Identification of
Trees and Shrubs in Winter
using Buds and Twigs

Identification of
Trees and Shrubs in Winter
using Buds and Twigs

Bernd Schulz

1900 drawings by Bernd Schulz

Kew Publishing
Royal Botanic Gardens, Kew

First published in 2018 by the
Royal Botanic Gardens, Kew,
Richmond, Surrey, TW9 3AB, UK
www.kew.org

Distributed on behalf of the Royal Botanic Gardens, Kew in North America by the University of Chicago Press, 1427 East 60th Street, Chicago, IL 60637, USA

ISBN 978-1-84246-650-6
eISBN 978-1-84246-651-3

British Library Cataloguing in Publication Data
A catalogue record for this book is available from the British Library

Originally published in German as:
Gehölzbestimmung im Winter mit Knospen und Zweigen
© 2013 by Eugen Ulmer KG, Stuttgart, Germany.

Translator: Monika Shaffer-Fehre
Scientific editor: Henk Beentje
Copy editor: Ruth Linklater
Proof reader: Sharon Whitehead
Design and page layout: Christine Beard
Production management: Georgina Hills

Printed and bound in Malta by Milita Press

For information or to purchase all Kew titles please visit
www.kewbooks.com or email publishing@kew.org

Kew's mission is to be the global resource in plant and fungal knowledge and the world's leading botanic garden.

Kew receives approximately one third of its running costs from government through Defra. All other funding needed to support Kew's vital work comes from members, foundations, donors and commercial activities including book sales.

Contents

Notes on translation

Each language has its own peculiarities, structure and vocabulary. Therefore, no text can be translated into another language without change. My wife, a linguist herself, likes to quote the Italian phrase 'traduttore, traditore' (translator, traitor). In the case of scientific texts, different traditions make adequate transmission more difficult.

For example, in English, three terms are used for the different kinds of armation: 'prickle', 'spine' and 'thorn'. In German, on the other hand, only 'dorn' (transformed organ) and 'stachel' (= prickle, emergence of the bark) are distinguished.

The difference between 'spine' (transformed leaf) and 'thorn' (transformed stem) can be represented in German by combining terms: spine = blattdorn, thorn = sprossdorn.

Over several centuries, German-speaking botanists have used the opportunity to assign new meanings to compound words. For example, the word blatt associated with prefixes or other words can be assigned to different versions of a leaf or even leaf parts: nebenblatt (stipule), oberblatt (petiole and lamina), unterblatt (leaf base and stipules), blattgrund (leaf base), keimblatt (cotyledon), niederblatt (cataphyll), vorblatt (bracteole in flower, prophyll in vegetative shoots), hochblatt, deckblatt (both bract), tragblatt (by flowers also bract, in the vegetative part without an English term, here simply called leaf, when fallen the leaf scar remains), hüllblatt (involucral bract), kelchblatt (sepal), kronblatt (petal), fruchtblatt (carpel) etc. The result of this sophisticated system of technical terminology is that for some of the terms I use in the German edition, there is no common English equivalent.

In his *Wörterbuch der Botanik* (botanical dictionary) Wagenitz gives for 'Vorblatt' 'prophyll, prophyllum or bracteole' as English botanical terms. But in English, the terms bract (deckblatt) and bracteole (vorblatt) are only used for leaves in inflorescences. A bract is the leaf that supports a single flower in its axilla; the bracteoles are the first two leaves on the pedicel (flower stalk). In German the first two leaves in lateral vegetative shoots (including buds) are 'vorblätter', too. They often significantly differ from the following scales or leaves of the bud. For a comprehensible description, a unique name is needed for the first two leaves in the vegetative part. Therefore, the publisher and the author have agreed to revive the term 'prophyll', though quoted as 'old-fashioned' by Henk Beentje.

New to the English botanical literature is the term 'enrichment bud'. In contrast to 'accessory buds', which emerge above or below the lateral bud from the same meristem, they stand in the axils of the scaly prophylls on the sides of the lateral bud. Enrichment buds and accessory buds are easily distinguishable in woody plants and provide good identification characteristics.

I thank everyone who contributed to the English edition. From the Royal Botanic Gardens, Kew: Gina Fullerlove and Lydia White, as well as Dr Henk Beentje, Dr Kevin McGinn, Ruth Linklater and Dr Sharon Whitehead. My greatest thanks go to Tony Kirkham, who recommended my book for publication by Kew, and to my friend Fred (Dr Friedrich Ditsch, Dresden), who invested a lot of time, helping me with the review of the translation.

Bernd Schulz,
Dresden, in the winter 2017/18

Foreword

Winter twigs and dormant buds — it's the identification test that all horticultural students dread and many gardeners shy away from! Why? Because there are no leaves, flowers, fruits or any other distinguishing features to help us with the identification, or so we think. But that's far from being true and if we take a closer look at a living, dormant twig, you couldn't be more wrong. By a simple process of elimination, observing whether the buds are alternate or opposite on the coloured or fissured stems, or whether there are lenticels and what is the colour, character and type of bud scales. With these simple steps, the positive identification of the different tree and shrub species can be narrowed down to a possible handful of taxa that fit the easy-to-follow key.

There are a few well-illustrated books with keys that help to guide us through this systematic approach to the identification of twigs and buds for a handful of common trees, but this monumental taxonomic work is one reference work that will help us all to be more confident in the identification of a comprehensive list of winter twigs and buds.

I first met Bernd Schultz in the fall of 2014 at a Polish Dendrology Society tree conference in Warsaw, Poland, where Bernd had an art exhibition featuring a selection of his most beautiful paintings. These paintings are used in this book to illustrate the winter twig of almost every temperate tree that we are likely to find growing in our gardens and arboretums. The artwork is so much better and more useful than a photograph, as it clearly shows the makeup and features of the naked twigs.

The original work was written in German which for English speaking horticulturists make the text, captions and keys difficult to understand and follow, so the user relied on matching the specimen to the respective images. I remember that moment when I first set eyes on this original book…. Quietly thinking to myself 'if only this book was written in English. How useful it would be to help further develop our tree identification skills in winter' and here it is. I am absolutely delighted with the outcome and I know just how useful this book will be to anyone wanting to be more accurate and effective with tree identification during the winter months.

Tony Kirkham
Head of Arboretum, Gardens
and Horticultural Services,
Royal Botanic Gardens, Kew

Preface

Not only in the description of the winter characteristics, no, generally in descriptive botany, the image must always be predominant to the word. It is of greatest importance to fix the recognised matter in the picture. When comparing sketches, differences, which can only be discerned with difficulty from the description, often become strikingly obvious.

Camillo Karl Schneider
Dendrologische Winterstudien 1903

When the leaves have fallen, the winter season begins. To many plant lovers this appears to be lost time, but a glance at the twigs of trees and shrubs shows that this need not be the case. In contrast to other life forms, trees and shrubs do not die above ground, but enclose their shoots for the next year in buds. These resting buds, initiated long before the leaves fall, show many characters which hardly change over time. Deciduous species of trees and shrubs can be classified reliably in winter by these twig and bud characters. For some genera they are so useful that these bud and twig features can also be used in summer, for example in ash or hickory (*Carya*) species.

The identification of woody plants without leaves is particularly important at planting time: gardeners, foresters, garden centres and their customers all have to recognise woody plants in winter. In areas of study like forestry and landscape architecture, the identification of woody plants is therefore part of the curriculum. When, during my studies, I was told to make a herbarium collection of winter twigs, I sketched my twigs, and instead of a thick folder, I handed in two A4 pages. Since then the subject has been of interest to me. I still investigate the structure of buds from woody plants and I observe and draw them from nature. And now the drawings fill several thick folders...

Again and again, I am surprised as to how little the complexity of bud structure is included in systematics or used for identification. Winter buds, as a complex adaptation to the season's climate, often reflect relationships of, and evolution within, a group of plants. The structure of winter buds can, particularly in wind-pollinated groups, often give more information on their relationships than the much-reduced flowers can.

In this second edition, a new main key leads the user to the genus; re-worked genus and family keys lead to the species. About 40 new genera and 60 new species have been included, and c. 10 species have been removed. As in the first edition, the species are presented in systematic order. Knowledge of the actual relationship of species has constantly improved in the past 15 years, thanks to the results of molecular biology, and (at least above the level of genera) has now stabilised. The sequence of the higher categories follows the currently recognised systems (APG III 2009, Strasburger 2008 and Stevens 2012).

Bernd Schulz
Dresden, January 2013

Introduction

I was often pleasantly surprised by the observation as to what excellent characters buds deliver, even in such species which in summer can be distinguished only with difficulty from one another.

J. G. Zuccarini
Charakteristik der deutschen Holzgewächse im blattlosen Zustande. 1829

History

To the present day the importance of bud morphology for systematics is not sufficiently recognised. Descriptions in systematic works limit themselves mostly to reporting whether there are naked or covered buds in a taxon. The specific architecture of buds and the comparison of their structure have only rarely been used in botanical works. The few older works dealing with the subject have, time and again, been forgotten by more recent botanists. In 1675–79 Marcello Malpighi showed buds and bud leaves

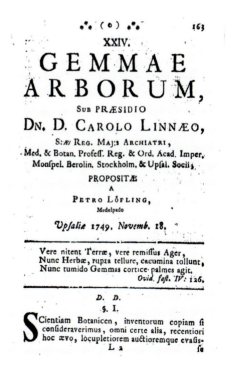

Title page of the oldest scientific treatise on the buds of trees (*Gemmae arborum*) by Pehr Löfling, 1749.

in his work on plant anatomy *Anatome Plantarum*. Seventy years later, in 1749, Pehr Löfling (1729–1756), a pupil of Linnaeus (1707–1778) introduced the buds of the trees in *Gemmae arborum*. He divided the buds of the 108 species investigated into several groups according to bud scales — their positional relationships and the morphological equivalent in the entire leaf. A century later, in 1847, a sizeable morphological study by Aimé C. F. Henry (1801–1875) was published in *Novorum Actorum Academiae Caesareae Leopoldinae-Carolinae Naturae Curiosorum*. In his *Knospenbilder* or *Bud pictures* Henry, a trained illustrator and lithographer, provided many excellent plates.

The structure of buds, with the role of stipules in bud protection, was investigated by John Lubbock. He published his results in his 1899 work *Buds and stipules*. Walter Sandt wrote *Zur Kenntnis der Beiknospen* in 1925.

Occasional studies of winter buds were shown in works such as *Die Anatomie der Knospenschuppen in ihrer Beziehung zur Anatomie der Laubblätter* by Eduard Brick, published in 1914. A few authors (e.g. Schulze (1934), Buchheim (1953), Schützsack (1965)) studied specific groups and investigated the anatomy, morphology and structure, sometimes including the structure and unfolding of winter buds.

Besides such anatomical and morphological works, little considered even by the scientific community, there were also early publications on the identification of the woody plants in winter. The first German book for identification came from Joseph Gerhard Zuccarini (1797–1848), Botany Professor in München. In 1829 he completed his *Charakteristik der deutschen Holzgewächse im blattlosen Zustande*. In his book with 18 hand-coloured lithographs — a bibliophilic treasure — he treated 34 of the most common native woody plant species.

Some decades later the forestry botanist Emil Adolf Rossmässler (1806–1867), with *Flora im Winterkleide* and *Der Wald*

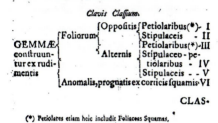

Classification of buds by Pehr Löfling, 1749.

and Moritz Willkomm (1821–1895) in his descriptive *Deutschlands Laubhölzer im Winter*, called attention to the variety of deciduous trees in winter. Willkomm for the first time gave an identification key, a systematic description in the form of a table as well as pen-and-ink drawings for each of the 102 species treated.

The *Dendrologische Winterstudien*, published in 1903 by Camillo Karl Schneider (1876–1951), far surpasses all works published before. The author dealt with 434 woody plants, with many illustrations and detailed descriptions. Later German authors no longer worked with such large numbers of species, usually treating only the most common native species that were important in forestry.

In England and North America many identification manuals appeared at the end of the 19th century, up to about 1930. For example, Marshall Ward published the first volume of his work *Trees*, called *Buds and Twigs* in 1904. After a valuable introduction to the morphology of buds, this book contains a special section in which, besides many species of the native flora, the most frequently planted woody plants are keyed out. But the highest number of species was treated in the 3rd edition of *Winter Botany* (1931) by the American Professor of Botany William Trelease (1857–1945): he keyed out over 1000 species. In spite of the number

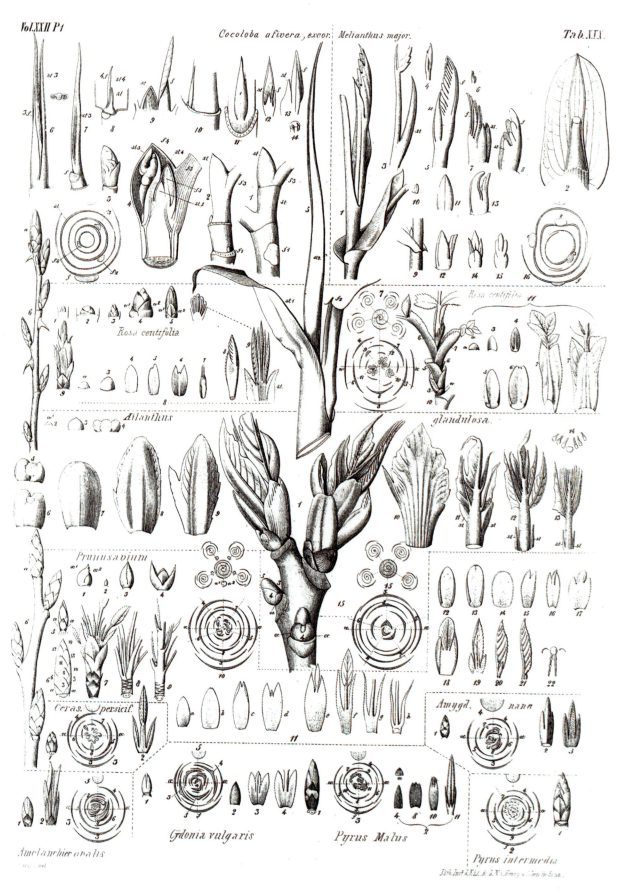

Plate from *Knospenbilder* by AIMÉ HENRY, 1847.

of species, the keys work well. The author did not give much in the way of description of the individual species and illustrated only selected species. In 1895, Shirasawa published *Die Japanischen Laubhölzer im Winterzustande*. In the Slavic countries, too, several interesting identification manuals of their deciduous tree floras in winter have been published.

Deciduous woody plants and their environment

The cycle of seasons determines our climate and plants have adapted to the seasons with various life forms. Raunkiaer, in 1904, established a system of life forms which he divided according to where the buds are placed during the adverse seasons. Annuals (*therophytes*) avoid the winter by a kind of suicide: after flowering and fruiting they scatter their seed and die. All other (perennial) plants protect their bud in one way or another. Herbaceous perennials (*geophytes*) hide them in the ground and the herbaceous *hemicryptophytes* use the cover of leaves and snow as a protective shield. By contrast, the buds of woody plants (*phanerophytes*) at least 30 cm above ground level are directly prone to wind and weather. The delimitation of the half- and dwarf shrubs (*chamaephytes*), standing between herbaceous perennials and woody plants, remains difficult.

Deciduous woody plants are found where summer temperature, rainfall maxima and winter frost periods make for a distinctly seasonal climate. In a vegetative period of at least 4 months duration, materials for flower and fruit formation and for growth, as well as reserves towards the new leaves of the next year, are formed.

In autumn the leaves are then shed, independent of weather conditions (**obligative**). The opposite is the **facultative** shedding of leaves in vegetation zones with irregularly occurring weather extremes (e.g. drought) but without a distinct seasonal climate.

Evergreen trees and shrubs retain their leaves during several periods of vegetative growth. They occur where there is no disadvantageous season requiring a break or pause in development.

LUBOCK, **1899**: 4 plates on bud unfolding.
When the buds unfold, the inner structure becomes visible.

1 guelder rose (*Viburnum opulus*)
2–4 Norway maple (*Acer platanoides*)
5 whitebeam (*Aria nivea*)

1–3 lime (*Tilia ×europaea*)
4–5 hornbeam (*Carpinus betulus*)

1–4 wych elm (*Ulmus glabra*)

1–7 beech (*Fagus sylvatica*)

Half evergreen woody plants stand in between the evergreen and deciduous woody plants. Last year's leaves fall only after the winter, with or shortly after the appearance of the new leaves.

In Europe, the vegetation zone of deciduous trees stretches from the Atlantic coast in the west to the Balkans in south-eastern Europe and even to Asia. Deciduous forests also predominate in eastern North America and in south-east Asia. From there come most of the ornamental woody plants planted in the temperate areas of Europe. In Asia Minor, the Caucasus and central Asia, as well as in the Southern Hemisphere in South America, there are further areas with deciduous vegetation. From these areas too, species have been introduced and are cultivated in Europe.

The European flora is relatively poor in woody plant species. During ice ages in the Quaternary, many woody genera died out in central Europe. Due to the mountain ranges of the Carpathians and Alps stretching from east to west, they could not escape the encroaching ice. Thus there are many genera, such as *Magnolia*, now native to the broad-leaved-forests of eastern Asia and North America, which also existed in central Europe before the ice ages. From the deciduous forests of the Northern Hemisphere also come most of the foreign species planted in open terrain, for instance: *Robinia*, naturalised from eastern North America; or many species of *Magnolia* from eastern Asia. Further species came to us from central Asia or south-eastern Europe, such as the horse chestnut. Very few species stem from the southern South America (*Nothofagus*).

The adaptation of the seasonal deciduous woody plants to the climate becomes particularly clear when vegetation types are compared: in the north, but also at high altitude in the mountains, deciduous forests are replaced by evergreen coniferous forests. These **boreal coniferous forests** occupy huge areas in the north of Eurasia. Under this climate insufficient reserves can be laid down for a,n annual leaf-fall and regrowth. On the other hand, the conifers' needle leaves are particularly adapted to low temperatures and **frost drought**. Only a few conifers, like larch, lose their needles in particularly cold and dry habitats.

In the Mediterranean south of Europe, the vegetation is characterised **by evergreen sclerophyllous forest.** Long dry periods in summer necessitate strong protection from evaporation. A thick epidermis with thick cuticle layers, often together with resin- and wax coverings and with deeply sunken stomata, characterises leathery leaves stabilised by supporting tissues. Leaves constructed with such lavish use of resources cannot be renewed completely during the short periods of rain maxima in spring and autumn. In a climate of mild winters, without the danger of winter frost, this is unnecessary.

Plates from ZUCCARINI 1829:
Characters of German woody plants without leaves.

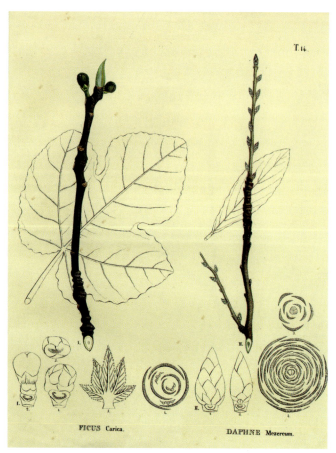

Fig (*Ficus carica*) and mezereon (*Daphne mezereum*)

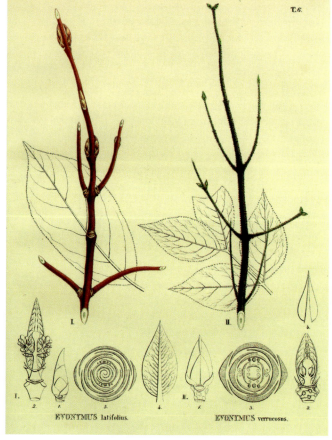

Broad leaved spindle (*Euonymus latifolius*) and warted spindle (*Euonymus verrucosus*)

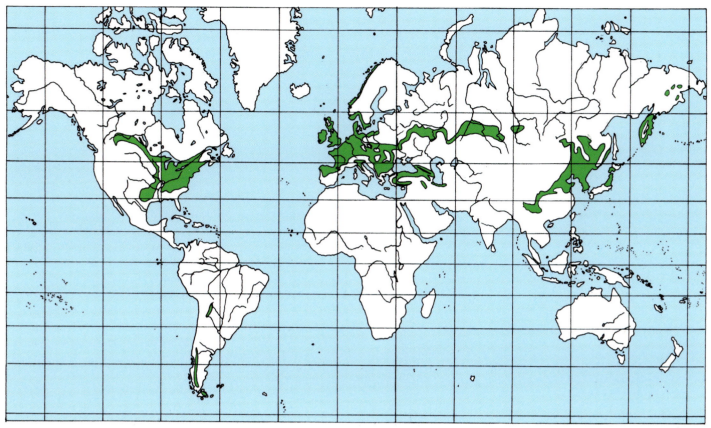

The worldwide distribution of deciduous forests.

Leaf fall

Before leaves fall, they undergo a natural ageing process (**senescence**) which is controlled by plant hormones: leaf matter is broken down and stored in the stem. During the breakdown of the green chlorophyll, yellow pigments that are already present become visible, and in some species red pigments develop through a complicated metabolic process. These pigments cause the change to autumn colours. Between the petiole and the twig a separation tissue of thin cell layers develops; here the leaf can break off easily. Below the abscission layer a closing tissue (the periderm) forms which, by incorporating cork and tannin, can isolate the site of the break from external dangers.

Frost

Growth stops in late summer. Towards autumn the twigs ripen and slowly harden in the gradually declining temperatures. In moist summer weather, the end of growth can be delayed. Exotic species from warmer climes can have more than one growth spurt; such plants are then not ready for winter and are particularly susceptible to frost. Native woody plants are better adapted to the rhythm of the seasons, and are less often damaged by deep winter frosts. The danger is particularly strong when frost begins unusually early in the autumn. In contrast to such early frosts, late frosts in spring are particularly dangerous for species that germinate, or come into leaf, early. Again, exotic species are particularly susceptible. Even species from Siberia can suffer from this effect in our climate. In their own area, with its long, hard winters and short growing periods, these species start their growth as soon as warmer temperatures signal winter's end. Under European conditions, this stage of development can be reached in February. Any cold weather after this can then easily lead to frost damage.

Another reason for plant damage is through **frost drought**. The plant dries out if it evaporates moisture on sunny winter days, because it loses more water than it can replace from the frozen ground.

Insects and fungi

Another danger is posed by living enemies such as insects and fungi. During the growth period, the healthy plant can defend itself actively by reactions such as the secretion of resin. In winter, when metabolic activity is low, an active defence against pests is not possible. Here barrier mechanisms prevent damage to the plant during winter. For instance, cork and tannins are stored in the outer tissues, such as in the leaf scar. Some species store poisonous cyanides (cyanic compounds) in their bud scales.

Despite this, there are some insect species which have specialised in feeding on the young and nutritional-rich shoot tips contained in the buds. For example, the caterpillars of ash-bud moth, *Prays fraxinella*, bore into the terminal buds of the ash in October and consume their content during winter. Later, in place of the emptied terminal bud, the upper two lateral buds germinate. In place of a straight growing stem, a forked branch develops.

How to use this book

Exclusively deciduous woody plants in their winter state are described. Included are all species of trees and shrubs native to Central Europe, and in addition those which have become naturalised and which are often cultivated. In this second edition, I have included nearly all genera of cultivated trees and shrubs, and I have added even those that are very rarely planted.

Keys lead to the species. Most useful for identification are first year's growth twigs with buds. In addition, growth form, bark and other characters can help in the identification of species. Information on where species come from is kept to a minimum; information on how common a species is applies to the use of species in gardens, landscaping, parks and forestry, and in species native to that region also to their natural frequency. In the keys to family and genera, more frequent species are emphasised in bold.

The figures

Not only do the architecture and shape of buds vary among species, but their size, too, can be very different: some Magnolias, e.g. *Magnolia fraseri*, have up to 50 mm long terminal buds, whereas the buds of species such as *Spiraea thunbergii* are less than 1 mm in length. Therefore it has been impossible to keep to a single illustration scale.

Measurements

For size orientation, most drawings are accompanied by scale bars. The scale bar with a double line corresponds to 1 cm, the scale bar with a single line represents 1 mm. The text gives further indications of size. The height or length is measured from the base of the bud to its tip. Thickness in terminal buds is given as the diameter; in lateral buds, thickness denotes the widening in the median direction (lengthwise), and width denotes widening in the transverse direction (across). This is often greater than the thickness.

Size of buds and twigs can vary. It can be influenced by habitat, position in the stand, competition with nearby plants and origin (provenance). Stool- or coppice shoots can be much stronger than "normal" twigs. Therefore it is best to use normal twigs to identify and compare species. In trees especially, buds at the base of branches are often very small and undeveloped (resting buds): such buds were not used for the measurements given in the text.

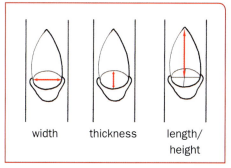

Orientation of measurements.

Definitions of place

We need to define the position of organs in the plant body and their spatial relationships. The **point of insertion** is the place at which one organ is affixed to another (where it is **inserted**). Lying at the base is called **basal**, lying at the tip **apical**. **Acropetal** describes the sequence from basal to terminal. **Axillary** shoots (buds, thorns, leafy shoots) stand in the axil of a leaf; **terminal** ones at the tip, at the free end of a shoot. The side away from the stem is called **abaxial** (as in the lower surface of a leaf), the side towards the stem **adaxial** (as in the upper surface of a leaf).

The **median plane** divides a structure into two symmetrical halves through the principle axis, the leaf and the lateral shoot; at right angles to this is the **transverse plane**. Organs away from the middle are **lateral**. According to their direction we distinguish: **orthotropous** (orientated upright) and **plagiotropous** (oblique or horizontal-growing plant parts, particularly shoots).

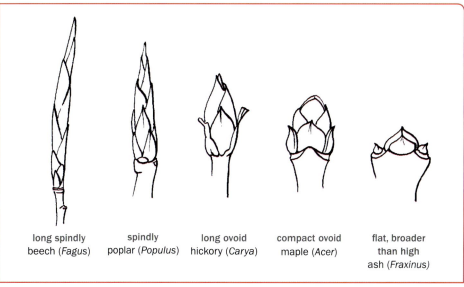

long spindly beech (*Fagus*) · spindly poplar (*Populus*) · long ovoid hickory (*Carya*) · compact ovoid maple (*Acer*) · flat, broader than high ash (*Fraxinus*)

Bud shapes (terminal bud).

Colours and descriptions of colours

Most twigs and buds are specifically coloured. Dependent on sunshine, wind and weather, such colouration can often vary on one side or the other. Anthocyanins in the cell sap of the epidermal vacuoles protect from solar radiation, and green twigs and buds often turn red in the sun. When a periderm (bark) replaces the epidermis, the twig turns brown and later grey. In many species, in addition, the epidermis dies, particularly on the side exposed to the weather: it dries out, the insides of cells fill with air, and finally the epidermis lifts off. What we see is a silvery grey coating which may look like bloom.

Descriptions of colour can only give an indication. They must be interpreted carefully because of the many nuances of colour as there are no equivalent technical terms.

Systematics and nomenclature

Every modern system tries to represent the natural relationships of plants. For a long time there was much disagreement among different authors concerning natural relationships. They accorded different degrees of importance to single characters and so the systems of different authors could differ markedly. In the past twenty years scientists, with the help of molecular-biological investigations, have deciphered the fundamental relationships of organisms. Because this process is not yet complete, many authors hold back from using this taxonomy. The relationships of the families introduced in this book are deemed safe and stable, so that no more fundamental changes are expected in the future. In lower categories like genus and species, however, many problems are not yet resolved. Independent of the knowledge concerning relationships, there also remains room for judgement; for example, an author might decide on large genera and use *Prunus* sensu lato to encompass all species with complete stone fruits, or he might prefer smaller genera and keeps under *Prunus* sensu stricto only plums and puts cherries (*Cerasus*), peach (*Amygdalus*), bird cherry (*Padus*) and other subgenera in their own, separate genera. Both these imterpretations are possible and logical. The current genus *Sorbus*, which for many authors includes whitebeam (*Aria*), wild service tree (*Torminalis*), dwarf whitebeam (*Chamaemespilus*), *Cormus* and *Sorbus*, is different. I do not see this as logical, because in between *Sorbus* and *Aria* there are several other genera of the tribe Pyreae.

The scientific naming of species follows international rules which are laid down in the International Code of Botanical Nomenclature. A species name consists of two parts, first the name of the genus, a noun with a capital letter, and second the species epithet, an adjective always written in lower case. Following the two-part (= binary) name of the species is the name of the author(s) who first validly published the name. Author's names are mostly abbreviated.

Basics of botany

Structure of woody plants

The vegetative body of higher plants is divided into three basal organs: leaf, stem, and root. Stems bear leaves, the roots never do. Leaves, as a rule, do not bear other leaves, stems or roots. Stems can bear subordinate lateral shoots and leaves, as well as adventitious roots, for instance aerial roots as holdfasts in climbing plants. Roots bear only shoots (stems with leaves) and subordinate lateral roots.

Modifications

Natural selection during the evolution of species led to adaptations to certain situations in life. Organs changed their form and appearance and took on several functions, say, as tendrils in a climbing plant, or as thorns in a species eaten by animals. When different organs take on the same role, they have the same function and are **analogous**. If, by contrast, a single organ fulfils different roles, its modifications are of similar derivation or **homologous**.

Leaf

Structure

The leaf is a lateral extension of the stem and is always situated at the node of a stem. Its function is mostly assimilation, but can take on other functions through various modifications; examples of this are bud scales, spines and leaf tendrils. The leaf consists, in its lower part, of the leaf base and (where present) the stipules; above this are the long thin stalk (the petiole) and the leaf blade (the lamina). In pinnate leaves, the leaf is divided into an axis, the rachis, and the leaflets. Stipules are not present in all groups; they consist, usually, of two tiny leaflets at the leaf-base. They can be free, or partly connate with the leaf-base or with each other.

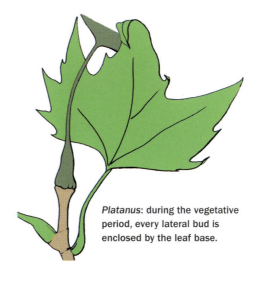

Platanus: during the vegetative period, every lateral bud is enclosed by the leaf base.

	leaf
	stipule
	leaf base and leaf scar
	prophyll
	stem
	accessory bud

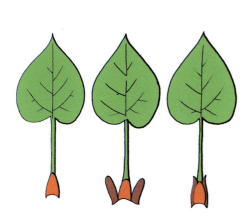

Structure of simple leaves of typical, true dicotyledonous Angiosperms (comparing different forms of stipules).

Platanus: a stipule envelops the lateral bud.

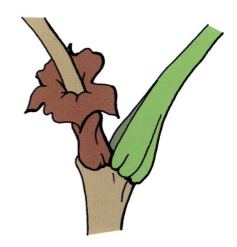

In summer, every leaf is accompanied by a stipule cover, which surrounds the shoot tip and leaves a ring-shaped scar at the shoot node.

lnterfoliar stipules or **interpetiolar stipules** are fused stipules of two neighbouring leaves on a node. An **ochrea** is a tubular stipule forming a sheath around the stem (as in *Platanus, Polygonum)*. **Median stipules** are connate on one side and are placed in the median plane of a leaf (e.g. *Ficus*).

Sequence of leaves

Plants are modular organisms. On the stem the leaves follow one another in indeterminate number, and in each leaf axil one leafy shoot can start growing. Among the following terms several can be applied to a leaf, as each term only applies to a single aspect of the position in the vegetative body. Thus a leaf can be a **prophyll** — i.e. the first or one of the first pair of a lateral branch; or it can, within an inflorescence, be an inflorescence bract.

The first or two first leaves emerging from the sprouting seed are **cotyledons**; transitional leaves lead to the fully differentiated **foliage leaves**, and in the inflorescence, there are various forms of **bracts**.

The first leaves of a lateral branch are **prophylls**. These are sometimes seen as homologous to the cotyledons. The monocotyledons and some basal angiosperms have a single prophyll. This is in the medial plane and has its dorsal side towards the principal axis ('addorsed'). Dicotyledonous angiosperms mostly have two prophylls in a transverse plane. Prophylls on the flower stalk are called bracteoles.

Scale-like leaves at the base of a shoot are called **scale leaves** or **cataphylls**. The first cataphylls of a lateral shoot can, at the same time, be prophylls. Every leaf serves as a subtending leaf for the shoot (e.g. lateral bud) in its axil. In the descriptions, they are simply named "leaf" or when fallen off "leaf scar". If a flower or fruit comes from the leaf axil, the leaf is referred to as a bract.

Diagram of transverse prophylls in *Spiraea*.

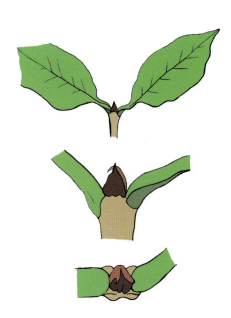

Coffee plant (*Coffea*) interfoliar stipule.

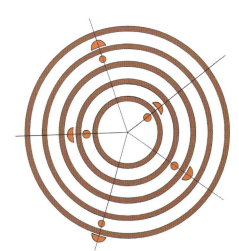

Completely connate stipules (ochrea) in the Polygonaceae.

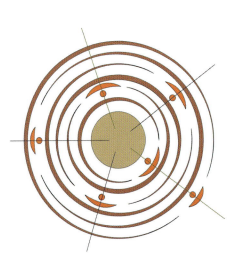

Unilaterally connate stipules in figs (*Ficus*).

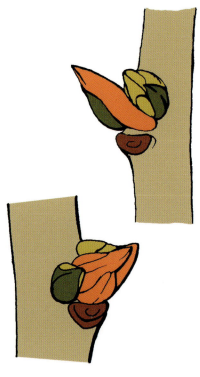

Spiraea: stipules and enrichment buds.

Catalpa : diagram of lateral bud.

Phyllotaxy

Phyllotaxy refers to the **position of leaves** on the stem. In winter this can be recognised from the position of leaf scars and the axillary buds. In **alternate phyllotaxy** there is only one leaf at each node; by contrast in **whorled** or **verticillate phyllotaxy** there are several leaves at each node.

Alternate or disperse phyllotaxy is differentiated into spiral and distichous phyllotaxy. In **spiral phyllotaxy** the leaves on sequential nodes stand at distinct angles to each other and form a spiral like a screw thread.

A frequent divergence in the vegetative state is the 2/5 divergence (144°) in which the screw line goes around the stem twice, before the sixth leaf stands exactly above the first leaf. If one connects the leaves exactly above each other with vertical lines, 5 longitudinal lines are obtained. In transverse section there are often 5 vascular bundles and in some species a distinct 5-radiate [rayed] pith is visible.

Catalpa: lateral bud.

Cercidiphyllum: addorsed prophyll.

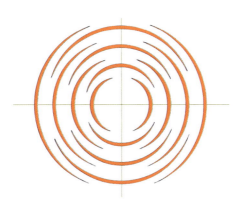

Opposite phyllotaxy with simple bud scales.

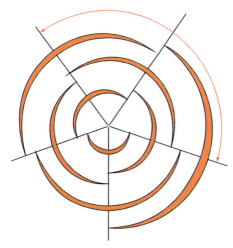

Spiral phyllotaxy with simple bud scales.

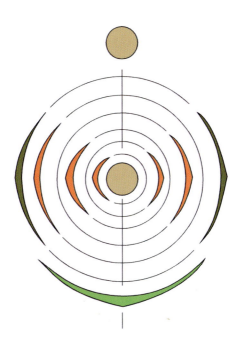

Change of phyllotaxy: opposite prophylls are followed by distichous leaves.

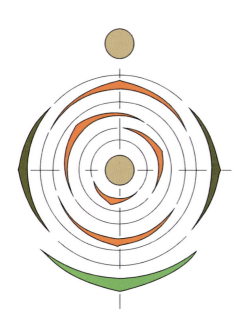

Change of phyllotaxy: opposite prophylls are followed by spirally inserted leaves.

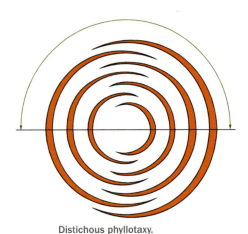

Distichous phyllotaxy.

Metamorphoses of the leaf

Leaves or parts of leaves can take up various functions. They can, for instance, be changed into tendrils, spines or bud scales. These metamorphoses can be recognised by their position as leaves: they are placed where leaves are situated ordinarily.

Bud scales or teguments: scale-shaped leaves or parts of leaves which serve as bud protection. According to their origin we differentiate between: stipule scales, simple scales (leaf base) or laminar scales (homologues of whole leaf).

Leaf tendrils: leaves or parts of leaves transformed into tendrils. For example, tendrils of stipules occur in *Smilax* and tendrils of the leaf rachis occur in *Clematis*. When the rachis turns into a tendril, the individual leaflets dehisce in winter and there will be no leaf scars.

Spines: any part of the leaf can turn into a spine: the rachis of pinnate leaves (e.g. *Caragana spinosa*), the paired stipules (e.g. *Robinia*) or the whole leaf (e.g. *Berberis*).

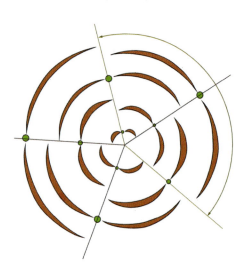

Spiral phyllotaxy with stipular bud scales.

In *Berberis*, the leaves on long-shoots become spines. During the summer the plant assimilates with the unfolded prophylls.

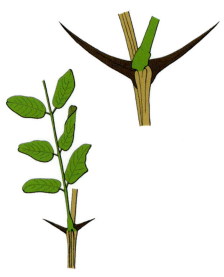

Only the stipules on *Robinia* become spines. The rest of the leaf is green and falls off in autumn.

The distichous position of leaves is a form of the alternate position, in which the leaves are in ½ divergence i.e. stand alternating in two rows.

In the verticillate position of leaves, several leaves are at one node. The leaves of one verticil stand at the same angle to one another (rule of equidistance) and the leaves of sequential nodes at the 'gap' (rule of alternation). In the vegetative realm the decussate position, in which two leaves stand at each node, is the most frequent form of the verticillate position of leaves. The dissolution of a verticil occurs during the transition of the basal verticillate position of leaves on opposing prophylls acropetally towards the alternate leaf position.

In *Caragana spinosa*, the green leaflets of the pinnate leaf assimilate during the summer and fall off in the autumn leaving a spiny leaf rachis.

Types of bud scales: simple (leaf base), simple with adnate stipules, laminar, stipular.

Leaf scar

After the leaves fall, scars remain, which are mostly sealed off by periderm. By the position of the leaf scar, one can tell the position of the leaves and therefore the phyllotaxy. On the leaf scar, **traces of the vascular bundles** usually remain, where metabolic exchange with the stem took place during the growing period. The number of traces is species-specific and, in winter, can be an important diagnostic character for deciduous woody plants. In many species, the traces in the leaf scars are clear, in others they are indistinct. Often traces that seem single at first glance are composed of several vascular bundles. Such groups are — as long as they look as if they belong together — counted as a single trace dutring identification. In cases of doubt, both possibilities (single or more than one trace) should be considered. Sometimes stipule scars can be found with the leaf scars; these are from free stipules leaving usually narrow scars.

Stem

The stem is composed of internodes and nodes. The leaves are attached at the nodes. An internode is the stem section located between two neighbouring nodes. Adventitious roots can arise everywhere, but develop preferentially in the vicinity of nodes.

Structure

Within the stem there is the **pith**, often a large-celled, loose tissue. At the border of the pith with the wood there is often the primary xylem, which sometimes is of a different colour or hue. The pith can tear transversally and is than called **chambered**. If the pith is entirely absent the result is **hollow twigs**, with the medullary cavity in place of the pith. Between pith and cambium is the **wood**. It offers some characteristics for identification (e.g. colour, surface}. Anatomical characters (such as ring-porous and diffuse-porous) are not used here. Outside the cambium is the bark.

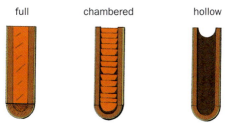

full   chambered   hollow

Pith schematic. Orange, pith; red-brown, wood; pale brown, bark.

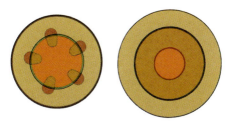

Transverse section of twig schematic: primary left, secondary right.

Bark

All tissues external to the cambium belong to the bark. Externally the bark of young twigs is surrounded by a single-layered cover, the epidermis. Above the epidermis cells, there is a wax-like, glossy layer, the **cuticle**. Within the epidermis there is sometimes — particularly in cells exposed to the sun — a red transparent colouring (anthocyanins in the cell sap of vacuoles). Underneath the epidermis there are the assimilating green tissues containing chlorophyll, especially bark parenchyma. This green tissue, which can be rather yellowish, causes the appearance of young twigs, either alone, or in combination with red colour in the epidermis. Patchy and mixed colours can develop between the complimentary colours of red and green, and in identification literature are unfortunately often vaguely described as "brown".

epidermis	cambium
wood	interfascicular cambium
pith	bast (phloem)
primary xylem	bark

      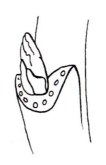

two traces: only ginkgo (*Ginkgo biloba*)

one trace: e.g. staff vine (*Celastrus*)

a composite trace: e.g. ash (*Fraxinus*)

three traces: e.g. nettle tree (*Celtis*)

three groups of traces: e.g. sweet chestnut (*Castanea*)

several traces: e.g. mulberry (*Morus*)

multiple traces: e.g. *Eleutherococcus*

Number of vascular bundle traces on the leaf scar.

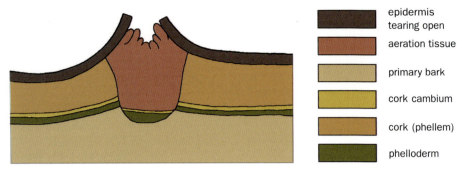

epidermis
tearing open

aeration tissue

primary bark

cork cambium

cork (phellem)

phelloderm

Section of lenticel (schematic).

In some species a **periderm**, consisting of several tissues, can replace the epidermis in the first growing period. From the outer bark tissue layers a new tissue arises: the cork-cambium or **phellogen.** On the outside this forms a cover of several layers of **cork** or **phellem** and sometimes (rarely) on the inside a cork-skin or **phelloderm.** As the epidermis usually does not continue growing, it is soon broken up by the expanding inner layers. The appearance of the newly formed surface peridermis (secondary dermal tissue) is distinctly different from that of an epidermal surface. Predominant colours are ochre-brown, brown and grey hues.

In the epidermis small stomata (slit-like openings) enable gas exchange. Macroscopically they are rarely visible. Only when they contain wax structures can they be seen, with a lens, as white dots.

Before a surface periderm comes into existence, the stomata in the epidermis are replaced by **lenticels.** The moment of their initiation can be long before the periderm is formed. In this process the cork cambium generates a mass of loose tissue which bursts the surface and enables gas exchange. At the same time cork (suberin) and tannins are laid down to protect the potential entrance portals against threats such as fungi. Externally, lenticels are visible as warty structures of a different colour.

Their form can vary greatly. Often lenticels on young twigs in the vicinity of nodes are round, but elongate on the internodes. The number, size and presence of lenticels are usually typical for species and genera.

If the first periderm (under the replaced epidermis) remains active for a longer time, the lenticels will determine the pattern of the bark in older branches and in trunks as well. When trunks become thicker, the lenticels are pulled laterally and a ringed bark arises with transversely placed lenticel bands, as in wild cherry (*Prunus avium*). Other barks with long active first periderm have rhomboid lenticels, as in aspen (*Populus tremula*) or remain entirely smooth as in common beech (*Fagus sylvatica*).

Secondary periderms, generated deeper within the bark, mostly succeed the first periderm. Now the outer layers die and tear open due to the increase in diameter. The woody bark is formed as the tertiary dermal tissue. The species-specific formation of the bark can be a helpful character when the choice of species is small: like, for instance, the flaking rind of the plane (*Platanus*) as opposed to the white smooth rind of the birch (*Betula pendula*). When there is a larger choice of species, identification by bark and habit is only rarely possible.

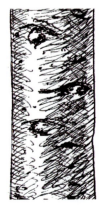

Bark remaining smooth
e.g. beech
(*Fagus sylvatica*)

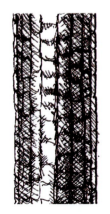

Bark tearing open
longitudinally and
horizontally giving rise
to scales, e.g. pear
(*Pyrus communis*)

Bark very scaly,
e.g. plane
(*Platanus ×hispanica*)

Bark rolling off obliquely
in thin layers, e.g. birch
(*Betula pendula*)

Bark tearing open
longitudinally and
splitting into fibres,
e.g. lilac
(*Syringa vulgaris*)

Short-shoots and long-shoots

Long-shoots are those with elongate internodes. They help with competition for space; particularly young plants form long-shoots.

Short-shoots are those with limited longitudinal growth. The internodes are very short and the leaf-bearing nodes follow each other closely. Short-shoots can take over particular functions: thus flowers are often formed only on short-shoots (e.g. *Malus)*. On short-shoots the plants can form, without much longitudinal growth, new foliage. The plant can adjust to unfavourable nutritional or spatial stress. When spatial expansion is impossible, the plant copes by forming predominantly short-shoots. This plasticity in the formation of shoots leads to individual differences, but it can also give hints as to the health of trees and can sometimes indirectly point to environmental stress.

Stem modifications

Parts of the stem can take over different functions, just like other basal organs can. The stem can, for instance, form thorns or tendrils. Such derived stem parts have their position in common: in the axil of a leaf, as a lateral shoot or at the free end of an axis as a shoot tip. Apart from the position, stem modifications always have remnants of leaves: leaf scars or scale leaves.

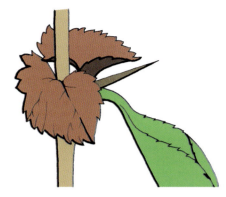

Thorn in flowering quince (*Chaenomeles*).

Thorn in hawthorn (*Crataegus*).

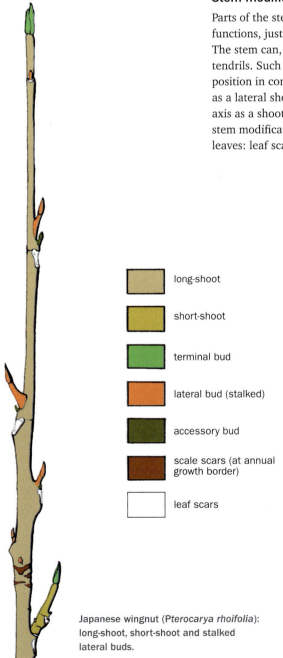

- long-shoot
- short-shoot
- terminal bud
- lateral bud (stalked)
- accessory bud
- scale scars (at annual growth border)
- leaf scars

Japanese wingnut (*Pterocarya rhoifolia*): long-shoot, short-shoot and stalked lateral buds.

Thorns are usually simple, more rarely they branch. Their position (as a lateral shoot or shoot tip) is species-specific.

Tendrils can originate from a shoot tip — when a lateral bud takes over the growth at the tip- or from lateral buds. They are usually branched.

Roots

Adventitious roots anchor the plant in the ground and serve in the supply of nutrients and water. They are hidden from view, and root modifications are much rarer than those in the other two basal organs, leaf or stem. Adventitious roots originating from the stem can serve as holdfasts in climbing plants, to anchor the plant to/in the support.

Prickles and hairs

Prickles

Prickles are pointed outgrowths of the bark. In contrast to the spines and thorns, they are not connected to the wood by vascular bundles; they are not organs themselves, but are scattered on the surface of stem or leaf. This irregular arrangement is typical for prickles. Sometimes, as when they occur in pairs near the nodes, they can resemble, and be confused with, stipular spines. In such cases a proper analysis will be necessary to determine what they are!

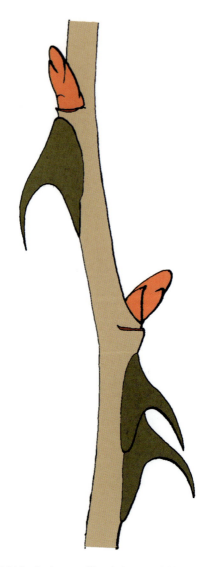

Prickles in the rose (*Rosa*). Green: prickles.

Hairs

Hairs (trichomes) are part of the epidermis. They can be single or multicellular. Dense hairiness slows the air current and breaks up the light rays, and in this way lowers evaporation. Many young shoots are hairy only during spring growth; here the hairs protect young organs whilst they break through the bud envelope. Such hairs are soon lost in many species. Completely hairless surfaces are called **glabrous**. Persistent hairiness differs from species to species, and can be dense and felty or loose, appressed or spreading, silky or rough. Especially naked buds, those not enveloped by bud scales, are often densely hairy.

Glandular hairs are frequent. These are ± long-stalked hairs with a globose tip, the gland. They exude different secretions during the growth period.

S**caly trichomes** or **scurfy scales** are also noticeable. These are multicellular and consist of a stalk with a disk-like or rayed top. They can give young twigs and buds a a silvery or bronzy sheen.

Stellate hairs are stalked hairs with the top branches radiating outward.

Fruits and infructescences

Fruits

Fruits develop from the carpels of the flower, and they enclose the seed(s). In some groups they are found in winter and can be useful in identification. Particularly suitable for this are dry fruits that remain on the plant and release their seed late. Other fruit forms, like those adapted to animal dispersal, in which the pericarp (ovary wall), or part of it, is juicy, rarely stay on the plant into winter. But even in such cases the form of infructescence, or the length of the fruit stalk, can give additional hints to species affiliation.

Frequent fruit forms:

1. Dry

1.1 pericarp lignified, not opening. . . . **nut**

1.2 deriving from one carpel, opening along two sides (dorsal and ventral). . **legume**

1.3 derived from one carpel, opening along one side only . . . . **follicle**

1.4 formed from several fused carpels . . . .**capsule**

2. Juicy

2.1 carpel separated into a hard inner endocarp and a soft outer exocarp . . . . **stone fruit**

2.2 seeds enclosed by juicy pericarp . . . . **berry**

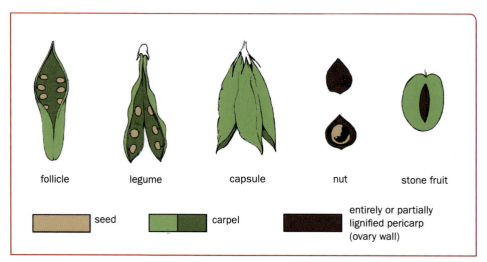

follicle   legume   capsule   nut   stone fruit

seed   carpel   entirely or partially lignified pericarp (ovary wall)

Most important fruit forms found in winter.

Infructescences

The infructescences are fruit-bearing shoot systems following the inflorescences. Terms that are frequently used to describe infructescences are listed below.

Spike: inflorescence/infructescence in which the single flowers/fruits sit on the main axis, without a stalk.

Cyme: inflorescence/infructescence in which each main axis ends in a flower, and is overtopped by one or several lateral axes.

Umbel: the single fruits are at the same level, while their the stalks originate from a single point.

Racemose corymb: the fruits are at the same level (as in the umbel), but their branched stalks originate at different levels, often paniculate.

Corymb: the fruits are at the same level (as in the umbel), but their unbranched stalks originate at different levels.

Panicle: flowers and fruits stalked, on branched lateral axes.

Raceme: fruits stalked, on unbranched main axis.

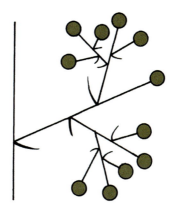

Cyme.

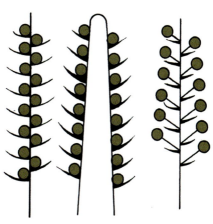

Left: spike; middle: spadix; right: raceme.

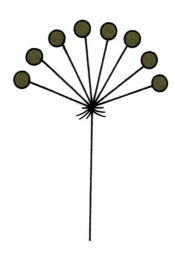

Above: umbel; below: capitulum.

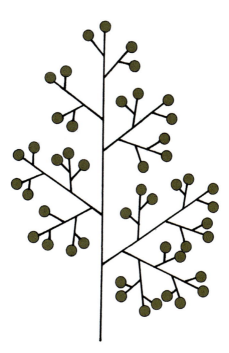

Panicle.

Growth

The 'look' of the plant, or habit, can be species-specific, because of branching patterns and the way growth takes place. The habit is modified by the history of each individual, depending on environment or habitat, and association with other plants nearby. Thus a tree grown in a closed stand of trees will be more slender than a free-standing one.

The growth of the shoot tip is called **monopodial** if it arises through a terminal bud; a continuing main axis, to which lateral axes are subordinate, is the result. Should the terminal bud fail, or if no terminal bud exists, one or more lateral buds will continue growth in a **sympodial** manner. Such sympodial growth can be differentiated according to the number of lateral buds maintaining growth. If a single sub-terminal lateral bud continues apical growth, this is called **monochasial**. A continuing axis can develop which is composed of many subsequent lateral axes (as in *Ulmus,* elm). But if two sub-terminal lateral buds continue the growth, a **dichasial** fork-like branching results (as in *Syringa vulgaris*). In many species with opposite leaves and branches, terminal inflorescences and infructescences 'use up' the shoot tip. In such cases, a dichasial growth continues the originally monopodial growth — as in the horse chestnut. When growth is continued by more than two lateral buds, this is called

Rhododendron: pleiochasial. Green: last year's persistent shoot tip (infructescence). Red-brown: terminal buds of lateral shoots which last year overtopped the shoot tip.

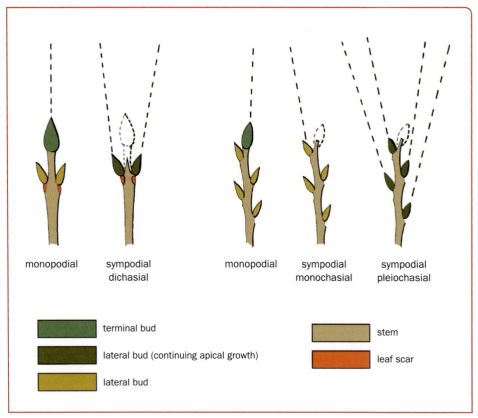

monopodial   sympodial dichasial   monopodial   sympodial monochasial   sympodial pleiochasial

■ terminal bud

■ lateral bud (continuing apical growth)

■ lateral bud

■ stem

■ leaf scar

Continuing growth.

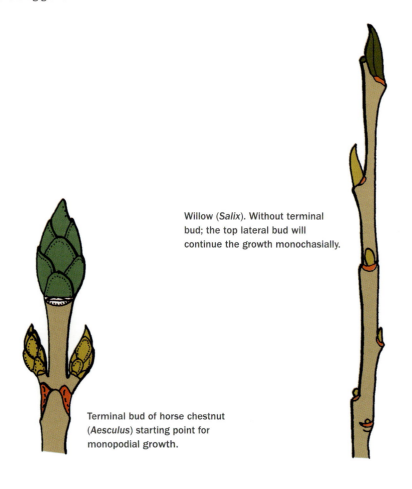

Willow (*Salix*). Without terminal bud; the top lateral bud will continue the growth monochasially.

Terminal bud of horse chestnut (*Aesculus*) starting point for monopodial growth.

pleiochasial; this can be observed regularly in *Rhododendron*. At the end of the shoot an inflorescence develops, which is overtopped by many lateral buds.

Branching — growth form

Depending on which mode of development predominates in new shoots of woody plants, two types of branching are distinguished. Trees have **acrotonic** branching: here buds or shoots near the top of the plant develop more than those near the base. Shrubs, on the other hand, have **basitonic** branching, where most growth takes place near the base, while upper branching is restrained.

Climbing shrubs or **lianas** are woody plants which use support to grow to considerable heights; they are not free-standing, like shrubs or trees. According to the manner of climbing we distinguish between climbers with tendrils and holdfasts, and rambling climbers. **Twiners** climb through their winding stems; depending on the direction of climbing there are right-handed and left-handed twiners. To determine the direction of twining, follow the direction of growth from below to above — if this is a constant left turn, then you are looking at a left-hand (anti-clockwise) twiner. **Ramblers** are the simplest climbers, they climb by growing through other plants. Spreading or recurved parts like spines and thorns prevent sliding back. Because of this most rose prickles are curved backwards. Occasionally there are such spreading parts in climbers as well, such as bent-back spiny prophyll scales in *Celastrus*, or prickly enations behind the leaf scar in species of *Wisteria*. Highly specialised are the **tendril climbers** in which the modified organs (stem, leaves or parts of leaves) are transformed into tendrils, by which they hold the growing plant secure. A speciality are holdfasts at the shoot tip of tendrils in *Parthenocissus*. **Root climbers** form adventitious roots as holdfasts on the climbing stems, with which they anchor to the substrate.

Buds

[…] The hopes of the forest, the small patient buds, are not so shapeless and similar, as one surmises – until one has looked at them better.

E. A. ROSSMÄSSLER
Flora im Winterkleide (1854)

The German word *knospe* (= bud) originates from the Old High German *knofsa* and is related to *knoten, knauf, knorren* and *knopf* as well as to the English *knob*. Very similar are the terms in the Scandinavian languages: Danish *knob*, Swedish *knopp* as well as Dutch *knop*. They have their common origin in the Germanic *knuppa*, which means clump. The word *knospe* was explained in the Germanic dictionary of the brothers Grimm, with the Latin words *nodus, tuber, gemma* (node, tuber, bud).

In the Slavic languages, the words for buds also relate to the compact shape: Czech *pupek*, Russian *potschek*. The French *bourgeon* means bud, but in medicine also means pimple; *bouton* can mean button as well as bud, the Latin *gemma* stands for both bud and jewel (or pearl), always items which are characterised by a rounded shape or closed-ness. In English, apart from bud, names such as burgeon have been used.

The verb deriving from the noun, budding, means opening: French *bourgeonner* and German *knospen*. In Polish the verb *pęcznieć* (swelling, billowing) or *pękać* (bursting) and even *począć* (begin) all have links with the words *pąk, pączek* (bud, little bud) and *pączkować* (budding, swelling). In different languages the content of the word root moves from the description of form to the meaning of the beginning, unfolding.

For the botanist a bud is an immature shoot with undeveloped internodes and, therefore, with leaves confined to the smallest space. The lowest bud leaves overtop the tip of the shoot and protect it. They are often in the shape of bud scales. At the tip of the shoot, enclosed by the basal leaves, is the apical meristem. It contains meristems ready to develop, and here the first new leaves are initiated, with more following after their unfolding. Another aspect is bud rest, a planned period of dormancy which can be very short or more prolonged.

Buds can be divided into lateral and terminal buds. **Lateral buds** (axillary buds) arise in leaf axils on the principal axis of a resting meristem; they are young lateral axes. **Terminal buds** terminate the stem and are capable of continuing growth with the meristems enclosed there.

Structure of buds

Bud scales and naked buds

The outer leaves protect the bud against environmental influences and against enemies like insects or fungi. They are ± strongly modified and adapted to the function of bud protection. Bud scales are completely reduced scale-shaped foliage leaves which only serve as bud protection.

A complete foliage leaf is divided into upper leaf and lower leaf. The upper leaf consists of stalk (petiole) and leaf blade (lamina); the lower leaf consists of the leaf base and, depending on the species, two stipules. Similarly, bud scales are differentiated according to their function into stipule homologues (**stipule**), leaf base homologues (**simple**) and leaf blade homologues (**laminar**) scales. These homologue relationships can be clearly distinguished when buds open in spring.

Stipule scales are placed in pairs where according to regular phyllotaxy one leaf is placed. Sometimes neighbouring stipules are fused. If there is more than one leaf at the node, the stipules of neighbouring leaves can be fused to interfoliar (or interpetiolar) stipules. Rarely are interfoliar stipules the only protection of the shoot tip, as in some Rubiaceae found as pot plants: *Coffea*, the coffee bush and *Gardenia*.

Simple and laminar scales cannot always be distinguished at first glance, and often there are transitions in one bud from the outer simple to inner laminar scales. The simple outer scales mostly have a very simple structure with a few strong veins which peter out. In the inner laminar scales, the veining is more strongly differentiated and corresponds to the veining of the foliage leaf. In pinnate-veined foliage leaves, it is pinnate-veined too.

Lateral bud of the rose (*Rosa*).

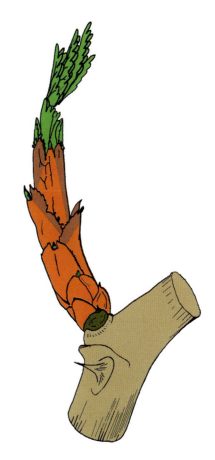

Opening lateral bud of the rose (*Rosa*).

Leaf of *Petteria ramentacea*.

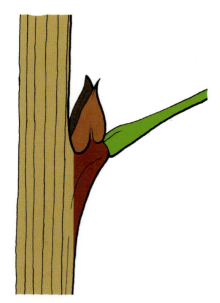

Detail: the stipules of a leaf remain in winter and cover the axillary bud.

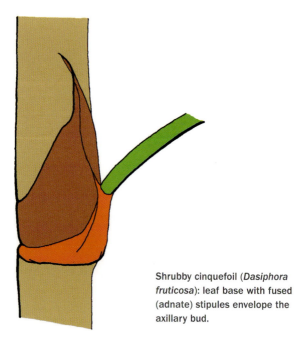

Shrubby cinquefoil (*Dasiphora fruticosa*): leaf base with fused (adnate) stipules envelope the axillary bud.

Sometimes outer laminar leaves are distinctly leaf-like. In other cases the bud leaves are undifferentiated and protected by a mostly quite dense indument. In both cases they develop during the unfolding into foliage leaves. True bud scales are missing and the buds are termed **naked**, in contrast to those covered by bud scales. Naked buds are, however, also adapted to the conditions of the cold season. The outermost leaves of large naked buds, as those of the wing nut (*Pterocarya*) are relatively firm. They contain much collenchyma: living support tissue capable of development. Almost all naked buds are densely hairy (e.g. *Rhamnus frangula*). In some species the lateral buds are sunk into the twig or surrounded by persistent parts of the leaf. The buds of the genera *Petteria* and *Berchemia* are enveloped by the persistent stipules of the leaf. In cinquefoil (*Dasiphora*) the buds are protected by thin membranous stipules, fused with the leafbase; they enclose the shoot node almost entirely and in this way also cover their axillary bud.

Terminal and lateral buds

It is fundamental to differentiate between end- or terminal buds and lateral or axillary buds.

A **terminal bud** ends the terminal end of a shoot, holds the tip meristem and is the requirement for monopodial growth. The bud leaves of the terminal buds follow the phyllotaxy of foliage leaves. When phyllotaxis is distichous, terminal buds are rarely developed.

A **lateral bud** sits in the axil of a leaf. It contains the resting meristem which is present (at least originally) in each leaf axil. In a resting bud, this can last several years as an organ reserve. A lateral bud is, most of all, a lateral branch initiation. Many species, however, only have lateral buds (they lack terminal buds). Here one or several lateral buds take over the growth of the tip — sympodially — in place of the terminal bud. Typical lateral buds often have few scales and are sometimes enveloped entirely by the prophyll scales. In some species the prophyll scales are partly or wholly fused (*Salix, Carya*).

Morphologically similar buds

In most species with terminal buds, the lateral buds hardly differ morphologically from these terminal buds. It appears that, as a rule, only one type of bud develops. Where these are lateral buds, the shoot tip dies regularly and only lateral buds remain. If terminal buds are present, lateral buds are always present too, but in their structure these are then mostly very similar to the terminal bud. As the axillary bud has a shoot tip too, it is not difficult to imagine that the terminal bud has here moved into the leaf axil. From the lateral bud often only the prophyll (sometimes not even this) remains and they are positioned in the leaf axil. A transitional form, from terminal buds on short-shoots to lateral buds homologous to terminal buds, is represented by stalked lateral buds: the first internode (hypopodium) of the lateral buds is distinctly lengthened. The subsequent internodes are, as in other buds, very short and covered by the outer bud leaves. Stalked buds are found mainly in species which also form terminal buds.

Walnut (*Juglans regia*): stalked lateral bud.

Lime (*Tilia*): opening lateral bud.

Lime (*Tilia*): infructescence.

Alder (*Alnus*): stalked lateral bud.

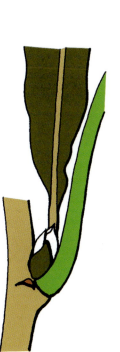

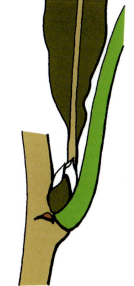

Lime (*Tilia*) detail: lateral bud on the infructescence.

Lime (*Tilia*): the first leaves of the lateral bud.

Morphologically different buds

Rarely does one species have two different types of buds. One example is the poplar (*Populus*): here the shoot tip is occupied by a terminal bud, enveloped by spirally positioned stipules; while the lateral buds have simple bud scales, which are distichous in the median plane. The first scale presumably consists of two partially fused prophylls. The closely related genus *Salix* (willows) has only lateral buds. Here we often find two separate axillary buds in the single, but fused all-round, bud scale. The bud scale has developed from two prophylls. The two genera have in common the position of the first scale (which is rather unobtrusive), as it is turned away from the principal axis.

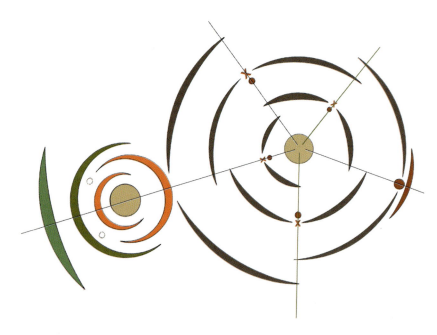

Poplar (*Populus*): diagram of terminal and lateral bud.

Twig of poplar (*Populus*) with terminal and lateral buds.

Lateral bud of poplar (*Populus*).

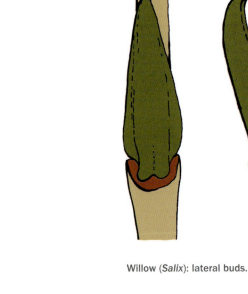

Willow (*Salix*): lateral buds.

Right: small lateral buds, placed basally on the twig in poplar (*Populus*); they are, as in willow (*Salix*), completely enclosed by prophyll scales.

Willow (*Salix*). Diagram of a lateral bud: two adnate simple prophylls form a single scale with two axillary buds.

Flower buds

Flower buds often differ externally from leaf buds. Particularly in species which flower early, the flowers and inflorescences are so far pre-formed that they influence the shape of the bud. A flower bud contains, as a rule, an inflorescence, which is entirely concealed in the winter bud. This inflorescence bud is often larger than a leaf bud. In winter identification manuals, inflorescence buds are abbreviated as flower buds and are compared with leaf buds (which do not contain a single leaf, but a leafy shoot). True flower buds in the strict sense do not contain an inflorescence, but a single flower enveloped by its own calyx lobes and the bracteoles. These are rarely seen in winter; over-wintering flower buds are found in *Asmina, Hamamelis, Paulownia tomentosa* and *Abeliophyllum distichum*. Here the single flower buds are united in inflorescences which are distinctly different from the vegetative shoot system.

Accessory and enrichment buds

In many species we find other buds above, below or next to the first-formed lateral bud. These buds serve as organ reserves (resting buds) or they can take the place of the main lateral bud should this be 'used up' as a thorn or tendril. Frequently these buds can develop as flower buds. Here accessory and enrichment buds can be differentiated. **Accessory buds** develop, additional to the first lateral bud, from the remnants of the same meristem. They are mostly smaller than the first lateral bud. If they are positioned transversely beside the primary lateral bud, then they are **colateral** accessory buds. These exist only in monocotyledons: the bulbils of garlic or the lateral fruits of a banana plant arise from colateral accessory buds. They are absent from all species treated in this book. But in woody plants there are frequently **serial accessory** buds: these occupy the median plane, either as **ascending accessory** buds (above the primary lateral bud) or as **descending accessory** buds (below the primary lateral bud). Because accessory buds arise from the same meristem as lateral buds, they do not differ from them, apart from their size.

Accessory lateral buds, diagrammatic. Black: primary lateral bud.

Serially ascending accessory buds, diagrammatic.

Serially ascending accessory buds (*Lonicera*).

Serially descending accessory buds, diagrammatic.

Serially descending accessory buds (*Neillia*).

Serially descending accessory buds (*Neillia*), detail.

Enrichment buds are, in many species, placed in the axils of the first bud scales of the lateral bud. These are mostly the axils of the first two leaves (prophylls) in the transverse plane. By this means, they are positioned (in contrast to the serial accessory buds) not over or below the first lateral bud, but beside it. As lateral ramifications, they are subordinate to the primary lateral bud. Through the scale-like leaves (prophylls) and their lateral position, they can be easily distinguished from the accessory buds.

Position of leaves in bud, and unfolding

Unfolding of buds

In areas with winter dormancy of the vegetation, the unfolding of leaves in spring is such a noticeable process, that it must have attracted attention even in earlier days. It seems likely to me that the term 'development', which we now use frequently in a figurative sense, originally comes from observations made on leaves, because many of them are enveloped in bud and then develop.

K. GOEBEL
Die Entfaltungsbewegungen der Pflanzen (1923).

Enrichment buds (*Prunus*).

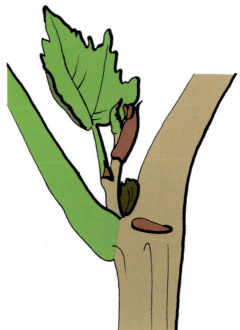

Grape vine (*Vitis*): enrichment buds.

Enrichment buds, diagrammatic. Green: prophylls and axillary buds = enrichment buds.

In spring the leafy tip of the young shoot frees itself from the strong embrace of the protecting bud scales. The ways that are followed in this process are different and dependent on the type of bud protection. Most frequently the bud scales elongate by the growth of intercalary meristems, often connected to an inside (adaxial) promotion. By these means, in firmly closed buds, the bud scales move strongly to the outside and free the shoot tip (*Aesculus*, *Magnolia*). Often also present, in the breaking open of the bud, are quite hairy transition leaves, located between the bud scales and the true leaves. These overlie the tender young foliage leaves and release them only after breaking through the outermost bud envelope.

This leads to the next form of unfolding, in which bud scales do not show any growth and the bud envelope is just burst by the leaves which gain in volume. This is noticeable in *Platanus*, where the bud envelope consists of a bud scale formed from fused stipules: this envelope is pierced by the shoot, which remains enveloped by the densely hairy stipules of the following leaves. In willow, which also has a single bud scale (here the product of fused prophylls) this tears off from the base during unfolding. It can then still cover the shoot for a while, like a hood. Other shoot-protective leaf organs are lacking here.

In the genus *Actinidia* the buds are enveloped by a persistent leaf base. During unfolding, the shoot tip is protected by short stubby and densely hairy leaves, which cover the centre of the bud. As in *Platanus*, the bud protection will simply be pierced.

Position of leaves in bud

Ptyxis, or folding of leaves in bud, indicates how the young leaves are folded within the bud. In this book aestivation is taken to mean the placing of leaves relative to each other, and vernation (ptyxis) the folding of a single leaf. Species-specific aestivation and vernation can help in the identification of species which are difficult to distinguish. It is easiest to observe leaf-folding during the unfolding of leaves; but it can also be seen in transverse section at any time when buds are present.

Vernation (ptyxis) is brought about by the forward or backwards bending of the leaves in the bud. In the bud there are folded and rolled leaves. Leaves in woody species are very rarely rolled up to the tip. These enrollings are an ancestral characteristic and can be frequently observed in ferns. In the woody species treated here it occurs only in the phylogenetically very old *Ginkgo biloba*: the leaves grow during development by tip-initials at the leaf margins, whilst the tissue from the base upwards changes to the permanent state. In most species, however, the leaves are folded laterally and rolled up. Here, intercalary extension growth begins during leaf development by means of a residual meristem.

In the placing of leaves relative to each other, three main types of aestivation are distinguished: overlapping (imbricate), twisted (contorted), and touching at the margins (valvate). Imbricate aestivation occurs mostly in species with distichous phyllotaxy, less often in those with alternate phyllotaxy. Here one leaf covers the following on both sides. The leaves in valvate aestivation touch at each side, but do not overlap. This occurs mainly with verticillate/opposite phyllotaxy. Contorted aestivation is found with alternate phyllotaxy. Here leaves cover one another on one side only. The direction of turning can be to the right or left. In terminal buds this follows the phyllotaxy of previous foliage leaves.

Elm (*Ulmus*): a developing lateral bud.

Hazel (*Corylus*): unfolding lateral bud and its first leaf.

Identification key

1 Leaves/leaf scars opposite or verticillate (2 or more leaf scars per node) . . . . 2 (Fig. 1)

1* Leaves/leaf scars alternate (only one leaf scar per node) . . . . 201 (Fig. 2)

Leaves/leaf scars opposite or verticillate

2 Leaf scars with one or three traces (of vascular bundles) 3

2* Leaf scars with more than three traces or leaf scars absent . . 101 (Figs 3 & 4)

3 Leaf scars with one trace (sometimes composed of many smaller traces) (if number of traces difficult to count, smooth surface by cutting across it with blade) . . . . 4 (Fig. 5)

3* Leaf scar with three traces . . 51 (Fig. 6)

4 Tree or liana 5

4* Shrub 11

5 Tree 6

5* Liana 37

6 Twigs with terminal bud 7

6* Twigs without terminal bud 8

7 Bud scales densely felty-hairy . . . .*Fraxinus* (Fig. 7)

7* Bud scales at most sparsely hairy . . . . *Chionanthus* (Fig. 8)

8 Every node with three leaf scars . . . . *Catalpa* (Fig. 9)

8* Leaf scars/buds opposite 9

9 Twigs with solid pith 10

9* Twigs with chambered pith . . . . *Paulownia tomentosa* (Fig. 10)

10 Buds at an angle of almost 90° to the red-brown branches . . . . *Metasequoia glyptostroboides* (Fig. 11)

10* Branches not red-brown, with latex . . . . *Broussonetia* (Fig. 12)

11 (4) Twigs grass-green all round, sometimes slightly reddened, lenticels absent 12

11* Twigs not grass-green, lenticels often present 14

12 Twigs without terminal bud 13

12* Twigs with terminal bud . . . . *Euonymus* (Fig. 13)

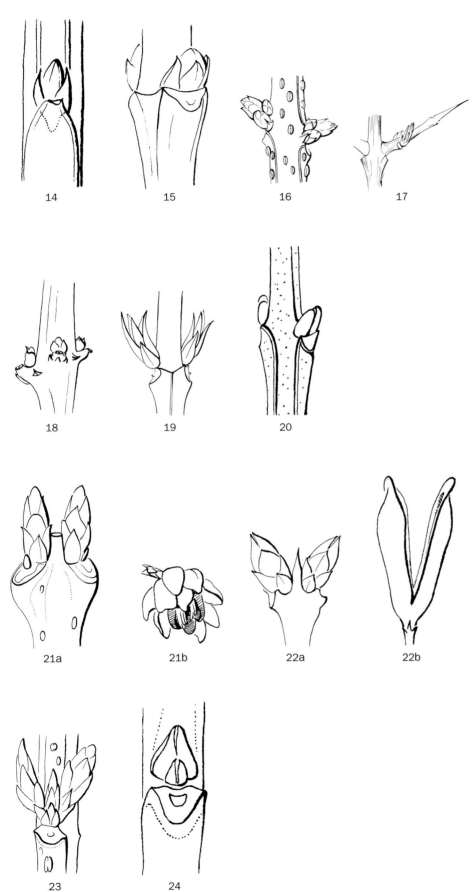

14 15 16 17

18 19 20

21a 21b 22a 22b

23 24

13 Buds almost hidden under projecting leaf cushion and stipules 36

13* Buds clearly visible. . . . . ***Jasminum*** (Fig. 14)

14 Twigs with solid pith 16

14* Twigs hollow or pith chambered . . . 15

15 Opposite leaf scars connected by a line . . . .***Symphoricarpos*** (Fig. 15)

15* Opposite leaf scars free, often slightly displaced towards each other 26

16 Twigs with terminal bud 29

16* Twigs without terminal bud 17

17 Buds with bud scales 18

17* Buds naked (without bud scales), active or hidden 38

18 Buds above the leaf scar solitary . . . 19

18* Buds above the leaf scar in groups, twigs with many large lenticels . . . . ***Coriaria*** (Fig. 16)

19 Twigs with latex . . . . ***Broussonetia*** (Fig. 12)

19* Twigs without latex 20

20 Twigs armed with thorns . . . . ***Punica granatum*** (Fig. 17)

20* Twigs unarmed 21

21 Buds ternate (in threes), twigs mostly dead. . . ***Fuchsia magellanica*** (Fig. 18)

21* Buds opposite 22

22 Twigs aromatic 27

22* Twigs not aromatic 23

23 Opposite leaf scars connected by a line, bud scales with pellucid dots, fruits with large sepals . . . . . ***Hypericum*** (Fig. 19)

23* Opposite leaf scars free, often slightly displaced towards each other 24

24 Twigs strongly 4-angular, thin, getting progressively thinner towards the tip. . . . . ***Fontanesia*** (Fig. 20)

24* Twigs relatively sturdy, with two terminal lateral buds 25

25 Flowering in winter, leaf cushion mostly larger than lateral bud . . . . ***Chimonanthus praecox*** (Fig. 21)

25* Not flowering in winter; when leaf cushions projecting more than bud, bark cross-hatched . . ***Syringa*** (Fig. 22)

26 (15) Buds more than 5 mm long; flower buds many, next to the leaf buds . . . . ***Forsythia*** (Fig. 23)

26* Buds under 5 mm long; inflorescences at shoot end, flower buds covered by dark purple sepals . . . . ***Abeliophyllum distichum*** (Fig. 24)

27 (22) Buds subopposite (opposite leaf scars slightly displaced towards each other)...... *Orixa japonica* (Fig. 25)

27* Buds exactly opposite, opposite leaf scars connected by a thin line; mostly half-shrubby................... 28

28 At least upper bud scales with stellate hairs .... *Perovskia* (Fig. 26)

28* Twigs and buds with simple glandular hairs... *Elsholtzia stauntoni* (Fig. 27)

29 (16) Twig tips and buds with metallic glossy lepidote scales ....
.... *Shepherdia* (Fig. 28)

29* Glabrous or with a different indument
.... 30

30 Buds naked 31

30* Buds with scales.. *Callicarpa* (Fig. 29)

31 Twigs green with warty lenticels or corky wings ..... *Euonymus* (Fig. 30)

31* Twigs different................. 32

32 Twigs with lenticels, buds less than 15 mm long 33

32* Twigs without lenticels, terminal buds fusiform, 1.5–2 cm long ....
.... *Euonymus* (Fig. 13)

33 Twigs hardly over 2 mm thick, lateral buds rarely over 2 mm long 34

33 Lateral buds larger or twigs thicker . 35

34 Tips of twigs about 1 mm thick, fruits black, in terminal panicles ....
.... *Ligustrum* (Fig. 31)

34* Tips of twigs less than 1 mm thick ...
.... *Forestiera* (Fig. 32)

35 Terminal buds compressed (less than 1.5 times longer than wide).........
.... *Chionanthus* (Fig. 8)

35* Terminal buds more elongated ......
.... *Syringa* (Fig. 33)

36 (13) Twigs rush-like, only partially opposite.. *Spartium junceum* (Fig. 174)

36* Twigs different, partly thorny.......
.... *Genista* (Fig. 34)

37 (5) Buds almost hidden below leaf remnants .... *Periploca* (Fig. 35)

37* Buds clearly visible; liana climbing with holdfast roots ..... *Campsis* (Fig. 36)

38 (17) Buds hidden or lateral twigs active
.... 39

38* Buds visible 40

39 Buds hidden, leaf scars mostly in whorls of 3 ....
.. *Cephalanthus occidentalis* (Fig. 37)

39* Twigs remain active and freeze or dry easily, phyllotaxy opposite ....
.... *Buddleja davidii* (Fig. 38)

25   26   27   28

29   30   31a   31b

32a   32b   33   34

35   36   37

38   39   40   41

42   43   44a   44b

45   46   47   48

49a   49b

40   Twigs thin, buds silvery-green hairy. . .
. . . . *Caryopteris* (Fig. 39)

40*   Twigs relatively thick, buds brown to
wine-red 41

41   Buds dark wine-red hairy . . . .
. *Clerodendrum trichotomum* (Fig. 40)

41*   Buds ochre-brown hairy . . . .
. . . . *Vitex agnus-castus* (Fig. 41)

Leaves/leaf scars opposite or verticillate/leaf scars with 3 traces

51 (3) All buds with bud scales. 64

51*   At least a part of the buds (terminal or
lateral) naked, or hidden 52

52   Buds visible 54

52*   Buds hidden under leaf scar, without
terminal bud 53

53   Twigs noticeably thickened at the
nodes; densely hairy around leaf scar,
buds often active . . . .
. . . . *Zabelia mosanensis* (Fig. 42)

53*   Leaf scar not hairy, twigs only slightly
thickened at the nodes. . . . .
. . . . *Philadelphus* (Fig. 43)

54   Free-standing shrubs or trees 55

54*   Liana climbing with holdfasts . . . .
. . . . .*Hydrangea* (*Decumaria*) *barbara*
(Fig. 44)

55   Opposite leaf scars touching, or
connected by a line 60

55*   Opposite leaf scars free, often somewhat
displaced towards each other 56

56   Single-stemmed trees 57

56*   Shrubs 58

57   Twigs with terminal bud . . . .
. . . . *Tetradium* (Fig. 45)

57*   Twigs without terminal bud . . . .
. . . . *Phellodendron* (Fig. 46)

58   Buds very dark-hairy 59

58*   Buds pale ochre-brown-hairy . . . .
. . . . *Vitex agnus-castus* (Fig. 41)

59   Twigs dark grey to red-brown . . . .
. . . .*Calycanthus* (Fig. 47)

59*   Twigs pale ochre-brown. . . . .
. . . . . . *Calycanthus chinensis* (Fig. 48)

60 (55) At least the lateral buds naked. . .61

60*   Only the terminal bud naked; twigs
mostly more than 3 mm thick . . . .
. . . .*Hydrangea* (Fig. 49)

61   Buds with simple hairs 62

61* Buds with stellate hairs . . . .
. . . .*Viburnum* (Fig. 50)

62 Buds finely hairy, the form of the
individual leaves can mostly be
recognised, twig bark not peeling . . 63

62* Terminal bud densely long-hairy, this
year's twigs with flaking bark. . . . . . .
. . . .*Jamesia americana* (Fig. 51)

63 Twigs stiff, 3–4 mm thick between
nodes . . . *Dipteronia sinensis* (Fig. 52)

63* Twigs thinner, occasionally luminescent
yellow or red, partly with larger globose
flower buds . . . . *Cornus* (Fig. 53)

64 (51) Liana 65

64* Tree or shrub 67

65 Liana with holdfast roots 66

65* Winding lianas with hollow twigs . . . .
. . . .*Lonicera* (Fig. 54)

66 Bark of young twigs grey-brown,
old bud scales remain at the growth
border . . .*Hydrangea (Schizophragma)
hydrangeoides* (Fig. 55)

66* Bark of young twigs red-brown . . . . . .
. . . . . .*Hydrangea petiolaris* (Fig. 56)

67 Leaf scar with stipule or stipule scars,
always without terminal buds 68

67* Leaf scar without separate stipule scars
. . . . 69

68 At least at the twig tip with small
stipule lobes at the leaf scar, twigs thin
. . . . . . *Rhodotypos scandens* (Fig. 57)

68* Only stipule scars present, twigs sturdy,
mostly more than 2 mm thick . . . .
. . . .*Staphylea* (Fig. 58)

69 Opposite leaf scars touching or
connected by a line 73

69* Opposite leaf scars free, often displaced
towards each other 70

70 Buds with only a single visible bud scale
. . . . 71

70* Buds with several (at least 2) visible
bud scales 72

71 Bud scale with dorsal suture on bud . .
. . *Cercidiphyllum japonicum* (Fig. 59)

71* Bud scale fused all round . . . .
. . . .*Salix* (Fig. 60)

72 Twigs with shoot thorns. . . . .
. *Rhamnus*, section *Rhamnus* (Fig. 61)

72* Twigs unarmed, flowering very early
(Jan/Feb) . . . .
. . . . . *Chimonanthus praecox* (Fig. 21)

73 (69) Shrub. 74

50   51   52a   52b

53a   53b   54   55

56a   56b   57   58

59   60   61   62

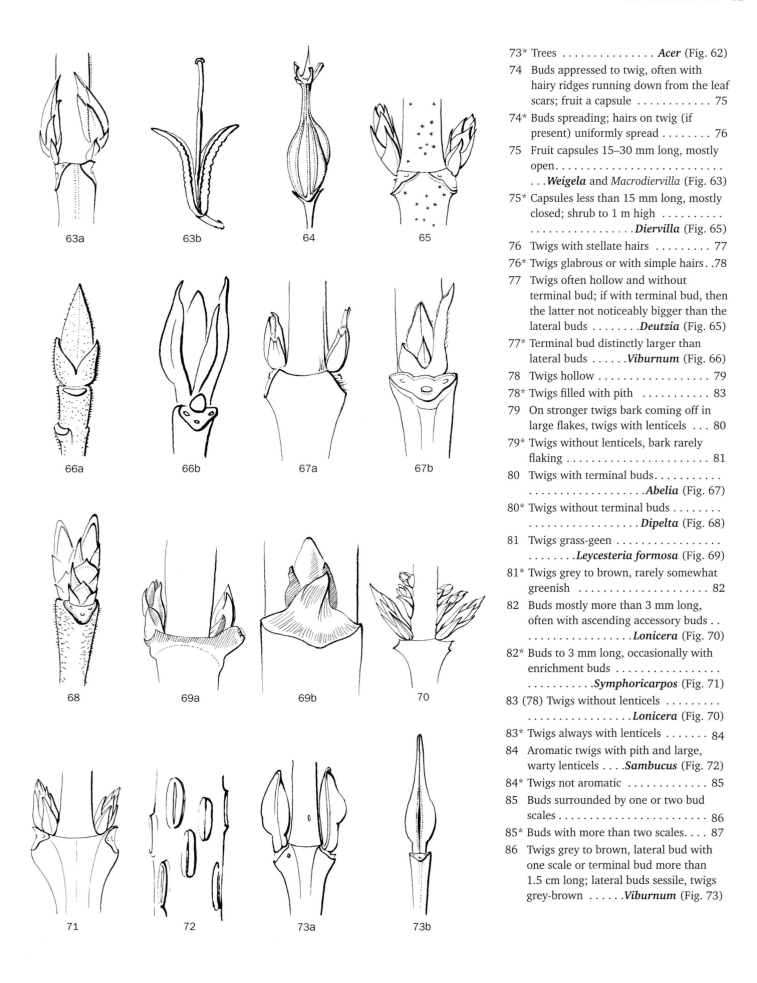

63a   63b   64   65

66a   66b   67a   67b

68   69a   69b   70

71   72   73a   73b

73* Trees . . . . **Acer** (Fig. 62)

74 Buds appressed to twig, often with hairy ridges running down from the leaf scars; fruit a capsule 75

74* Buds spreading; hairs on twig (if present) uniformly spread 76

75 Fruit capsules 15–30 mm long, mostly open. . . . . **Weigela** and *Macrodiervilla* (Fig. 63)

75* Capsules less than 15 mm long, mostly closed; shrub to 1 m high . . . . **Diervilla** (Fig. 65)

76 Twigs with stellate hairs 77

76* Twigs glabrous or with simple hairs. .78

77 Twigs often hollow and without terminal bud; if with terminal bud, then the latter not noticeably bigger than the lateral buds . . . . **Deutzia** (Fig. 65)

77* Terminal bud distinctly larger than lateral buds . . . . . . **Viburnum** (Fig. 66)

78 Twigs hollow 79

78* Twigs filled with pith 83

79 On stronger twigs bark coming off in large flakes, twigs with lenticels . . . 80

79* Twigs without lenticels, bark rarely flaking 81

80 Twigs with terminal buds. . . . . **Abelia** (Fig. 67)

80* Twigs without terminal buds . . . . **Dipelta** (Fig. 68)

81 Twigs grass-geen . . . .**Leycesteria formosa** (Fig. 69)

81* Twigs grey to brown, rarely somewhat greenish 82

82 Buds mostly more than 3 mm long, often with ascending accessory buds . . . .**Lonicera** (Fig. 70)

82* Buds to 3 mm long, occasionally with enrichment buds . . . .**Symphoricarpos** (Fig. 71)

83 (78) Twigs without lenticels . . . .**Lonicera** (Fig. 70)

83* Twigs always with lenticels 84

84 Aromatic twigs with pith and large, warty lenticels . . . .**Sambucus** (Fig. 72)

84* Twigs not aromatic 85

85 Buds surrounded by one or two bud scales 86

85* Buds with more than two scales. . . . 87

86 Twigs grey to brown, lateral bud with one scale or terminal bud more than 1.5 cm long; lateral buds sessile, twigs grey-brown . . . . . .**Viburnum** (Fig. 73)

86* Lateral buds with two scales, twigs mostly reddish to greenish .... **Acer** (Fig. 74).

87 Buds whorled in 3 .... **Hydrangea paniculata** (Fig. 75)

87* Buds opposite 88

88 Two basally connate densely bristly indehiscent fruits with long-extending 5-lobed calyx tube .... **Kolkwitzia amabilis** (Fig. 76)

88* Fruits absent or different 89

89 Leaf scars entirely surrounding lateral buds 90

89* Leaf scars hardly surrounding buds ..91

90 Shoot tips often with erect infructescence axis. Shrub with curved branches growing upwards .... **Aesculus parviflora** (Fig. 77)

90* Habit different .... **Acer** (Fig. 62)

91 Bark of twig flaking off .... **Hydrangea** (Fig. 78)

91* Bark of twig not flaking off 92

92 Twigs with terminal bud .... **Viburnum** (Fig. 79)

92* Twigs dying off slightly at the tip, without terminal buds ....**Heptocodium miconioides** (Fig. 80)

Leaves/leaf scars opposite or verticillate; leaf scars with more than 3 traces or leaf scars absent

101 (2) Leaf scars with more than 3 traces 102

101 Leaf scars not present 118

102 Buds with bud scales 107

102* Buds without bud scales 103

103 Twigs with terminal bud 105

103* Twigs without terminal bud 104

104 Leaf scar surrounding the bud .... **Phellodendron** (Fig. 46)

104* Bud above the leaf scar .... **Clerodendrum trichotomum** (Fig. 40)

105 Tree .... **Fraxinus** (Fig. 7)

105* Shrub 106

106 Bark of twig longitudinally fissured and flaking ..... **Hydrangea** (Fig. 78)

106* Bark of twig not flaking .... **Dipteronia sinensis** (Fig. 52)

107 (102) Tree or shrub 109

107* Liana climbing with holdfasts 108

108 Twigs with terminal buds .... **Hydrangea** (Schizophragma) **hydrangeoides** (Fig. 55)

108* Twigs without terminal buds .... **Campsis** (Fig. 36)

74   75   76   77a

77b   78   79a   79b

80a   80b   81   82

83   84   85   86

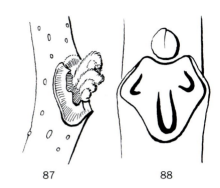

87   88

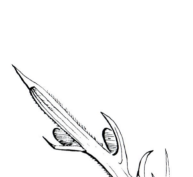

89

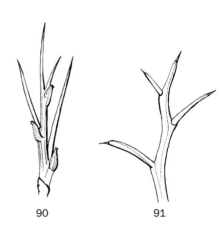

90   91

Phyllotaxy alternate

Leaves/leaf scars alternate, twigs armed with prickles, spines or thorns

216 Twigs without terminal buds, flower buds closely set, relatively large: 4–6 mm long, bronze-brown. . . . .
. . . . *Hippophae* (Fig. 92)

216* Flower buds smaller or twigs silver-white or with terminal bud . . . .
. . . . *Elaeagnus* (Fig. 93)

217 Twigs with terminal bud. 218

217* Twigs always without terminal bud . .
. . . . 221

218 Thorns sharply pointed 219

218* Thorny tip of branches not very sharp, tip of branches hairy . *Malus* (Fig. 199)

219 Tips of twigs appressed-hairy or glabrous 220

219* Twig tips spreading long-hairy . . . . .
. . . . *Crataegus* (*Mespilus*) *germanica* (Fig. 197)

220 Buds pointed, matt. . *Pyrus* (Fig. 198)

220* Buds rounded, glossy . . . .
. . . .*Crataegus* (Fig. 94)

221 Thorns unbranched, buds visible . . 222

221* Thorns branched, buds sunk into tissue. . . . . *Gleditsia* (Fig. 95)

222 Thorns thicker in middle, thinner towards base . . . .
. . . . . . *Hemiptelea davidii* (Fig. 96)

222* Thorns thickest at base. 223

223 Youngest twigs with latex 224

223* Twigs without latex 225

224 Young twigs greenish . . . .
. . . .*Maclura pomifera* (Fig. 97)

224* Young twigs brown. . . . .
. . . *Maclura* (*Cudrania*) *tricuspidata* (Fig. 98)

225 Cushion shrub with thorny inflorescence axis. . . . . *Alyssum spinosum* (synonym of *Ptilotrichum spinosum*)

225* Plant different 226

226 Stipule scar surrounds twig at every node. Small shrub, hardly over 50 cm high . . . *Atraphaxis spinosa* (Fig. 99)

226* Mostly larger shrubs. 227

227 Leaf scar with one trace 228

227* Leaf scar with three traces 230

228 Thorns pointed, twigs round 229

228* Thorns not pointed, twigs edged, pale ochre . . . . *Lycium* (Fig. 100)

229 Thorn above the bud (bud is accessory bud to thorn); pith chambered. . . . . .
. . . .*Prinsepia* (Fig. 101)

229* Buds laterally on the thorn; pith solid
. . . . *Chaenomeles* (Fig. 102)

230 Buds solitary . . . .
. . . . *Sophora davidii* (Fig. 103)

92   93   94   95

96   97   98a   98b

99   100   101   102

103   104

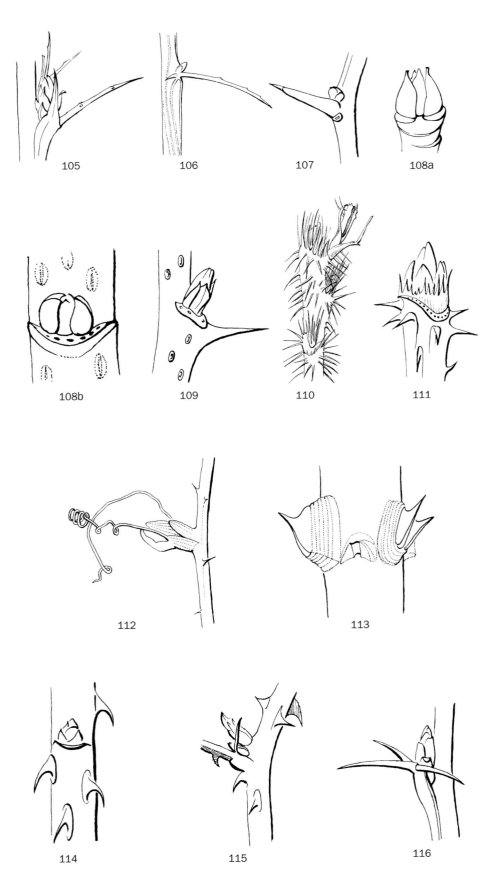

105   106   107   108a

108b   109   110   111

112   113

114   115   116

230* Next to the leaf buds often larger flower buds......*Prunus* "plums" (Fig. 104)

231 (211) Spines simple 232

231* Spines several-parted 235

232 Spines (=leaf rachis) with tiny opposite scars of dehisced leaflets 233

232* Spines without scars........... 234

233 Stipules flattened or thorns less than 1.5 cm long ....*Caragana* (Fig. 105)

233* Stipules round in section, spines more than 2 cm long..... *Halimodendron halodendron* (Fig. 106)

234 Twigs and spines brown 245

234* Twigs and spines green .... *Citrus trifoliata* (Fig. 107)

235 Spines 2-parted 246

235* Spines of 3-parted or more 245

236 (210) Leaf scars wide and thin, with many traces, coarse-branched shrubs or small tree 237

236* Leaf scars with one to three traces, or liana 240

237 Lateral buds semi-globose with dark wine-red glossy scales, small tree.... .. *Kalopanax septemlobus* (Fig. 108)

237* Shrubs 238

238 Twigs to 5 mm thick.............. .... *Eleutherococcus* (Fig. 109)

238* Twigs thicker 239

239 Twigs c. 10 mm thick, prickles numerous, covering buds .... *Oplopanax horridus* (Fig. 110)

239* Twigs thicker, prickles shorter than the buds ....*Aralia* (Fig. 111)

240 Liana, of the leaf the leaf base remains with two thin stipule-tendrils....... ....*Smilax* (Fig. 112)

240* Erect shrubs or trees 241

241 Ground bark of twig aromatic, older prickles on cork cushions .... *Zanthoxylum* (Fig. 113)

241* Twigs not so................. 242

242 Prickles irregularly distributed .. 243

242* One to three prickles per node .. 244

243 Leaf scars wide and thin .... *Rosa* (Fig. 114)

243* Leaf remnants formed by a projecting leaf cushion or the whole petiole .... *Rubus* (Fig. 115)

244 Two prickles/spines at the node . 246

244* One, three or more prickles/spines on the node 245

245 (234/244) Above every spine a short-shoot with distinct leaf scars at the base .... *Berberis* (Fig. 116)

245* Above the prickles a distinct leaf scar/ above that (on this year's long-shoots) buds with bud scales, but no further scars . . . . **Ribes** (Fig. 117)

246 (235/244) At each node there is always one larger and one smaller, often curved, stipule spine. 247

246* Prickles/spines at the node of almost equal size 248

247 Twigs grey-brown, spines mostly under 1 cm long . . . . **Paliurus spina-christi** (Fig. 118)

247* Twigs wine-red; longer spine mostly more than 1 cm long . . . . **Ziziphus jujuba** (Fig. 119)

248 Buds visible, shrubs 249

248* Buds hidden, often a tree, with paired stipule-spines . . . . **Robinia** (Fig. 120)

249 Twigs with prickles that are easily broken off. 243

249* With membranous stipule-spines connected firmly to the twig. . . . .**Caragana** (Fig. 121)

Leaves/leaf scars alternate, lianas

251 (201) Liana winding by its twigs . 252

251* Liana holding with tendrils 264

252 Buds clearly raised above the twig and visible, with bud scales. 253

252* Buds hidden (at least not raised above the twig surface) or without bud scales 261

253 Leaf scars with one trace, but never with a scar circling the twig 254

253* Leaf scar with more than one trace, or with a line (=stipule scar) circling the twig 257

254 Buds globose and compact, spreading from the twig 255

254* Buds elongate, lying along the twig . . . . **Wisteria** (Fig. 122)

255 Twigs not green, pith solid or, more rarely, chambered 256

255* Twigs greenish and hollow, half shrubby . . . . . **Solanum dulcamara** (Fig. 123)

256 Twigs grey-brown **Celastrus** (Fig. 124)

256* Twigs red-brown, with many lenticels . . . .**Tripterygium wilfordii** (Fig. 125)

257 Each node with a line encircling twig. . . .**Fallopia baldschuanica** (Fig. 126)

257* Nodes without lines encircling twig 258

258 Buds with many bud scales 259

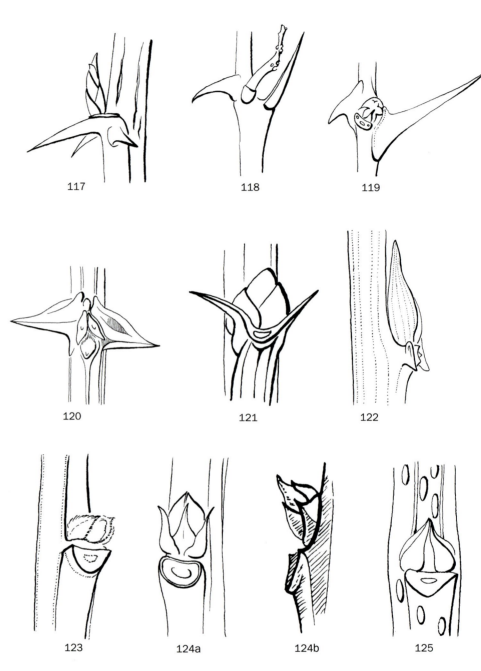

117   118   119

120   121   122

123   124a   124b   125

126

127a 127b 128 129

130 131 132 133

134 135 136a 136b

137 138

258* Buds along the twig, enclosed by two scales which are leaf stipules . . . . *Berchemia* (Fig. 127)

259 Twigs less than 3 mm thick 260

259* Twigs thick (more than 5 mm) . . . . *Sinofranchetia chinensis* (Fig. 128)

260 Leaf cushion clearly projecting. . . . . *Akebia* (Fig. 129)

260* Leaf scar relatively flat on the twig . . . . *Schisandra* (Fig. 130)

261 (252) Buds concealed or not raised above the twig surface, pith chambered or solid 263

261* Buds well visible, pith never chambered 262

262 Leaf scar narrowly U-shaped, encircling the buds . . . . . *Aristolochia* (Fig. 131)

262* Leaf scar round-ovate. . . . . *Cocculus* (Fig. 132)

263 Buds visible only when twig is cut, without scales, pith often chambered. . . . . *Actinidia* (Fig. 133)

263* Pith never chambered, buds with scales . . . . . *Menispermum* (Fig. 134)

264 (251) Tendrils opposite the leaf scars 265

264* Tendrils always formed from 2 leaf stipules. . . . . *Smilax* (Fig. 112)

265 Twigs with lenticels, pith white . . 266

265* Twigs without lenticels, pith brown . . . . *Vitis* (Fig. 135)

266 Buds hidden or longer than 8 mm, tendrils without holdfasts. . . . .*Ampelopsis* (Fig. 136)

266* Buds less than 8 mm long, tendrils often with holdfasts . . . . *Parthenocissus* (Fig. 137)

Leaves/leaf scars alternate, twigs with terminal buds, buds without bud scales

271 (204) Leaf scars with three traces . 272

271* Leaf scars one- or multi-traced . . 283

272 Twigs, buds with stellate hairs . . 278

272* Twigs, buds glabrous or with other kinds of hairs 273

273 Twigs with chambered pith 277

273* Twigs with uninterrupted pith . . . 274

274 Trees, lateral buds with accessory buds, often with small peltate hairs . . . . *Carya* (Fig. 138)

274* Lateral buds without accessory buds 275

275 Large leaf cushions, clearly distinct from the twig; partly with stipule lobes, buds silvery pilose . . . . **Laburnum** (Fig. 139)

275* Without distinct leaf cushions, brownish hairy 276

276 Terminal bud wider than long . . . . **Picrasma quassioides** (Fig. 140)

276* Terminal bud longer than wide . . . . **Rhamnus**, subgenus **Frangula** (Fig. 141)

277 (273) Terminal bud longer than 1.8 cm, bud leaves spreading . . . . **Pterocarya** (Fig. 142)

277* Bud leaves appressed, buds shorter than 1.5 cm . . . . . . **Juglans** (Fig. 143)

278 (272) Twigs and buds matt 279

278* Twigs glossy, sticky . . . . **Chamaebatiaria millefolium** (Fig. 144)

279 Inflorescences/infructescences terminal; flowers in winter to the end of March. Lateral buds with enrichment buds or bark of twig flaking off in scales. . 280

279* Inflorescences/infructescences axillary. Flowering later; lateral buds partially with accessory buds 282

280 Inflorescences black-brown, cylindrical . . . . . . **Sinowilsonia henryi** (Fig. 145)

280* Flower buds compact-globose . . . 281

281 Lateral buds distinctly stalked, bud leaves appressed. Large shrub to 7 m high . . . **Parrotiopsis jaquemontiana** (Fig. 146)

281* Lateral buds sessile or shortly stalked, outermost bud leaves mostly with spreading tip. Small shrub, 1–2 m high . . . . **Fothergilla** (Fig. 147)

282 Flower buds 4–5 mm thick. Flowers without petals, bark on older branches and on stems flaking . . . . **Parrotia persica** (Fig. 148)

282* Flower buds smaller, in dense groups, flowers with 4 narrow linear petals, leaf buds long-stalked with accessory buds . . . . **Hamamelis** (Fig. 149)

283 (271) Leaf scar with one trace . . . 284

283* Leaf scar with multiple traces. . . 288

284 Twigs with stellate hairs or lepidote scales 285

284* Twigs with simple hairs or glabrous 291

285 Twigs with lepidote scales . . . . **Elaeagnus** (Fig. 150)

139

140a

140b

141

142

143

144a

144b

145

146

147

148

149

150

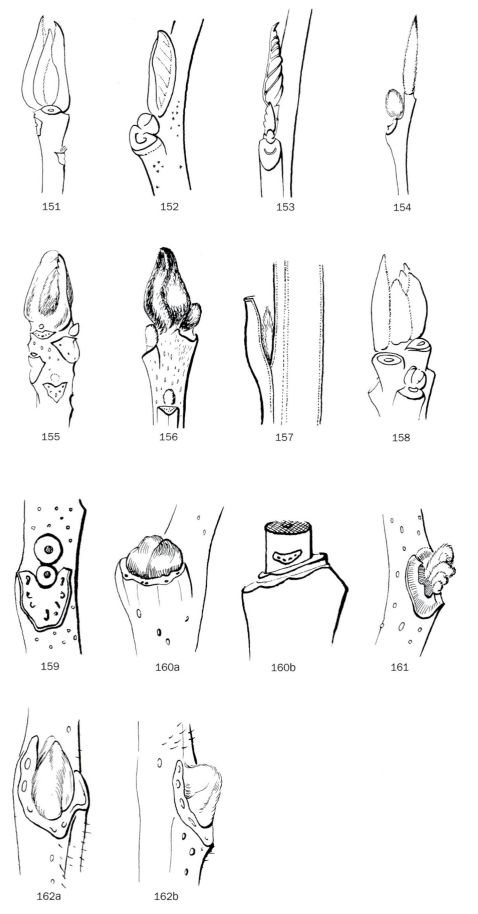

151   152   153   154

155   156   157   158

159   160a   160b   161

162a   162b

285* Twigs with stellate hairs 286

286   Lateral buds short-stalked, mostly with accessory buds 287

286* Lateral buds singly, sessile. . . . . *Clethra* (Fig. 151)

287   Lateral buds to 8 mm long, brown. . . . . ***Sinojackia*** (Fig. 152)

287* Lateral buds longer than 10 mm, outer scales easily dehiscing, then greenish. . . . .*Pterostyrax* (Fig. 153)

288 (283) Terminal bud long (at least 3 times as long as wide) 289

288* Terminal bud ovoid, not very long. . 290

289   Buds with simple black-brown hairs, flower buds globose . . . . ***Asimina triloba*** (Fig. 154)

289* Lateral buds often with accessory buds, twigs and buds often with peltate hairs . . . . ***Carya*** (Fig. 138)

290   Leaf scar with 5 traces . . . . ***Toona sinensis*** (Fig. 155)

290* Leaf scar with multiple traces, twigs with latex (attention! contact poison!) . . . . ***Toxicodendron*** (Fig. 156)

291 (284) Leaf scar below the bud, trace of vascular bundle distinct 292

291* Leaf cushion extends at least to above base of bud, often with stipule lobes . . . .***Chamaecytisus*** (Fig. 157)

292   Buds dark violet-brown . . . .*Daphne* (Fig. 158)

292* Buds greenish . . . . ***Edgeworthia***

Leaves and leaf scars alternate, twigs without terminal buds, buds without bud scales or concealed

301 (205) Buds visible 302

301* Buds hidden under leaf remnants . 312

302   Leaf scar with a single trace 315

302* Leaf scar with several traces 303

303   Leaf scar with 3 traces 307

303* Leaf scar with more than 3 traces . 304

304   Twigs more than 8 mm thick . . . .***Gymnocladus dioicus*** (Fig. 159)

304* Twigs thinner 305

305   Leaf scars encircling the node; shrub . . . .***Dirca palustris*** (Fig. 160)

305* Leaf scars not encircling the node; usually tree-like 306

306   Buds with several accessory buds. . . . . ***Cladrastis*** (Fig. 161)

306* Buds with 1 or without accessory buds . . . . ***Alangium*** (Fig. 162)

307 (303) Leaf scars encircle the visible buds in a U-shape 308

307* Leaf scars not encircling buds, sometimes buds under the leaf scar or in bark tissue 310

308 Lateral buds without accessory buds, twigs brown .... *Rhus* (Fig. 163)

308* Lateral buds with accessory buds . 309

309 Twigs green .... ***Styphnolobium japonicum*** (Fig. 164)

309* Twigs brown, up to 3 accessory buds below lateral bud .... *Cladrastis* (Fig. 161)

310 Buds with accessory buds sunk in tissue of twig above leaf scar .... *Gleditsia* (Fig. 165)

310* Buds hidden under leaf scar..... 311

311 Twigs round .... ***Ptelea trifoliata*** (Fig. 166)

311* Twigs angular .... *Robinia* (Fig. 167)

312 (301) Twigs light grey-brownish; shrubs 313

312* Twigs green or dark brown 314

313 Twig internodes less than 1 cm long .... ***Ononis fruticosa*** (Fig. 168)

313* Twig internodes longer........... .... ***Hibiscus syriacus*** (Fig. 169)

314 Twigs yellowish-green to grass-green; shrubs................319

314 *Twigs dark greenish to brown; often a tree 310

315 (302) Twigs and buds with stellate hairs 316

315* Twigs glabrous or with simple hairs... 317

316 Buds sessile .... ***Buddleja alternifolia*** (Fig. 170)

316* Buds stalked, with accessory buds ... . *Styrax* (Fig. 171)

317 Twigs green 319

317* Twigs grey-brown 318

318 Tree .... ***Meliosma dilleniifolia*** subsp. ***tenuis*** (Fig. 172)

318 Shrub; buds and lateral twigs often clearly distichous .... *Cotoneaster* (Fig. 173)

319 (314) Twigs ± round, finely grooved, rush-like, buds not visible.......... ...... ***Spartium junceum*** (Fig. 174)

319* Twigs different 320

320 Twig nodes thickened by the leaf bases, which cover the buds........ .... ***Genista pilosa*** (Fig. 175)

320* Twig nodes not thickened......... .... *Cytisus* (Fig. 176)

163 164 165 166

167 168 169 170

171 172 173 174

175 176

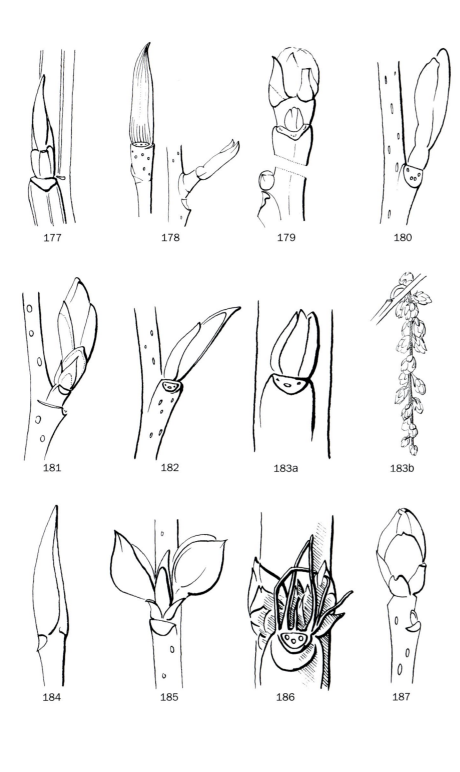

177 178 179 180

181 182 183a 183b

184 185 186 187

Leaves/leaf scars alternate, twigs with terminal buds, buds with bud scales, leaf scars with three traces

331 (206) Leaf buds and flower buds distinctly different 332

331* Leaf buds and flower buds externally no different 341

332 First bud scale of lateral bud dorsal . . . . *Populus* (Fig. 177)

332* First scales of lateral buds lateral to the bud 333

333 Flower buds: catkins without bud scales 334

333* Flower buds different from leaf buds, either buds larger or smaller 338

334 Pith of twig chambered 335

334* Twig with solid pith 336

335 Terminal buds more than 20 mm long . . . . . *Pterocarya rhoifolia* (Fig. 178)

335* Buds compact . . . .*Juglans* (Fig. 179)

336 Lateral buds stalked. .*Alnus* (Fig. 180)

336* Lateral buds sessile 337

337 Terminal buds (only on short shoots) with 4 or more bud scales . . . .*Betula* (Fig. 181)

337* Terminal buds with 2–3 outer scales . . . .*Alnus alnobetula* (Fig. 182)

338 (333) Flower buds different, numerous in elongated racemes . . . .*Stachyurus* (Fig. 183)

338* Flower buds on same twig part as the leaf buds 339

339 Flower buds as enrichment buds beside a leaf bud 340

339* Flower buds often solitary; more bulging than the slim leaf buds . . . . *Corylopsis* (Fig. 184)

340 Flower buds slightly stalked, scales greenish to red; lowest scale of a bud almost as long as the bud . . . .*Lindera* (Fig. 185)

340* Buds sessile, brownish . . . . *Prunus* (Fig. 186)

341 (331) Twigs with solid pith 343

341* Pith chambered 342

342 Shrub . . . .*Oemleria cerasiformis* (Fig. 187)

342* Tree 335

343 Small shrubs, rarely over 2 m high; older twigs rarely more than 2 cm thick, bark often coming off in large plates 344

343* Trees or larger shrubs; older twigs and stems mostly significantly thicker . . 347

344 Bark flaking irregularly and longitudinally 345

344* Bark not flaking or (only) transversely 346

345 Leaf scar flaking off together with the bark of the twig, only the three traces visible..*Sibiraea altaiensis* (Fig. 188)

345* Leaf scar distinct, lateral buds slightly stalked ....*Ribes* (Fig. 189)

346 Multiple fruits numerous in corymbs, with 2–5 dry follicles ....*Physocarpus* (Fig. 190)

346* Fruits lacking or different 347

347 Buds brownish 348

347* Twig bark without transverse stripes or buds green to red 349

348 Twig bark of two-year-old and older twigs with transverse stripes and buds brownish, often with enrichment buds ....*Prunus* (Fig. 191)

348* Buds solitary, twigs with large warty lenticels ....*Photinia* (Fig. 192)

349 Terminal buds slender and spindle-shaped 350

349* Terminal buds more compact, at most elongate-ovoid 353

350 Buds with 3–5 bud distichous scales . ...*Aronia* (Fig. 193)

350* Buds with more than 5 visible bud scales 351

351 Less than 10 bud scales 352

351* Buds with more than 10 ochre-brown bud scales, in four rows. Trees with smooth grey bark ... *Fagus* (Fig. 194)

352 The first bud scale of the lateral buds points outward... *Populus* (Fig. 177)

352* The first bud scales are lateral on the lateral buds .. *Amelanchier* (Fig. 195)

353 Twigs hairy or pruinose 354

353* Twigs glabrous 360

354 Twigs pruinose, wood yellow....... ....*Cotinus* (Fig. 196)

354* Twigs hairy 355

355 Twigs towards the tip spreading hirsute .... *Crataegus* (*Mespilus*) *germanica* (Fig. 197)

355* Twigs appressed hairy 356

356 Twig tips with patches of felty hairs.. 357

356* Twig tips evenly and sparsely hairy..358

357 First bud scale of lateral buds points outward .... *Populus* (Fig. 177)

357* First bud scales of lateral buds are lateral to bud ....*Pyrus* and *Malus* (Figs 198 & 199)

188   189   190a   190b

191   192   193   194

195   196   197   198

199

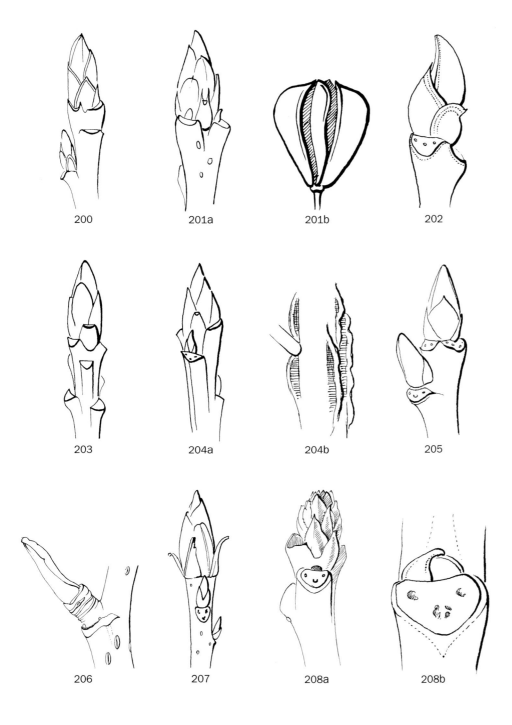

200   201a   201b   202

203   204a   204b   205

206   207   208a   208b

358 Bud scales dorsally hairy 359

358* Bud scales at most ciliate 360

359 Leaf cushion projecting from twig, upper bud scales greenish, bud densely silvery-hairy; fruits are legumes . . . . *Laburnum* (Fig. 139)

359* Bud scales not densely hairy, reddish or green with brownish margin . . . . *Aria* and others (Fig. 200)

360 Terminal buds to 5 mm long 361

360* Terminal buds longer 362

361 Terminal buds with more than 5 scales, shrub, fruits with 5 woody follicles . . . . *Exochorda* (Fig. 201)

361* Terminal buds with 3–4 scales, tree . . . . *Nyssa sylvatica* (Fig. 202)

362 The first bud scale of the lateral buds points outwards. . .*Populus* (Fig. 177)

362* The first bud scales are lateral on the lateral bud 363

363 Terminal bud glossy, red to green. . 364

363* Terminal buds mostly brown to grey-brown; if greenish, then not glossy 367

364 Twigs same colour as bud . . . . *Cornus alternifolia* and *C. controversa* (Fig. 203)

364* Twigs different in colour from bud 365

365 Buds green to red, to 8 mm long; twigs frequently with cork ridges . . . . *Liquidambar* (Fig. 204)

365* Terminal buds longer 366

366 Bud scales with brown margin . . . . *Sorbus* and others (Fig. 200)

366* Buds wine-red all round. . . . . *Davidia involucrata* (Fig. 205)

367 Leaf scars very narrow, encircling the twig, bud enveloped by a scale. . . . . *Tetracentron sinense* (Fig. 206)

367* Buds with more than one scale. . . 368

368 Leaf scars shield-shaped, with 3 groups of traces, lateral buds spreading, often with accessory buds . . . . *Carya* (Fig. 207)

368* Leaf scars obliquely triangular to rounded-ovate; lateral buds never with accessory buds 369

369 Buds ovoid, scales spreading at the tip . .*Xanthoceras sorbifolium* (Fig. 208)

369* Buds conical, scales appressed, lateral buds frequently with enrichment buds . . . . *Pyrus* (Fig. 198)

Leaves/leaf scars alternate, twigs with terminal buds, buds with bud scales, leaf scars not with 3 traces

381 (206) Leaf scars present 382

381* Leaf scars absent 416

382 Leaf scars with 2 traces . . . . *Ginkgo biloba* (Fig. 209)

382* Leaf scars with one or with many traces 383

383 Leaf scars with one trace 384

383* Leaf scars with many traces 402

384 Leaf bases of the fallen leaves running downward on the twig. Many more leaf scars than lateral buds; mostly one-stemmed trees 385

384* Most leaf scars on the long-shoot have axillary buds; often shrubs 387

385 Short shoots present on older twigs 386

385* Large scars present from the short shoots which fall off in their entirety . . . . *Taxodium* (Fig. 210)

386 Terminal bud at the base with thread-like scales, overtopping the bud . . . .*Pseudolarix amabilis* (Fig. 211)

386* Scales different. . . . .*Larix* (Fig. 212)

387 Lateral buds covered by leaf remnants at least at the base 388

387* Lateral buds free above the leaf scar. 391

388 Lateral buds covered entirely by leaf remnants 389

388* Lateral buds only about half-covered 390

389 Twigs red-brown, terminal bud with a ball of bud leaves . . . . *Dasiphora* (Fig. 213)

389* Twigs grey-brown, terminal bud normal . . . .*Calophaca wolgarica* (Fig. 214)

390 Twigs round . . . . *Chamaecytisus* (Fig. 215)

390* Twigs angular. . .*Caragana* (Fig. 216)

391 Several lateral twigs overtop the original shoot tip, which was used up by an inflorescence. 392

391* Growth is continued mostly by a single terminal twig 395

392 Terminal buds thicker than the twig. 393

392* Terminal buds narrower than the twig, to 5 mm long. . . . .*Daphne* (Fig. 217)

393 Buds rarely more than 6 mm long, glabrous or coarsely ciliate. 394

393* Buds mostly larger and often hairy . . . .*Rhododendron* (Fig. 218)

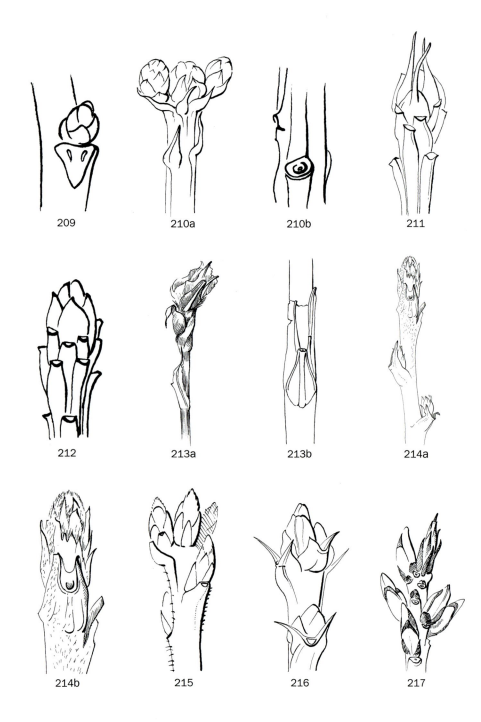

209   210a   210b   211

212   213a   213b   214a

214b   215   216   217

218

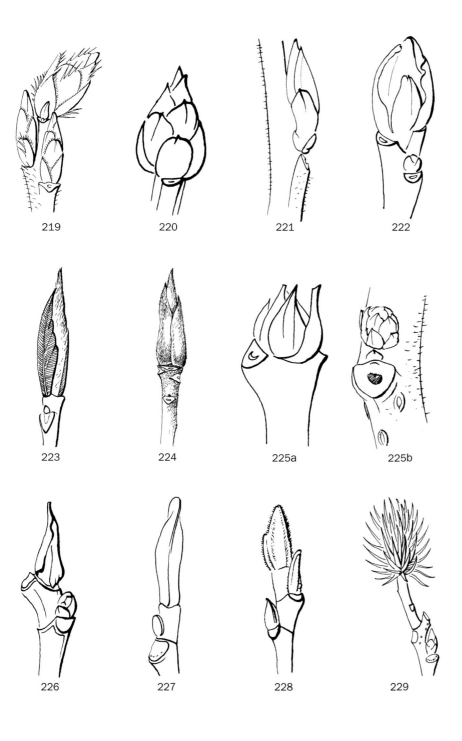

219   220   221   222

223   224   225a   225b

226   227   228   229

394   Buds dark, brownish. . . . .
. . *Rhododendron menziesii* (Fig. 219)

394*   Buds greenish to reddish . . . .
. . . . **Enkianthus** (Fig. 220)

395   Twigs with chambered pith . . . .
. . . . *Halesia* (Fig. 221)

395*   Pith not chambered 396

396   Tips of twigs and buds with metallic
shimmering lepidote scales . . . .
. . . . **Elaeagnus** (Fig. 150)

396*   Twigs glabrous or differently hairy. . 397

397   Terminal buds more than 8 mm long . .
. . . . 398

397*   Terminal buds shorter 400

398   Terminal buds ovoid, green, pruinose,
with more than 3 scales . . . .
. . . . *Sassafras albidum* (Fig. 222)

398*   Terminal buds elongate-pointed, with
2 scales. 399

399   Scales red-brown, spreading, loosely
hairy (stem bark with flat scales) . . . .
. . . . *Stewartia* (Fig. 223)

399*   Scales appressed, violet-brown, dense
with short white hairs . . . .
. . . . *Franklinia alatamaha* (Fig. 224)

400   Terminal buds less than 2 mm long,
brown . . . . *Ilex* (Fig. 225)

400*   Terminal buds larger and/or greenish
. . . . 417

402 (383) At each node a scar (of leaf
stipules) encircling the twig 403

402*   Nodes without a line encircling the
twig 405

403   Twigs with latex . . . .
. . . . *Ficus carica* (Fig. 226)

403*   Twigs without latex 404

404   Terminal bud glabrous, duckbill-shaped
. . . . *Liriodendron* (Fig. 227)

404*   Terminal bud hairy or glabrous, when
glabrous often very large . . . .
. . . . *Magnolia* (Fig. 228)

405   With fruit cones or naked inflorescence
. . . . 406

405*   Fruits, if present, different 407

406   Lateral buds stalked . **Alnus** (Fig. 180)

406*   Lateral buds sessile. . . . .
. . . . *Platycarya strobilacea* (Fig. 229)

407   Terminal buds longer than 12 mm . . 408

407*   Terminal buds smaller 412

408   Leaf scars on differently coloured leaf
cushion . . . . **Sorbus** (Fig. 200)

408*   Leaf cushion, if present, not distinct. .
. . . . 409

409   Buds slim and spindle-shaped, with
more than 15 scales. Tree. . . . .
. . . . **Fagus** (Fig. 194)

409* Buds with fewer than 10 scales or
    ovoid 410
410 Shrubs to 2 m high 411
410* Larger shrubs or trees 412
411 Bud scales greenish, twigs decumbent,
    rising to 1 m. . . . .
    . . . . *Xanthorhiza simplicissima*
    (Fig. 230)
411* Bud scales brown when dry, erect
    shrub . . . . *Paeonia* (Fig. 231)
412 (407) Leaf scars broad, with more than
    8 traces. . . *Eleutherococcus* (Fig. 232)
412* Leaf scars different 413
413 Dense shrub with thin twigs and dry
    fruits . . . . *Physocarpus* (Fig. 190)
413* Large shrubs or trees 414
414 Large lateral buds flanking the
    terminal bud. Buds with many scales,
    the lowermost sometimes with thread-
    like tip . . . . *Quercus* (Fig. 233)
414* Terminal bud solitary, at most flanked
    by much smaller lateral buds . . . 415
415 Terminal bud as broad as high . . . . . .
    . . . . *Idesia polycarpa* (Fig. 234)
415* Terminal buds noticeably longer than
    broad (ovoid) . . . . . . *Carya* (Fig. 207)
416 (381) Leaves becoming scale-like, small
    aromatic half-shrubs. . . . .
    . . . . *Artemisia* (Fig. 235)
416* Decumbent shrub, leaves drying up . .
    . . . . *Arctostaphylos alpina* (Fig. 236)
417 (400) Twigs green 418
417* Twigs brown 419
418 Terminal buds with more than 3 scales
    . . . .*Euonymus nanus* (Fig. 237)
418* Terminal buds with 2–3 scales . . . . . .
    . . . . . . *Helwingia japonica* (Fig. 238)
419 Margins of bud scales with brownish
    glands . .*Escallonia virgata* (Fig. 239)
419* Bud scales different, buds over 4 mm
    long . . . .*Daphne* (Fig. 217)

Leaves/leaf scars alternate, buds without terminal buds, buds with bud scales, leaf scars with 3 traces

420 (207) Buds with a maximum of 5
    scales 421
420* Buds with 5 or more scales. 437
421 Buds distichous on twig 422
421* Buds in spirals 424
422 Buds solitary in the leaf axil 423
422* Buds with accessory buds. . . . .
    . . . . *Cercis* (Fig. 240)

230

231

232

233

234

235

236

237

238

239a

239b

240

241

242

243

244a

244b

245

246a

246b

247

248

249

250

251

252a

252b

253

423 Buds mostly> 4 mm, (yellow) green to wine-red . . . . *Tilia* (Fig. 241)

423* Buds < 4 mm, brownish. . . . . *Celtis* (Fig. 242)

424 Buds with enrichment or accessory buds 425

424* Buds solitary 427

425 Shrub 426

425* Tree . . . . . . *Hovenia dulcis* (Fig. 243)

426 Bud with accessory buds (not branching from main bud). . *Amorpha* (Fig. 244)

426* Bud with enrichment buds (in axil of scale of primary bud). . . . . *Neviusia alabamensis* (Fig. 266)

427 At each node with stipules or stipule scar encircling the twig, small shrub 428

427* Without stipule scar surrounding twig 429

428 Stipules (fused) appressed to twig . . . . *Atraphaxis* (Fig. 245)

428* Stipules separate, spreading from the twig or deciduous, buds densely short white-hairy. . . . . *Hedysarum multijugum* (Fig. 245)

429 Buds with a single bud scale. . . . *Salix*

429* Buds with more than one bud scale 430

430 Twigs angular 431

430* Twigs round 435

431 Tree . . . *Albizia julibrissin* (Fig. 247)

431* Shrubs 432

432 Leaf stipules enclosing bud. . . . . *Petteria ramentacea* (Fig. 248)

432* Bud scales independent of leaf. . . .433

433 Buds 2–3 mm long 434

433* Buds about 1 mm long, fruits many small terminal legumes (nuts) . . . . *Amorpha* (Fig. 244)

434 Leaf scar with stipule remnants, bud scales acute . . . *Indigofera* (Fig. 249)

434* Scales hardly acute. . . . . *Holodiscus discolor* (Fig. 250)

435 Young twigs thinner than 3 mm. . .436

435* Twigs more than 3 mm thick . . . . *Maackia amurensis* (Fig. 251)

436 Small shrubs; or trees with straight stem (bark of stem often transversely striped and whitish) . . . . *Betula* (Fig. 252)

436* Large shrub or small tree with quickly branching stem. . . *Cydonia* (Fig. 253)

437 (420) Buds on twig distichous 438

437* Buds spiral 446

438 Buds with accessory buds or accessory shoots; shrubs .... **Neillia** (Fig. 254)

438* Buds without accessory buds 440

440 Buds to 2 mm long, partly with enrichment buds. . .**Zelkova** (Fig. 255)

440* Buds more than 2 mm long 441

441 Bud scales distichous 442

441* Bud scales in four rows or spiral ..443

442 Buds mostly > 4 mm, oblique above the leaf scar .... **Ulmus** (Fig. 256)

442* Twigs at the tip less than 1 mm thick, buds < 4mm ....
...... **Aphananthe aspera** (Fig. 257)

443 Buds elongate-ovate to short fusiform
.... 444

443* Buds squat, ovoid . . **Corylus** (Fig. 258)

444 Buds shortly fusiform, with more than 15 bud scales ... **Carpinus** (Fig. 259)

444* Buds with distinctly fewer scales . 445

445 Twigs densely covered with small, pale lenticels ....
... **Disanthus cercidifolius** (Fig. 260)

445* Lenticels few, relatively inconspicuous
.... **Ostrya** (Fig. 261)

446 (437) Twigs sticky or aromatic. ... 447

446* Twigs neither sticky nor aromatic ..
.... 449

447 Flower buds larger than leaf buds 448

447* Flower buds not larger than leaf buds; mostly with old, dry leaves.........
.... **Baccharis** (Fig. 262)

448 Without stipule scars ....
.... **Myrica** (Fig. 263)

448* With stipule scars ....
.... **Comptonia peregrina** (Fig. 264)

449 Twigs angular 450

449* Twigs round 452

450 Twigs grass green................
.... **Kerria japonica** (Fig. 265)

450* Twigs with different colour 451

451 Buds with enrichment buds ....
.... **Neviusia alabamensis** (Fig. 266)

451* Inflated fruits present.............
.... **Colutea** (Fig. 267)

452 Twig with solid pith 453

452* Twigs with chambered pith ....
.... **Itea virginica** (Fig. 268)

453 Twigs thick with wide pith, buds spreading, globose-ovoid ....
.... **Sorbaria** (Fig. 269)

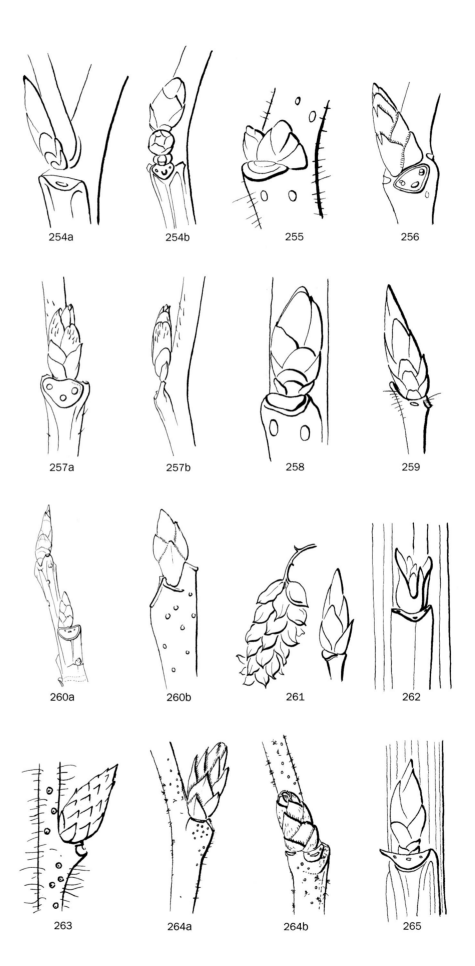

254a 254b 255 256

257a 257b 258 259

260a 260b 261 262

263 264a 264b 265

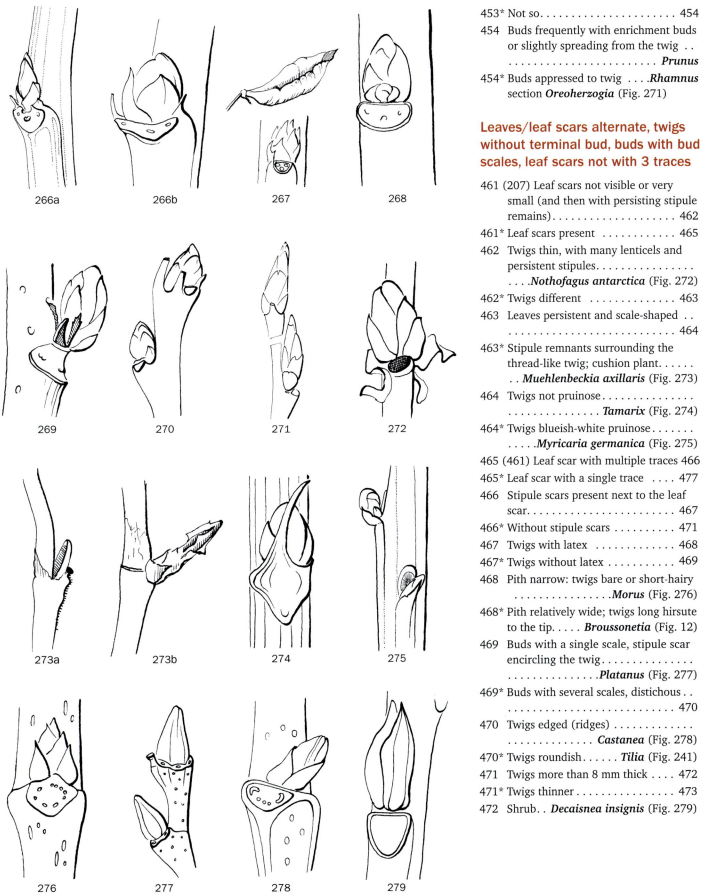

266a   266b   267   268

269   270   271   272

273a   273b   274   275

276   277   278   279

453* Not so. 454

454 Buds frequently with enrichment buds or slightly spreading from the twig . . . . ***Prunus***

454* Buds appressed to twig . . . .***Rhamnus*** section ***Oreoherzogia*** (Fig. 271)

Leaves/leaf scars alternate, twigs without terminal bud, buds with bud scales, leaf scars not with 3 traces

461 (207) Leaf scars not visible or very small (and then with persisting stipule remains). 462

461* Leaf scars present 465

462 Twigs thin, with many lenticels and persistent stipules. . . . .***Nothofagus antarctica*** (Fig. 272)

462* Twigs different 463

463 Leaves persistent and scale-shaped 464

463* Stipule remnants surrounding the thread-like twig; cushion plant. . . . . ***Muehlenbeckia axillaris*** (Fig. 273)

464 Twigs not pruinose. . . . . ***Tamarix*** (Fig. 274)

464* Twigs blueish-white pruinose. . . . .***Myricaria germanica*** (Fig. 275)

465 (461) Leaf scar with multiple traces 466

465* Leaf scar with a single trace 477

466 Stipule scars present next to the leaf scar. 467

466* Without stipule scars 471

467 Twigs with latex 468

467* Twigs without latex 469

468 Pith narrow: twigs bare or short-hairy . . . .***Morus*** (Fig. 276)

468* Pith relatively wide; twigs long hirsute to the tip. . . . . . ***Broussonetia*** (Fig. 12)

469 Buds with a single scale, stipule scar encircling the twig. . . . .***Platanus*** (Fig. 277)

469* Buds with several scales, distichous 470

470 Twigs edged (ridges) . . . . ***Castanea*** (Fig. 278)

470* Twigs roundish. . . . . . . ***Tilia*** (Fig. 241)

471 Twigs more than 8 mm thick 472

471* Twigs thinner 473

472 Shrub. . ***Decaisnea insignis*** (Fig. 279)

472* Tree . . . . *Ailanthus* (Fig. 280)

473 Sub-shrub or shrub. 476

473* Tree-shaped 474

474 Leaf scars on distinct cushions with many traces (> 8) or in 3 groups of traces 475

474* Scars with 5–7 traces, flower buds distinctly larger than leaf buds. . . . . .
. . . . *Euptelea* (Fig. 281)

475 Traces in 3 groups . . . .
. . . . . *Poliothyrsis sinensis* (Fig. 282)

475* Scar with many traces, with old bud scales at base of shoot . . . .
. . . *Koelreuteria paniculata* (Fig. 283)

476 Shoot ends with many fruits (dry capsules) or round fruit-stalk scars. . .
. . . . *Hibiscus syriacus* (Fig. 169)

476* Buds narrow and pointed . . . .
. . . . . . *Ceratostigma willmottianum*
(Fig. 284)

477 (465) Twigs green on all sides, angular
. . . . 478

477* Twigs different 483

478 Leaf scar with stipule lobes and/or buds partially hidden under the leaf cushion. 479

478* Leaf scars without stipule lobes . . 481

479 Buds with enrichment buds . . . .
. . . . . . *Hippocrepis emerus* (Fig. 285)

479* Buds solitary 480

480 Buds brown; stipule lobes thin and acute . . . . *Genista* (Fig. 286)

480* Buds greenish; stipule lobes flat, triangular . . . . *Cytisus* (Fig. 176)

481 Buds appressed to the twig 482

481* Buds spreading from the twig . . . . . .
. . . . *Jasminum* (Fig. 287)

482 Buds elongate. Twigs deeply furrowed, apparently distichous. Shrub to 50 cm high . . . . *Vaccinium* (Fig. 288)

482* Shrubs larger and buds compact, or twigs more rounded and only slightly furrowed . . . . *Cytisus* (Fig. 289)

483 (477) Small shrub with extensively dying, slightly angular twigs, buds often in pairs . . . *Lespedeza* (Fig. 290)

483* Characteristics different 484

484 Small shrubs with shoot tips dying back quite far and with round twigs. . . 485

484* Characteristics different 487

485 Scales completely enclosing buds. Living twigs with latex 486

485* Bud scales spreading, inner leaflets without bud scales, hairy . . . .
. . . . *Ceanothus* (Fig. 291)

280   281   282a   282b

283   284   285   286

287a   287b   288   289a

289b   290   291

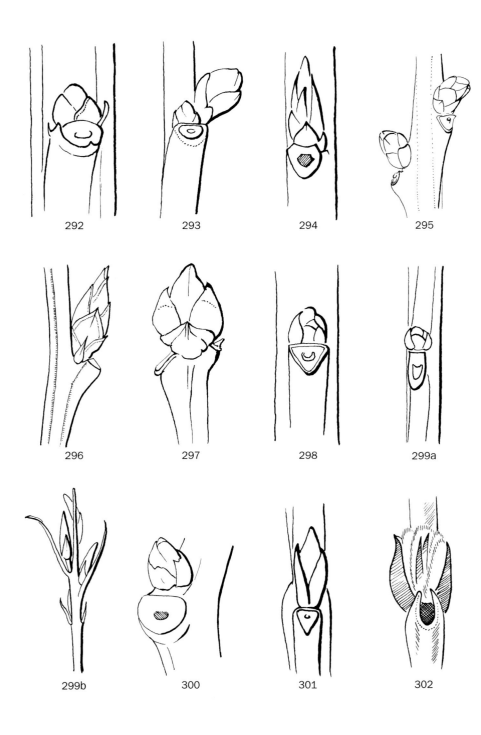

292   293   294   295

296   297   298   299a

299b   300   301   302

486 Twigs red brown, about 2 mm thick at base . . . . ***Flueggea*** (*Securinega*) ***suffruticosa*** (Fig. 292)

486* Twigs grey-brown, thinner than 2 mm . . . . . . ***Leptopus chinensis*** (Fig. 293)

487 Twigs without lenticels, mostly angular 488

487* Twigs with lenticels, often round. . 498

488 Single-stemmed tree, the number of leaf scars exceeding the number of buds . . . . ***Taxodium*** (Fig. 210)

488* Mostly shrubs 489

489 Leaf with stipule remnants 490

489* Stipules absent 491

490 Lateral buds covered by leaf stipules . . . . ***Dasiphora*** (Fig. 213)

490* Buds clearly visible. . . . . ***Indigofera*** (Fig. 249)

491 Twigs thin, brown to grey, fruits of 5 free follicles, pith narrow to normal. . . . . ***Spiraea*** (Fig. 294)

491* Fruits capsules or berries, of several fused carpels. Pith often relatively wide 492

492 Twigs not pruinose. 493

492* Twigs orange-ochre, blueish pruinose . . . . ***Zenobia pulverulenta*** (Fig. 295)

493 Buds > 3 mm. 494

493* Buds < 3 mm. 495

494 Twigs with hairy ridges . . . .***Vaccinium corymbosum*** (Fig. 296)

494* Twigs glabrous . . . .***Lyonia mariana*** (Fig. 297)

495 Leaf buds flat, half-globose; twigs green to wine-red. 496

495* Buds ovoid to elongate, brown. . . 497

496 Twig bark tearing open, later with deeply furrowed bark. . . . .***Oxydendrum arboreum*** (Fig. 298)

496* Twigs different . . . . ***Lyonia ligustrina***

497 Buds spreading off twig . . . . ***Eubotrys racemosa*** (Figs 299 & 300)

497* Buds appressed . . . . ***Vaccinium*** (Fig. 301)

498 (487) Small shrubs, rarely more than 2 m 499

498* Large shrubs or trees 500

499 Buds appressed to twig, with 2 scales, inner leaves densely hairy . . . . ***Cotoneaster*** (Fig. 302)

499* Buds with several scales, spreading off twig, on large leaf cushions . . . . ***Symplocos paniculata*** (Fig. 303)

500 Bark of twig with sticky gum or latex 501

500* Without latex or gum 502

501 Bark of twig with sticky gum . . . . ***Eucommia ulmoides*** (Fig. 304)

501* Bark of twig with latex . . . . ***Broussonetia*** (Fig. 12)

502 Buds to 2 mm long with more than 3 visible scales, often with enrichment buds . . . . ***Zelkova*** (Fig. 255)

502* Buds with 2 visible scales, usually more than 2 mm long . . . . ***Diospyros*** (Fig. 305)

303

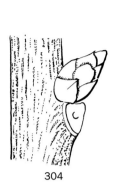

304

305a

305b

Systematic arrangement of the featured woody plants

(following APG 2009 and Stevens 2012)

* paraphyletic

Subdivision Spermatophytina, Seed plants

1. Class Ginkgoopsida, Gymnosperms

Order Ginkgoales
    Family Ginkgoaceae

2. Class Coniferopsida, Gymnosperms

Order Pinales
    Family Pinaceae
    Family Cupressaceae
Order Gnetales (Class Gnetopsida)
    Family Ephedraceae

3. Class Magnoliopsida, Basal angiosperms

Basal groups*

Order Austrobaileyales
    Family Schisandraceae

Magnoliids

Order Magnoliales
    Family Magnoliaceae
    Family Annonaceae
Order Laurales
    Family Calycanthaceae
    Family Lauraceae
Order Piperales
    Family Aristolochiaceae

Monocotyledons

Order Liliales
    Family Smilacaceae

Eudicots

Basal groups*

Order Ranunculales
    Family Eupteleaceae
    Family Lardizabalaceae
    Family Menispermaceae
    Family Ranunculaceae
    Family Berberidaceae
Order Sabiales
    Family Sabiaceae
Order Proteales
    Family Platanaceae
Order Trochodendrales
    Family Trochodendraceae

Core dicots

Order Saxifragales
    Family Paeoniaceae
    Family Altingiaceae
    Family Hamamelidaceae
    Family Cercidiphyllaceae
    Family Iteaceae
    Family Grossulariaceae

Rosids

Order Vitales
    Family Vitaceae
Order Celastrales
    Family Celastraceae
Order Malpighiales
    Family Salicaceae
    Family Hypericaceae
    Family Phyllanthaceae
Order Cucurbitales
    Family Coriariaceae
Order Fabales
    Family Fabaceae
Order Fagales
    Family Nothofagaceae
    Family Fagaceae
    Family Juglandaceae
    Family Myricaceae
    Family Betulaceae
Order Rosales
    Family Rosaceae
    Family Rhamnaceae
    Family Elaeagnaceae
    Family Ulmaceae
    Family Cannabaceae
    Family Moraceae
Order Myrtales
    Family Lythraceae
    Family Onagraceae
Order Crossosomatales
    Family Staphyleaceae
    Family Stachyuraceae
Order Malvales
    Family Thymelaeaceae
    Family Malvaceae
Order Sapindales
    Family Anacardiaceae

Family Sapindaceae
Family Simaroubaceae
Family Meliaceae
Family Rutaceae

Order Caryophyllales
Family Tamaricaceae
Family Plumbaginaceae
Family Polygonaceae

Asterids

Order Cornales
Family Cornaceae
Family Hydrangeaceae

Order Ericales
Family Ebenaceae
Family Theaceae
Family Symplocaceae
Family Styracaceae
Family Actinidiaceae
Family Clethraceae
Family Ericaceae

Order Garryales
Family Eucommiaceae

Order Gentianales
Family Rubiaceae
Family Apocynaceae, Subfamily Asclepiadoideae

Order Lamiales
Family Oleaceae
Family Scrophulariaceae
Family Lamiaceae
Family Paulowniaceae
Family Bignoniaceae

Order Solanales
Family Solanaceae

Order Aquifoliales
Family Helwingiaceae
Family Aquifoliaceae

Order Asterales
Family Asteraceae/Compositae

Order Escalloniales
Family Escalloniaceae

Order Dipsacales
Family Adoxaceae
Family Caprifoliaceae

Order Apiales
Family Araliaceae

Descriptions of the
tree and shrub species

Gymnosperms

Ginkgoaceae
maidenhair tree family

Only one species of this evolutionarily ancient group has survived to modern times — a "living fossil". An archaic characteristic is the forked branching of the main vascular bundles in the leaf, just as it occurs in ferns. This is reflected in the two vascular traces of the leaf scar.

Ginkgo biloba L., ginkgo, maidenhair tree

Buds conical ovoid, 4–5 mm long, simple bud scales arranged spirally, the alternate lateral buds with two additional small and narrow prophyll scales. Bud scales red-brown, partly glossy. **Twigs** (long-shoots): smooth, but grooved irregularly, first year's growth ochre to dark brown, with silver-grey parts. Older parts: grey to grey-brown with oblong irregular tears. Branches distinctly thickened at annual growth border. Long-shoots of several years' growth have many short-shoots, as in larch. These short-shoots are robust and lengthen distinctly over the years. **Pith** relatively narrow, light-coloured when fresh, turning orange-brown when exposed to air. **Leaf scars** with two traces. To more than 30 m in height. Frequently planted, originally from China.

Pinaceae
pine family

Woody plants of the Northern Hemisphere, mostly evergreens with needle leaves in a spiral arrangement. Leaf scar always with one trace, because only one vascular bundle enters the needle. Cones which become woody, with many seed scales and ± developed narrow bract scales which partly emerge from under the seed scales.

Key to Pinaceae

1 Terminal buds at the base with thread-like scales which are longer than the bud . . . . *Pseudolarix amabilis*

1* Terminal buds without thread-like scales
. . . . *Larix*

Larix MILL., larch

Tall trees with a mostly ± straight stem. Young tree narrowly pyramidal; later with an irregular open crown. From a few large branches slender lateral branches hang down. Leaves spirally inserted.

The branches are differentiated into long-shoots and short-shoots of a few millimetres (on last year's and older long-shoots). One year-old long-shoots, still without lateral short-shoots, have different colours depending on the species: yellow, brown or ± reddish. Their surface is covered in cushion-like swellings which end in slightly diverging, adaxial leaf scars with a single trace. The needles are shed in autumn.

Cones and branches offer the most important means for differentiation. The cones differ among the species by their size, the number and shape of the seed scales as well as the shaping of the bract scales.

The genus is distributed in the cooler latitudes and in the mountains of the Temperate Zone. It reaches the tree line of vegetation.

Key to *Larix*

1 First year's growth twigs straw-coloured, grey when older; glabrous and without bloom . . . . ***Larix decidua***

1* First year's growth twigs reddish or brown, partly hairy and/or with waxy bloom 2

2 Bract scales in closed cones invisible, twigs slender to medium thick 3

2* Bract scales longer than seed scales, twigs thick . . . . *Larix occidentalis*

3 Bark with rough structure and cones larger than 10–15 mm 4

3* Bark first smooth, later fine-scaly; twigs slender; cones small (15 mm) with few scales . . . . *Larix laricina*

4 Twigs mostly glabrous, cone scales reflexed. . . . . ***Larix kaempferi***

4* Twigs mostly hairy, cones relatively small, very variable species. . . . . . *Larix gmelinii*

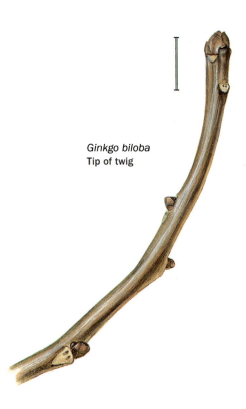

Ginkgo biloba
Tip of twig

Ginkgo biloba
Short-shoot

Larix decidua
Section of two-year-old long-shoot with short-shoots

Larix decidua
Tip of twig

Larix decidua
Twig with cones

Larix gmelinii
Tip of twig

Larix sibirica
Cone

Larix gmelinii
Cone

Larix decidua MILL., European larch
[*Larix europaea* LAM. & A. DC.]

Buds red-brown to dark brown, partly with light wax-like bloom. **Terminal buds** of long-shoots half-globose to ovoid, those of the short-shoots ± sunken below, surrounded by the light ochre-yellow scales of the short-shoots and the revolute grey scale leaves from buds of earlier years. **Lateral buds** flattened hemispherical. **Twigs** of lowest order (youngest) long, hanging down, with ochre-yellow needle cushions, glabrous, without bloom. **Cones** pale brown, narrowly ovoid, 3–3.5 cm long with 45–70 seed scales with rounded margin. Seed scales glabrous or slightly hairy. Margins not inrolled or revolute. Only in the lower part are the bract scales larger than seed scales. Tree to 45 m high. Occurring naturally in European mountains and frequently planted in forestry. *Larix decidua* var. *polonica* (RACIB. ex WÓYCICKI) OSTENF. & SYRACH has smaller often slightly hairy cones.

Larix sibirica LEDEB., Siberian larch
[*Larix russica* (ENDL.) SABINE ex TRAUTV.]

Similar to the European larch, differing mainly in the 2.5–3 cm long, dark red-brown, ovoid **cones** which are abruptly blunt, with 20–40 seed scales which are very hairy especially at the cone base. Margins of the seed scales slightly thickened and bent inwards. Bark of stem rough and fissured. Slender tree. The species, rarely planted, comes into leaf about two weeks earlier than the European larch. Origin: north and east Russia.

Larix gmelinii (RUPR.) KUZEN., Dahurian larch
[*Larix dahurica* TURCZ.]

A very variable species, which shows similarity to the Siberian larch, but has distinctly divergent forms. **Twigs** from (reddish) yellow and glabrous (typical), to red-violet-brown and ± hairy in var. *japonica* (MAXIM. ex REGEL) PILG.; densely hairy in var. *olgensis* (HENRY) OSTENF. & SYRACH. **Cones** relatively small, 15–25 mm long. Seed scales (10–30) opening wide, at their tip wavy-rounded or slightly notched. Tree to 30 m high, only in specialised collections. Origin: northern east Asia.

Larix kaempferi (LAMB.) CARRIÈRE, Japanese larch
[*Larix leptolepis* (SIEBOLD & ZUCC.) GORDON]

Buds red-brown, glossy, only slightly resinous. **Twigs** reddish, with blue-white waxy bloom, mostly glabrous or slightly hairy. **Cones** cannot be confused with any others: almost globose, 20–30 mm diameter with thin seed scales that are strongly bent outwards. Tree, 30–35 m high. Origin: Japan.

Larix × marschlinsii COAZ,
[*Larix × eurolepis* HENRY], hybrid larch
[*L. decidua* × *L. kaempferi*]

This hybrid is found especially in forestry plantings, as it is a fast-growing and highly productive timber tree.

Larix laricina (DU ROI) K. KOCH, American larch
[*Larix americana* MICHX.]

Buds small and dark black-brown. **Twigs** glabrous, slender, matte red-brown with a waxy bloom. Bark when young, grey and smooth, later with small scales. **Cones** appearing early, very small, 10–15 mm long, hardly opening. Few (10–20) seed scales, with the margins bent slightly inwards. Tree to 25 m high. The only larch to inhabit moist habitats. Origin: northern North America.

Larix occidentalis NUTT., western larch

Twigs thick, light orange-brown, initially slightly hairy in the furrows. **Cones** 20–35 mm long. Seed scales bending strongly on opening, bract scale long and acute, even in the closed cone emerging distinctly from under the seed scales. Tree 25–30 m high in our area; in Western North America, its home, it can grow to over 70 m high.

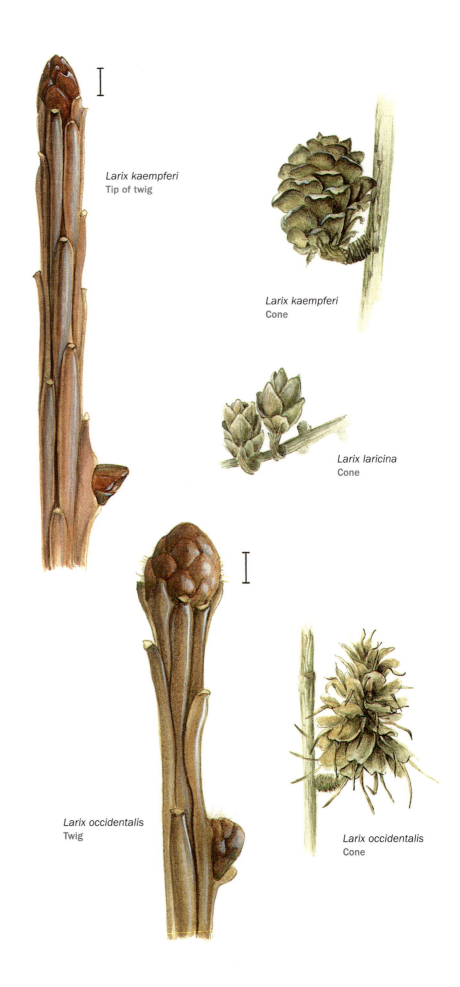

Larix kaempferi
Tip of twig

Larix kaempferi
Cone

Larix laricina
Cone

Larix occidentalis
Twig

Larix occidentalis
Cone

Pseudolarix amabilis
Tip of twig

Pseudolarix amabilis
Short-shoot

Metasequoia glyptostroboides
Opposite lateral buds

Metasequoia glyptostroboides
Tip of twig

Pseudolarix GORDON, golden larch

Differs from the genus *Larix* by cones that disintegrate when ripe, by short-shoots which increase in length and by long-acute lower bud-scales which overtop the bud. A single species.

Pseudolarix amabilis (J. NELSON) REHDER, golden larch

Terminal buds of long-shoots globose-ovoid, c. 3 mm long, ochre to brown-violet, lowest bud scales acute and overtopping the bud, on short-shoots covering the bud completely. **Lateral buds** on long-shoots in the first year hemispherical to ovoid, with a broad base on the twig. **Branches** as in *Larix* wrinkled by the leaf cushions on the twig, ochre-brown (more so on the base of the twig and on the shaded side) to brown-violet (especially towards the shoot tip and in the light), partly with a light waxy bloom. The cones, to 7 cm long, disintegrate on the tree. A tree 30–40 m high, quite rare in Europe. Origin: Mid- to Eastern China.

Cupressaceae
cypress family

Mainly evergreen trees and shrubs with scale- or needle-shaped leaves. In some basal genera, in the past included in Taxodiaceae, the needles are united on short-shoots; these are abscised in one unit in deciduous species.

Key to Cupressaceae

1 Buds opposite . . . .
. . . . *Metasequoia glyptostroboides*
1* Buds alternate . . . . *Taxodium distichum*

Metasequoia glyptostroboides HU & W. C. CHENG, dawn redwood

Partly with terminal **buds**, lateral buds oblique opposite, ovoid, 2–4 mm long, narrow at the base (and due to this appearing slightly stalked) and ± spreading at right angles to the twig. With many (12–14) bud scales: the lower ones dark red-brown, the upper light to yellowish-brown. Above, or more rarely below the buds, single trace scars of short-shoots are often present. The male floral buds are arranged in dense axillary racemes, to 30 cm long towards shoot end. The female floral buds are mostly solitary at the ends of thin stalks. **Twigs** initially red-brown, later dark violet-brown, with a slight waxy bloom, with a bark that peels in long thin strips. **Stem** with longitudinally scaly bark. Major branches often come out of the trunk in deep furrows. **Cones** globose, 1.5–2 cm in diameter, on to c. 4 cm long stalks. Tree conical, to 35 m high, frequently encountered. Origin: China.

Taxodium distichum (L.) RICH.
swamp cypress

Buds alternate, very small: c. 1 mm diam, rounded and sparsely scaly. **Branches** orange-brown to red-brown when young, glabrous and slender with many very small raised leaf scars which make the twig slightly angular. The short-shoots fall off and leave slightly larger scars with a single trace. **Bark** red-brown, coming off in fibres. Base of the stem often thickened. In wet habitats breathing roots are formed, the so-called 'cypress knees' with their tips 5–15 cm above the swamp surface. **Cones** short-stalked and ± globose, 2.5 cm thick. Conical tree 30 m high (40 m in its native habitat) with horizontal branches. Frequent. Origin: south-eastern USA.

Taxodium distichum
Long-shoot

Taxodium distichum
Long-shoot: section with short-shoot scar

Ephedra distachya
Branches

Ephedra distachya
Part of twig

Ephedra distachya
Lateral buds

Ephedra distachya
Lateral buds

Ephedraceae
jointfir family

A genus with 40 species from dry areas of the Northern Hemisphere and South America. Low shrubs, with some species from southern middle and east Europe to the Mediterranean. The leaves are reduced and scale-shaped; the plants photosynthesise through the green stems.

Key to *Ephedra*

1 Branches round, more than 1.5 mm thick
. . . .*Ephedra distachya*
1* Branches partly 4-angular, mostly less
than 1.5 mm thick. . . . . .*Ephedra majo*r

Ephedra distachya L., jointfir, joint pine

Buds (and branches) alternate, often not visible, concealed below the scale-shaped leaves, flat-appressed to the twig, with 2–3 green to ochre-brown pairs of scales. Buds at the base of the twig are often more easily visible and globose. **Twigs** green when young, delicately narrowly striate with 1.5–5 cm long and c. 2 mm thick internodes, older twigs grey-brown. Scars of scales not visible. Shrub to 50 cm high with prostrate main axis and erect twigs. Origin: from south and southern middle Europe to Asia.

Ephedra major HOST is similar. Plant 1–2 m high, the twigs are initially only 1–1.5 mm thick, with internodes about 2 cm long. Origin: Mediterranean area to western Asia.

Angiosperms — Basal Groups

Schisandraceae
star-anise family

Schisandra chinensis (TURCZ.) BAILL.,
Chinese magnolia-vine

Buds alternate, narrowly ovoid, to 5 mm
long, spreading from the twig; laterally
often with smaller, enrichment buds. Bud
scales glabrous, orange-red-brown to dark
red-brown. **Branches** glabrous, slightly
angular, red-brown to ochre-brown, grey-
brown where exposed to sun; with many
narrowly ovate protuberant and slightly
paler lenticels. Liana to 7 m high, planted
occasionally. Origin: eastern Asia.

Magnoliaceae
tulip tree family

Within the basal orders of the Angiosperms,
this is the only family presented here in
which stipules protect the bud. They form
an envelope which leaves a linear scar at
every node, encircling the twig, next to the
base of the peduncle. In the magnolias,
the leaf base, too, is enclosed in this cover.
Every cover includes the subsequent leaf
with its stipules which, again, form an
envelope. By this means, the tip of the shoot
is protected even during the growth phase,
by a series of envelopes of stipules following
each other in sequence.

Key to Magnoliaceae

1 Buds ± acute, tip in transverse section ±
  round, often hairy. . . . .***Magnolia***
1* Buds broad, rounded toward the tip and
  flattened (duck-bill-shaped) . . . .
  . . . .***Liriodendron***

Magnolia L., magnolia

The magnolias have about 300 species. They
occur in South East Asia, from northern
South America via central America to eastern
North America and on the Antilles. Most
species are evergreen species of the Tropics.
Only about 30 of the species reach temperate
latitudes and are deciduous. The section
Yulania includes many early-flowering

garden varieties; among these many double
or poly-hybrids which are almost impossible
to place taxonomically. The scars of the
leaves have several traces, the scars of the
stipules encircle the branch in a line. A small
scar on the bud scale marks the place of the
stalk of the mostly reduced leaf.

Key to *Magnolia*

1 Terminal floral buds narrowly ovoid,
  distinctly to very hairy; flowering before
  the leaves or when the leaves appear. . 7
1* Terminal buds oblong, often acute,
  glabrous or slightly hairy; flowering after
  the leaves appear 2

Subgenus *Magnolia*

2 Flower buds glabrous, more than 3 cm
  long; club-shaped, thickened and slimmer
  portions of twigs alternate 4
2* Buds slightly hairy; without distinctly
  thickened portions of twigs 3
3 Young twigs and buds pure green, largest
  diameter of the buds in the lower half . .
  . . . .*Magnolia virginiana*
3* Young twigs and buds with brown hues,
  flower buds tubular, largest diameter
  about in the middle 6
4 Buds mostly less than 50 mm long, twigs
  grey-brown to greenish 5
4* Largest terminal buds more than 50 mm
  long, to 12–16 mm thick, young twigs
  red-brown. . . . .*Magnolia fraseri*
5 Buds tubular, bluntly acute, often
  greenish . . . .*Magnolia obovata*
5* Buds slim, long-acute, mostly blue-violet
  . . . .***Magnolia tripetala***
6 Shrub . . . .***Magnolia sieboldii***

Subgenus *Yulania*

6* Tree . . . .*Magnolia acuminata*
7 (1) Flower buds densely hairy 10
7* Buds slightly hairy or relatively finely so
  8
8 Twigs distinctly red-brown. 9
8* Twigs grey-brown to greenish 10
9 First internode below the flower bud ±
  glabrous . . . .***Magnolia denudata***
9* First internode below the flower bud still
  distinctly hairy . . . .
  . . . .***Magnolia ×soulangeana***

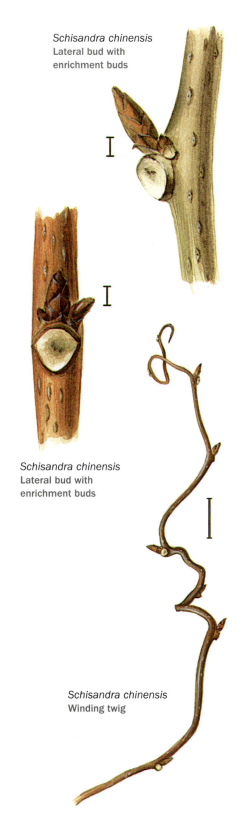

Schisandra chinensis
**Lateral bud with
enrichment buds**

Schisandra chinensis
**Lateral bud with
enrichment buds**

Schisandra chinensis
Winding twig

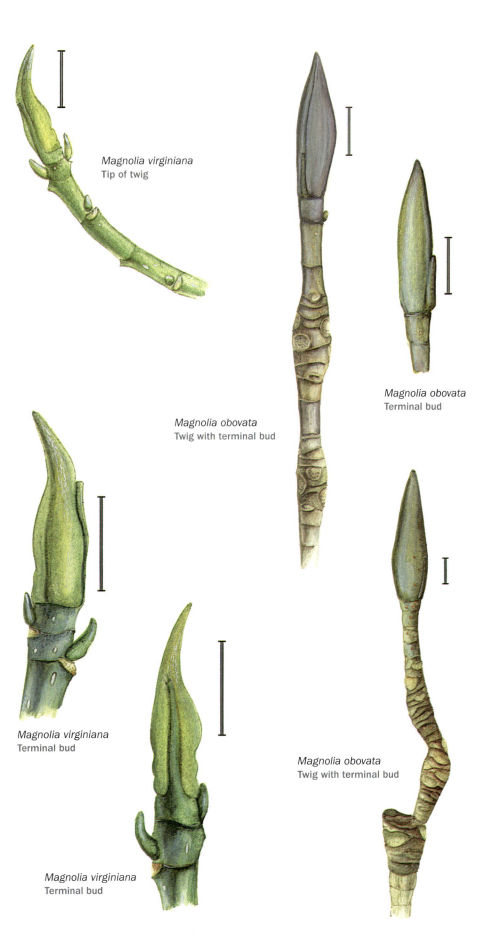

Magnolia virginiana
Tip of twig

Magnolia obovata
Twig with terminal bud

Magnolia obovata
Terminal bud

Magnolia virginiana
Terminal bud

Magnolia virginiana
Terminal bud

Magnolia obovata
Twig with terminal bud

10   Plants shrub-like, flower buds less than 2 cm long 12

10* Large shrubs or small trees, flower buds more than 2 cm long 11

11   Below the buds ± glabrous, only top internode occasionally slightly hairy . . . . **Magnolia kobus**

11* First internode below the bud still ± densely hairy **Magnolia ×soulangeana**

12   Several internodes below the flower buds ± densely hairy . . . .**Magnolia stellata**

12* Internodes below the buds ± glabrous, only top internode occasionally slightly hairy . . . .**Magnolia kobus** and **Magnolia liliiflora**

Subgenus Magnolia

Section Magnolia

Buds slightly hairy or glabrous. The ovoid to cylindrical fruit 'cones' are slightly asymmetrical; they consist of many carpels in a spiral arrangement.

Magnolia virginiana L., sweetbay magnolia

Terminal bud 10–20 mm long, green, hairy towards the tip with fine silky appressed hairs. Lateral buds condensed to oblong, mostly relatively small. **Twigs** green, towards the tip also blue-green, with slight waxy bloom and with a few white lenticels which have a dark brown margin; older parts of twig red-brown. **Flowers** simultaneously with appearance of leaves or following it. Deciduous, occasionally also half evergreen shrub, 3–5 m high. Origin: east- or south-east USA.

Section Rhytidospermum

Subsection Rhytidospermum, umbrella magnolias

Buds and twigs glabrous. Thick twig sections with short internodes and large leaf scars alternate with thin sections with small leaf scars and longer internodes (like the last section before the terminal bud). Flowering when in full leaf. Fruits symmetrical, conical, rose-red to wine-red, becoming woody. Medium-sized to large trees with distant branching.

Magnolia obovata THUNB.,
Japanese big-leaf magnolia

[*Magnolia hypoleuca* SIEBOLD & ZUCC.]

Buds glabrous, violet to (only in shade) matte green, with waxy bloom. Terminal buds 35–45(–50) mm long, cylindrical and with a blunt point. **Twigs** glabrous and smooth, green to light brown and violet brown with pale, raised lenticels. **Bark** light brown. A broad, pyramidal tree 15–20 m high (to 30 m in Japan), the stem to 60 cm thick. Planted occasionally. Origin: Japan and Kurile Islands.

Magnolia tripetala (L.) L.,
umbrella magnolia

Buds, particularly terminal buds, long, acute and relatively slender, to 40 (rarely 50) mm long, mostly blue-violet, less often greenish. **Twigs** grey-brown to greenish with few small lenticels. **Bark** smooth and pale grey. Tree 9–15 m high, with open round crown; planted frequently. Origin: south-east and east USA.

Subsection Oyama, summer magnolias

Magnolia sieboldii K. KOCH,
Siebold's magnolia

Buds indistinct, finely appressed-hairy, olive-green to green-brown. **Terminal buds** oblong, 1.5–2 cm, lateral buds mostly only 5–6 mm long. **Twigs** 3–4 mm diameter, light ochre-brown to greyish (resembling a waxy bloom) sparsely hairy, with sparse slightly oblong light grey lenticels. The most frequent species of a group including *Magnolia sinensis* (REHDER & E.H. WILSON) STAPF and *Magnolia wilsonii* (FINET & GAGNEP.) REHDER among others, small shrubs, to 3 m high, summer-flowering, planted occasionally. Origin: eastern Asia.

Section Auriculata

Magnolia fraseri WALTER,
Fraser's magnolia

Buds very large, dark violet-blue, often more than 50 mm long and particularly in flowering buds to 16(–20) mm thick. **Twigs** initially with strong, red-brown surface and light ochre to grey-brown leaf scars as well as many lenticels. **Bark** initially brown,

Magnolia tripetala
Twig with terminal bud

Magnolia tripetala
Twig with terminal bud

Magnolia sieboldii
Twig

Magnolia fraseri
Terminal bud

Magnolia fraseri
Terminal bud

becoming grey later. Tree often has several stems, is laxly branched, 9–18 m high; rarely planted. Origin: mountains of the USA.

Subgenus Yulania

Buds mostly densely hairy. **Fruiting cone** often twisted asymmetrically, with only a few seed-bearing carpels enlarging; development of those without seeds stops. Shrubs and trees from Eastern Asia are parents of many garden hybrids.

Section Yulania

Magnolia denudata DESR., Yulan magnolia

Buds: terminal flowering buds to 30–40 mm long and 8 mm thick. Lateral buds generally smaller, 10 mm long and up to 3 mm thick. Bud scales brown to dirty olive green with appressed short white hairs. **Twigs** below the tip olive green and slightly hairy, lower on the branch red to violet-brown and glabrous. Lenticels few, pale and rounded. **Flowers** with 9 cream-white tepals of equal shape, bell-shaped, later bowl-shaped, 12–15 cm diameter, flowering in April, long before the leaves appear. Tree to 10 m high. Origin: central China.

Magnolia kobus DC., northern Japanese magnolia

Buds: flowering buds pale grey to silvery-green, densely hairy and velvety to slightly shaggy. Leaf buds mostly smaller and less hairy. **Twigs** ± glabrous below the buds, only the first internode sometimes slightly hairy, greenish to brown-grey. **Flowers** appear before the leaves: fully opened to 10 cm diameter, with 6–9 petaloid tepals; these are obovoid, white, occasionally pale red-violet at their base and with 3 deciduous sepaloid tepals. Shrub or tree to 10 m high with slender twigs; frequently planted. Origin: Japan.

Magnolia denudata
Twig with terminal bud

Magnolia kobus
Long-shoot with short-shoots

Magnolia kobus
Short-shoots with flowering bud

Magnolia kobus
Tip of twig with terminal bud

Magnolia stellata (SIEBOLD & ZUCC.) MAXIM., star magnolia
[*Magnolia kobus* var. *stellata* (SIEBOLD & ZUCC.) BLACKBURN]

Buds silvery-green with dense long hairs, which cover the contours of the bud and extend down the twig over several internodes, only gradually falling. **Flowers** with 12–18 irregularly reflexed narrow white tepals; flowers appear very early, March to April, long before the leaves. Frequently planted shrub to 3 m high. Origin: from the mountain forests of the Japanese Island of Honshu (Hondo).

Magnolia ×*loebneri* KACHE
Loebner's magnolia
[*Magnolia kobus* × *Magnolia stellata*]

This hybrid, planted quite frequently, is intermediate between its parents. It differs from the star magnolia in size and gets to 6–8 m high, often being rather tree-like. In contrast to the northern Japanese magnolia, the twigs below the buds are hairy, although not as densely so as in the star magnolia. Early flowering with 12–15 petals. Some cultivars are planted more frequently: **'Leonard Messel'** with pink and **'Merrill'** with white flowers.

Magnolia 'Susan'
[*Magnolia stellata* × *Magnolia liliiflora*]

Also fairly common are these triploid hybrids. 'Susan' is the most frequent among several magnolias bearing female names. The 2–3 m high shrub has, as Loebner's magnolia, the buds with a covering of hairs, and the first internode below the bud is still densely hairy, while the second has deciduous hairs. Below this the twigs are glabrous and glossy dark brown with pale lenticels. Early flowering (together with the appearance of the leaves) **flowers** wine-red with (6–)8–15(–10) tepals.

Magnolia liliiflora DESR.

Buds greenish to grey-green, with quite tightly appressed silky grey-white hairs. Terminal buds 2–2.5 mm long. **Twigs** olive-brown to violet-brown, glossy, greenish towards the tip, with some pale grey-white lenticels. Only slightly hairy below the terminal bud, the second internode mostly completely hairless. **Flowers** mostly after leaf appearance. Frequently planted shrub, 3–5 m high. Origin: central China.

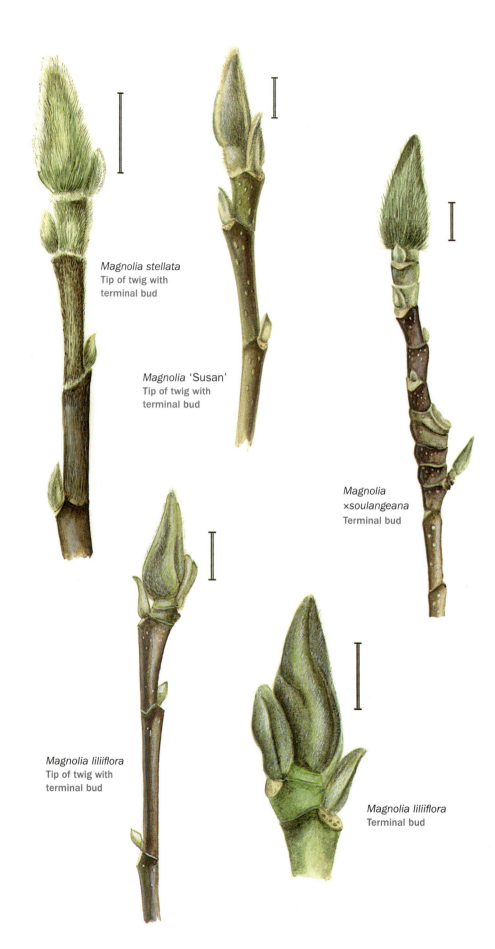

Magnolia stellata
Tip of twig with terminal bud

Magnolia 'Susan'
Tip of twig with terminal bud

Magnolia ×*soulangeana*
Terminal bud

Magnolia liliiflora
Tip of twig with terminal bud

Magnolia liliiflora
Terminal bud

Magnolia ×soulangeana
Terminal bud

Magnolia acuminata
Long-shoot with
terminal bud

Magnolia acuminata
Short-shoot with
terminal bud

Liriodendron tulipifera
Long-shoot

Liriodendron tulipifera
Opening of bud

Liriodendron tulipifera
Tip of twig with
terminal bud

***Magnolia* ×*soulangeana* SOUL.-BOD.,** saucer magnolia
[*Magnolia denudata* × *Magnolia liliiflora*]

Buds with rather dense appressed to shaggy grey-green to silvery-green hairs. **Twigs** red-brown to brown-grey, with paler raised lenticels. **Flowers** before the leaves appear, usually with 9 tepals. The colour varies from white via pink to a strong wine-red. The most frequently planted magnolia with very many forms which are differentiated predominantly by flower colour and shape. Shrub or small tree 3–6 m high.

Section Tulipastrum

Magnolia acuminata (L.) L., cucumber tree

Buds 10–15 mm long and 3–4 mm thick, cylindrical, with a blunt tip and with silvery, light greenish to brownish appressed velvety hairs. **Twigs** smooth, glossy olive-brown to red-brown, with few paler lenticels; later red- to violet-brown, finely oblong fissured at the lenticels. **Flowers** with the leaves or after these appear. Rarely seen, the young trees rather narrow, later broadening out and 15–20 m high. Origin: eastern USA.

Liriodendron L., tulip tree

In the tulip tree, the two stipules are not fused with the leaf base. At their tip they are broadly rounded and in the bud they lie with their margins flat on top of one another, so that the buds resemble small duckbills. There is a smooth transition between sessile lateral buds, short- to long-stalked lateral buds and terminal buds on short-shoots. Two species, one each in North America and in China.

Liriodendron tulipifera L., tulip tree

Buds duckbill-shaped, 6–10 mm long, dark wine-red to violet, in shadow also greenish, often with a slightly whitish bloom. **Twigs** olive-brown to red-brown, glabrous and glossy, towards the tip with a light bloom. Lenticels few, very pale. **Pith** compact, white, interrupted by horizontal woody diaphragms visible in longitudinal section. **Leaf scars** grey-brown, large and rounded with several indistinct traces. Scars of stipules encircling the twig at every node and meeting at a point on the opposite side of the leaf scar. Frequently planted tree reaching over 40 m. Origin: North America.

Annonaceae
soursop family

Asimina triloba (L.) DUNAL, pawpaw

Buds naked, distichous, variable. **Floral buds** axillary, globose to obovoid, c. 3 mm long and 2–3 mm thick; greenish and covered in black hairs; surrounded by two transverse bracteoles and by the 3 sepals. **Leaf buds** when terminal 6–8 mm long, 2 mm wide and 1–2 mm thick; when axillary smaller, densely red-brown to dark brown hairy. Bud leaves distichous, each surrounding the next (equitant). **Twigs** grey to grey-brown, ± red-brown hairy below terminal bud. Leaf scar with 5 to 7 traces, central trace the largest one. At the top margin of the leaf scar there often is a densely hairy remnant of the leaf base. **Flowers** together with or shortly before the leaves appear. Occasionally planted as 'exotic fruit', shrub or small tree to 5 m high. Origin: west or south-west USA.

Calycanthaceae
spicebush family

Medium-sized shrubs originating from Asia and North America. Twigs mostly aromatic, tetragonal. In cross section with green spots concentrated in the corners. Leaves obliquely opposite, lateral buds in the axils; terminal buds not formed.

Key to *Calycanthaceae*

1 Buds naked, flowering in the summer . . . . ***Calycanthus***
1* Buds with bud scales, flowering in the winter. . . . . *Chimonanthus praecox*

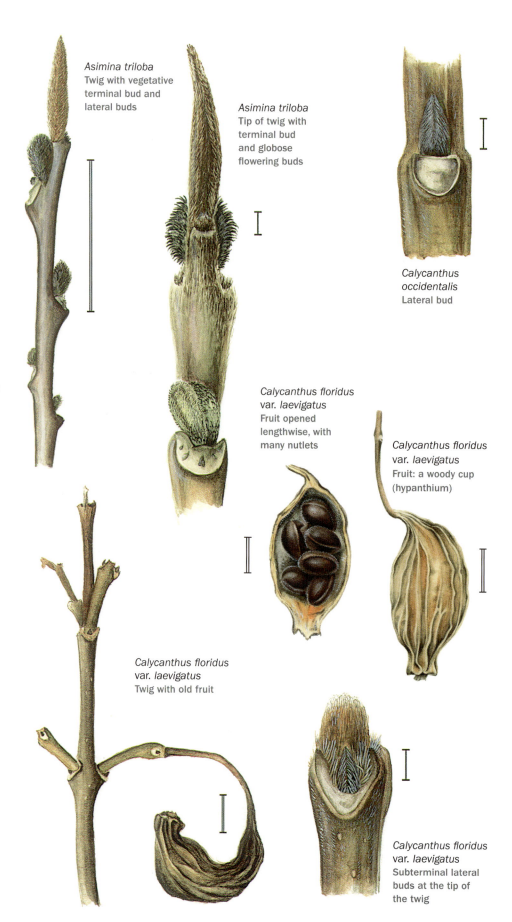

Asimina triloba
Twig with vegetative terminal bud and lateral buds

Asimina triloba
Tip of twig with terminal bud and globose flowering buds

Calycanthus occidentalis
Lateral bud

Calycanthus floridus var. *laevigatus*
Fruit opened lengthwise, with many nutlets

Calycanthus floridus var. *laevigatus*
Fruit: a woody cup (hypanthium)

Calycanthus floridus var. *laevigatus*
Twig with old fruit

Calycanthus floridus var. *laevigatus*
Subterminal lateral buds at the tip of the twig

Calycanthus occidentalis
Fruiting twig

Calycanthus occidentalis
Fruit with wide opening

Calycanthus L., spicebush

Buds naked, with very dark hairs. Leaf scars with 3 traces, on raised cushions, not connected to each other (sub opposite). Twigs and other parts often with an aromatic scent. Flowers long after the leaves appear. Fruit, a multiple fruit similar to a capsule, and persistent through the winter; the woody base of the flower (receptaculum) forms the fruit 'cup' (hypanthium), which holds the nutlets. It also bears the spiralling scars of tepals and has an opening at the tip. Three good species; there are a few hybrids, rare in cultivation.

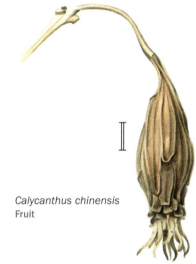

Calycanthus chinensis
Fruit

Calycanthus chinensis
Subterminal lateral buds

Calycanthus chinensis
Lateral bud

Calycanthus chinensis
Opened fruit

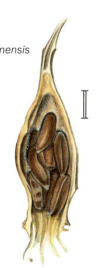

Key to *Calycanthus*

1 Buds encircled by the leaf scar 2
1* Buds oblong, 2–4 mm long, leaf scar does not encircle bud; fruit cup squat, with wide opening at the tip . . . .*Calycanthus occidentalis*
2 Twigs pale grey to ochre-brown, relatively stout, the opening of the fruit surrounded by long tips . . . .*Calycanthus chinensis*
2* Twigs dark grey to red-brown, no distinct tips around the narrow opening of the fruit . . . . **Calycanthus floridus**

Calycanthus floridus L., Carolina allspice

Buds small, blackish hairy, surrounded by the leaf scars and partly by remains from leaf stalks. **Twigs** very finely densely hairy, with few, small and rounded lenticels. **Bark** especially of the young branches strongly aromatic. **Fruits** narrowly ovoid, towards the tip narrowed and distinctly neck-like. Small 1.5–3 m high shrub. Origin: southeast USA.

The variety *Calycanthus floridus* var. *laevigatus* (WILLD.) TORR. & A. GRAY [*Calycanthus fertilis* WALTER] is often planted; in winter it can only be distinguished from the main variety with difficulty, it has more fruits, glabrous twigs and a weaker scent.

Calycanthus occidentalis HOOK. & ARN., California allspice

Buds clearly visible, not surrounded by the leaf scar, dark hairy. Fruits bell-shaped, with wide opening, hardly narrowing to the tip. Rarely planted 1.5–3 m high shrub. Origin: California.

Calycanthus chinensis (W. C. CHENG & S. Y. CHANG) P. T. LI, Chinese spicebush
[*Sinocalycanthus chinensis* W. C. CHENG & S. Y. CHANG]

Buds compact, naked, 2–4 mm long, surrounded by leaf scar, densely hairy, with black hairs at the base, towards the tip yellowish-green. **Twigs** glabrous, relatively stiff and of medium thickness, only slightly angular; surface light brown to ochre-brown, partly with yellow-green shades, glabrous, with many lenticels. Outside botanical gardens still rarely planted, open shrub growing to 2.5 m high. Origin: eastern China.

A hybrid, **Calycanthus × raulstonii** (F. T. Lass. & Fantz) F. T. Lass. & Fantz ex Bernd Schulz, exists between the North American species *C. floridus* and the Chinese species. Several, currently only rarely planted forms, which show great promise, belong to this group.

Chimonanthus praecox (L.) Link, wintersweet

Buds ovoid, 3–4 mm long, with small red-brown to grey-brown bud scales, densely white hairy around the margin, sparsely so on the outside. Occasionally with descending accessory buds. **Flower buds** more globose with ochre-coloured bud scales/outer tepals. **Twigs** grey-brown, scattered with lenticels of the same colour, hardly distinguishable or a little paler. **Leaf scars** on relatively large cushions. **Flowers** often as early as January or February, strongly scented. Tepals many in a spiral arrangement: the outermost longer, pale yellow, the inner ones shorter and dark purple, with transitions between the two. Occasionally planted, shrub 1–2 m high. Origin: China.

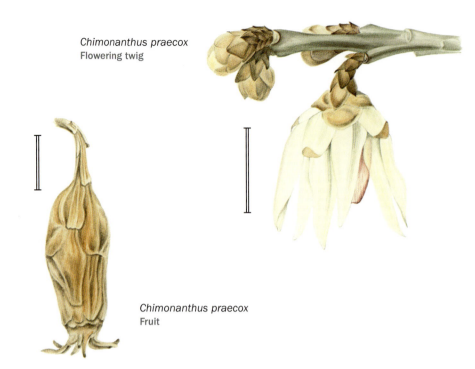

Chimonanthus praecox
Flowering twig

Chimonanthus praecox
Fruit

Lauraceae
bay-laurel family

Mostly evergreen species in the subtropics and tropics. Often with aromatic essential oils.

Key to Lauraceae

1   With large terminal buds, lateral buds solitary . . . . *Sassafras albidum*
1* Lateral buds with enrichment buds . . . . *Lindera benzoin*

Lindera benzoin (L.) Blume, Benjamin bush

Buds glabrous, yellowish green partly with reddish hue. **Flower buds** mostly 2, more rarely 1 or 3 from the sides of the prophylls of the slender vegetative lateral buds; with a short stalk, globose, slightly acute and c. 2 mm long. **Twigs** glabrous, olive-brown–brown, slightly glossy. **Flowers** greenish-yellow, 2–5 in small, to 5 mm broad, sessile heads before leaf appearance in March to April. **Bark** grey-brown. Seldom planted, broad shrub 3–6 m high. Origin: south-east USA.

Of the about 100 species of the genus, occasionally other species are cultivated

Chimonanthus praecox
Lateral buds

Chimonanthus praecox
Subterminal lateral buds

Lindera benzoin
Many flower buds as enrichment buds beside the lateral buds

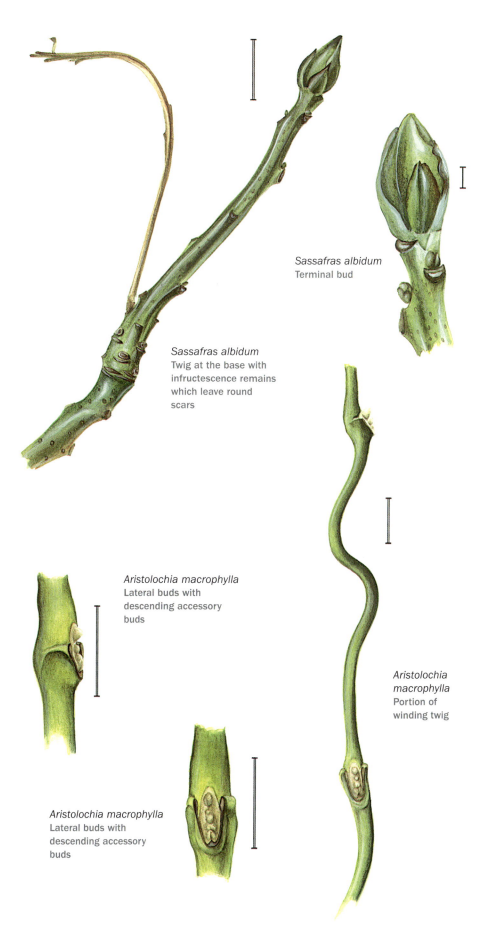

Sassafras albidum
Terminal bud

Sassafras albidum
Twig at the base with
infructescence remains
which leave round
scars

Aristolochia macrophylla
Lateral buds with
descending accessory
buds

Aristolochia
macrophylla
Portion of
winding twig

Aristolochia macrophylla
Lateral buds with
descending accessory
buds

in botanical collections, such as the east-Asian *Lindera umbellata* Thunb. and *Lindera obtusiloba* Blume.

Sassafras albidum (Nutt.) Nees, sassafras

Buds greenish. **Lateral buds** in a spiral, mostly relatively small. **Terminal buds** ovoid, 8–10 mm long with few bud scales, the outer ones occasionally with remnants of the leaf. **Twigs** grass green to yellowish green; towards the tip with a slight waxy bloom, this easily wiped off; distinctly thickened at the year's growth margin and thinnest below the terminal bud. Older twigs reddish-green on the side exposed to light. **Lenticels** many: initially flat, later raised with a dark margin. Leaf scars half round to round, at the upper edge flat or notched, with a central furrow. At the base of the twig, in axils of narrow leaf scars (last year's bud scales) there are round scars of the infructescences. **Bark** deeply fissured. Occasionally planted, tree to 15(–20) m high, with runners. Origin: North America.

Aristolochiaceae
birthwort family

Aristolochia L., birthwort

The only prophyll of the lateral buds is turned towards the principal axis (addorsed). This is more characteristic for monocotyledons (e.g. *Smilax*), but it also occurs in this genus of basal angiosperms.

Aristolochia macrophylla Lam., Dutchman's pipe

Buds: alternate lateral buds, ± without bud scales, silvery grey-green hairy; with many descending accessory buds. Some accessory buds are larger than the primary lateral bud, due to the differentiation of these buds as flower buds. **Twigs** winding to the left, anti-clockwise, green, smooth, without lenticels, only the older twigs turn grey-brown. **Leaf scar** horseshoe-shaped, largely enveloping the buds, with three traces, on large cushions. Frequently planted, winding shrub to 10 m high. Origin: eastern North America.

More rare are the species distinguished by hairy twigs: *Aristolochia manshuriensis* Kom., from eastern Asia, or *Aristolochia tomentosa* Sims from North America.

Angiosperms — Monocotyledons

Smilacaceae
catbriar family

Smilax L., catbriar

Only **lateral buds**, these with an addorsed prophyll, i.e. the form of prophyll typical for the monocotyledons. Twigs ± green, often with prickles. Climbing shrubs in the temperate to tropical zones. Often evergreen, like the species *Smilax aspera*, occurring from Southern Europe to India.

Key to *Smilax*

1 Older twigs round 2
1* Twigs ± angular. . . . . . *Smilax bona-nox*
2 Shoots with many thin prickles, plant without runners . . . . . .*Smilax tamnoides*
2* Shoots with few, thick prickles, plant with runners. . . . . *Smilax rotundifolia*

Smilax rotundifolia L., common catbriar

Buds alternate, at almost right angles to the twig; on the underside covered by the persistent leaf base and on the upper side covered by the addorsed prophyll. The leaf base usually is dry and brown and has two thread-like stipular tendrils. **Twigs** green, initially angular, older twigs round; with few thick prickles between the nodes. Strong liana, climbing to 7–10 m high and forming runners. Origin: North America.

Other North American species are seldom cultivated: *Smilax bona-nox* L. with 4-angular very prickly twigs, that are basally beset with bristly stellate hairs; as well as *Smilax tamnoides* L., a shrub climbing to over 10 m high; it rarely forms runners and has round twigs that basally have dense straight, thin, dark prickles.

Smilax rotundifolia
Twig

Smilax
The stipular tendrils on the leaf-remains (leaf base) and the addorsed prophyll of the lateral bud are typical

Angiosperms — true Dicotyledons

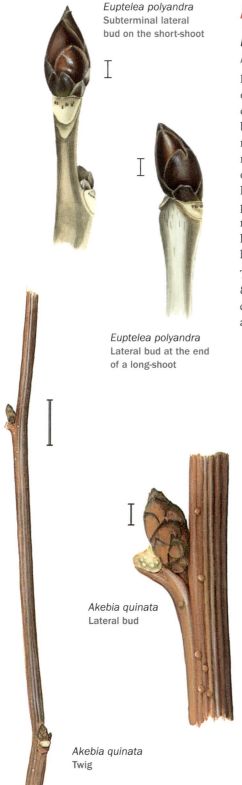

Euptelea polyandra
Subterminal lateral bud on the short-shoot

Euptelea polyandra
Lateral bud at the end of a long-shoot

Akebia quinata
Lateral bud

Akebia quinata
Twig

Eupteleaceae
Asian elm family

Euptelea polyandra Siebold & Zucc., Asian elm

Buds exclusively alternating lateral buds, ovoid and acute, c. 6 mm long and 3 mm diameter, dark red-brown glossy with c. 10 bud scales, the lowest one finely hairy, the remnant only with some small hairs on the margins. **Twigs** red-brown, with waxy bloom on one side, caused by the lifting epidermis, later dark grey-brown with many dot-shaped pale lenticels. **Leaf scars** pale ochre with more than five dark brown stripes. Below the leaf scar with a paler margin. Tree to 7 m high. Origin: Japan.

The species ***Euptelea pleiosperma*** Hook.f. & Thomson is very similar and cannot be distinguished in winter. Origin: Himalaya and central China.

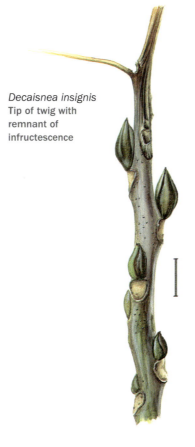

Decaisnea insignis
Tip of twig with remnant of infructescence

Lardizabalaceae
Zabala-fruit family

No terminal buds, leaf scars with many traces. Lianas, one shrub (*Decaisnea*), in Asia, few in South America.

Key to Lardizabalaceae

1 Erect shrub with large branches, laxly branched . . . . ***Decaisnea insignis***
1* Winding lianas 2
2 Twigs brown, leaf cushions protruding, leaf scar with many traces . . . .*Akebia quinata*
2* Branches green to violet-brown, with waxy bloom, leaf scars broad with 3 traces . . . . *Sinofranchetia chinensis*

Decaisnea insignis (Griff.) Hook.f. & Thomson

Buds: only spirally arranged lateral buds, to c. 20 mm long, surrounded by two bud prophyll scales. The outer bud scale surrounds the inner at the base, each one ends in a small tip, so the buds are two-tipped. Scales fleshy, thick, externally green to olive brown, bluish-white with waxy bloom, with a slightly bumpy surface. **Twigs** very thick (in the first year to 10 mm), rigid, not branching much, green to olive, with a bluish-white waxy layer which can be wiped off. **Lenticels** many, small, pale brown, prominent. **Leaf scars** large, mostly with 5 inconspicuous (groups of) traces, the central ones consisting of several small single traces. Remnants of the **infructescences** persist in winter. Common, erect 3 m high shrub. Origin: western China.

Akebia quinata (Houtt.) Decne., chocolate vine

Buds lateral, alternate, broadly ovoid, 4–7 mm long and 3 mm thick, often with enrichment buds. Lateral buds with many brown bud scales; these are a little darker towards the margin and slightly keeled and spreading at the tip. **Twigs** red- to olive-brown, ± furrowed, with few to many lighter lenticels. The leaf base remains and resembles a prominent leaf cushion. Leaf scar with more than three traces of vascular bundles. Frequently planted, winding liana. Origin: Japan and central China to Korea.

Sinofranchetia chinensis (FRANCH.) HEMSL.
Branches differentiated into winding long-shoots and short-shoots on older twigs. **Lateral buds** on young long-shoots, standing away from the twig, c. 6 mm long. **Terminal buds** at the end of a short-shoot, ovoid, to c. 10 mm long. **Bud scales** glabrous, fleshy on indistinctly green base, with a dark wine-red-violet surface hue with a lighter, dirty ochre to cinnamon-brown margin. Especially the basal bud scales with a distinctly spreading leaf stalk remnant. **Branches** c. 3–5 mm diameter, glabrous, violet-brown to greenish in shadow, often appears lighter through a waxy bloom on the surface; with a few warty, pale ochre-brown lenticels. **Leaf scars** shield-shaped, 5–10 mm broad, distinctly broader than internodes, therefore making the nodes wider. Rarely planted, liana climbing to 10 m. Origin: central to western China.

Menispermaceae
moonseed family

Anti-clockwise winding lianas, without terminal buds: branching sympodial. Lateral buds alternate, without bud scales and hairy; mostly with descending serial accessory buds.

Key to Menispermaceae

1 First year's twigs slender (<2.5 mm) and hairy. . . . .*Cocculus orbiculatus*
1* Twigs thicker (mostly >3 mm), glabrous or hairy. . . . .
. . . .***Menispermum canadense***

Cocculus orbiculatus (L.) DC.,
queen coralbead
[**Cocculus trilobus** (THUNB.) **DC.**]

Buds to 2 mm, globose, densely pale grey-brown hairy, often with a second, relatively large descending accessory bud. **Twigs** 1–2 mm diameter, green and cylindrical; densely fine pale grey-brown hairy. **Leaf scar** rounded, obliquely ovoid on a large leaf cushion. Shrub, climbing to 4 m, rarely cultivated. Origin: Japan to Himalaya and the Philippines.

Sinofranchetia chinensis
Lateral bud on long-shoot

Sinofranchetia chinensis
Short-shoot with terminal bud

Cocculus orbiculatus
Winding twig

Cocculus orbiculatus
Lateral bud with descending accessory bud

Menispermum dauricum
Winding twig

Menispermum canadense
Small lateral buds with descending accessory buds, almost hidden by the large leaf scar

Xanthorhiza simplicissima
Tip of twig with terminal bud

Menispermum canadense L., Canada moonseed

Lateral buds small, 2–3 mm thick, flat to conical, surrounded by two outer dark violet-brown bud scales; with small, descending accessory buds. **Twigs** initially finely hairy, glossy green and ± becoming glabrous, later brown. **Leaf scar** large and rounded, on a large leaf cushion, deeply notched on upper rim and enclosing the accessory buds situated below the primary bud. Above the lateral bud is occasionally a further small scar from the fruiting heads. Occasionally planted, compact liana, climbing to 4 m high. Origin: North America.

Less common is *Menispermum dauricum* DC.,with entirely glabrous twigs. Origin: eastern Asia.

Ranunculaceae
buttercup family

A species rich family. The main bulk are annual and perennial herbaceous plants, with only a few woody genera. Phyllotaxy alternate, more rarely, as in *Clematis*, opposite. Origin: Northern Hemisphere.

Key to Ranunculaceae

1 Buds opposite, liana with leaf rachis-tendrils . . . . *Clematis*
1* Buds alternate, small prostrate to ascendant shrubs . . . . *Xanthorhiza simplicissima*

Xanthorhiza simplicissima MARSHALL, yellowroot

Buds: terminal buds 1.5–2 cm long, fusiform and lateral buds alternate, 2–4 mm long and appressed. **Bud scales** of the terminal buds from a parallel-veined leaf base with remnants of the lamina: spreading tips to persistent small leaflets; greenish, violet-wine-red hue, glabrous, only margins inconspicuously ciliated. **Twigs** slender and glabrous, smooth, grey-olive, inner surface yellow. Lenticels few, round. **Leaf scars** broad and encircling the twig with 7–9 traces side by side. Little-branched shrub, forming runners, prostrate to ascending, 0.6–1 m high, rarely planted. Origin: North America.

Clematis L.

Climbing with opposite leaf rachides, or more rarely erect perennials or shrubs, the leaf rachides persistent even after leaflets have fallen. Due to the persistent leaf rachides, there are no leaf scars. A useful characteristic is the head-like multiple fruits with style topped nutlets which persist for the whole winter. The intensive work on the genus by breeders, with the resulting hybrids, makes the identification of these plants in winter very difficult. This is why only a few of the more frequent species are dealt with here.

Key to *Clematis*

1 Fruits many in axillary panicles, twigs sturdy, most frequent species in Europe . . . . ***Clematis vitalba***

1* Fruits solitary or few, mostly with long stalks 2

2 Fruits solitary, at the end of short-shoots ***Clematis alpina*** and
C. macropetala

2* Without short-shoots 3

3 Twigs almost entirely glabrous . . . . ***Clematis orientalis***

3* Twigs hairy at least at the nodes 4

4 Twigs dainty, when young c. 2 mm diameter, from southern Europe . . . . ***Clematis viticella***

4* Twigs mostly thicker 5

5 Fruitlets: c. 4 mm long nut with a finely hairy style to 4 cm long . . . . ***Clematis tangutica***

5* Fruitlets: more than 5 mm long nut with long-hairy style to 3 cm long . . . . ***Clematis montana***

Subgenus Atragene

Fruits long-stalked, on axillary short-shoots.

Clematis alpina (L.) MILL., Alpine clematis

Buds ovoid to narrowly ovoid, acute, to c. 10 mm long, surrounded by a few dark grey-brown bud scales which are slightly hairy right to the tip. **Twigs** grey-brown to black-brown, ± glabrous. Frequently planted liana, to 2 m high. Origin: from Europe to western Asia, with the typical subspecies in the Alps, from south-eastern France to Austria, in the Apennines and in Carpathian mountains.

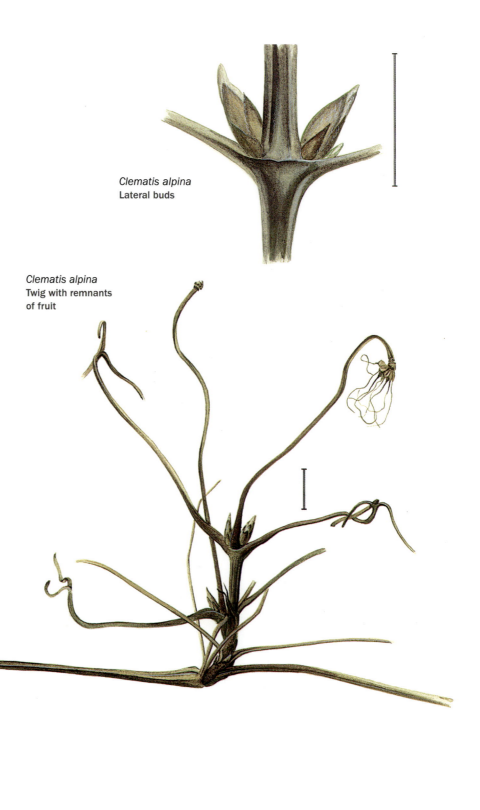

Clematis alpina
Lateral buds

Clematis alpina
Twig with remnants of fruit

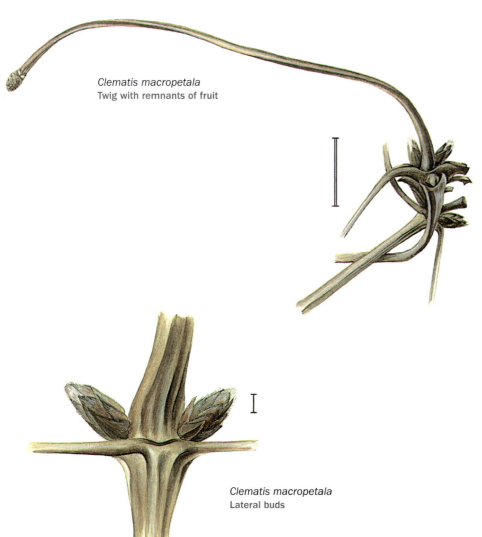

Clematis macropetala
Twig with remnants of fruit

Clematis macropetala **Lᴇᴅᴇʙ.,**
downy clematis

Buds 3–5 mm long, ovoid, with several outer pairs of scales. Lowest bud scales dark grey-brown, almost glabrous, the upper ones lighter, ± densely long-hairy. **Twigs** grey-brown, relatively slender, angular especially on the nodes, shortly hairy. With many short-shoots, which end in a 12–15 cm long fruit stalk. Base of leaf rachis persistent. Frequently planted, shrub climbing to 3 m high. Origin: Asia from Manchuria and Siberia to northern China.

Subgenus Flammula

Section Viticella

Strong climbing lianas. Infructescences at the tip of last year's shoots.

Clematis viticella L., purple clematis

Buds small, globose to ovoid, with few bud scales. **Twigs** slender, angular-furrowed, red-brown. **Fruits** with short, glabrous style. Climbing shrub to 4 m high. Origin: southern Europe to Asia Minor.

A frequently planted hybrid is *Clematis ×jackmanii* T. Mᴏᴏʀᴇ [*Clematis lanuginosa × Clematis viticella*]. **Buds** globose-ovoid, 2–4 mm long, surrounded by a few grey-brown bud scales. **Twigs** red-brown, grooved, hairy on the nodes, slender (c. 3 mm) and furrowed. Leaf base of the opposite leaves meet and extend prominently, almost wing-like. Liana climbing to 3–4 m high.

Clematis macropetala
Lateral buds

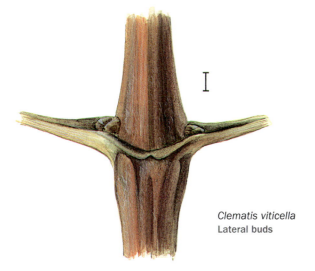

Clematis viticella
Lateral buds

Clematis viticella
Lateral buds

Subgenus Cheiropsis

Section Montanae

Clematis montana BUCH.-HAM. ex DC., mountain clematis

Buds 5–7 mm long, outer bud scales sparsely hairy to almost glabrous, subsequent scales densely hairy. **Twigs** ochre-brown to dirty red-brown, slightly grooved, of medium thickness (of first year's growth 2–4 mm). **Fruits** on two-year-old shoots, nutlets 5–6 mm long, flattened, circular, red-brown and glabrous, style 25–30 mm long and densely hairy (hairs 5–6 mm long, shorter towards the style's tip). Strongly growing liana, climbing to 8 m high. Origin: central to western China and Himalaya.

Clematis montana
Nutlet

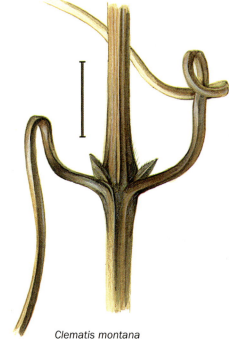

Clematis montana
Twig with tendrils
of leaf rachides

Clematis montana
Lateral buds, leaf
rachides removed

Clematis vitalba
Twig with infructescences

Clematis vitalba
Twig with leaf rachis-tendrils

Subgenus Clematis

Section Clematis

Clematis vitalba L.,
traveller's joy, old man's beard

Twigs mostly strong and angular-furrowed.
Main branches to several cm thick. **Fruits**
with hairy styles, these can be formed
variously in different individuals: from
relatively long to short. Liana of strong
growth, climbing to 15 m, more rarely 30 m
high, and often covering large areas. The
most frequent species in Europe, extending
to the Mediterranean area and to Asia Minor.

Subgenus Campanella

Section Meclatis

Clematis orientalis L., Oriental clematis

Twigs relatively slender, only very slightly
angular, grey-brown, towards the nodes
mostly darker brown, longitudinally striped,
glabrous, or at most with a few small hairs
on the nodes. **Fruits** with feathery styles.
Liana, frequently planted, climbing 3–5 m
high. Origin: Himalaya.

Clematis tangutica (Maxim.) Korsh.,
golden clematis

Twigs initially hairy, persistent hairy around
the nodes. Fruiting heads stalked for 8–15
cm long, **fruits** pale ochre-brown, c. 4 mm
long nutlets with 30–40 mm long finely
hairy style (hairs c. 4 mm long, shorter
to glabrous at the tip). Frequent liana,
climbing to 3 m high. Origin: Mongolia.

Clematis orientalis
Twig with leaf
rachis-tendrils

Clematis orientalis
Lateral buds

Clematis tangutica
Nutlet

Berberidaceae
barberry family

As well as evergreen representatives from the genera *Nandina* and *Berberis* (including *Mahonia*), within *Berberis* there are some common deciduous species.

Berberis L., barberry

These shrubs can be easily recognised from the spine-forming leaves. Assimilation is carried out by foliage leaves in the axils of these spines, on short-shoots; leaf scars remain. At the tip of the current year's short-shoot, there is usually a terminal bud. But when an infructescence ends the shoot, a lateral bud will continue growth. The wood and the inner bark are yellow.

Useful characters are the shape of the spines (1-, 3- or many-parted), the bark colour of younger twigs (yellow, brown, grey), covering of hairs (indumentum) and grooves on the twigs as well as the height of growth and, as long as these remain, the infructescences.

Key to *Berberis*

1 Spines mostly 3-parted (or more) 2
1* Spines predominantly simple 8
2 Shrubs to about 1 m high 3
2* Shrubs usually over 1.50 m high 4
3 Twigs initially hairy, red-brown 6
3* Twigs glabrous, yellowish-brown 7
4 (2) Twigs grey-brown, grey when older . .
. . . . *Berberis vulgaris*
4* Twigs brown 5
5 Spines thin, mostly 3-parted; twigs not pruinose . . . . *Berberis aggregata*
5* Spines sturdy, lower on the twig widened and leaf-like, twigs initially pruinose . . .
. . . . *Berberis koreana*
6 (3) Spines thin, 3-parted . . . .
. . . . . . *Berberis wilsoniae* (and hybrids)
6* Spines sturdy, lower on the twig widened and leaf-like . . . . *Berberis koreana*
7 (3) Infructescences up to 5 cm long panicle . . . . *Berberis×carminea*
7* Infructescences umbels or fascicles . . . .
. . . . *Berberis ×ottawensis*
8 (1) Twigs yellowish, ochre-brown . . . . . .
. . . . *Berberis ×ottawensis*
8* Twigs red-brown . . *Berberis thunbergii*

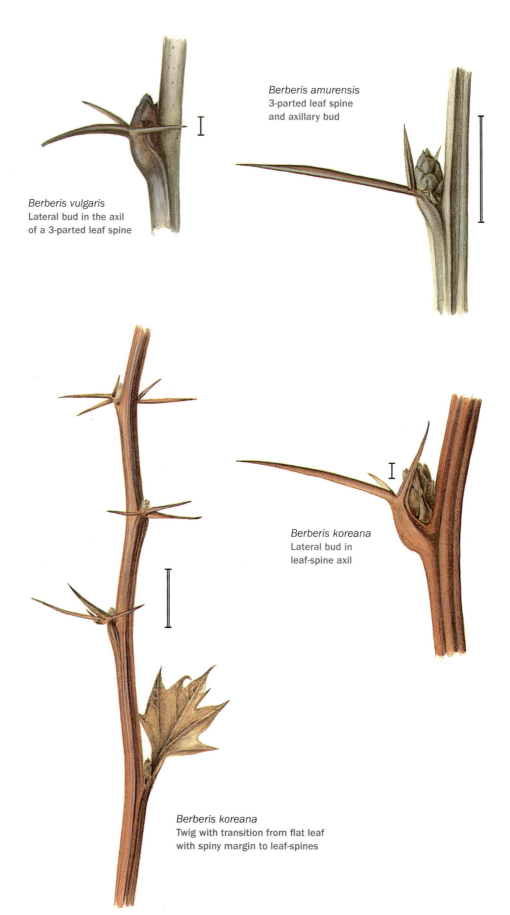

Berberis vulgaris
Lateral bud in the axil of a 3-parted leaf spine

Berberis amurensis
3-parted leaf spine and axillary bud

Berberis koreana
Lateral bud in leaf-spine axil

Berberis koreana
Twig with transition from flat leaf with spiny margin to leaf-spines

Berberis thunbergii
Twig with simple leaf-spine

Berberis ×*ottawensis*
**Lateral bud in the axil of a
simple leaf-spine**

Berberis wilsoniae
**Three-pointed ochre-
brown leaf-spine**

Series Berberis

Berberis vulgaris L., common barberry

Buds at the end of short axillary shoots, grey-brown, ovoid, 4–5 mm long: basally enclosed by leaf remains (leaf base with round petiole scars) and the subsequent bud scales. Where the tip of the short-shoot turns into an infructescence, a lateral bud often develops in a leaf axil. **Twigs** furrowed, with mostly 3-parted, 1.5–2 cm long, grey-brown to red-brown spines. Shrub to more than 2.5 m, often planted. Native to west and south Europe to the Crimea and the Caucasus.

Berberis amurensis RUPR. cannot be distinguished with certainty from *Berberis vulgaris*; this comes from east Asia and is also widely planted.

Series Sinensis

Berberis koreana PALIB., Korean barberry

Buds ovoid, 3-4 mm long, grey-brown, with some bud scales. The leaves modified to leaf-spines are often still leafy near the base of the branch; otherwise mostly 3–5(7)-parted, but near the tip sometimes simple; parts of the spine sturdy and to 1.5 cm long. **Twigs** orange-brown to dark red-brown, strongly grooved, occasionally somewhat pruinose; at two years old dark brown. **Pith** white, wide and full. **Fruits** ovoid, to 8 mm long, in 4–5 cm long, hanging racemes. Erect shrub, 1.5–2 m high, occasionally planted. Native to Korea.

Berberis thunbergii DC., purple barberry

Buds ovoid, 2–3 mm long; upper bud scales grey-brown and pointed, lowermost initially reddish. **Twigs** glabrous, thin, strongly red-brown and densely furrowed. **Leaf spines** thin, simple and relatively short: 5–15 mm, those towards the shoot tip occasionally upright and at a sharp angle to the twig. **Fruits** few: 1–2(–5) in sessile fascicles; c. 8 mm long, ellipsoid and without style. Small shrub to 1 m high, densely branched with bow-shaped twigs; frequently planted. Origin: Japan.

Similar and difficult to differentiate is: *Berberis* ×*ottawensis* C. K. SCHNEID. [*Berberis thunbergii* × *Berberis vulgaris*]. **Twigs** yellowish-brown to ochre-brown, thin and furrowed, but more robust than *Berberis thunbergii*. **Spines** simple, thin, 8–12 mm long. **Fruits** around 5–10 in fascicles or short racemes, ovoid and red. Shrub to 1.3 m high. Some common varieties with red foliage and with ± red brown twigs can only be distinguished with the help of fruit remnants.

Series Polyanthae

Berberis wilsoniae HEMSL., Wilson's berberis

Buds ovoid, 2–3 mm long; bud scales brown, scattered short-glandular hairy and ciliate. **Twigs** angular, red-brown, initially hairy (glabrous in some varieties), later only with black dot-like remains of hair. **Spines** 3-parted, 1–2 cm long, thin and pointed, yellowish-brown. **Fruits** 2–6, salmon-pink, globose, with short style. Dense shrub to 1 m high. Native to China: west Sichuan.

Two frequently planted garden hybrids of *B. aggregata* with Wilson's berberis are **Berberis ×rubrostilla** CHITT., a shrub to 1.5 m high with red-brown, initially hairy, angular twigs and ovoid 1.5 cm long fruits without styles in corymbs; and **Berberis ×carminea** CHITT. ex AHRENDT, a 1 m high shrub with yellowish-brown, glabrous, angular twigs and c. 8 mm long fruits hanging in up to 8 cm long panicles.

Berberis aggregata C. K. SCHNEID., salmon barberry

Buds short-ovoid, c. 3 mm long; red-brown to grey-brown, hairy. With many spiral bud scales, the lowest one with distinct petiole scars, the top-one± pointed. **Twigs** angular, rather thin, matte ochre-brown to grey-brown, finely hairy. **Spines** ochre-brown to orange-brown, mostly 3-parted, less often simple, 1–2(–3) cm long, thin and pointed. **Fruits** remain on the bush for extended periods, firmly attached, in short, almost sessile panicles to 3 cm long: red pruinose globose berries, with short style, c. 6 mm long. Shrub 1.5–2 m high. Origin: China.

Var. **prattii** (C. K. SCHNEID.) C. K. SCHNEID. [*Berberis prattii* C. K. SCHNEID.] **Spines** 1–3 -parted. **Fruits** numerous in 10(–20) cm long panicles.

Sabiaceae
pao-hua family

Meliosma dilleniifolia subsp. *tenuis* (MAXIM.) BEUSEKOM

Buds naked, with several densely hairy, separated bud leaves. **Terminal buds** 5–7 mm long, the elongate cylindrical bud leaves partly twisted around one another. Lateral buds smaller, appressed to the twig, on noticeable leaf-cushions. **Twigs** robust, pale brown, finely hairy at the tip, lower down glabrous, with numerous pale ochre lenticels. The current year's long-shoot at the base with variably-sized short-shoots. **Pith** full, white with greenish margin. **Leaf scars** with one trace. Rarely planted tree, 10–15 m high. Origin: western China.

Berberis aggregata var. *prattii*
Lateral bud and several-parted leaf-spine

Berberis aggregata var. *prattii*
Twig with persistant remnants of infructescences

Meliosma dilleniifolia subsp. *tenuis*
Twig tip

Meliosma dilleniifolia subsp. *tenuis*
Twig tip

Platanaceae
plane-tree family

Platanus L., plane

Large trees with distinct irregularly descaling bark. **Buds** enclosed by fused stipules forming an envelope (ochrea). The leaf scar of the leaf which envelops the bud with its hood-shaped leaf base during the growing season, has many traces and surrounds the bud almost entirely. From its upper margin a fine, linear scar, pertaining to fused stipules, encircles the twig. A very distinct genus. But, it is more difficult to identify the species: the number of globose fruit clusters, the growth form and the bark are helpful.

Key to *Platanus*

1 Infructescences of 2 or more globose fruit clusters 2
1* Single globose fruit cluster, bark coming off in small scales . . *Platanus occidentalis*
2 Infructescences of 3 or more globose fruit clusters . . . . *Platanus orientalis*
2* Mostly two globose fruit clusters per infructescence. . . . *Platanus ×hispanica*

Platanus ×hispanica MÜNCHH., London plane
[*Platanus orientalis* × *Platanus occidentalis*]

Buds: only lateral buds, alternate, rounded-angular, bluntly conical, enveloped by a single bud scale. Terminal lateral buds to 10 mm long and at the base 6 mm thick; buds lower on the twig distinctly smaller. **Twigs** rigid, olive-brown to red-brown, with many roundish pale ochre-coloured lenticels; older grey-brown to grey. **Leaf scar** entirely surrounding the bud, with many traces of vascular bundles; stipule scar surrounding the twig as a narrow strip. The **bark** dehisces in large plates and lends a typically patchy look to the trunk. **Fruits** in globose clusters, persistent in winter, mostly in pairs on a hanging stalk. Often planted tree to 35 m high, trunk continuing almost to the top.

The parent species are less common: *Platanus orientalis* L., the oriental plane, mostly with the trunk growing right to the top and with distinctly spreading branches. The bark dehisces in relatively large plates and the infructescence consist of 3 or more fruit clusters. Tree to 30 m high, native from the Balkan and Asia Minor to Western Asia. In *Platanus occidentalis* L., the American or occidental plane, the lateral branches grow upwards and the trunk ends well below the top of the tree. The bark dehisces in small scales and the infructescences consist mostly of a single globose fruit cluster. The tree grows to 40 m high. Origin: North America.

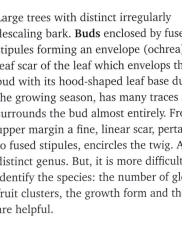

Platanus ×hispanica
Twig with short-shoots

Platanus ×hispanica
Short-shoot with terminal lateral bud and (on the left) the dead shoot tip

Platanus orientalis
Lateral bud

Platanus orientalis
Long-shoot

Trochodendraceae
wheel-tree family

Tetracentron sinense OLIV., spur leaf

All parts glabrous. **Buds** mainly ochre-green, with ± strongly red tints, long and narrow, entirely enclosed by a single scale which overlaps at its margin. Terminal bud and subterminal lateral bud appressed to each other, lower lateral buds spreading, the tip slightly bent back towards the twig, c. 12 mm long and at the base 2 mm diameter. Lateral buds on the long-shoot alternate distichous. **Twigs** (long-shoot) relatively slender, in the light dark wine red, in shade red-brown, with very pale lenticels. Pith solid. **Leaf scar** narrow with 3 traces, enveloping half of the bud base, elongated on both sides by small scars of stipules so that the entire scar encircles three-quarters of the twig. Older plants with many long-lived short-shoots which have only a single terminal bud. This unmistakable tree grows to 20 m high and is only rarely cultivated. Origin: northern India and south-west China.

Paeoniaceae
peony family

Paeonia L., peony

Mostly herbaceous perennial plants, more rarely little-branched low shrubs with thick branches and large buds. Lateral buds alternate. Old bud scales are retained like the leaves are, and both fall off late, leaving irregular abscission points. **Fruits**: 2–5 several-seeded follicles.

Key to Paeonia

1 Young twigs strongly wine-red to violet-brown, plants with stolons . . . .
. . . .**Paeonia delavayi** var. **delavayi**
1* Twigs grey-brown. 2
2 Shrubs to 2 m high, with little branching . . . .**Paeonia rockii** and **P. ×suffruticosa**
2* Shrubs to I m high, almost without any branching. .**Paeonia delavvayi** var. **lutea**

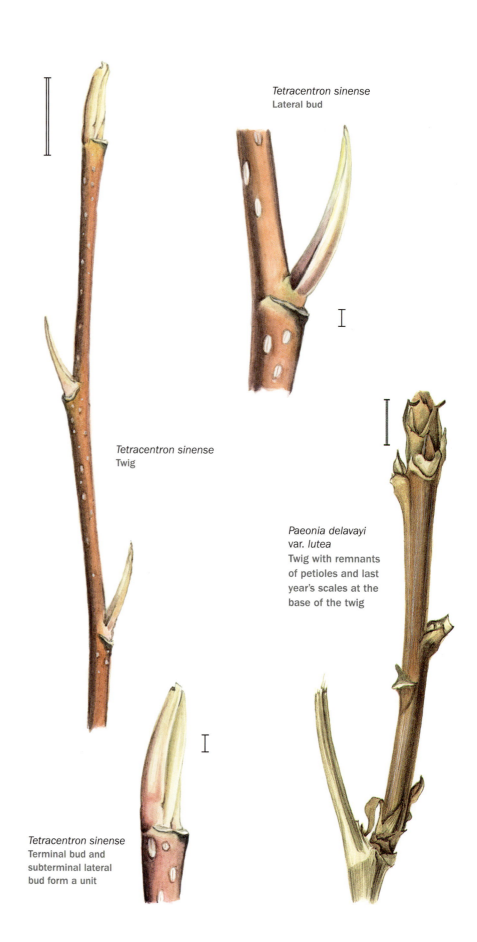

Tetracentron sinense
Lateral bud

Tetracentron sinense
Twig

Paeonia delavayi
var. lutea
Twig with remnants
of petioles and last
year's scales at the
base of the twig

Tetracentron sinense
Terminal bud and
subterminal lateral
bud form a unit

Paeonia ×suffruticosa
Twig with dry remnants
of leaves and terminal
multiple fruit

Paeonia delavayi
Twig with distinctive
reddish-brown surface

Paeonia rockii (S. G. HAW & LAUENER) T. HONG & J.J. LI and ***Paeonia ×suffruticosa*** ANDREWS, tree peonies

Buds of several scales, to over 2 cm long, ovoid and acute, with grey-brown loose bud scales which spread at the margins. **Twigs** 8–10 mm diameter, above the base of the first year's growth twigs matte grey-brown with many dot-like small lenticels. **Leaf scars** on distinct leaf cushions: large, half-round with several traces (vascular bundles) arranged in a semicircle. **Multiple fruit** consisting of 5 densely hairy c. 2 cm long follicles. Frequently planted, sparsely branching shrub to 2 m high, with many garden hybrids united under *P. ×suffruticosa*. Origin: China.

Paeonia delavayi FRANCH.

Buds covered by loosely fitting brown to grey-brown dried remnants of (parts of) leaves, usually the leaf base and stalk, but near the base of the bud with entire dead leaves. Terminal buds mostly longer than 2 cm, lateral buds mostly smaller. **Twigs** initially wine-red-brown to violet-brown, matte-glossy, later grey-brown with peeling bark and warty lenticels. Frequent, shrub to 1.5 m high. Origin: China.

Paeonia delavayi* var. *lutea (DELAVAY ex FRANCH.) FINET & GAGNEP.
[*Paeonia lutea* DELAVAY ex FRANCH.]

Buds with loosely appressed pale brown to grey-brown bud scales. Terminal buds to 1.5 cm, ovoid, blunt; lateral buds smaller, ovoid and blunt, enclosed by two bud scales. **Twigs** sturdy, 6–7 mm thick at the base of one-year-old shoots, matte ochre-brown (from the yellow-flowered variety red pigments are absent), finely striated longitudinally; older branches pale grey. Frequently planted, hardly branched shrub, to 1 m high. Origin: China.

Altingiaceae
sweetgum family

Liquidambar L., sweetgum

Few species, distributed from Asia Minor to Asia and in North America. **Buds** with bud scales. **Branches** often with corky wings. Striking in winter are the 2–3.5 cm thick, compound globose fruit clusters which consist of many woody 2-locular capsules.

Key to *Liquidambar*

1 Terminal buds c. 8 mm long; tree mostly one-stemmed . .*Liquidambar styraciflua*
1* Terminal buds mostly smaller; large shrub often with several stems . . . .
. . . .*Liquidambar orientalis*

Liquidambar styraciflua L., sweetgum

Buds ovoid and acute, with 5–6 bud scales, glossy, green to reddish with thin ciliate margins. Lateral buds mainly below the tip of the twig, very small. **Twigs** often with irregular cork ridges. Young twigs 3–4 mm diameter, olive-green to aubergine-coloured, slightly glossy, with some warty lenticels; two year old twigs initially brown, later grey due to dead epidermis, older twigs entirely grey with dark grey lenticels. **Leaf scars** on distinct cushions with 3 vascular bundle traces of similar size. **Bark** deeply furrowed. Frequently planted, tree 10–20(–45) m high. Origin: North America.

Much more rare — mostly encountered in botanic gardens — is *Liquidambar orientalis* MILL., a large shrub or up to 20 m high tree from Syria and southern Asia Minor. It differs by smaller, to 6 mm long and 3 mm thick terminal buds as well as by more slender twigs, 2–2.5 mm thick, compared to the preceding species.

Hamamelidaceae
witch hazel family

Deciduous and evergreen woody plants. **Buds** naked and then often with stellate hairs or with bud scales which are glabrous or which will turn glabrous. Lateral buds alternate, and as the part of the stem after the first or second prophyll is distinctly elongated, these may appear stalked. **Leaf scars** with 3 vascular traces. **Flowers** sessile in heads or spikes, either on axillary short-shoots in winter or to early spring (in *Hamamelis, Parrotia, Corylopsis*) or terminal on the new shoot following the appearance of the leaves. The two fused carpels end in free styles and develop into two-loculate woody capsules which, at least basally, are enclosed by the persistent flower cup (hypanthium). The persistent capsules tear open through the styles (resulting in four points when open) and a little less deeply between the carpels, thereby ejecting the two seeds. Helpful characters in winter are the position of the flower buds, or the buds of the inflorescence, and the fruits arising from them.

Key to Hamamelidaceae

1 Buds with bud scales, flowers and fruit axillary 2
1* Buds without bud scales, mostly densely covered in stellate hairs 3
2 Twigs with terminal buds, upper bud scales green-red (prophyll- and stipule scales), the yellow-green flowers in March in axillary spikes . . . . . **Corylopsis**
2* Twigs without terminal bud, bud scales distichous, flowers from enrichment bud, starting in autumn, two flowers each from one bud, with 5 dark wine-red petals . . . . *Disanthus*
3 Inflorescences as cylindrical buds at the shoot tip, fruits in more than 10 cm long pendulous spikes, lateral buds with two prophylls . . . . *Sinowilsonia*
3* Flower buds of inflorescences globose-ovoid, or invisible in winter, lateral buds with one prophyll[1] 4

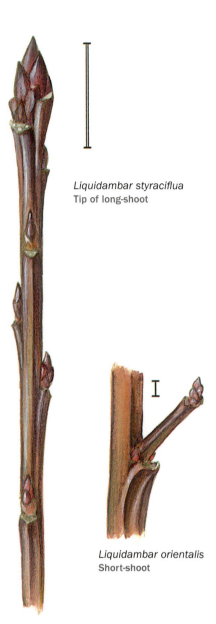

Liquidambar styraciflua
Tip of long-shoot

Liquidambar orientalis
Short-shoot

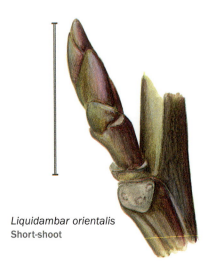

Liquidambar orientalis
Short-shoot

---

[1] The prophylls are here simple scales without stipules. They are distinct from the subsequent leaves, which are already formed like foliage leaves and have paired stipules.

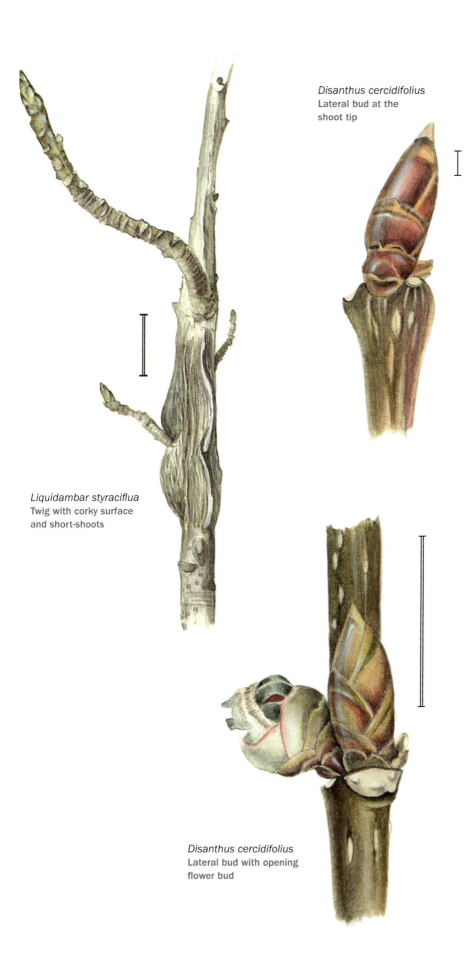

Liquidambar styraciflua
Twig with corky surface
and short-shoots

Disanthus cercidifolius
Lateral bud at the
shoot tip

Disanthus cercidifolius
Lateral bud with opening
flower bud

4   Flower buds and fruits in terminal spikes
    or heads 5
4* Flower buds and fruits in axillary spikes
    6
5   Large shrub over 2 m high, fruits in
    compact heads . . . . *Parrotiopsis*
5* Small, densely branching shrubs, rarely
    attaining 2 m high, fruits in upright
    spikes . . . . **Fothergilla**
6   Several globose to 3 mm thick flower
    buds on the short-shoot. Flowers often
    in February with 4 linear yellow to red
    petals . . . . **Hamamelis**
6* Actual flower buds concealed until
    anthesis in a larger black-brown
    inflorescence bud (4–5 m diameter),
    flowers from March without petals,
    conspicuous with its red anthers. Bark
    peeling as in a plane tree . . . . . **Parrotia**

SUBFAMILY Disanthoideae

Disanthus cercidifolius Maxim., long-stiped disanthus

Without terminal buds. **Lateral buds**
covered by distichous leaf-base scales. The
bud scales are violet-brown at base, the
margin ochre brown, lighter inside, darker
brown outside. The conical leaf buds are
5–6 mm long and have 6–10 visible bud
scales. Flower buds solitary or at the tip of
the long-shoots as enrichment bud(s) of the
leaf buds; they are ± globose, to 4–5 mm
long, with up to 11 bud scales. **Twigs** with
pale lenticels. Shrub attaining over 2 m in
height. Origin: Japan.

SUBFAMILY Hamamelidoideae

TRIBE Corylopsideae

Corylopsis Siebold & Zucc.

Buds fusiform: relatively narrow at the
base, acute at the top, in between slightly
(leaf buds) to strongly (flower buds)
convex-thickened. Terminal buds covered
by one to two pairs of stipules, according to
the leaf arrangement of the plant. Lateral
buds with two simple deciduous prophylls
(persisting below the shoot tip), followed by
several pairs of stipules, the first (belonging
to the third leaf) covering and encircling
the bud at least half-way (or smaller buds

completely), eventually the second pair of stipules (belonging to the fourth leaf), enclosing the bud entirely. **Twigs** with many small, pale lenticels. **Flowers** very early, March to April, in axillary hanging spikes, yellowish-green and partly scented. Where hairs occur on the calyx, they will remain until the fruits are ripe. **Fruits** are bifid capsules sunk into a persistent flower cup.

About 12 species in Eastern Asia (from the literature 7–29 species). In botanical collections there are many unrecognised hybrids. Their identification, even at flowering time in late winter, is almost impossible. Relatively safe for identification is *Carylopsis pauciflora* with few flowers, *Carylopsis spicata* which is propagated vegetatively in tree nurseries and the group of *Corylopsis sinensis*.

Key to Corylopsis

1 Erect shrub, buds often more than 7 mm long, twigs glabrous or hairy, more than 5 flowers fruit in one spike, flowers with 10 (5 2-parted) staminodia 2

1* Flat spreading shrub, buds 5–7 mm long, twigs glabrous, two to three flowers/fruits in each spike, flowers with 5 undivided staminodia. . . . . . . **Corylopsis pauciflora**

2 Up to 10 flowers and fruits in distinctly stalked spikes, anthers mostly dark red* . . . .*Corylopsis spicata*

2* More than 10 flowers and fruits in spikes, anthers yellow, without any hint of red . . . . *Corylopsis sinensis*

Note: It is not clear whether the pale red anthers often seen in collections are a characteristic of hybrids.

Corylopsis pauciflora SIEBOLD & ZUCC., buttercup winter hazel

Buds 5–7 mm long, fusiform to globose and acute. Bud scales red-brown to green, in the sun also to wine-red. **Twigs** slender and glabrous, with many lenticels. **Inflorescences/infructescences** on the shrub many, with 2–3 flowers/fruits in short spikes. Flowers March to April, faintly yellowish-green with yellow anthers, five undivided staminodia and reddish bracts. Frequently planted, densely branched small shrub to 2 m high. Origin: Japan.

Corylopsis spicata SIEBOLD & ZUCC., spike winter hazel

Buds 7–9 mm long, acute, thin at the base, leaf buds fusiform. Flowering buds more bulgy, later in winter, appearing shortly before flowering, ovoid-globose. Bud scales thin orange-red-brown and ochre-brown to pale green finely ciliate; outermost scales sometimes deciduous. **Twigs** olive and red-brown to grey-brown, with many fine lenticels and ± hairy, particularly towards the tip of the twig. **Flowers** March to April, 6–11 crowded at the tip of the to 4 cm long spikes. Petals cuneate at base, pale yellowish-green. Anthers dark red, filaments tinged with red. **Fruits** also in the upper $^2/_3$ of the spike, hairy. Frequent, ornamental shrub to 2 m high. Origin: Japan.

Also originating from Japan, the up to 6 m high **Corylopsis glabrescens** FRANCH. & SAV., the 'fragrant winter hazel' distinct by its glabrous twigs and fruits.

Corylopsis sinensis HEMSL., Chinese winter hazel

Different from *Corylopsis spicata* only by flowers and fruit: up to 18 flowers distributed evenly along a 3–5 cm long spike, the petals clawed and upright, anthers yellow and fruits hairy. Strong shrub to 5 m high; in its native country also a tree. Origin: central to western China.

A related species is *Corylopsis willmottiae* REHDER & E. H. WILSON. A shrub to 4 m high, petals rounded, reflexed, fruits glabrous. Origin: western China.

TRIBE Eustigmateae

Sinowilsonia henryi HEMSL.

Buds without bud scales, densely stellate-haired. Lateral buds 6–12 mm long, appearing stalked, at the base of the stalk initially with 2(!) opposite scale-like prophylls. Later on, mostly prophyll scars and their axillary buds are visible. **Inflorescences** overwinter as naked, felty black-brown hairy catkins (10 mm long and 3 mm wide) and begin flowering shortly before the leaves appear.
Twigs towards the tip (the upper 1–2 internodes) covered with felt due to stellate hairs, lower on the one year old twig less dense; with some ochre-whitish lenticels. One year old twigs greenish, darker on one

Corylopsis pauciflora
Twig with bulbous flower buds

Corylopsis pauciflora
Twig with fusiform leaf buds

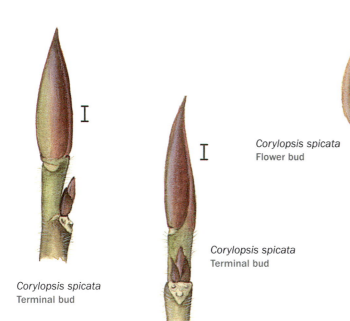

Corylopsis spicata
Terminal bud

Corylopsis spicata
Flower bud

Corylopsis spicata
Terminal bud

Corylopsis spicata
Infructescence

Sinowilsonia henryi
Twig with naked
inflorescence bud

side; two years old — completely glossy, grey-brown. Phyllotaxy distichous, twigs zig-zag between the nodes. **Pith** solid, green. **Fruits** many in 12–20 cm long spikes, to 9 mm long and 7–8 mm diameter. The floral cup encloses the capsule for more than half to almost completely. Rarely planted shrub to 8 m high. Origin: central to western China.

TRIBE Hamamelideae (including Fothergilleae)

Lateral buds appearing stalked: the second internode above the first, often deciduous, prophyll, is often lengthened; the second leaf is then inserted distinctly higher than the first and is similar to the subsequent bud leaves.

Hamamelis L., witch hazel

Buds 8–15 m long, ± naked, oblong, densely covered with stellate hairs, pale ochre to grey-brown, lateral buds appear stalked, often with descending accessory buds. Flower buds globose-ovoid, 2–3 mm diameter, grouped on axillary short-shoots. **Twigs**, dependent on species, either persistently stellate hairy or glabrescent. **Flower** in late autumn (*Hamamelis virginiana*) or in late winter into early spring. The 4-merous flowers have long, linear ± strongly yellow, reddish or red petals. Small staminodia above the petals, widened at the tip, spatula-like, except in *Hamamelis vernalis*. Due to the unusual flowering time, these are frequently planted shrubs.

Key to *Hamamelis*

1 Twigs pale grey, buds more than 15 mm long, calyx externally dark brown . . . . **Hamamelis japonica**

1* Twigs brown to ochre, buds mostly smaller 2

2 Twigs thick (at the tip c. 3 mm diameter), densely felty, buds thick, flowers with almost flat petals . . . . **Hamamelis mollis**

2* Twigs more slender, without felty hair, flowers with strongly crumpled petals . 3

3 Lateral buds slim, distinctly stalked, stipules of the outermost leaf rudimentary. Flowering Oct.–Nov., staminodia spatulate. . . . . **Hamamelis virginiana**

3* Lateral buds more squat, some almost without stalk, entirely surrounded by the stipules of the outermost leaf, flowering Jan.–Mar., staminodia narrow and not widened . . . . **Hamamelis vernalis**

Hamamelis virginiana L., Virginia witch hazel

Lateral buds stalked, upright. Stipules of the outermost leaf reduced, not surrounding the bud. **Flowers** October to November, rarely into early winter; after flowering the calyx is persistent, there are no more flower buds. The old **fruits** of the previous year long-persistent, widely enclosed by the flower-cup. Broad shrub, 4–5 m high. Origin: North America.

Hamamelis mollis OLIV., Chinese witch hazel

Lateral buds shortly stalked, upright. The outmost lamina entirely enclosed by its stipules; these later often tear off at base and can sit hood-like on the bud for a while. Buds and **twigs** brownish, densely covered in stellate hairs. First year's shoot completely hairy. **Flowers**: petals smooth and flat, hardly twisted, deep yellow, basally slightly tinged with red; calyx ± upright. In all parts more condensed than *Hamamelis japonica*. Shrub to 3 m high, in its native area also as a tree. Origin: central China.

Hamamelis japonica SIEBOLD & ZUCC., Japanese witch hazel

Lateral buds stalked and slightly curved. As in *Hamamelis mollis* at least initially surrounded by the stipules. **Twigs** pale grey, finely fissured longitudinally. In the fissures partly corky lenticels, initially with stellate hairs, soon becoming glabrous; and at least the base of the last shoot of the year ± glabrous. **Flowers** with reflexed calyx lobes which are wine-red on the inside. In the typical form, the crumpled petals are pale yellow, in var. *flavopurpurascens* (MAKINO) REHDER at the base they are dark red to yellow-orange, turning lighter towards the tip. Frequent, broad and stiff shrub, 2–3 m high. Origin: Japan.

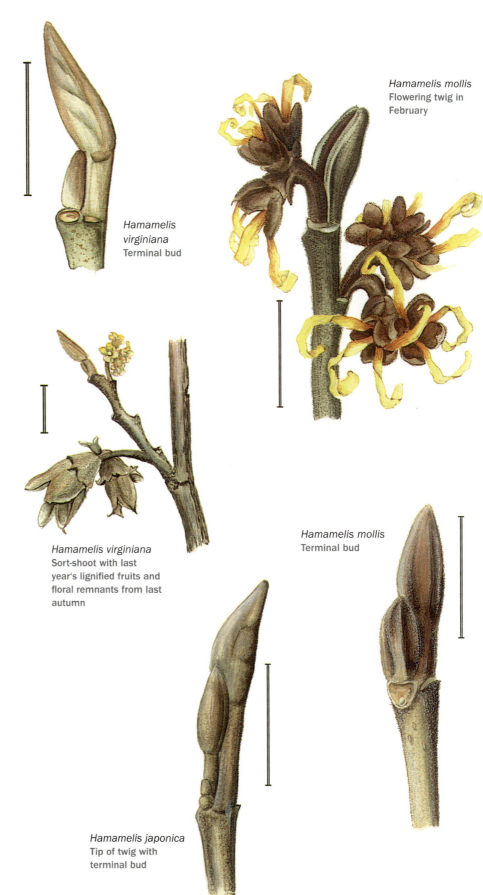

Hamamelis virginiana Terminal bud

Hamamelis mollis Flowering twig in February

Hamamelis virginiana Sort-shoot with last year's lignified fruits and floral remnants from last autumn

Hamamelis mollis Terminal bud

Hamamelis japonica Tip of twig with terminal bud

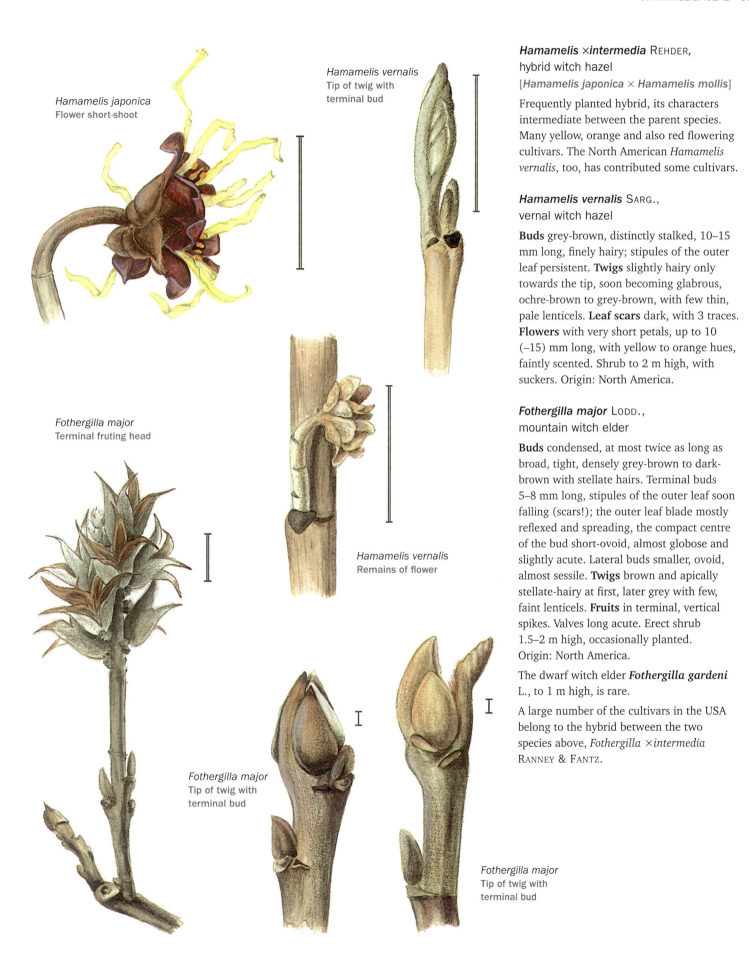

Hamamelis japonica
Flower short-shoot

Hamamelis vernalis
Tip of twig with
terminal bud

Fothergilla major
Terminal fruting head

Hamamelis vernalis
Remains of flower

Fothergilla major
Tip of twig with
terminal bud

Fothergilla major
Tip of twig with
terminal bud

Hamamelis ×intermedia REHDER,
hybrid witch hazel

[*Hamamelis japonica × Hamamelis mollis*]

Frequently planted hybrid, its characters
intermediate between the parent species.
Many yellow, orange and also red flowering
cultivars. The North American *Hamamelis
vernalis*, too, has contributed some cultivars.

Hamamelis vernalis SARG.,
vernal witch hazel

Buds grey-brown, distinctly stalked, 10–15
mm long, finely hairy; stipules of the outer
leaf persistent. **Twigs** slightly hairy only
towards the tip, soon becoming glabrous,
ochre-brown to grey-brown, with few thin,
pale lenticels. **Leaf scars** dark, with 3 traces.
Flowers with very short petals, up to 10
(–15) mm long, with yellow to orange hues,
faintly scented. Shrub to 2 m high, with
suckers. Origin: North America.

Fothergilla major LODD.,
mountain witch elder

Buds condensed, at most twice as long as
broad, tight, densely grey-brown to dark-
brown with stellate hairs. Terminal buds
5–8 mm long, stipules of the outer leaf soon
falling (scars!); the outer leaf blade mostly
reflexed and spreading, the compact centre
of the bud short-ovoid, almost globose and
slightly acute. Lateral buds smaller, ovoid,
almost sessile. **Twigs** brown and apically
stellate-hairy at first, later grey with few,
faint lenticels. **Fruits** in terminal, vertical
spikes. Valves long acute. Erect shrub
1.5–2 m high, occasionally planted.
Origin: North America.

The dwarf witch elder *Fothergilla gardeni*
L., to 1 m high, is rare.

A large number of the cultivars in the USA
belong to the hybrid between the two
species above, *Fothergilla ×intermedia*
RANNEY & FANTZ.

Parrotiopsis jacquemontiana (DECNE.) REHDER

Buds narrowly ovoid, terminal buds to 10 mm long, densely grey-green to pale ochre-grey stellate hairs. Lamina of the first leaf mostly appressed to the ovoid centre of the bud. Lateral buds shortly stalked. **Twigs** olive to grey, with dense stellate hairs towards the tip. **Bark** not peeling. **Fruits** in terminal capitula, the valves shortly acute. Rarely planted, stiffly erect shrub to 7 m high. Origin: western Himalaya.

Parrotia persica C.A.MEY., Persian ironwood

Buds dark grey-brown to brown, partly pale stellate-haired. Terminal buds 6–8 mm long, acute; lateral buds smaller, 4–6 mm long, hardly stalked. **Inflorescence buds** black-brown, solitary on long-shoots or several on short-shoots, shortly stalked, globose and 5 mm across (without stalk), flowering in late winter or early spring. **Twigs** grey-brown to grey-green, slightly hairy only towards the tip. Bark peeling in flat pieces; the different colours of differently aged bark plates give a typical variegated surface (like the bark of a plane tree). **Flowers:** mid-February to April, without perianth, but with 10–15 red, to 15 mm long pendulous anthers, and 2 inconspicuous carpels. Often planted, large shrub or medium-sized several-stemmed broad tree to 10 m high. Origin: northern Iran.

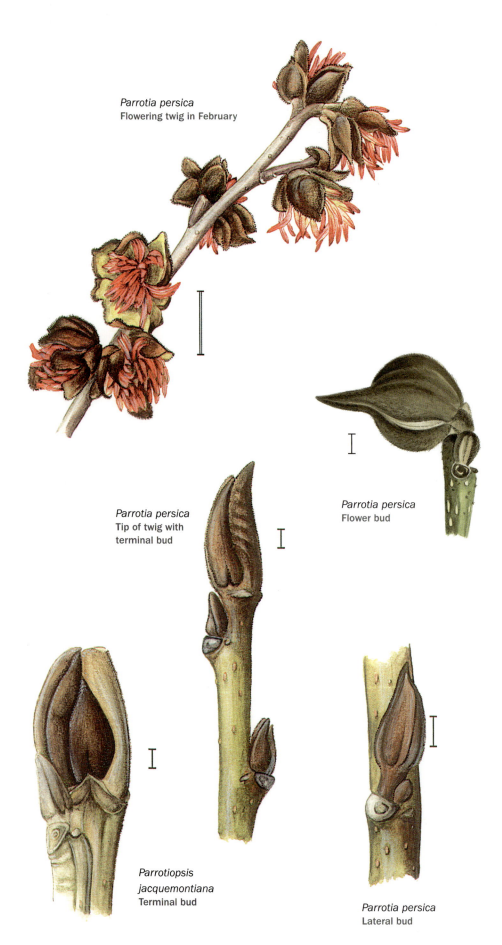

Parrotia persica
Flowering twig in February

Parrotia persica
Flower bud

Parrotia persica
Tip of twig with terminal bud

Parrotiopsis jacquemontiana
Terminal bud

Parrotia persica
Lateral bud

Cercidiphyllum japonicum
Short-shoots with terminal lateral buds and remnants of old inflorescence

Cercidiphyllum japonicum
Tip of long-shoot without terminal bud

Cercidiphyllum japonicum
Lateral buds on a long-shoot

Itea virginica
Lateral bud

Itea virginica
Lateral bud

Cercidiphyllaceae
caramel-tree family

Only **lateral buds** occur. The first 4 leaves in the bud are distichous, while the leaves on long-shoots are opposite. The outermost bud scale is positioned adaxially (on the 'inner' side) and encloses the bud almost completely, its margins overlap on the abaxial ("outer") side. Two further scale leaves in the median plane are followed by the only foliage leaf preformed in the bud. On the short-shoots this remains the only leaf; and in its axil the new lateral bud is formed, ending the short-shoot. The tip of the shoot can grow into an inflorescence, or it can die off, or grow into a long-shoot.

Cercidiphyllum japonicum SIEBOLD & ZUCC. ex J. J. HOFFM. & J. H. SCHULT., caramel- or katsura-tree

Buds sub-opposite on long-shoots or solitary at the end of short-shoots: slender, 4–5 mm long, tightly appressed to the twig, glabrous, glossy, in the light strongly violet-wine-red, in the shade glowing red. **Twigs** slender, glabrous, on the shaded side pale olive-green to olive-grey, on the sun side deep red-brown to grey-brown. **Lenticels** many, rounded and warty. **Leaf scars** small, grey-brown with 3 traces, on distinct cushions. **Bark** flat and irregularly furrowed. **Flowers** of the dioecious trees inconspicuous, March to April: in heads, with 8–13 red anthers or with 4 purple-red style branches. Frequently planted, tree to 30 m high. Origin: Japan.

Iteaceae
sweetspire family

Itea virginica L., Virginia sweetspire

Buds alternate, flattened half-globose to blunt-conical, 2–3 mm long lateral buds, mostly with small, descending accessory buds. Bud scales: the first reddish, the next ochre-brown. **Twigs** slender, slightly zig-zag, mostly glabrous, wine-red-brown to brownish-green on the shaded side. **Pith**: chambered. **Leaf scars** semicircular, dark grey, with 3 small conical traces. Rarely planted, shrub c. 1 m high. Origin: North America.

Grossulariaceae
gooseberry family

Ribes L., currants and gooseberries

Genus distributed in the northern temperate zone and in the mountains of South America. It contains many fruit shrubs and frequently planted ornamental shrubs.

Buds alternate, occasionally slightly stalked. Bud scales in a spiral, simple, sometimes with leaf remnants. Some species with prickles. Often with remnants of racemose inflorescence. The **flowers** appear shortly before or with the leaves, while before that the flower buds are visible for a long time: they are broadly globose. In most species the **bark** peels off from the second year onward in ± large sheets.

Key to *Ribes*

1 Shrubs without prickles 2
1* Shrubs with prickles on the nodes . . . . .
. . . . ***Ribes uva-crispa***
2 Buds and twigs strongly aromatic. . . . . .
. . . . ***Ribes nigrum***
2* Plants not aromatic 3
3 Terminal buds to 6 mm long 4
3* Terminal buds over 7 mm long. 6
4 Twigs less than 3 mm thick, buds pale or red-brown. 5
4* Twigs more than 3 mm thick, buds very dark grey-brown. . . . . ***Ribes rubrum***
5 Twigs red-brown. . . . . ***Ribes aureum***
5* Twigs grey to grey-brown. . . . .
. . . . ***Ribes alpinum***
6 Terminal buds over 12 mm long; twigs often with remnants of glandular hairs . .
. . . . ***Ribes sanguineum***
6* Terminal bud to 10 mm long 7
7 Twigs over 3 mm thick. . ***Ribes petraeum***
7* Twigs more slender . . . . . ***Ribes aureum***

Ribes alpinum L., Alpine currant

Buds oblong-acute, 4–6 mm long. Bud scales spiral, pale ochre-yellow to light grey-brown, slightly darker towards the tip and often with a dark-brown remnant of a lamina. **Twigs** unarmed, pale grey, partly almost white to ochre-brown, below the leaf scars and tip of the twig darker brown to violet-brown. Older twigs grey-brown

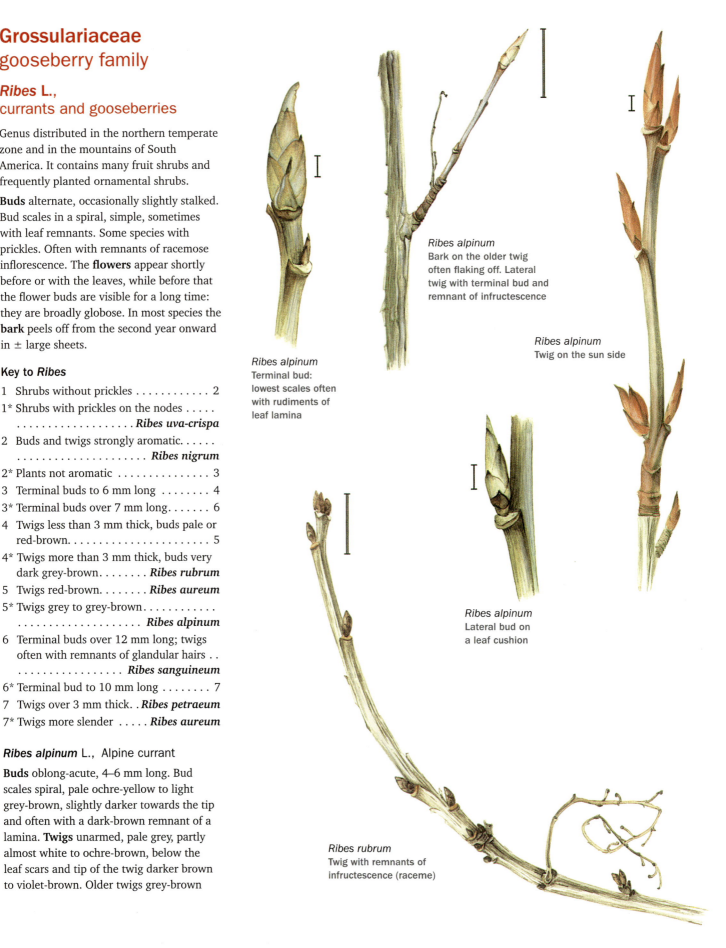

Ribes alpinum
Terminal bud: lowest scales often with rudiments of leaf lamina

Ribes alpinum
Bark on the older twig often flaking off. Lateral twig with terminal bud and remnant of infructescence

Ribes alpinum
Twig on the sun side

Ribes alpinum
Lateral bud on a leaf cushion

Ribes rubrum
Twig with remnants of infructescence (raceme)

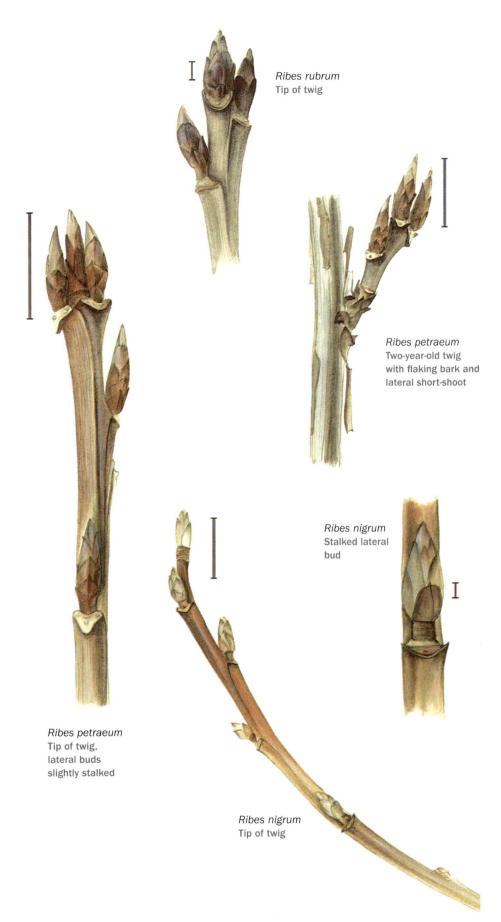

Ribes rubrum
Tip of twig

Ribes petraeum
Two-year-old twig
with flaking bark and
lateral short-shoot

Ribes nigrum
Stalked lateral
bud

Ribes petraeum
Tip of twig,
lateral buds
slightly stalked

Ribes nigrum
Tip of twig

to violet-brown with thin, flaking bark. **Lenticels** slightly darker than the young twigs, later tearing longitudinally. **Leaf scars** narrow and 3-traced. Very frequently planted, shrub 1–2 m high, finely branching. Origin: mountains of Europe and Siberia.

Ribes rubrum L., redcurrant

Buds conical-ovoid, 4–6 mm long bud scales dark grey-brown to red-brown, lower ones ciliate, upper ones short pale hairy. **Twigs** glabrous, strong, pale grey and ochre-white to dark grey-brown, finely fissured longitudinally and flaking. **Leaf scars** with 3 traces, grey to ochre-brown with red-brown margins. Unarmed shrub to 2 m high. Origin: western central Europe and western Europe. Most of the fruit varieties with red and white fruits belong to this frequent species.

Closely related is *Ribes petraeum* WULFEN, rock redcurrant: **buds** oblong-acute, 7–9 mm long, with 8–12 visible, brown to grey-brown, ± strongly keeled and very finely hairy bud scales. **Twigs** relatively thick and stiff: below the tip 3 mm diameter, when one year old matte, pale grey-brown, very finely fissured longitudinally, glabrous; fine ridges run down the stem from the leaf scars on both sides. **Bark** in two-year-old twigs strongly fissured and thinly peeling off in rolls, grey below, with few faint ochre-coloured lenticels. **Leaf scars** pale ochre to ochre-grey, prominent, with three prominent traces of vascular bundles of equal size. Frequent shrub to 2 m high. Origin: moist slopes of the mountains in western and central Europe.

Ribes nigrum L., blackcurrant

Buds ovoid, slightly stalked, 6–8 mm long. Bud scales: lowest one reddish, otherwise light grey-green to pale yellowish green, slightly acute and only slightly keeled. **Twigs** dark red-brown towards the tip with blackish remnants of hairs. Towards the base of the first year's growth shoot lighter: ochre to orange-brown, ± becoming glabrous and glossy. **Leaf scars** narrow, with 3 small traces and with ridges running down the stem, but disappearing before reaching the next node. All parts aromatic. Frequently planted, shrub to 2 m high. Cultivated for many years with natural origin in central Europe to Siberia and Asia; naturalised in western Europe.

Ribes aureum PURSH, golden currant

Buds 6–8 m long, oblong-acute, with red-brown to wine-red, very finely velvety hairy bud scales. **Twigs** glossy, red-brown to orange-brown, due to tearing epidermis also silvery grey, towards the tip velvety grey hairy; relatively slender, below the tip less than 2 mm thick, slightly zig-zag. **Leaf scars** narrow, red-brown to dark-brown, with 3 traces on a weak leaf cushion. Frequently planted, ornamental shrub 1.5–2.5 m high. Origin: western- and central North America.

Ribes sanguineum PURSH, flowering currant

Buds large: terminal buds 1.2–1.2(–2) cm long, fusiform with ochre-brown acute bud scales which are darker towards the tip. Lateral buds smaller, almost appressed to the twig. **Twigs** aromatic, initially hairy, red-brown with fine black points (remnants of hair). **Leaf scars** with 3 distinct traces. Frequently planted, shrub 2(–4) m high. Origin: North America.

Ribes uva-crispa L., gooseberry

Buds to 5–6 mm long, narrowly ovoid, with red-brown bud scale, keeled with ciliate margin. Lateral buds standing slightly away from the twig, or terminal buds on spreading short-shoots. **Twigs** initially hairy, later becoming glabrous and finely spotted dark due to hairs, with silvery-grey fissuring and peeling epidermis. **Leaf scars** with 3 traces. Below the buds mostly 3 prickles, connected firmly to the stem. Shrub 1–1.5 m high. Natural or widely naturalised in parts of Europe; distributed to North Africa, the Caucasus and China. Very commonly cultivated is var. **sativum** DC. [*Ribes grossularia* L.]

Vitaceae
grapevine family

Shrubs climbing with shoot tendrils. Tendrils and inflorescence emerge from the tip of the shoot and lateral shoots repeatedly continue the growth of the tip. As in most climbers, no terminal buds are formed. The buds are then enrichment buds of the lateral shoots which continue growth. As a rule, regular lateral buds are found on every third node without tendrils.

Ribes aureum
Tip of twig with terminal bud and stalked lateral buds

Ribes sanguineum
Tip of twig with large terminal bud

Ribes uva-crispa
Lateral bud. Frequently a 3-parted prickle below the bud

Vitis vinifera
Lateral buds

Vitis vinifera
Lateral buds

Ribes uva-crispa
Part of a twig

Vitis vinifera
Longitudinally cut open at node

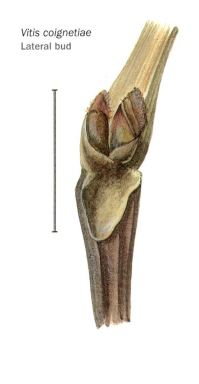

Vitis coignetiae
Lateral bud

Key to Vitaceae

1 Twigs with lenticels, pith white 2
1* Twigs without lenticels, pith brown. . . .
. . . .*Vitis*
2 Tendrils with adhesive discs . . . .
. . . . **Parthenocissus**
2* Tendrils without adhesive discs 3
3 Buds visible 4
3* Buds not visible . . . . **Ampelopsis**
4 Buds to 10 mm long, pyramidal and
slightly 3-angular. . . . .
. . . .*Ampelopsis megalophylla*
4* Buds smaller. . . . . .*Parthenocissus inserta*

Vitis L., grapevines

Lateral buds in older plants are mostly enrichment buds in the axils of the first prophylls of sterile side shoots which develop slightly and mostly die off. Of the dead side shoot, a scar remains opposite the scale leaf which envelops the bud. The enrichment buds develop later on into strong shoots, so-called canes. **Twigs** without lenticels, the bark disintegrating in long vertical fibres. The species can hardly be distinguished from one another in winter. Useful characters are the thickness of the transverse wall in the nodes (in longitudinal section) which separate the pith, whether the shoots are round or angular, as well as the regular appearance, including the lack of tendrils on the nodes.

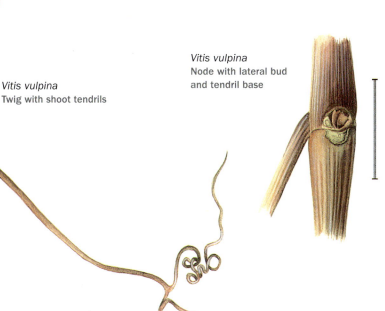

Vitis vulpina
Twig with shoot tendrils

Vitis vulpina
Node with lateral bud
and tendril base

Key to *Vitis*

1 Buds rusty red hairy. . . .*Vitis coignetiae*
1* Buds different or not hairy 2
2 Stem at the thinnest parts less than 6 mm
diam. . .*Vitis amurensis* and *Vitis vulpina*
2* Stem often thicker . . . .*Vitis vinifera*

Vitis vinifera L., grape vine

Buds flattened-globose to conical, up to 6–7 mm high. Particularly outer bud scales loosely appressed. **Twigs** very thick, one year-olds between the nodes often 8–10 mm and at the nodes 12–15 mm diameter; closely furrowed and, depending on the cultivar, glabrous to sparsely floccose-hairy. **Tendrils** or remnants of infructescences, or their scars, mostly on 2 out of 3 stem nodes. Walls in the nodes relatively thick, 2–3 mm. **Leaf scars** irregular, pale-edged with narrow scars of stipules which surround the twig. Traces of vascular bundles indistinct. 10–20 m high plant, climbing by tendrils, widely distributed and in cultivation since time immemorial. The original species *Vitis vinifera* subsp. *sylvestris* (C. C. GMEL.) HEGI, the 'wild grape' is rarely encountered. Daintier in all parts, this can climb to 20–30 m high. It grows wild from southern central Europe to Asia minor and in North Africa. Similar and difficult to distinguish is the occasionally planted Amur grape, *Vitis amurensis* RUPR., a strong liana from central- to eastern Asia.

Vitis coignetiae PULLIAT ex PLANCH., crimson glory vine
[*Vitis labrusca* auct. mult. non L.]

Buds globose-ovoid, 3–5 mm long, tomentose rust-coloured at the tip, felty hairy. **Twigs** of medium thickness, finely furrowed, initially rusty-red and with tomentose hairs. **Tendrils** mostly lacking on each third node. Occasionally planted, strongly growing vine. Origin: Japan, Sakhalin and Korea.

Vitis vulpina L., chicken grape

Buds hemispherical, 3–5 mm high and broad. **Twigs sturdy**, at one-year old c. 4 mm thick and at the nodes to 8 mm thick; pale brown to red-brown with longitudinal stripes and furrows, without fibres when young. **Pith** diaphragms distinct, 2–5 mm thick and even below the tip of the stem still almost 1 mm. **Tendrils** not on every node, mostly absent from every third node. Rarely planted liana. Origin: North America.

Parthenocissus PLANCH.

Buds visible, with brown bus scales. **Twigs**
with lenticels and white pith. **Tendrils** mostly
with developed adhesive discs, with which
it grows over vertical walls that otherwise
afford little support. Useful identification
characters are short-shoots, the development
of adhesive discs and the colour of the new
shoot.

Key to Parthenocissus

1 Tendrils with distinctly developed
   adhesive discs. New shoot reddish 2
1* Tendrils without adhesive discs or only
   slightly developed ones, new shoot
   green . . . . **Parthenocissus inserta**
2 Tendrils mostly shorter than 3 cm, shoot
   violet-red. Old plants with series of short-
   shoots (yearly growth here c. 1 cm) . . . .
   . . . .**Parthenocissus tricuspidata**
2* Tendrils longer, shoot rose-red. Short-
   shoots poorly developed . . . .
   . . . . **Parthenocissus quinquefolia**

Parthenocissus tricuspidata (SIEBOLD & ZUCC.) PLANCH., Boston ivy

Buds surrounded by brown scales; the
outermost on top of and enfolding the next,
like a hood. New shoot coloured a strong
red. **Twigs** brown to grey-brown, initially
with small roundish to oblong lenticels
which later tear open longitudinally and
lead to a rhombic-structured bark. Older
plants, covering a surface, develop in time a
chain of short-shoots from growing about 1
cm each year. **Tendrils** short and branched,
to 3 cm long with round adhesive disks that
hold fast to the base (the plane). **Fruits** in
panicles, blue-black berries, with a slightly
waxy bloom. Frequently planted, broadly
branched liana. Climbing to 20 m high on
vertical walls; spreading broadly by almost
horizontal veering main branches. Origin:
Eastern Asia.

Parthenocissus quinquefolia (L.) PLANCH., Virginia creeper

Buds: shoot initially dirty rose-red, first
leaves greenish. **Twigs** relatively strong.
Tendrils over 3–7 cm long; some branchlets
ending in adhesive disks. **Fruits** in terminal
panicles, on 2–4 cm long short-shoots,
blue-black berries, with hardly any waxy
bloom, deciduous in the course of winter.
Frequently planted, liana climbing to 15 m
high. Origin: North America.

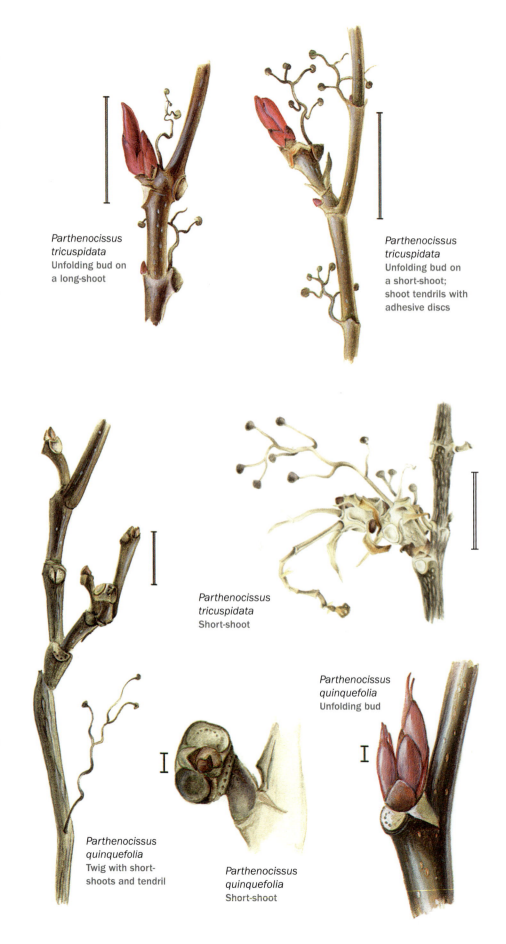

Parthenocissus tricuspidata
Unfolding bud on
a long-shoot

Parthenocissus tricuspidata
Unfolding bud on
a short-shoot;
shoot tendrils with
adhesive discs

Parthenocissus tricuspidata
Short-shoot

Parthenocissus quinquefolia
Unfolding bud

Parthenocissus quinquefolia
Twig with short-
shoots and tendril

Parthenocissus quinquefolia
Short-shoot

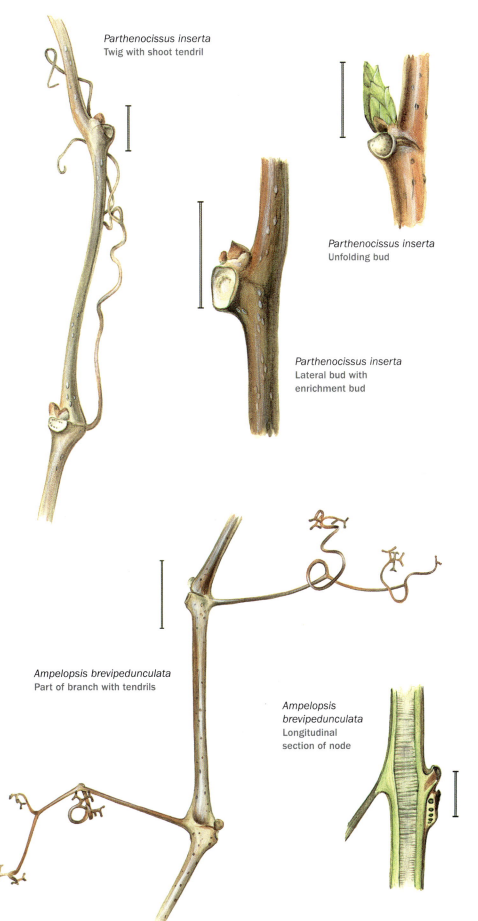

Parthenocissus inserta
Twig with shoot tendril

Parthenocissus inserta
Unfolding bud

Parthenocissus inserta
Lateral bud with
enrichment bud

Ampelopsis brevipedunculata
Part of branch with tendrils

Ampelopsis
brevipedunculata
Longitudinal
section of node

Parthenocissus inserta (A. KERN.) FRITSCH, thicket creeper

[*Parthenocissus vitacea* (KNERR) HITCHC.]

Buds on large leaf cushions, broad, short-conical with matte red-brown bud scales; sometimes with ovoid enrichment buds. Young shoots initially ± green, only rarely slightly flushed red. **Twigs** grey-brown, with relatively many pale lenticels. **Tendrils** long and branched, without or with only slightly developed adhesive disks. Planted occasionally, a weak liana climbing to 8 m high. Origin: North America.

Ampelopsis MICHX.

Buds in many species covered by the leaf scar and not visible, but very large in *Ampelopsis megalophylla.* **Twigs** strongly thickened at the nodes, with white pith and distinct lenticels. **Tendrils** long and branched, opposite the leaf scars, without adhesive disks.

Key to *Ampelopsis*

1 Buds large, tendrils more than 15 cm long . . . . *Ampelopsis megalophylla*
1* Buds not visible, tendrils shorter . . . . *Ampelopsis brevipedunculata* and ***Ampelopsis aconitifolia***

Ampelopsis brevipedunculata (MAXIM.) TRAUTV., porcelain berry

Buds hidden below the leaf scar with many small descending accessory buds. **Twigs** grey-brown to red-brown, initially hairy, but soon becoming glabrous, angular with many small, round, warty lenticels. **Pith** white, solid to slightly transversely fissured. Thickened on the node and distinctly thinner towards the internodes. **Leaf scars** grey, roundish and multi-traced. Stipular scars narrow and oblong, lateral at the nodes. Shrub up to 8 m high, with tendrils. Origin: from the Ussuri region across China, Manchuria and Korea to Japan.

Similar, and likewise frequently seen is ***Ampelopsis aconitifolia*** BUNGE which cannot be recognised with certainty in winter. Origin: northern China and Mongolia. In contrast to ***brevipedunculata***, its branches are glabrous from the start.

Ampelopsis megalophylla DIELS & GILG

Buds developed particularly on nodes without tendrils, large: c. 10 mm long, 3-angular-pyramidal, with 3 outer ochre-red-brown glabrous bud scales. **Twigs** round to slightly angular; smooth, glossy ochre-brown to red-brown, with many small lenticels. **Tendrils** branching and long, reaching to over 20 cm, red- to dark brown. Rarely encountered, to 10 m high climbing, liana. Origin: western China.

Celastraceae
spindle family

Only three *Euonymus* species are native to Central Europe from among the c. 100 genera, mostly distributed in tropical areas. Species from two more genera are planted frequently. The species mentioned here have buds surrounded by simple bud scales and leaf scars with a single, clearly demarcated vascular bundle.

Key to Celastraceae

1 Lateral buds opposite or whorled. Terminal buds mostly present. Mostly erect shrubs . . . .*Euonymus*
1* Lateral buds alternate. 2
2 Climbing liana, terminal buds absent. . 3
2* Small, upright, partly prostrate shrub, to 80 cm high with terminal buds. . . . .
. . . . *Euonymus nanus*
3 Twigs strongly angular 4
3* Twigs round . . . .*Celastrus*
4 Pith solid, twig covered densely in lenticels, fruits three-winged . . . .
. . . . *Tripterygium wilfordii*
4* Pith with chambers, fruits globose . . . . .
. . . . *Celastrus angulatus*

Euonymus L., spindle tree

Leaves/buds apart from few exceptions [*Euonymus nanus*] sub-opposite. The treated deciduous species are ± erect shrubs, besides these there also evergreen species, climbing with adventituous roots, which are cultivated. In early winter usually still with **fruits**: 3–5-valvate capsules, seeds hanging on funicles from the compartments and surrounded by orange to red arils ("seed covers").

About 175 species in Europe, Asia, Australia, north and central America with high diversity in the Himalayas and eastern Asia.

Key to *Euonymus*

1 Buds sub-opposite 2
1* Buds alternate or spiral . . . .
. . . . *Euonymus nanus*
2 Buds long, fusiform, terminal buds more than 15 mm long 3
2* Buds smaller, rarely wine-red 4
3 Buds matte brownish-red . . . .
. . . . *Euonymus latifolius*
3* Buds at least on the lighter side glossy wine-red . . . . *Euonymus planipes*
4 Twigs with distinctive cork ridges. 5
4* Twigs without ridges 7
5 Cork ridges very broad (wing-like), bud scales brownish 6
5* Ridges only weak, bud scales green . . . 8
6 Buds with more than 4 pairs of bud scale pairs . . . . *Euonymus alatus*
6* Buds with 3 bud scale pairs . . . .
. . . . *Euonymus phellomanus*
7 Twigs densely beset with warty lenticels
. . . .*Euonymus verrucosus*
7* Twigs not warty 8
8 Terminal buds to 5 mm long, with 3(–4) visible bud-scale pairs . . . .
. . . .*Euonymus europaeus*
8* Terminal buds 6–8 mm long, with 4–5 visible bud scale pairs. . . . .
. . . . *Euonymus hamiltonianus*

Subgenus Euonymus

Euonymus alatus (THUNB.) SIEBOLD, winged spindle tree

Buds globose-ovoid, with 4–7 externally visible bud scale pairs. Scales acute, ochre-cork coloured, darker brown towards the margin and the tip. **Twigs** green with thin, wing-like corky ridges. Frequently planted, dense shrub 1.5–3 m high. Origin: North Eastern Asia to Central China.

Superficially observed, the more rarely planted **Euonymus phellomanus** LOES. ex DIELS appears similar, a species near *Euonymus hamiltonianus*. **Twigs** with still more strongly developed corky wings; on older cork ridges growth zones are visible. **Buds** with distinctly fewer scale pairs of short-ovate shape. Shrub to 3 m high. Origin: China.

Ampelopsis megalophylla
Node with lateral bud

Euonymus alatus
Twig with thin corky wings

Euonymus alatus
Terminal bud flanked by two lateral buds

Euonymus phellomanus
Tip of shoot with terminal bud

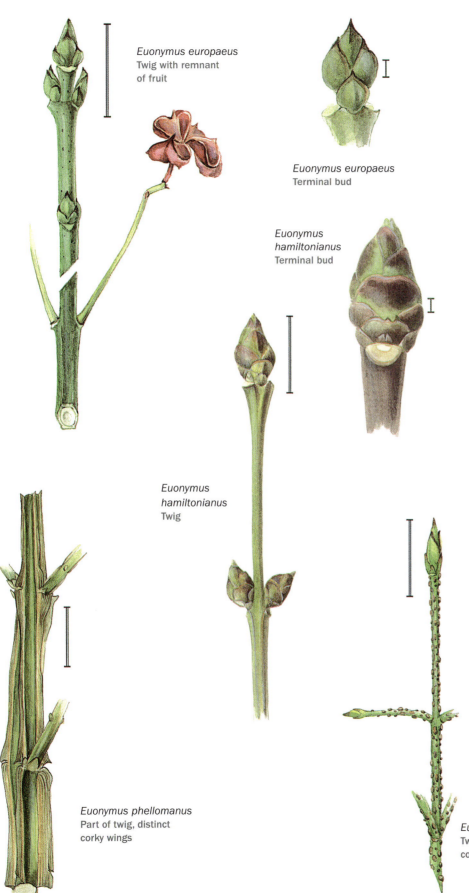

Euonymus europaeus
**Twig with remnant
of fruit**

Euonymus europaeus
Terminal bud

*Euonymus
hamiltonianus*
Terminal bud

*Euonymus
hamiltonianus*
Twig

Euonymus phellomanus
**Part of twig, distinct
corky wings**

Euonymus verrucosus
**Twig with dense, warty
cork pores**

Euonymus europaeus L., spindle tree

Buds with 3, rarely 4 pairs of externally visible bud scales. Terminal buds 4–6 mm long and 3–4 mm thick. Lateral buds sometimes just as large, but mostly smaller, with 2 pairs of scales. Bud scales matte green with brown-grey margin. **Twigs** green when young with very fine pale dots, round to 4-angular, sometimes slightly winged, with few dark lenticels. Erect shrub 1.5–7(–7) m high. In Central Europe growing naturally in herb-rich broad-leaved forests. Species frequently planted as an ornamental. Similar is *Euonymus hamiltonianus* WALL., with several cultivars occasionally planted, a shrub 3–4 m high. **Buds** larger, to 6–8 mm long and 5 mm thick. Bud scales 4–5 visible pairs, green to reddish, with a fine-grained cover, the lowest scale with dark brown, the upper with paler, narrow margin. First year's growth **twigs** green, on the lighter side also reddish, when several years old orange-brown to grey and moderately 4-angular. Origin: eastern Asia.

Euonymus verrucosus SCOP., warted spindle tree

Buds opposite, with mostly three pairs of bud scales. These are visible, green, with a dark grey-brown margin. **Twigs** initially green, very dense, covered in wart-like (hence the name!) large dark grey-brown **lenticels**. Shrub to 2 m high, only occasionally planted. Origin: eastern Europe to western Asia.

Euonymus verrucosus
Terminal bud

Euonymus nanus M. BIEB.,
dwarf spindle tree

Buds very small, lateral buds 1–2.5 mm
long, 3–4 in a spiral or ± alternate,
appressed to the twig or slightly spreading,
with few bud scales. Terminal buds slightly
larger, globose, sometime surrounded by
few remaining leaves. **Twigs** glabrous,
green, finely furrowed. Planted occasionally,
to 80 cm high shrub. Origin: central Asia.

Subgenus Kalonymus

Euonymus latifolius (L.) MILL.,
broad-leaved spindle tree

Buds matte brownish-red, olive green on
the shaded side, fusiform, with 4–6 visible
pairs of scales. Terminal buds over 2 cm
and lateral buds to 1.5 cm long. **Twigs**
when young slightly flattened, but not
angular; basic colour green, on the sun side
with a reddish hue. Frequently used as an
ornamental, shrub 1.5–6 m high. Origin:
southern central Europe and southern
Europe to western Asia as well as north
Africa and Asia Minor.

Euonymus planipes (KOEHNE) KOEHNE, the
flat-stalked spindle tree, is very similar. The
buds on the light side are lustrous wine-red.
Frequently planted shrubs from East Asia.

Celastrus L., staff vine

Climbing/twining, more rarely erect shrubs
with alternate leaves. Without terminal bud,
lateral buds only. Fruits: globose, yellow to
orange capsules opening by 3 valves. The
seeds inside are covered entirely by the red
aril. Origin: 35 species in North America,
eastern and southern Asia and Australia.

Key to *Celastrus*

1 Twigs strongly angular. Buds with many
   scales, pith chambered. . . . .
   . . . . *Celastrus angulatus*
1* Twigs roundish, the first two bud scales
   acute and the buds mostly overtopping,
   pith solid 2
2 Fruits in axillary cymes. . . . .
   . . . . **Celastrus orbiculatus**
2* Fruits in terminal panicles . . . .
   . . . . **Celastrus scandens**

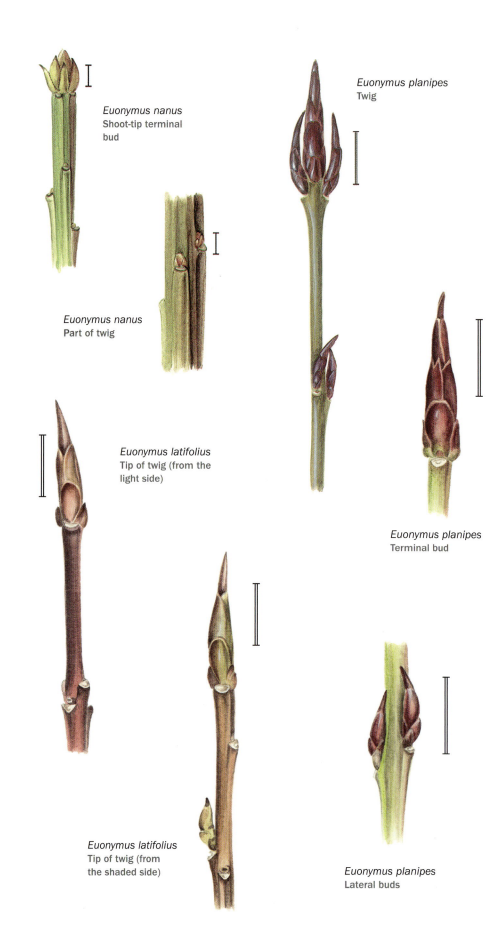

Euonymus nanus
Shoot-tip terminal bud

Euonymus nanus
Part of twig

Euonymus planipes
Twig

Euonymus latifolius
Tip of twig (from the light side)

Euonymus planipes
Terminal bud

Euonymus latifolius
Tip of twig (from the shaded side)

Euonymus planipes
Lateral buds

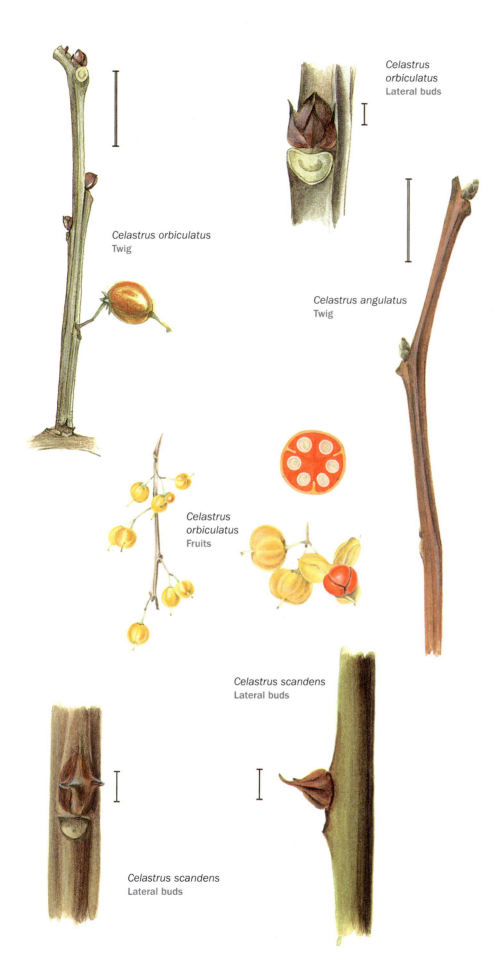

Celastrus orbiculatus
Twig

Celastrus orbiculatus
Lateral buds

Celastrus angulatus
Twig

Celastrus orbiculatus
Fruits

Celastrus scandens
Lateral buds

Celastrus scandens
Lateral buds

Celastrus angulatus
Lateral buds

Celastrus orbiculatus THUNB., Oriental bittersweet or staff tree

Buds grey to brown, globose, 2–3 mm long, first scale pair (prophylls) opposite, particularly in buds close to the tip of the twig, ± long acute and reflexed, extending beyond the bud and serving as hooks in climbing. **Twigs** grey-brown to grey, round with white pith. **Leaf scars** with one trace/vascular bundle. **Fruits** in axillary cymes, found in only some of the plants because the sexes are separate (dioecious). Frequently planted, liana to 12 m high. Origin: eastern Asia.

Also encountered frequently is the American bittersweet, *Celastrus scandens* L. It is mainly distinguished from the Oriental bittersweet by the fruits in 5–10 cm long terminal panicles.

Celastrus angulatus MAXIM., Chinese bittersweet

Buds dark, 3–4 mm long, ovoid, with some imbricate bud scales: black-brown, slightly lighter towards the margin and partly silvery-grey due to dead epidermis. **Twigs** 5-angular, strongly orange-red-brown to olive-brown with chambered pith. **Lenticels** few, orange-brown. **Leaf scar** on distinct cushion, with one trace. **Fruit** in 10–15 cm long terminal panicles; dioecious. Rarely planted, liana winding to 7 m high. Origin: central to north-west China.

Tripterygium wilfordii HOOK. f.,
Thunder-god vine
[*Tripterygium regelii* SPRAGUE & TAKEDA]

Buds matte red-brown, lateral buds in a spiral: 3–4 mm broad and high, rounded-ovate, surrounded by the two outer acute prophyll scales. Mostly solitary, sometimes with ascending accessory buds when the primary branching in the inflorescence comes to an end. **Twigs** matte red-brown, slightly 5-angular, covered in many warts of the same colour. One-trace **leaf scars**, semi-circular on distinct cushion. **Fruits** mostly persistent in winter: 3-winged, one-seeded nuts without arils in up to 20 cm long terminal panicles. Rarely planted, slightly climbing shrub. Origin: eastern Asia, Manchuria to Japan.

Salicaceae
willow family

Additional to the two classic genera of Salicaceae, the willows (*Salix*) and poplars (*Populus*) there are many genera that have come out of the Flacourtiaceae, so that Salicaceae has grown from two genera with c. 350 species to 55 genera with 1000 species. The closely related genera *Salix* and *Populus* have in common that both prophylls of lateral buds have fused into one scale. Prophylls of the genus *Salix* unite completely, those of *Populus* only on the abaxial side; prophylls on the lateral buds of the new family members do not fuse, and they are positioned to the side of (transversely to) the lateral bud.

Many members of the dioecious genera *Salix* and *Populus* flower ahead of the appearance of their leaves; they have many inconspicuous flowers with a reduced floral envelope in oblong spikes/catkins. The flowers of *Idesia* and *Poliothyrsis* appear after the leaves, in large panicles.

Key to Salicaceae

1 Trees with terminal buds, terminal buds
  with many scales 2
1* Shrubs or trees without terminal buds. . 3
2 The first two bud scales of the lateral
  buds lateral (transversal) . . . . . . *Idesia*
2* The first bud scale of the lateral buds
  turned away from the principal axis. . . .
  . . . . *Populus*
3 Lateral buds with a single, all-round
  fused bud scale. . . . . *Salix*
3 Lateral buds with two separate, slightly
  spreading bud scales . . . . *Poliothyrsis*

Idesia polycarpa MAXIM., wonder tree

Lateral buds alternate, mostly relatively small, surrounded by two prophyll scales and one to two sequential bud-scale pairs. **Terminal buds** short-conical, 5–6 mm long and just as thick with some shiny wine-red-brown bud scales hairy to the tip. Distinctive are three-tipped transition leaves: consisting of a centrally longer tip (remnant of leaf) and two neighbouring shorter tips (remnants of stipules). **Twigs** olive green to ochre-brown, glabrous and glossy, with many oblong ochre-brown to grey-brown lenticels, twigs thickest at the annual border and below the terminal buds. **Bark** smooth, greyish-white. **Leaf scars** on cushions, almost parallel to the twig surface, roundish to ovate, with many, U-shaped arranged, traces of vascular bundles. Distinctive in early winter are the many orange-red berries in hanging 10–25 cm long panicles. Frequently planted, laxly branched tree to 15 m high.
Origin: southern Japan and central and western China.

Poliothyrsis sinensis OLIV.,
Chinese pearl-bloom tree

No terminal buds (rarely the leaf-like shoot tip may resemble a bud), **lateral buds** compact, c. 3 mm long, surrounded by two slightly spreading, pale brown, finely hairy prophyll scales; in between, densely hairy tips visible. **Twigs** slender, distinctly thickened at the node (strong leaf cushions), glabrous, pale grey with brownish and greenish shades, finely fissured lengthwise (under the lens showing dark streaks), later joining and forming rhombic shapes, with few oblong lenticels, initially light ochre yellow, brown on older twigs. **Pith** solid, white, faintly angular. Leaf scars on leaf cushions surrounding the bud for $2/3$, with 3 groups of traces. Very rarely cultivated tree, to 15 m high. Origin: Central China.

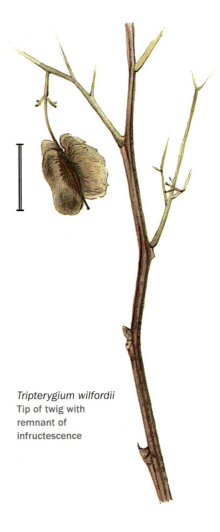

Tripterygium wilfordii
Tip of twig with
remnant of
infructescence

Tripterygium wilfordii
Lateral bud

Idesia polycarpa
Tip of twig

Idesia polycarpa
**Infructescence and,
right, seed and berry
(cross section)**

Poliothyrsis sinensis
Lateral bud

Poliothyrsis sinensis
Tip of twig

Populus L., poplar

The structure of the terminal buds is different from that of the lateral buds. The terminal buds are surrounded by adaxial stipule scales in a spiral (like the leaves). It is noticeable that the stipules of leaves situated relatively far below the terminal bud take part in the protection of buds.

In the **lateral buds** the first bud scale is opposite the principal axis (median-abaxial). The subsequent bud scales are distichous in the median plane and gradually change to a spiral phyllotaxy and thence to foliage leaves.

The species are sometimes difficult to tell apart. The identification is made more complex by many natural and artificial hybrids.

Key to *Populus*

1 Lateral buds very thick (> 10 mm) and condensed . . . .***Populus lasiocarpa***

1* Buds thinner. 2

2 Terminal buds rarely longer than 10 mm 7

2* Terminal buds mostly longer than 10 mm 3

3 Twigs in transverse section ± round or occasionally slightly angular 4

3* Twigs ± distinctly angular 5

4 Buds sticky, but not aromatic 6

4* Buds with an aromatic, sticky covering; twigs dark. . . . .***Populus balsamifera***

5 Twigs ochre-brown, basally grey; lateral buds appressed to twig . . . . ***Populus simonii***

5* Twigs red-brown, lateral buds standing slightly away from twig . . . .***Populus trichocarpa***

6 Tips of lateral buds spreading, twigs grey-brown to ochre-brown . . . . *Populus nigra*

6* Lateral buds facing towards twig, twigs red-brown. . . . . . ***Populus ×berolinensis***

7 (2) Buds and twigs densely white-felty-hairy. . . . .***Populus alba***

7* Buds glabrous, glossy . .***Populus tremula***

Section Populus, aspens and white poplars

Flower buds of the very early flowering species are often rather swollen and thicker than the leaf buds.

Populus alba L., white poplar

Buds and tips of twigs densely white-tomentose hairy. **Buds** ovoid, slightly acute, terminal buds on average 4–5 mm, lateral buds 2–4 mm long. **Twigs** initially smooth and pale grey-green with large grey rhomboid lenticels, forming only later an dark bark fissured lengthwise. Common, up to 30 m high tree with a broad crown. Distributed naturally from North Africa and Southern Europe across Central Europe to Central Asia.

Populus tremula L., aspen

Buds red-brown, narrowly ovoid, attenuate, glabrous and slightly sticky. Terminal buds 7–9 mm long. Flowering buds globose with acute tip, often breaking open very early, freeing the as yet undeveloped, stunted, densely grey-hairy catkins. **Twigs** glabrous and initially brown. **Bark** ochre to grey-green and, when older, fissured black-grey. **Flowers**: in initially silvery-grey, 10 cm long catkins. Common, to 30 m high tree with stolons. Origin: Europe to Japan.

In between the parent species of *Populus alba* × *Populus tremula* stands the fairly common grey poplar, **Populus ×canescens** (AITON) SM.: **twigs** grey-green, sparsely tomentose hairy; never with continuous hairy cover, but also never glabrous.

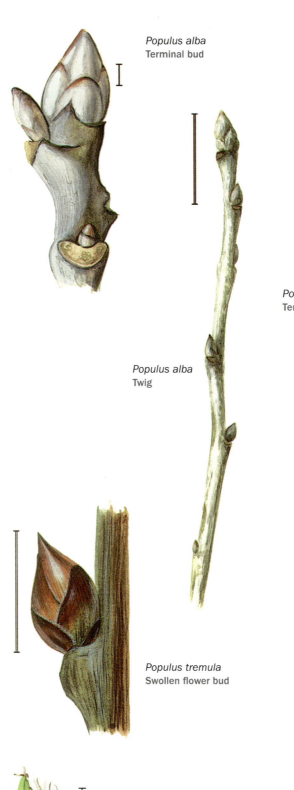

Populus alba
Terminal bud

Populus alba
Twig

Populus tremula
Terminal bud

Populus tremula
Swollen flower bud

Populus tremula
Flowering twig at the end of March

Populus tremula
Branch with young infructescences and young fruit with dry bract

Section Aigeiros, black poplars

Populus nigra L. and **hybrids**, black poplars

Buds sticky, slender conical to spindle-shaped, lateral buds pointing outwards. **Bud scales** red-brown to ochre-brown and olive to green, mostly with darker brown margin, the lowermost often keeled. **Twigs** round, pale orange-brown to grey-brown, with raised, oblong pale brown lenticels. **Bark** deeply fissured, black-grey. Tree to 30 m high. Origin: from North Africa and Europe to Central Asia.

Quite common is the columnar, pyramidal poplar, ***Populus nigra*** 'Italica', growing very upright.

More common than the pure species is the hybrid with the American black poplar *Populus deltoides* W. BARTRAM ex MARSHALL, the so-called Canadian poplar, ***Populus ×canadensis*** MOENCH. It differs from the black poplar by angular, darker olive green to brown twigs; in winter it is difficult to distinguish this hybrid from the parents.

A further hybrid between the pyramidal poplar *Populus nigra* 'Italica' and *Populus laurifolia* LEDEB. from the section of the balsam poplars, is the Berlin poplar ***Populus ×berolinensis*** K. KOCH, often planted; it has dark brown and almost round twigs.

Populus nigra
Male flowers

Populus nigra
Male catkin

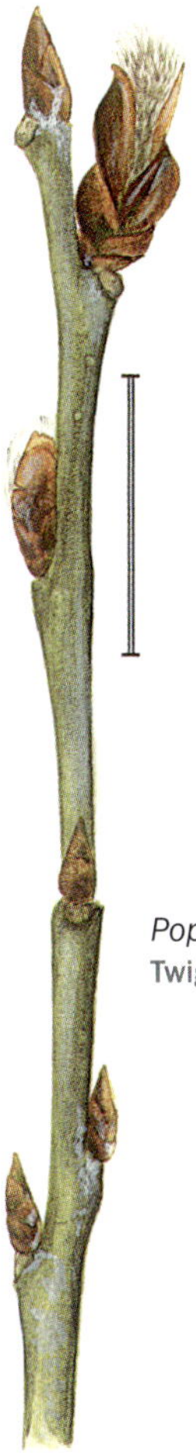

Populus nigra
Branching

Populus ×canescens
Twig with flowering buds

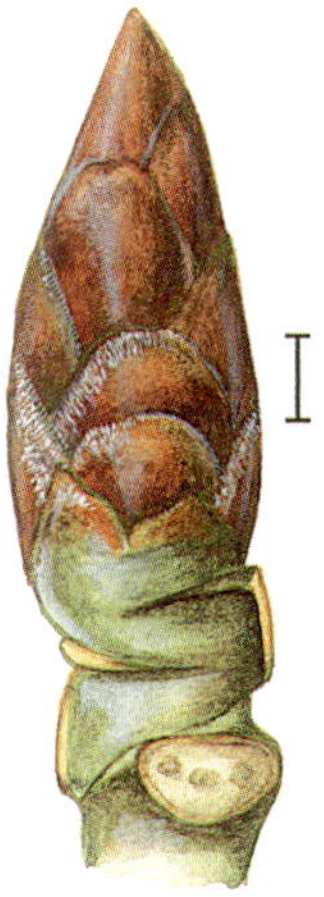

Populus ×canescens
Terminal bud

Populus ×berolinensis
Twig tip

Section Tacamahaca, balsam poplars

Populus simonii CARRIÈRE, Simon's poplar

Buds small, conical. **Lateral buds** 8–10 mm long, appressed to the twig, slightly reflexed at tip, with three outer bud scales. **Terminal buds** 10–12 mm long with many bud scales. **Bud scales** slightly sticky, dark violet-brown, occasionally ochre- to red-brown, margins slightly membranous (particularly basal ones). **Twigs** long, virgate, vertically upright, cylindrical near the base and, like the several year old twigs, distinctly pale, greyish with light brown- or greenish hues, upper epidermis densely fine oblong-fissured. Tips of twigs relatively slender, angular and ochre-brown to olive-brown. Slim, frequently planted, tree to 15 m high coming into leaf early. Origin: North China.

Populus trichocarpa TORR. & A. GRAY, western balsam poplar

Buds long, fusiform. Bud scales sticky, aromatic, brown or greenish with a brown margin. Terminal buds 12–15 mm long. Lateral buds slightly smaller, appressed to the twig to slightly spreading. **Twigs** angular, glabrous or hairy, olive-grey. Leaf scars pale, with three traces. common, tree to 30 m high. Origin: North America.

Populus balsamifera L., balsam poplar

Buds very large, to 25 mm long and especially the terminal buds relatively thick (to 8 mm), unlike in other species of the section; bud scales red-brown and olive-brown to yellowish-green, covered by aromatic, sticky resin. **Twigs** olive-brown to dark brown, in places also ochre-brown, glabrous, ± round, only towards tip slightly angular and slightly knotty due to leaf insertion points. **Lenticels** pale orange-ochre, oblong to dot-like. **Leaf scars** pale ochre, with three traces of vascular bundles. Frequently planted tree to 30 m high. Origin: North America. The occasionally planted var. *subcordata* HYLAND differs from the species by slightly hairy twigs.

Populus simonii
Twig: insertion of branching and tip

Populus trichocarpa
Terminal bud

Populus simonii
Lateral bud

Populus trichocarpa
Tip of twig

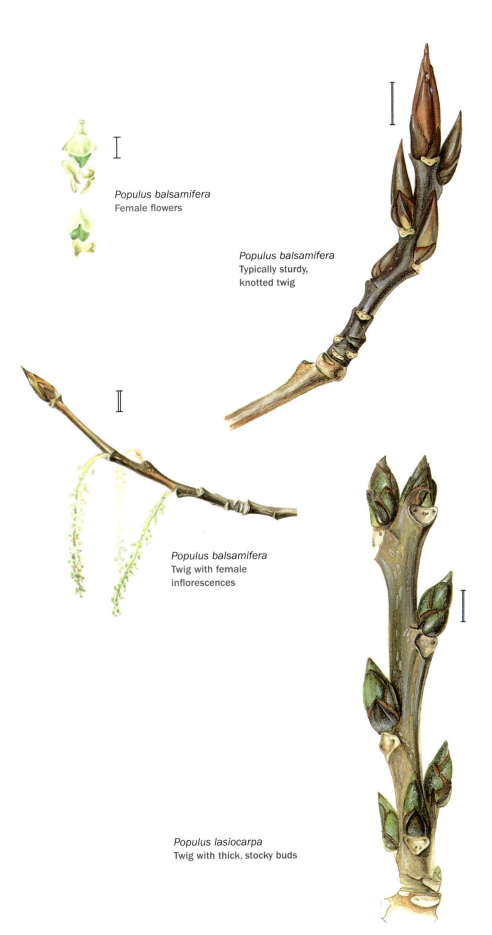

Populus balsamifera
Female flowers

Populus balsamifera
**Typically sturdy,
knotted twig**

Populus balsamifera
**Twig with female
inflorescences**

Populus lasiocarpa
Twig with thick, stocky buds

Section Leucoides, large leaf poplars

In comparison to other poplars, slow growing and coarsely branched.

Populus lasiocarpa OLIV., Chinese necklace poplar

Buds condensed: over 2 cm long and often more than 1 cm thick. Bud scales green, margins dry, brown, with a cuticle and, most of all lower scales, with long, shaggy hairs. **Twigs** divaricate and thick, c. 1 cm diameter, olive- to ochre-brown, tomentose-hairy, when older grey-brown and longitudinally fissured. **Lenticels** pale, ochre-brown, mostly oblong or round. **Leaf scars** with three groups of traces, the central one noticeably of several parts. Stipules partly drying up and persistent. The 20 m high tree with sturdy branches and a laxly branching crown is sometimes planted. Origin: south-west China.

Salix L., willow

This genus contains c. 400 species. Centres of distribution are the cooler, temperate zones of Central Europe, Eastern Asia and North America. The habit spans from arctic small creeping willows to shrubs to large trees.

In contrast to the poplars, the willows form no terminal buds and their lateral buds are completely enclosed in a single bud scale. This bud scale is a fusion product of both prophylls: within the bud scale there are often two axillary buds.

The leaf scars are mostly alternate but occasionally show a tendency to be opposite (*Salix purpurea*). They are relatively narrow, have three traces and surround the base of their axillary buds. Next to the leaf scars are often small separate stipule scars. Flowers often come early, before the leaves, in spike-like catkins. Being dioecious, male catkins are often yellowish due to pollen, female ones greenish. Each flowers has nectar glands.

There are many hybrids (both natural and artificial). Identifying these in winter is mostly impossible.

Three main forms can be distinguished:

1. Tree willows. These have buds of the same kind (no difference between leaf- and flower buds), which gradually become smaller towards the base of the stem. The flowers arise only after the leaves.

2. Shrub willows. These often have very different leaf- and flower buds. The flower buds are mostly larger and different in shape compared to the leaf buds. They are positioned near the shoot tip; but the upper one or two buds are mostly vegetative. Many species flower before the leaves appear.

3. Alpine willows. These are mostly relatively low and often ± creeping shrubs. Leaf- and flower buds are not very different. At the stem base there are buds, different in both shape and size, which do not expand in spring but form resting buds, building a reserve for new stems.

Key to *Salix*

1 Trees usually over 10 m high; flowers catkins either with leaves or later. 2
1* Mostly shrubs, more seldom small trees up to 10 m high, usually flowering early before the leaves 14

Tree willows

2 Buds and twigs glabrous or only the tips of young twigs slightly hairy 3
2* Twigs distinctly hairy 11
3 Bark on older twigs and stems lifting off in scaly plates. Buds blunt . . . .*Salix triandra*
3* Bark not scaly-deciduous. Buds mostly distinctly acute 4
4 Twigs easily break at base 5
4* Twigs (in absence of frost) not easily breaking. 7
5 Twigs hanging down . . . **Salix ×blanda**
5* Twigs not hanging. Crown widely branching. 6
6 Twigs and buds glabrous. Scars of stipules large . . . . **Salix fragilis**
6* Twigs and buds slightly hairy to tip. Scars of stipules smaller . . . . . .**Salix ×rubens**
7 (4) Buds with spreading, often darker tip. Fruit catkins of the previous year long-persistent . . . . **Salix pentandra**
7* Fruit catkins of previous year deciduous, buds mostly ± appressed to twig 8
8 Twigs straight. 10
8* Twigs sinuous. 9
9 Twigs yellow-orange to red with hairy tips . . . . **Salix ×erythroflexuosa**
9* Twigs glabrous, greenish ochre-yellow, strongly twisted . . . . **Salix babylonica 'Tortuosa'**
10 Twigs hanging down, very long reddish-brown at least on the sun side . . . . **Salix babylonica**

10* Twigs shorter, greenish-brown . . . . **Salix ×salamonii**
11 (2) Twigs a strong yellow. 12
11* Twigs of different colour 13
12 Twigs hanging . .**Salix alba** var. **vitellina** '**Tristis**' and hybrids
12* Twigs not hanging . . . .*Salix alba* var. *vitellina*
13 Twigs orange-red to red. . . . .*Salix alba* '**Chermesina**'
13* Twigs red-brown to brown. . **Salix alba**
14 (1) Dwarf shrubs to 30 cm high. 31
14* Shrubs more than 50 cm high, but rarely tree-like 15

Shrub willows

15 Twigs with waxy bloom, at least at the base 19
15* Twigs not with waxy bloom 16
16 Buds and twigs glabrous, at most at the base of twigs with some hairs 17
16* Buds or twigs hairy 20
17 Buds globose-ovoid, c. 3 mm long . . . .*Salix caesia*
17* Buds slightly oblong. 18
18 Buds sub-opposite . . . . **Salix purpurea**
18* Buds alternate . . . .*Salix glabra*
19 (15) Twigs greenish, flower buds more than 15 mm long, often a small tree . . . . **Salix daphnoides**
19* Twigs brown-red, flower buds to 15 mm long . . . .**Salix acutifolia**
20 (16) With floccose hairs that can be wiped off, internodes very short, the narrowly lanceolate leaves often ± persistent . . . .*Salix eleagnos*
20* Indument (hairiness) of leaves different. 21
21 Indument on the back of the buds distinctly stronger. Flower buds with duckbill-like flattened tip . . . . *Salix nigricans*
21* Indument and flower buds different . .22
22 Wood of young twigs with distinct streaks 23
22* Wood (under the bark) without streaks 25
23 Twigs slender, buds shining red, becoming glabrous. . . . . . .*Salix aurita*
23* Buds different 24
24 Twigs with persistent velvety hair . . . .*Salix cinerea*
24* Twigs becoming glabrous at base. . . . . *Salix appendiculata*

25 Stipules large, long-persistent . . . . **Salix lanata**
25* Stipules deciduous. 26
26 Shrub to 50 cm high . . . . **Salix repens**
26* Shrub higher or small tree 27
27 Buds thick. Inside of peeled bark becoming orange-brown . .**Salix caprea**
27* Inside of bark not changing colour. . 28
28 Buds round in cross-section. 30
28* Buds with distinct ridges 29
29 Scars of stipule indistinct . . . . **Salix helvetica**
29* Scars of stipules distinct . . . . **Salix viminalis**
30 Buds narrowly ovoid with straight spreading tip . . . . *Salix hastata*
30* Buds appressed to twig, tip bent outwards . . . .*Salix phylicifolia*

Alpine dwarf shrub willows

31 (14) Buds glabrous 32
31* Buds at least initially hairy. 33
32 Shoot base often with persistent last year's bud scales . . . . **Salix retusa**
32* Twigs partly underground, with pale runners . . . . *Salix herbacea*
33 Buds of two types. . . . . *Salix alpina*
33* Buds similar. 34
34 Buds mostly over 3 mm long, ovoid to obo-void. Twigs reddish. . . .**Salix reticulata**
34* Buds to 3 mm, ovoid. Twigs with short internodes, greenish yellow . . . . **Salix serpyllifolia**

Subgenus Salix, tree willows

Section Salix

Salix alba L., white willow

Buds flattened, appressed to the twig, c. 5 mm long, red to dark-brown, hardly drying at tip. Leaf- and flower buds the same size. Buds toward shoot tip densely and finely hairy, matte, becoming glabrous at base and more strongly coloured and glossy. Inside the bud scale follow, without transition, the first foliage leaves. **Twigs slender**, round and flexible, in the end hanging down, dirty-red at the tip and later green-brown. With round, raised, brown lenticels. **Bark** rough, longitudinally fissured. Tree to 30 m high with broad-spreading branches spreading at a sharp angle. Crown densely branched. Common willow. Origin: Eurasia.

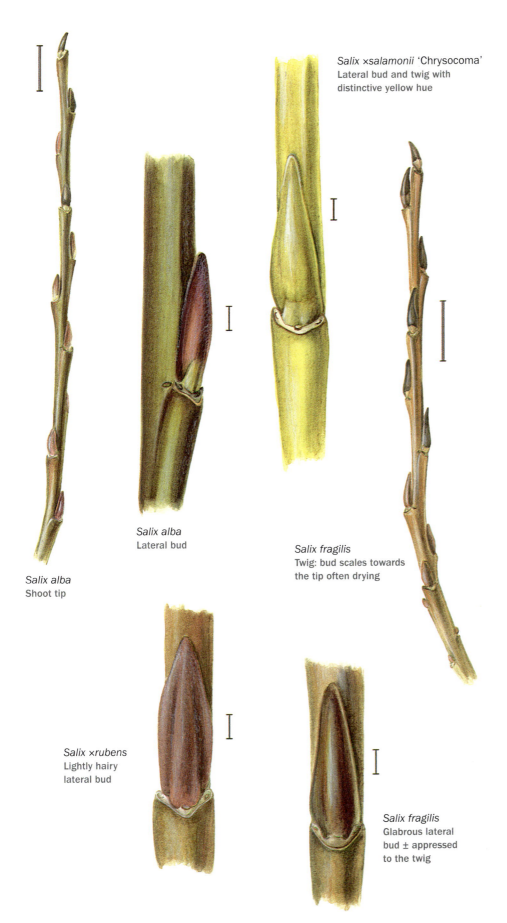

Salix ×salamonii 'Chrysocoma'
Lateral bud and twig with
distinctive yellow hue

Salix alba
Lateral bud

Salix fragilis
Twig: bud scales towards
the tip often drying

Salix alba
Shoot tip

Salix ×rubens
Lightly hairy
lateral bud

Salix fragilis
Glabrous lateral
bud ± appressed
to the twig

In **Salix alba** var. **vitellina** (L.) STOKES, the golden willow, young twigs are a strong yellow, later yellow-brown, slightly short-hairy. The scarlet willow, **Salix alba 'Chermesina'** [*Salix alba* var. *britzensis* SPÄTH], has bright orange-red twigs.

Weeping willows

The cultivar **Salix alba 'Tristis'** is a weeping form with vertically hanging down, slender and yellowish twigs which are slightly hairy towards the tip. Very similar and difficult to distinguish in winter is **Salix ×salamonii** CARRIÈRE **'Chrysocoma'** [*Salix chrysocoma* DODE] a hybrid with the Babylonian weeping willow [*Salix alba* 'Tristis' × *Salix babylonica*]. This is the most frequently planted weeping willow in central Europe: leaves long persistent, often with witches brooms and twigs becoming glabrous. **Salix ×blanda**, the hybrid between *Salix babylonica* and *Salix fragilis*, has more dirty green, easily breaking twigs.

Salix fragilis L., crack willow

Buds glabrous, 4–9 mm long, flat-appressed to the branch, only the flower buds slightly spreading. Bud scales ochre-brown, olive-brown or red-brown, often from the tip slightly dried up before bud expansion, and upper part a strong blackish-brown. Inside the bud scale is lodged a second scale-like leaf which is followed by a transitional leaf before the real leaves. **Twigs** slightly angular, ochre, grey-green to reddish-brown, glossy, spreading at almost right angles and bent upward. Easily breaking at the base (this character only useful when it is not frosty). Bark rough, oblong-fissured and dark grey. Frequent, tree 10–15(–25) m high. Origin: Eurasia.

Salix ×rubens SCHRANK, hybrid crack willow
[**Salix alba** × **Salix fragilis**]

Intermediate between the parents. **Buds** ochre-yellow to reddish, appressed to the twig, only tips of buds slightly spreading. **Twigs** glossy, ochre-yellow, never hanging down. Natural in central Europe, the area of distribution of the parents.

Section Subalbae

Salix babylonica L., weeping willow

Buds and twigs initially silky hairy, becoming glabrous. **Twigs** brown. Tree to 10 m high with twigs drooping. Origin: Transcaucasus to Eastern Asia. The weeping willow planted most frequently in southern Europe, needs a lot of warmth.

Salix babylonica 'Tortuosa', corkscrew willow

[**Salix matsudana** KOIDZ 'Tortuosa']

Buds oblong, to c. 6 mm long, flat-appressed to the twig or slightly spreading at the tip, ochre-yellow to grey-brown, sparsely finely hairy or glabrous. **Twigs** greenish or yellow-ochre to ochre-brown, glossy, finely hairy only initially, strongly twisted in corkscrew fashion. Frequently planted, tree 10(–15) m high.

Similar is the hybrid with the pendulous cultivar of the silver willow, *Salix alba* 'Tristis': **Salix ×erythroflexuosa** RAGONESE. **Twigs** reddish-orange-yellow, finely hairy at the tips and slightly less twisted.

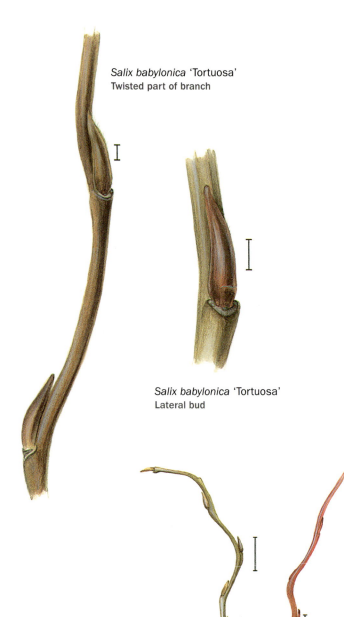

Salix babylonica 'Tortuosa'
Twisted part of branch

Salix babylonica 'Tortuosa'
Lateral bud

Salix ×erythroflexuosa
Lateral bud

Salix babylonica 'Tortuosa' and
Salix ×erythroflexuosa
**Twigs of both species: clay-yellow
to grey-green in S. babylonica &
reddish in S. ×erythroflexuosa**

Series Pentandrae

Salix pentandra L., bay willow

Very similar to the crack willow, *Salix fragilis*. Buds glabrous and glossy, 5–7 mm long, slightly spreading; tip straight or slightly inclined towards twig, wine-red to red-brown or often of two colours: basally greenish-ochre and upper part sharply different, reddish-brown to dark violet-brown. **Twigs** glabrous, flexible, red- to ochre-brown, glabrous and glossy. Flowers catkins only long after leaves appear, old fruit catkins of the year before persist for a long time. **Bark** initially smooth and grey, later longitudinally fissured. Large shrub or small tree to 12 m high. Origin: Eurasia.

Series Amygdalinae

Salix triandra L., almond willow
[*Salix amygdalina* L.]

Buds glabrous, reddish to clay-yellow, slender, 5–6 mm long, appressed to the twig, only the tip slightly spreading. **Twigs** glabrous, slender, ochre grey-green to brown, glossy. **Bark** peeling off in scaly plates. 1–4 m high shrub or to 10 m tall tree; many cultivars exist. Origin: from Europe to Asia including Japan.

Salix pentandra
Part of twig

Salix triandra
Lateral bud

Salix triandra
Lateral bud

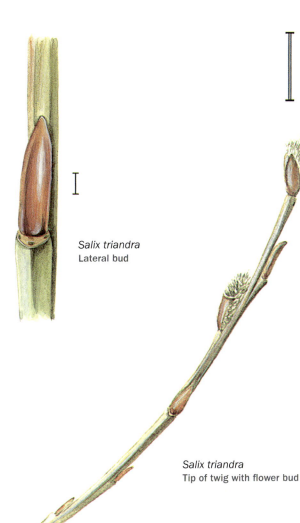

Salix triandra
Tip of twig with flower bud

Subgenus Vetrix

Series Vetrix

Salix caprea L., goat willow, great sallow

Buds ovoid to conical, acute, spreading from the twig, 4–8 mm long, dirty yellowish green to orange-brown, basally slightly lighter; initially hairy, turning glabrous. Flower buds distinctly larger than leaf buds. **Twigs** thick, initially short-hairy, soon becoming glabrous, olive, red-brown on the sun side. Pith brown, wood without streaks. Bark grey, tearing in rhomboid shapes, developing into a net-like bark. Frequent, large shrub or small tree of 8(–12) m. Origin: Europe to north-east Asia.

Salix cinerea L., grey willow

Buds alternate, occasionally opposite, appressed to the twig or slightly spreading, flat to broadly conical, 4–5 mm long. Flower buds slightly larger, conical-ovoid, short velvety grey hairy, matte orange-brown to grey-brown, sometimes becoming glabrous in places and glossy yellow-green to orange-red. **Twigs** yellowish green to dark grey, matte short velvety grey-hairy, later glabrous and glossy. Wood with distinct streaks, branches often furrowed. **Bark** smooth glossy and grey. Frequent broad shrub 4–6 m high. Origin: Europe to North Africa and to Kamchatka.

Salix caprea
Tip of twig with lateral buds

Salix caprea
Lateral bud

Salix caprea
Catkin emerging from the axil of fused prophylls

Salix caprea
Lateral bud

Salix caprea
Twig with male catkins and male single flowers

Salix caprea
Female catkin and female single flowers

Salix cinerea
Lateral bud,
side view

Salix aurita
Lateral bud,
dorsal view

Salix aurita
Lateral bud,
side view

Salix eleagnos
Tip of twig

Salix appendiculata
Part of twig with
flower buds

Salix eleagnos
Part of twig

Salix aurita L., eared willow

Buds 3–4 mm long, appressed to twig or slightly spreading, thick, seen from above triangular, glossy, dark wine red to coral-red, turning glabrous. **Twigs** slender, dirty wine-red and on the shaded side olive-green to yellowish green, towards the tip more densely hairy. **Pith** white. **Wood** young with short, sharp streaks. **Leaf scars** on thick cushions, with distinct traces of stipules. Frequently planted, compact shrub 1–2(–3) m high, forming stands. Origin: Europe to western Asia.

Salix appendiculata VILL., large-leaved willow

Buds oblong, to more than 10 mm long, basally appressed to twig, occasionally towards the tip flattened and duckbill-shaped, matte grey-green to ochre-green on the side of the light, slightly reddish brown, densely grey-white hairy, becoming glabrous in patches. **Twigs** sturdy, slightly angular, matte grey-green to yellowish-green or brown, short velvety hairy, becoming glabrous, at the base of the shoot with slightly shaggy hairs. **Pith** white. Older branches with smooth grey bark and wood with streaks. **Leaf scars** on cushions with roundish traces of stipules. Large shrub to 4(–6) m high or small tree. Native to the lower slopes of the Alps, in the Black Forest and the Bohemian Forest.

Series Canae

Salix eleagnos SCOP., olive willow
[*Salix incana* SCHRANK]

Buds close together due to short internodes, flat-appressed to twig, 4–7 mm long. On the light side of twig ± deep red, on the shaded side yellowish-green to olive-green, most of all towards the tip of the twig a little felty-hairy turning glabrous. **Twigs** growing straight up, due to the dense nodes with large leaf cushions initially very angular, reddish to yellowish green and grey-green, with felty indumentum which can be wiped off. Leaves mostly long-persistent: narrowly lanceolate, 7–12 cm long and to 2 cm wide, with inrolled margin. **Bark** grey and remaining smooth. Shrub or tree to 20 m high. Origin: the mountains of central and southern Europe, to Asia Minor.

Frequently planted is *Salix eleagnos* subsp. *angustifolia* (CARIOT) RECH. f. [var. *lavandulifolia* LAPEYR]. A shrub to only 3 m high. The leaves are narrow, c. 15 cm long and 5 mm wide and tightly inrolled at the margins. Origin: southern Europe.

Series Daphnella

Salix daphnoides VILL., violet willow

Buds finely hairy, with a clear difference between fat, almost 20 mm long flower buds and the to 8 mm long flat leaf buds. **Flower buds** fat and acute, dark black-brown and yellowish red towards the tip, bud scale of male catkins early deciduous. Flower catkins very early, before leaves appear. **Leaf buds** flat, appressed to twig, wine-red, in the shade also greenish. **Twigs** round, glabrous and glossy wine-red to greenish, at the base white with waxy bloom and slightly brittle. **Bark** grey, smooth for a long time, when older slightly longitudinally fissured and yellow. Tree to 15 m high. Origin: in the mountains of Europe.

Salix acutifolia WILLD., Siberian violet willow

[*Salix daphnoides* subsp. *acutifolia* (WILLD.) DAHL]

Buds rather long, 4–6 mm long, appressed to twig, wine-red, glabrous, and partly with waxy bloom. **Twigs** red-brown glossy and whitish-blue with waxy bloom layer which can be wiped off. **Leaf scars** mostly without separate stipule scars, with 3 traces. Shrub to 4 m high. Origin from eastern Europe to central Asia.

Series Glabrella

Salix glabra SCOP.

Buds 6–9 mm long, at the base rather flat-appressed to twig, towards tip often slightly spreading, glabrous, glossy, olive- to yellowish-green, partly reddish-brown, especially flower buds flattened at the tip. **Twigs** relatively thick and stiff, slightly angular, slightly glossy, green to brown, mostly completely glabrous and only at the base of the twig with long, shaggy hair. Mostly solitary compact shrub to 1.5(–2) m high. Grows on chalky soils in the Alps.

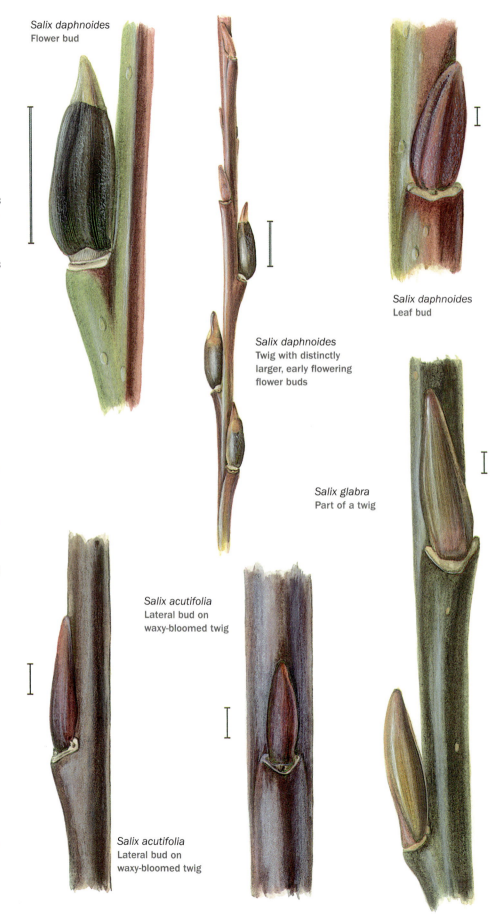

Salix daphnoides
Flower bud

Salix daphnoides
Twig with distinctly
larger, early flowering
flower buds

Salix daphnoides
Leaf bud

Salix glabra
Part of a twig

Salix acutifolia
Lateral bud on
waxy-bloomed twig

Salix acutifolia
Lateral bud on
waxy-bloomed twig

Salix purpurea
Lateral bud

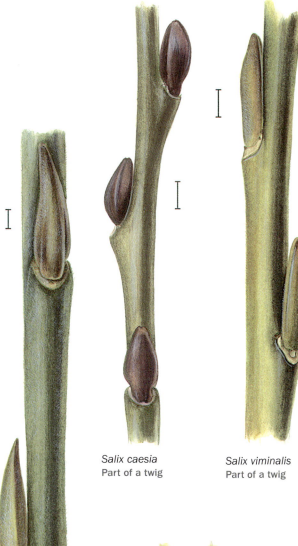

Salix ×sericans
Part of a twig

Salix caesia
Part of a twig

Salix viminalis
Part of a twig

Salix viminalis
Lateral bud

Salix purpurea
**Sub-opposite
lateral buds**

Series Helix

Salix purpurea L., purple osier

Buds glabrous, glossy dark red, mostly sub-opposite and only some alternate, 4–6 mm long, flat-appressed to twig, tips of buds also appressed or slightly bent outwards. **Twigs** very slender, glabrous, glossy, on the light side purple to red-brown, in shade slightly greenish, when older yellow-grey to green. Lenticels few. **Leaf scars** without stipule scars. Frequent much-branched shrub to 6 m. Cannot be confused as the only local species of willow with mostly opposite phyllotaxy. Origin: from Europe to north Africa, Asia minor and eastern Asia.

Salix caesia VILL., blue willow

Buds small, 2–3 mm long, alternate or occasionally sub-opposite, short-ovoid, usually with a slight notch at the tip, glabrous, dark brown-red, on the shaded side also yellowish-green to ochre. **Twigs** slender, glabrous, matte olive-green and grey-green to brown. Rare small shrub to 1(–1.5) m high, growth erect to ascending, forming dense stands. Origin: separate areas in the Alps and in central Asia.

Series Vimen

Salix viminalis L., common osier

Buds oblong, with broad obtuse tip, to 7 mm long, appressed to twig or slightly spreading, towards tip of twig matte and hairy, lower down turning glabrous and glossy, ochre olive-green to orange yellow. **Twigs** initially soft-felty silver grey hairy, later turning glabrous and glossy, yellowish-green to brownish-yellow. No stipule scars. Shrub or small tree, 2–6(–10) m high. Many cultivars, frequent. Origin: from western central Europe to eastern Asia.

Salix ×sericans TAUSCH ex A. KERN., **broad-leaved osier** [*Salix ×smithiana* WILLD.] is an old hybrid with *Salix caprea*. **Buds** similar to *Salix viminalis*, slightly acute but darker: yellowish-green to grey-green with reddish hues, appressed to twig and initially densely hairy. **Twigs** thick, when young green to greenish grey-brown, densely felty hairy, later becoming glabrous and reddish-brown. Frequent shrub, to 9 m high.

Series Villosae

Salix helvetica VILL., Swiss willow

Buds narrowly ovoid-cylindrical, 4–6 mm long, matte to glossy yellowish green to ochre-red-brown, woolly-hairy, turning partly glabrous in places, more rarely entirely glabrous, appressed to twig and only the tip slightly spreading. **Flower buds** thicker, more distinctly standing away from twig. **Twigs** olive-green on all sides, to matte to glossy brown, towards tip of twig ± hairy. Lenticels few. **Leaf scars** without distinctive stipule-scars. Frequent, shrub to 1.5 m high. Origin: central Alps from sub-alpine to alpine levels.

Series Incubaceae

Salix repens L., creeping willow

Buds alternate, occasionally sub-opposite, flat-appressed to twig or slightly spreading in upper part, c. 3 mm long, dark black-brown and ± dense, short-appressed, silvery-white hairy. Flower buds shorter and thicker. **Twigs** slender, hairy, dark brown, more rarely yellow- to red-brown. **Leaf scars**, particularly the stipule scars, almost hidden behind the dense hairs. Small shrub up to 50 cm high, with ascending branches. Origin: Europe to central Asia and Siberia.

Series Hastatae

Salix hastata L., halberd willow

Buds 6–7 mm long, long-ovoid, ± appressed to the twig (only the tip slightly spreading), red-brown, slightly glossy, especially the buds near the tip long-shaggy hairy. **Flower buds** larger and more oblong, the tip frequently flattened like a duckbill and shaggy-hairy. **Twigs** matte grey-brown to grey-green, towards the tip mostly thick and shaggy-hairy, lower down becoming glabrous, only at the base of first year's growth shoots with long persistent remnants of hair. Pith white. Shrub, upright and much branching, to 1.5(–2) m high. Origin: Northern Europe as well as in the mountains of Central- and Southern Europe (e.g. sub-alpine level of the Alps) and on to north-east Asia.

Salix helvetica
Lateral bud

Salix helvetica
Twig with flower buds

Salix repens
Lateral buds, dorsal view

Salix hastata
Part of a twig

Salix repens
Lateral buds, lateral view

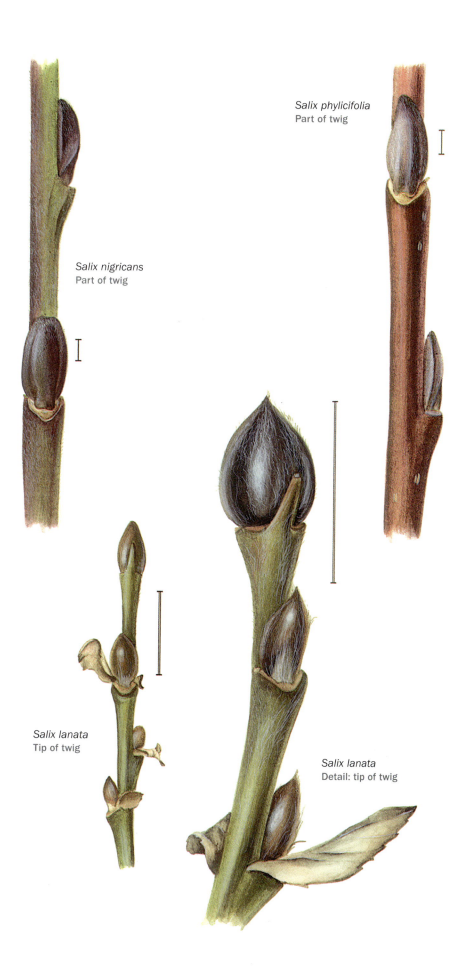

Salix phylicifolia
Part of twig

Salix nigricans
Part of twig

Salix lanata
Tip of twig

Salix lanata
Detail: tip of twig

Series Nigricantes

Salix nigricans Sm., dark-leaved willow
[*Salix myrsinifolia* Salisb.]

Buds dark brown-red, to 5–6 mm long, appressed to twig, towards the tip often flattened and densely hairy. Flower buds more globose, to 9 mm long and towards the tip flattened like a duckbill. **Twigs** green to blackish-grey, round, short-hairy, later becoming glabrous. Lateral twigs branching off at right- or blunt angles. **Leaf scars** with stipule scars. Shrub 2–5 m high, more rarely small tree to 10 m high; rarely planted. Origin: in Europe from the high north- to the south-west of the Alps, especially on high plateaux and low mountain ranges.

Series Arbuscella

Salix phylicifolia L., tea-leaved willow
[*Salix bicolor* Ehrh. ex Willd.]

Buds flat-appressed to twig, 4–5 mm long, dark brown and sparsely hairy. **Twigs** mostly entirely glabrous, slightly glossy with light, small lenticels. Occasionally planted shrub, 1–2 m high. Origin: the Pyrenees, the Vosges and the Harz, as well as in northern Europe and central Asia.

Series Lanatae

Salix lanata L., woolly willow

Buds slightly spreading from the twig, mainly in the upper half; long, close to the twig felty-hairy, frequently flanked by persistent dry stipules. **Leaf buds** 6–7 mm long, ovoid, blunt to acute, slightly glossy, red to olive-brown. **Flower buds** dark, black-brown, slightly larger than the leaf buds: 7–9 mm long and almost globose. **Twigs** relatively thick, slightly glossy green and mainly towards the tip with remains of long, felty indument. Leaf cushion relatively large, with large stipules or their scars. Frequently planted, small shrub. Origin: northern Eurasia.

Subgenus Chaetia

Series Myrtosalix

Salix alpina Scop., alpine willow

Buds yellow-ochre to olive-brown, ± glabrous, 3–4(–6) mm long, slender, narrowly ovoid. Flower buds fat, the tip relatively long, spreading from twig. **Twigs** slender, green to olive-green, initially finely white-hairy, later easily tearing off and rough. Shrub 10–20 cm high. Origin: eastern Alps.

Series Retusae

Salix retusa L., blunt-leaved willow

Buds in transverse plain ± round, obovoid to cylindrical, appressed to twig, 2–4 mm long. **Twigs** prostrate on the ground, glabrous, light red or the upper side red to dark brown, the shaded-side yellowish-green, glossy. At the base of the shoot often with remnants of last-year's bud scale. Of stronger growth than *Salix reticulata*. Shrub 5–30 cm high, planted occasionally. Origin: high mountains of Europe and the Balkans.

Salix serpyllifolia Scop.
thyme-leaved willow

Buds small, ovoid, 1.5–3 mm long, initially hairy but soon becoming glabrous scales, these olive-green to brown. **Twigs** slender, yellowish green to olive green, sometimes also slightly reddish, prostrate on the ground with many adventitious roots. Very flat shrub, often spreading wide and carpet-forming. Twigs partly underground with only 3–5 cm emerging above ground. On chalk-rich soils of the alpine and sub-alpine level in the higher mountains of central and southern Europe.

The dwarf willow, **Salix herbacea** L., is similar. **Buds** obovoid, glabrous (as are the twigs), ochre to pale brown. In its carpet-like growth only the youngest shoots extend above ground. Origin: on acid soils, species of the European high mountains and circumpolar.

Series Chamaetia

Salix reticulata L., net-leaved willow

Buds dark wine-red, alternate, particularly towards the bud's tip shortly white-hairy;

Salix alpina
Lateral bud

Salix alpina
Lateral bud

Salix serpyllifolia
Lateral bud

Salix serpyllifolia
Lateral bud

Salix serpyllifolia
Branching twig with roots

Salix reticulata
Lateral bud

Salix reticulata
Part of a twig

Hypericum androsaemum
Fruiting twig

Hypericum androsaemum
Lateral buds

Hypericum androsaemum
Fruit

differentiated into small 1–2 mm long leaf buds flat-appressed to the twig and spreading ovoid to 4 mm long flower buds. **Twigs** towards the tip finely appressed-hairy, becoming glabrous lower down, red-brown to grey-green to blackish. In its natural habitat this dwarf shrub is hardly 5 cm tall, but it is cultivated as 5–30 cm high ground cover. Origin: circumpolar and in the mountains of central and southern Europe.

Hypericaceae
St John's wort family

Hypericum L., St. John's wort

Several mostly herbaceous species native, and several shrubby species, mostly evergreen and common in cultivation. The few deciduous species often have remnants of the dead opposite leaves. These, as well as the bud scales, have dot-shaped oil glands visible against the light. Leaf scars with one trace. Terminal buds absent. Lenticels absent.

Key to *Hypericum*

1 Fruits berry-like, ovoid, blackish . . . . . .
. . . . ***Hypericum androsaemum***
1* Fruits woody capsules that break open. .
. . . . ***Hypericum kouytchense***

Hypericum androsaemum L., tutsan

Buds 2–4 mm long, lateral buds spreading from the twig, the bud scales with twig-coloured, ochre-brown to orange- or red-brown, long mostly ± spreading tips. **Twigs** glabrous, finely longitudinally striped, just below the nodes sharply two-angular, orange-brown and dark violet-brown to red-brown. **Leaf scars** dark, with one trace on small cushions, the opposite ones connected by a line. **Fruits** in pseudo-umbellate terminal racemes, persistent on the shrub during the winter. Fruit superior, berry-like, ovoid, to 10 mm long, with persistent sepals. Frequently planted, shrub to 70 cm high. Origin: from western to southern Europe to the Caucasus, Asia minor and north Africa.

Hypericum kouytchense H.Lév.,
large-flowered St. John's wort
[***Hypericum patulum*** Thunb. var.
grandiflorum hort.]

Buds 3–4 mm long, ovoid, spreading from
the twig. Bud scales sometimes spreading,
long acute. **Twigs** ochre-brown to red-
brown, at the tip 2–4 angular, at the base
± round. Older twigs with smooth, flaking
bark. **Leaf scars** one indistinct trace, very
dark. **Fruits** united in racemes or panicles,
5-valvate, the woody capsules opening; fruit
valves ending in long thread-like remnants
of styles. Frequently planted shrub, often
more than 1 m high. Origin: western China.

Phyllanthaceae
leafflower family

The centre of distribution for this family
(formerly a subfamily of the Euphorbiaceae)
lies in the tropics of the Southern
Hemisphere. Only a few, rarely planted
deciduous shrubs reach the temperate
latitudes of the Northern Hemisphere.

The family frequently has latex, and
terminal buds are absent; lateral buds are
alternate, more rarely opposite; leaf scars
with one trace.

Key to Phyllanthaceae

1 One year old twigs basally more than
   2 mm thick; buds without scales, more
   broad than high . . ***Flueggea suffruticosa***
1* Twigs more slender, buds longer than
   thick, with more than two bud scales. . .
   . . . . ***Leptopus chinensis***

Leptopus chinensis (Bunge) Pojark.
[***Andrachne colchica*** Fisch. & C. A. Mey. ex
Boiss.]

Buds small, globose-ovoid, to 2 mm long,
with some red-brown, basally greenish,
slightly ciliate bud scales. **Twigs** close to
each other, slender, glabrous, olive-brown to
red-brown, towards the tip often dead and
dry: pale grey-brown. **Leaf scar** with one
trace, above this frequently small scars of
the fallen fruits. Rarely planted, shrub to 50
cm high. Origin: Asia minor and Caucasus.

Hypericum kouytchense
Lateral buds, bud scales
have oil glands that are
visible as transparent little dots

Hypericum kouytchense
Twig with fruits: 5-valvate
capsules

Leptopus chinensis
Twig with dry tip

Flueggea suffruticosa
Twig: tips mostly
dying for quite
some distance

Leptopus chinensis
Lateral bud with
enrichment bud

Flueggea suffruticosa
Lateral bud

Coriaria myrtifolia
Twigs angular with warty
lenticels; lateral buds with
many enrichment buds

Coriaria myrtifolia
Twigs angular with
warty cork pores;
lateral buds with many
enrichment buds

Flueggea suffruticosa (PALL.) BAILL.
[*Securinega suffruticosa* (PALL.) REHDER]

Buds: small, appressed lateral buds, broadly rounded, 1–1.5 mm high and 2–2.5 mm broad, protected by two glabrous, red-brown, glossy bud scales; margins of bud scales and point of bud dark black-brown. **Twigs** long, rod-like, slightly angular, the tips drying straw-coloured; living parts of twig glabrous, glossy-orange-brown to red-brown, initially finely pale-dotted (lens!) and with many distinctly visible paler, warty lenticels. Two-year-old twigs wine-red to violet-brown. Bark tearing longitudinally into ochre-brown, corky strips. **Leaf scars** with one trace, mostly relatively pale. Shrub to 2 m high, rarely planted. Origin: Mongolia to northern China.

Coriariaceae
tanner bush family

Coriaria myrtifolia L.,
redoul or tanner bush

Terminal buds not formed. **Lateral buds** sub-opposite, mostly in small clusters, with many enrichment buds in the axils of the primary lateral bud; globose-ovoid, 2–5 mm long, with some loose brown bud scales. **Twigs** 4-angular, long, curving upwards, basally sturdy, more slender towards tips which die down over long distances, green to olive-grey. **Lenticels** many, warty, ochre-brown. **Leaf scar** with one trace. 1–3 m high shrub. Origin: south-west Europe and north Africa.

Fabaceae
pea family

The family, on a world-wide scale containing almost 20,000 species, includes represent-atives of all life forms: herbs, perennials, shrubs, climbers and trees. German dendrofloras often include the sub-shrubs restharrow (*Ononis*) and canary clover (*Dorycnium*); these sub-shrubs stand between perennial herbs and shrubs, and often only basal branch sections survive in winter.

Buds alternate, rarely opposite (in part of *Genista*), with bud scales or without. **Leaf scars** mostly with 3 traces, more rarely one or multi-traced. **Fruits** of a single carpel, mostly called legumes, but apart from true legumes opening along both sutures, there are also follicles, nuts and special forms like lomentums.

Key to Fabaceae

1 Buds without scales or hidden below leaf scar/leaf base, but never with terminal bud 8
1* Buds with bud scales 2
2 Plants with thorns/spines. 3
2* Plants not armed 15
3 With terminal buds 11
3* Without terminal buds 4
4 Plant with thorns, occasionally additional leaf-spines 5
4* Only leaf spines present (on stipules and rachis) 7
5 Twigs and thorns/spines green, twigs finely longitudinally furrowed or angular. . . . . **Genisteae**
5* Thorns/spines ochre to grey-brown or red-brown, twigs almost round or slightly 5-angular 6
6 Thorns mostly branched, red-brown, glossy, mostly more than 3 cm long . . . . *Gleditsia triacanthos*
6* Thorns/spines not branched, ochre to grey-brown . . . . *Sophora davidii*
7 Twigs initially brown, at least stipules with thorny tips . . . . *Caragana* and *Halimodendron*
7* Twigs initially green, stipules weak and deciduous. . . . . **Genisteae**
8 (1) Buds hidden under remnants of leaf 9
8* Buds visible 10
9 Buds hidden below the three-parted leaf scar, next to the leaf-scar often with paired stipule spines . . . . *Robinia*
9* Buds hidden below persistent leaf base, twigs mostly greenish. . . . . . **Genisteae**
10 Buds distinctly emerging above surface of twig, with several accessory buds . . . . *Cladrastis kentukea*
10* Buds almost level with twig 13
11 Twigs round, stipules awl-shaped or caducous 12
11* Twigs slightly angular, stipules flat-broad, partly covering lateral buds . 27
12 Shrubs, rarely higher than 1 m, fruit remnants in fascicles . . *Chamaecytisus*
12* Trees to 7 m high, fruits in racemes . . . . *Laburnum*
13 Twigs thick, grey-brown to grey-green 14

SUBFAMILY Cercidoideae

Cercis L., Judas tree

Shrubs and medium-large trees in Eurasia and America. It is remarkable that the flowers grow on the older wood (cauliflory): the plants flower and fruit from long time resting buds on older twigs and stems. **Fruits** are legumes which do not open; they are situated on the trunk and on sturdy branches during the whole winter. **Buds** are exclusively lateral buds with descending accessory buds. Bud scales distichous, prophyll scales from the leaf base, subsequent scales formed by stipules. Apical lateral bud pointing obliquely beyond the dying tip of shoot. Twigs zigzag.

Key to *Cercis*

1 Buds longer than 3 mm . . . . *Cercis siliquastrum*

1* Buds less than 3 mm long . . . . *Cercis canadensis*

Cercis siliquastrum L., Judas tree

Buds only alternate, lateral buds on first year's growth twigs, oblong, acute, to 8 mm long, smaller towards the base of the twig, flat and surrounded externally by only two simple bud scales. Bud scales dark red-brown, towards the tip black-brown. With 1–2 descending accessory buds which often remain for many years. From these and from the lateral buds at the base of the twig, the flower buds will develop at a later stage. They are multi-scaled, large and ± globose: 8–10 mm long. **Twigs** dark red-brown, with few small grey-brown lenticels. From the leaf scars two ridges extend down the stem, these soon lose their definition. **Pith** narrow and white. **Wood** greenish-white. **Leaf scars** on cushions; with three traces, the central trace the largest. Common, tree to 12 m high. Origin: Mediterranean region to Asia minor.

Cercis siliquastrum
Subterminal lateral bud with descending accessory buds

Cercis siliquastrum
Twig

Cercis canadensis
Flower bud on older wood

Cercis canadensis
Twig

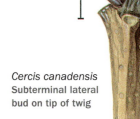

Cercis canadensis
Subterminal lateral bud on tip of twig

Gymnocladus dioicus
Twig: above the large leaf scars, the tiny buds are immersed in the tissue

Gymnocladus dioicus
Tip of twig

Gleditsia triacanthos
Twig with branching shoot thorns

Gleditsia triacanthos
Lateral bud small, hardly distinctive below thorn

Cercis canadensis L., forest pansy

Buds roundish-ovoid c.1(–2) mm, mostly with a slightly sunken, very small accessory bud. Flower buds like in *Cercis siliquastrum* on old wood, but smaller: 2–3 mm long here, rarely bearing fruit in our area. Bud scales deep lilac wine-red, glabrous, only at the margin slightly whitish ciliate. **Twigs** round, brown to brown-violet and glossy. **Lenticels** many, but less distinctive than in the Judas tree. Older **twigs** finely fissured, grey. Rarely planted, large shrub or small 6–10 m high tree. Origin: North America, south to Mexico.

SUBFAMILY Caesalpinioideae

Gymnocladus dioicus (L) K. KOCH

Buds relatively small, without bud scales, lateral buds sunk in the bark: visible as a round, olive-brown swelling with at the centre a circular, c. 2 mm large part of the grey-brown-hairy bud. With descending accessory buds. **Twigs** very thick, becoming glabrous, dark wine-red to brown, initially with waxy bloom, later covered almost completely by a grey-white dying epidermis. Many small light orange-brown to white lenticels. **Leaf scars** large, shield-shaped, with 5 large traces of vascular bundles. **Pith** broad, ochre-brown. Frequently planted, laxly branched tree with thick branches. Origin: North America.

Gleditsia triacanthos L., honey locust

Buds: only small, lateral buds surrounded by few bud scales, with descending accessory buds. Lateral buds below thorns are the accessory buds of the thorns. The buds are sunk in tissue with a red-brown bark so that frequently only the uppermost bud is visible. **Twigs** olive-green to brownish-green, glossy and mostly glabrous. Thorns long and mostly branched, dark wine-red to violet-brown, at the base red-brown. **Leaf scars** with 3 traces. Tree to 30 m high. Particularly the form without thorns (f. *inermis* WILLD.) frequently planted. Origin: North America.

Albizia julibrissin Durazz., pink silk tree

Buds only lateral buds: 1–2 mm long, surrounded by two dark brown, sparsely hairy bud scales, slightly paler at the tip; the outer partly surrounding and overtopping the inner (equitant). With tiny descending accessory buds. **Twigs** glabrous, zig-zag, more slender towards the shoot tip; slightly angular/longitudinally furrowed, pale grey-green; with many pale ochre-coloured lenticels. **Leaf scars** on thick cushions, black-brown, with 3 indistinct traces. Large shrub or small, to 6 m high tree with a broad crown. Origin: SW and E Asia.

SUBFAMILY Faboideae (Papilionoideae)

Cladrastis group

Cladrastis kentukea (Dum.-Cours) Rudd
[*Cladrastis lutea* (F. Michx.) K. Koch]

Buds: only lateral buds, these without bud scales, dense ochre-olive-grey hairy, with 3 descending accessory buds. Sizes: c. 4 mm, that of the 3 accessory buds 2.2, 2 and 1.5 mm. **Leaf scar** which encloses the buds horse-shoe-shaped, with 5 traces; it shows the protection of the axillary buds by the leaf base during the previous growth period. Between the leaf scar and the buds is a small red-brown ridge of hair. **Twigs** 3–4 mm thick, initially glossy, wine-red-violet, with few very much paler lenticels; later matte and grey. **Lenticels** oblong and running into each other. Tree to 10 m high, planted frequently. Origin: North America.

Styphnolobium japonicum (L.) Schott, Japanese pagoda tree
[*Sophora japonica* L.]

Buds: very small lateral buds, 1.5–2 mm thick and high; often, particularly on long-shoots, with descending accessory bud; without dedicated bud scales, surrounded by dark hairy prophylls. **Twigs** differentiated into short and long short-shoots, smooth, green and matte, glossy: with sparse small, grey, dot- to line-shaped **lenticels**; on the short-shoots the nodes are thickened due to the strongly developed leaf-cushions, on the long-shoots, buds on smaller leaf cushions.

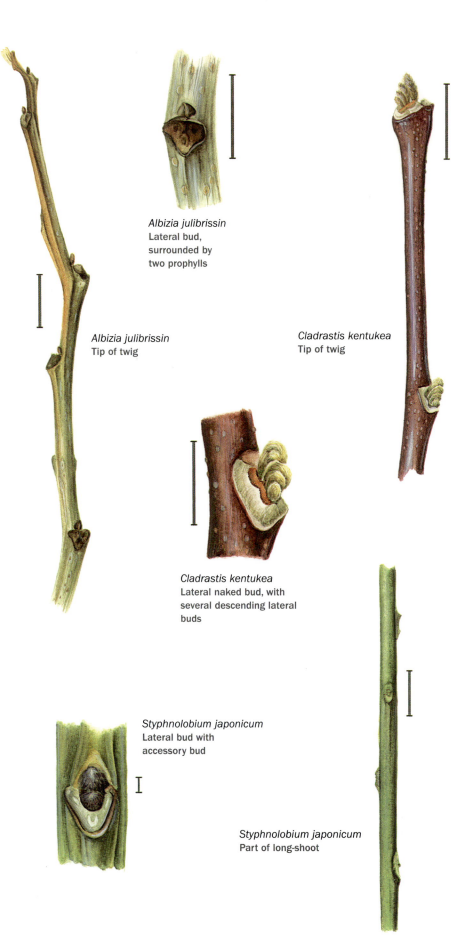

Albizia julibrissin
Lateral bud, surrounded by two prophylls

Albizia julibrissin
Tip of twig

Cladrastis kentukea
Tip of twig

Cladrastis kentukea
Lateral naked bud, with several descending lateral buds

Styphnolobium japonicum
Lateral bud with accessory bud

Styphnolobium japonicum
Part of long-shoot

Sophora davidii
Lateral bud next to thorn

Sophora davidii
Lateral bud beside thorn

Styphnolobium japonicum
Short-shoot

Maackia amurensis
Tip of twig

Maackia amurensis
Lateral bud with two scales

Leaf scars semi-circular to U-shaped with three traces of vascular bundles: the central one larger, semi-circular; the two lateral ones small, line-shaped. Tree to 20 m high. Origin: eastern Asia.

Sophora davidii (FRANCH.) SKEELS

Buds small, c. 3 m long, ovoid to compact, densely whitish-grey hairy lateral buds without bud scales. First bud scales with delicate stipule-spines, as beside the leaf scar. **Twigs** slender, grey-green, fine and relatively densely hairy. In the leaf axils with **thorns**. Leaf scar with a large central trace and two smaller ones at the margin. Shrub to 3 m high. Origin: western China.

Maackia amurensis RUPR., Amur maackia

Buds: only lateral buds: broad, ovoid, acute, flat, 5–7 mm long and 5–6 mm broad, but only 3–4 mm thick, with two bud scales. Scales dark violet-brown, at the base and in shadow also greenish, dorsally and towards the tip very delicately hairy; scale margin partly dry and paler brown. **Twigs** finely hairy only at the very tip, otherwise glabrous, sturdy: 3–4 mm thick, round, grey-green to grey-brown with small, roundish-oblong lenticels. **Leaf scars** with three traces, central trace large and acute to conical, adjacent traces small, hardly noticeable. **Fruits** flat to 5 cm long, legumes in terminal racemes. Particularly the infructescence axes persistent in winter. Rare, tree to 15 m tall. Origin: Manchuria and China.

TRIBE Genisteae

Often with green twigs, assimilating entirely or partly with the stem; shrubs, more rarely small trees (*Laburnum*).

Key to Genisteae

1 Branching alternate 2
1* Branching opposite. . . . . *Genista*
2 Twigs round, often with terminal buds. 3
2* Twigs angular or furrowed, without terminal buds 4
3 Fruits in long racemes, trees or shrubs to 7 m high . . . . *Laburnum*
3* Fruits in fascicles, shrubs rarely more than 2 m high . . . . *Chamaecytisus*
4 Plants thorny 5
4* Unarmed 7

5 Buds hidden under persistant bladder-like enlarged leaf-base . . . . **Erinacea anthyllis**

5* Buds visible or leaf base different . . . 6

6 All twigs thorny and all leaves spiny . . . .**Ulex europaeus**

6* Short-shoots turning thorny present on long-shoots, flower shoots often unarmed . . . . **Genista**

7 (4) Twigs glabrous, bud cover relatively large: composed of leaf base and stipules, hairy on the inside with spreading tips. . . . . *Petteria ramentacea*

7* Buds smaller, often hairy or hidden under the leaf base. 8

8 Twigs round, finely furrowed, long and rush-like. . . . . *Spartium junceum*

8* Twigs different. 9

9 Buds ovoid, not covered by leaf base . . . . **Chamaecytisus**

9* Buds small and concealed at least basally by the leaf base. 10

10 Twigs 5-angular or 8-furrowed, mostly with fine white dots (lens) . . . . .*Cytisus*

10* Twigs furrowed, with more than 5 angles, very slender twigs occasionally 4-angular . . . . **Genista**

Genista group

Closely related to the genus *Genista* are, among others, *Spartium* and *Ulex*. In future they might become part of the genus *Genista*, as *Echinospartum* and *Cytisanthus* already have.

Genista L., broom

Deciduous, more rarely evergreen (*Genista sagittalis*) or semi-evergreen (*Genista anglica, Genista germanica*). Shrubs armed with thorns or unarmed.

Key to *Genista*

1 Buds and branching alternate. 3

1* Branching opposite. 2

2 Twigs unarmed. . . . . . . **Genista radiata**

2* Twigs thorny. . . . . *Genista horrida*

3 Twigs thorny. 4

3* Twigs unarmed. 6

4 Thorns at least 1.5 cm long and leafy (scars), often leafy in winter 5

4* Thorns shorter, leaves absent, deciduous . . . . *Genista hispanica*

5 Twigs glabrous . . . .*Genista anglica*

5* Twigs hairy. . . . . *Genista germanica*

6 Leaf base (scar) with stipule lobes . . . 7

6* Stipules absent, strongly gnarled twigs due to large leaf cushions *Genista pilosa*

7 Twigs blue-green, ± 4-angular. . . . .*Genista lydia*

7* Twigs yellowish-green with more angles . . . . **Genista tinctoria**

Subgenus Genista

Genista tinctoria L., dyer's greenweed

Buds largely covered by leaf bases, but mostly visible, small, brownish; bud scales mostly glabrous, with sparse hairs. **Twigs** glossy green, angular, with many shallow furrows; initially sparsely hairy, towards winter mostly glabrous. Persistent leaf bases with brown drying tips of stipules. Frequently planted, erect shrub to 1 m high. Origin: Europe to Asia minor.

Genista lydia Boiss., Lydian broom

Twigs more or less 4-angular, blue-green to dark green, more slender at tip but not thorn-acute. Frequently planted, to 50 cm high, prostrate-ascendent shrub. Origin: south-eastern Europe.

Genista pilosa L., hairy greenweed

Buds initially not visible, hidden below leaf cushion, the persistent leaf base. When the buds develop they become densely pale hairy, globose, and visible under the leaf cushion. **Twigs** green, appressed white hairy, slender, slightly angular; due to densely following leaf cushions with knobbly nodes. Older twigs grey-brown and hairy, with sessile short-shoots and persistent brown prophylls. Twigs spreading from branches at an acute angle. Prostrate-ascending, 10–30 cm high, densely branching shrub. Frequently planted, orginally from central and western Europe.

Genista tinctoria
Lateral bud

Genista tinctoria
Lateral bud

Genista lydia
Part of twig

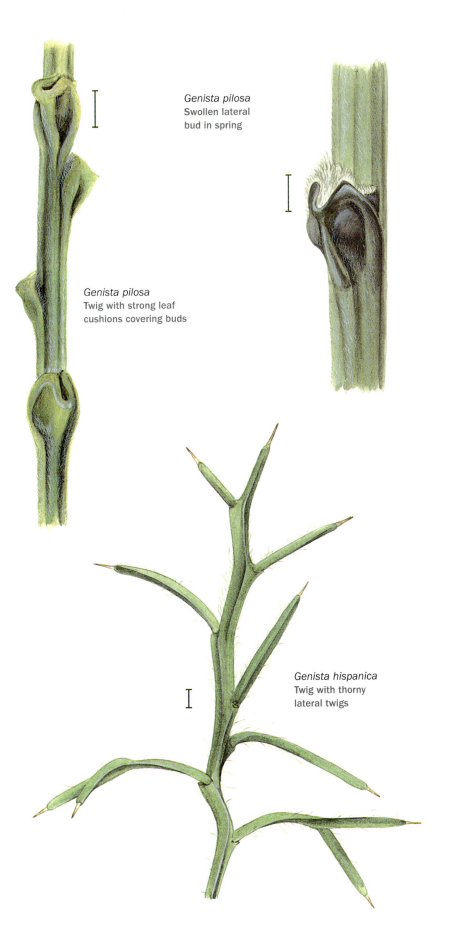

Genista pilosa
Swollen lateral
bud in spring

Genista pilosa
Twig with strong leaf
cushions covering buds

Genista hispanica
Twig with thorny
lateral twigs

Subgenus Phyllobotris

Genista anglica L., petty whin

Buds small, globose-ovoid, below the axillary thorns. **Twigs** green, glabrous, relatively slender. The c. 2 cm long, slender, leafy shoot thorns sit on long-shoots. Low shrub, half-evergreen, rarely reaching 80 cm high. Central and western Europe.

Closely related is **Genista germanica** L., the German broom. A 30–50 cm high half evergreen shrub also in Central and Western Europe. It differs from the English broom by having slightly sturdier, hairy twigs.

Genista hispanica L., Spanish gorse

Buds globose-ovoid, 1–2 mm long, with some brownish-green bud scales, largely hidden between shoot thorns and the leaf base; they are descending accessory buds formed below the proleptic thorns. **Twigs** green with very fine white dots, furrowed and 4-angular, hairy. Lateral shoot thorns short, 7–13 mm long, also 4-angular, but flattened, c. 1 mm wide with a distinct very slender brown tip. Very dense shrub, 30–70 cm high. Origin: southern Europe from Spain to northern Italy.

Genista hispanica
Bud as accessory bud
below a shoot thorn

Subgenus Echinospartum

Genista horrida (VAHL.) DC.,
large-spined broom
[*Echinospartum horridum* (VAHL.) ROTHM.]

Buds opposite, small, densely white hairy, hidden below leaf remnants (leaf base). **Twigs** blueish-green, finely furrowed, initially sparsely hairy, older twigs grey-brown. All tips of twigs thorny. **Leaves** with broad base and persistent stipule lobes and remnant of brown petiole. Small, densely branching, cushion-shaped shrub. Origin: southern central France to northern Spain.

Subgenus Asterospartum

Genista radiata (L.) SCOP.,
rayed broom
[*Cytisanthus radiatus* (L.) O. LANG.]

Buds not visible. **Side twigs** opposite, longitudinally furrowed; first year twigs dark green, finely appressed-hairy, older ones pale grey-brown. Unarmed, flat and broad shrub, 20–50(–100) cm high. Due to the forked branching of twigs, relatively dense. Frequently planted. Origin: southern Alps to southern Europe.

Spartium junceum L., Spanish broom

Buds hidden below leaf cushions. **Twigs** glabrous, green, rush-like: long and narrow, round, finely furrowed. **Leaf cushion** only slightly protuberant from the twig, ending in a small slender stalk. Shrub 2–3 m high. Origin: from the Mediterranean area to south-western Europe and the Canary Islands.

Genista horrida
Twig with opposite thorns

Genista horrida
Lateral buds below the thorny tip of twig

Genista radiata
Branching

Genista radiata
Tip of twig

Genista radiata
Opposite branching

Spartium junceum
Branching in detail

Ulex europaeus L., common gorse

Buds visible, globose, c. 3 mm long, without bud scales, densely yellowish to pale grey-green hairy. **Twigs** green, furrowed and, like the leaves, terminating in thorny/spiny tips. With spreading hairs. Without leaf scars, because the leaves will become spines. Frequent, shrub over 1 m high. Origin: western and southern Europe, locally naturalised in northern Europe.

Erinacea anthyllis LINK, hedgehog broom
[**Erinacea pungens** BOISS.]

Buds not visible below the membranous leaf bases, which are expanded and look like bladders, yellowish-green to brown; buds small and densely hairy, without bud scales. **Twigs** all ending in thorny tips, green, finely furrowed and short-appressed, silvery hairy. Densely branching, globose, small shrub to 30 cm high. Origin: the Mediterranean region.

Ulex europaeus
Twigs with shoots and
leaves ending in thorns

Spartium junceum
Branching

Erinacea anthyllis
Lateral bud hidden under
leaf base: an accessory
bud of a shoot thorn

Ulex europaeus
Thorn-tipped shoot with
lateral buds which sit in
the axils of spiny leaves

Erinacea anthyllis
Tip of a thorn twig

Erinacea anthyllis
Branching

Cytisus and relatives

More closely related to *Cytisus* are *Chamacytissus* and *Laburnum*. In winter, the two genera are distinct from *Cytisus* in their round twigs, the formation of short-shoots and often the presence of terminal buds.

Cytisus DESV., broom

[including *Sarothamnus* WIMM.]

Unarmed, with angular or furrowed long-shoots; small to medium large shrubs without terminal buds. Origin: from the Mediterranean region across central Europe to western Asia.

Key to Cytisus

1   Twigs with five wings or angles 2
1* Twigs with 8(–10) furrows 6
2   To 20 cm high, branches very slender, soon glabrous . . . . . . **Cytisus decumbens**
2* Taller, twigs glabrous or hairy 3
3   Buds visible 4
3* Buds mostly hidden completely below the persistent leaf base 5
4   Twigs with 5 narrow wings, mostly glabrous or ± glabrous . . . . **Cytisus scoparius**
4* Twigs slightly 5-angular, buds densely hairy . . . . **Cytisus nigricans**
5   Twigs green all around . . . . **Cytisus sessilifolius**
5* At least the angles reddish in the sun . . . . **Chamaecytisus purpureus**
6   Twigs when young c. 2 mm thick and relatively stiff . . . . **Cytisus purgans**
6* Twigs clearly less than 2 mm thick 7
7   Shrub to 30 cm high 8
7* 2–3 m high shrub . . . . **Cytisus ×praecox**
8   Twigs curving downwards . . . . **Cytisus ×kewensis**
8* Twigs growing upwards . . . . **Cytisus ×beanii**

Cytisus cultivars: selections and hybrids

Cytisus cultivars can be divided into two large groups, distinct even in winter: the *Scoparius* group and the *Praecox* group. The *Scoparius* group are probably derived from broom, *Cytisus scoparius*, or more rarely closely related hybrids. The twigs have, as a rule, 5 distinct angles, more than 5 is much less common. The second group, the *Praecox* group, derive predominantly from the species grouped here under the section

Spartothamnus. A common character is that twigs are round in this section, with 8 furrows. Definite hybrids between broom and species of the section *Spartothamnus* are called *Cytisus ×dallimorei* ROLFE (*C. multiflorus*? × *C. scoparius* 'Andreanus').

Section Sarothamnus

Cytisus scoparius (L.) LINK, common broom, Scotch broom [*Sarothamnus scoparius* (L.) WIMM. ex W. D. J. KOCH]

Buds to 2 mm long, round when seen from above, flat-appressed to twig, with 2–4 outer, almost glabrous bud scales. **Twigs** green, 5-angular with prominent edges and in-between deeply cut furrows, mostly glabrous. Erect shrub, curving upwards and 0.5 to 2 m high. In cultivation since ancient times. Origin: central to southern Europe.

Section Corothamnus

Cytisus decumbens (DURANDE) SPACH

Buds very small, to c. 1 mm long, partly hidden below the leaf base, surrounded by two slightly hairy bud scales. **Twigs** green, very slender, angular, sparsely hairy or glabrous. Flat shrub to 20 cm high with decumbent, often rooting twigs. Origin: southern to south-west Europe.

Cytisus scoparius
Lateral bud with few scales sunk in a furrow

Cytisus scoparius
Lateral bud with few scales sunk in a furrow

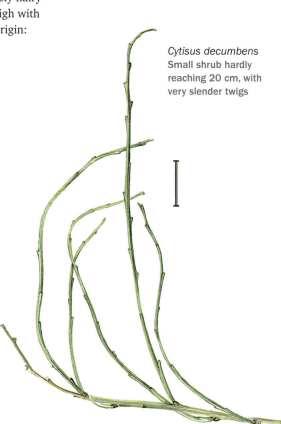

Cytisus decumbens
Small shrub hardly reaching 20 cm, with very slender twigs

Cytisus decumbens
Lateral bud

Cytisus purgans
Part of twig

Cytisus ×kewensis
Lateral bud

Cytisus purgans
Tip of twig

Cytisus ×praecox
Lateral bud

Cytisus nigricans
Lateral bud

Cytisus nigricans
Lateral bud

Section Spartothamnus

Twigs in most species 8-furrowed (finely striped longitudinally), but in **Cytisus multiflorus**, which is not hardy, 5-angular.

Cytisus ×praecox BEAN, ivory broom
[*Cytisus multiflorus* × *Cytisus purgans*]
Dense shrub 0.7–1.5(–2.5) m high with long down-curving green twigs. Many forms differing in type of growth. In the shape of the twig close to *Cytisus purgans*, twigs always 8-furrowed. **Cytisus purgans** (L) BOISS., the Spanish gold broom, is uncommon: twigs green, stiff, when young c. 2 mm thick, round, striped, initially sparsely rough-hairy. Densely branching shrub 0.2–1 m high. Origin: south-western Europe, Spain to southern France and north Africa. The second parent which also comes from Spain and north Africa: *Cytisus multiflorus* (AITON) SWEET is not hardy.

Cytisus ×kewensis BEAN and **Cytisus ×beanii** G. NICHOLSON
[*Cytisus ardoinii* × *Cytisus multiflorus* or *Cytisus purgans*]
One parent of both hybrids is the southern French species **Cytisus ardoinii** E. FOURN.: a 10–20 cm high shrub with shaggy-haired branches. The hybrids, growing to 30–40 cm high, link with their 8-furrowed twigs to the *Praecox* Group. The twigs of **Cytisus ×kewensis** curve downwards, whereas the short branches of **Cytisus ×beanii** grow strongly upwards.

Section Phyllocytisus

Cytisus nigricans L., black-rooted broom
Buds small, but not hidden, 1 mm high and 1–2 mm broad, naked, densely long-hairy, bud leaves globular-bulgy. **Twigs** green, erect, round and finely furrowed, appressed, short, yellowish-white hairy; older twigs grey-green to brown-olive. Branches and trunks with dark brown bark. **Leaf cushions** relatively narrow, prominent, with an indistinct trace. Frequent, shrub with many shoots, 0.5–2 m high. Origin: central to south-east Europe.

Cytisus sessilifolius L.

Buds very small, grey-brown hairy, often completely hidden below the persistent leaf base. **Twigs** green, long and glabrous; initially distinctly 5-angular, later ± round with fine ridges running down from the leaf scars. **Leaf scars** dark, small and round with an indistinct trace, in the leaf axils often with stalked remnants of the infructescence. Frequently planted shrub, much branched, upright 1(–2) m high. Origin: South and South-West Europe to North Africa.

Chamaecytisus LINK

The fruits grouped in axillary short-shoots or a shoot tip. Old bud scales often persistent. Twigs with terminal buds.

Key to *Chamaecytisus*

1 Twigs round, often hairy 2
1* Twigs slightly angular, glabrous . . . .
. . . . *Chamaecytisus purpureus*
2 Fruits or their remnants grouped at tip of twig 3
2* Fruits or their remnants evenly spread on last year's shoots . . . .
. . . . *Chamaecytisus hirsutus* and *Chamaecytisus ratisbonensis*
3 Half-evergreen: some leaves persistent in winter. . . . .*Chamaecytisus supinus*
3* Deciduous, losing all leaves . . . .
. . . .*Chamaecytisus austriacus*

Chamaecytisus purpureus (SCOP.) LINK, purple broom

[*Cytisus purpureus* SCOP.]

Buds densely hairy. Lateral bud on young lateral shoots almost hidden below the leaf cushion. **Long-shoots** slightly 5-angular, only initially hairy, mostly becoming glabrous, green and grey-green to wine-red, the latter predominantly on the ridges and the leaf cushions; finely pale-dotted. **Leaf scar** with one trace. Shrub to 60 cm high, prostrate to upright, spreading by runners. Origin: south-eastern Europe. Frequently cultivated.

Chamaecytisus hirsutus (L.) LINK

[*Cytisus hirsutus* L.]

Lateral buds clearly visible, ovoid, to 5 mm long, lowest bud scales almost glabrous, greenish to wine-red-violet, upper scales greyish, more densely fine white-hairy.

Cytisus sessilifolius
Branching

Cytisus sessilifolius
Part of branch

Chamaecytisus purpureus
Lateral buds concealed under leaf remnant (leaf base and petiole scar)

Chamaecytisus hirsutus
Part of branch

Chamaecytisus hirsutus
Tip of twig

Chamaecytisus ratisbonensis
Short-shoot with small terminal bud

Chamaecytisus ratisbonensis
Lateral bud

Chamaecytisus ratisbonensis
Lateral bud

Chamaecytisus ratisbonensis
Tip of twig

Chamaecytisus ratisbonensis
Lateral bud

Chamaecytisus supinus
Lateral bud: the lowermost bud leaves with developed lamina

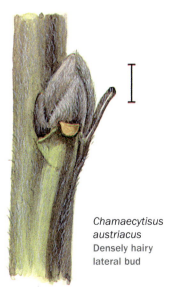

Chamaecytisus austriacus
Densely hairy lateral bud

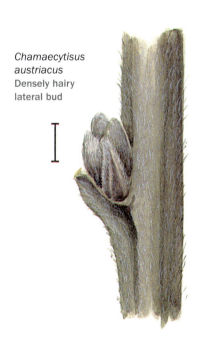

Chamaecytisus austriacus
Densely hairy lateral bud

Twigs greenish to red-brown, round, hardly angular, with warty dots, shaggy hairy, but at two years old completely glabrous. **Leaf scars** with one trace on small, but distinct leaf cushions (much smaller than the buds). Frequent, shrub 0.3–0.6(–1) m high, mostly prostrate with upwards arching branches. Origin: southern Europe to the Caucasus.

Very similar is ***Chamaecytisus ratisbonensis*** (Schaeff.) Rothm. [*Cytisus ratisbonensis*]

Buds silver-grey to grey-blue hairy. **Lateral buds** on first year long-shoots, ovoid with few outer bud leaves, 3 mm long. **Terminal buds** of long-shoots to 6 mm long and those of the short-shoots smaller, 2–3 mm long. **Twigs** roundish, dark olive-grey, finely appressed-hairy; differentiated in long and short-shoots. Short-shoots compressed, on older long-shoots and placed on very large leaf cushions together with old grey-brown bud scales. **Leaf scars** small, dark and indistinctly 3-traced. Occasional, prostrate or upright shrub 0.5–1(–2) m high. Origin: central Europe to western Siberia and Caucasus.

Chamaecytisus austriacus (L.) Link [*Cytisus austriacus* L.]

Lateral buds hardly surrounded by small leaf cushions, completely visible, ovoid, 2–3 mm long and densely hairy. The outer leaves of the bud develop into normal foliage leaves, but are shed in autumn, so that only the leaf base remains. **Twigs** slightly angular, towards the tip densely appressed-hairy, basally becoming glabrous, greenish to wine-red or brownish. Rarely planted erect shrub, 0.3–1 m high. Origin: from southern central Europe to eastern Europe.

The buds of clustered broom, the frequently planted half evergreen ***Chamaecytisus supinus*** (L.) Link, are similar. Here the leaves of lateral buds, develop into normal foliage leaves and remain throughout the winter. **Twigs** round, spreading with strong shaggy hair, glabrous much later. A tall, broad, upright, rarely decumbent shrub, 0.2–0.6(–1) m high. Origin: central to southern Europe to central France.

Laburnum FABR.

Persistent leaf base (leaf cushion) with small awl-shaped stipules or their scars, and a dark leaf scar with 3 indistinct traces of vascular bundles. Fruits: flat, 4–6 cm long legumes, slightly constricted between the seeds, mostly persistent in winter. Frequently planted small trees or shrubs with branches hanging down. In Eurasia represented by 2 or 3 species.

Key to *Laburnum*

1 Leaf cushion hardly prominent, silvery hairy . . . . **Laburnum anagyroides**

1* Leaf cushion glabrous . . . . **Laburnum alpinum**

Laburnum anagyroides MEDIK., common laburnum, golden rain

Buds surrounded by densely hairy silver-white scales; terminal buds hemispherical to ovoid, 3–4 mm long, lateral buds more narrowly ovoid, slightly spreading from the twig. **Twigs** differentiated into short and long-shoots, dark green to paler grey-green or silvery grey, appressed-hairy mainly towards tip of the twig. **Leaf cushion** strongly developed and often clustered below terminal bud, densely appressed silvery-white hairy. Shrub to 7 m high. Origin: southern Europe to southern central Europe and sometimes naturalised in the rest of central Europe.

Laburnum alpinum (MILL.) BERCHT & J. PRESL, Scotch laburnum

Buds similar to the common laburnum, but particularly the lowest bud scales only thinly hairy. **Twigs** glabrous, olive-green to grey-brown. **Leaf cushions** glabrous, glossy green to brownish. Frequently planted shrub, 5–7 m high. Origin: widely distributed in southern central Europe.

The frequently planted hybrid **Laburnum ×watereri** (WETTST.) DIPPEL (*Laburnum alpinum* × *Laburnum anagyroides*) is intermediate between its parents. The leaf cushions are not densely hairy as in the common laburnum, but also not completely glabrous as in the Scotch laburnum.

Petteria ramentacea (SIEBER) C. PRESL, Dalmatian laburnum

Buds very small, without bud scales and densely hairy, completely enclosed by

Laburnum anagyroides
Infructescence: hanging racemes of legumes

Laburnum anagyroides
Tip of twig with terminal bud

Laburnum alpinum
Tip of twig

Petteria ramentacea
Lateral bud enclosed by stipules of leaf

Petteria ramentacea
Lateral bud enclosed by stipules of leaf

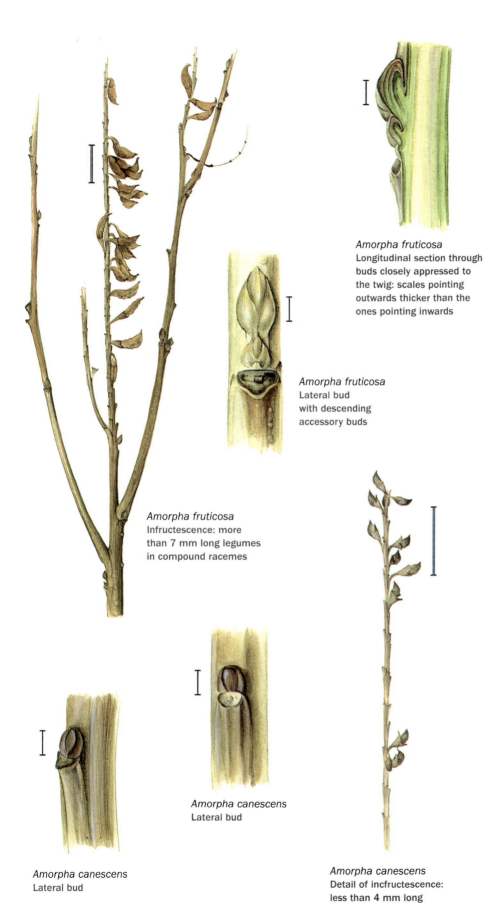

Amorpha fruticosa
Longitudinal section through
buds closely appressed to
the twig: scales pointing
outwards thicker than the
ones pointing inwards

Amorpha fruticosa
Lateral bud
with descending
accessory buds

Amorpha fruticosa
Infructescence: more
than 7 mm long legumes
in compound racemes

Amorpha canescens
Lateral bud

Amorpha canescens
Lateral bud

Amorpha canescens
Detail of incfructescence:
less than 4 mm long
legumes in raceme

stipules of the persistent leaf bases. **Twigs** glabrous, pale grey-brown, slightly angular and furrowed. Stipules and base of leaves with very distinct colour, outside dark violet-brown and glabrous, inside densely hairy, ochre-brown. **Leaf scar** roundish with 3 traces. Shrub, 2 m high, rarely planted. Origin from Istria to northern Albania.

TRIBE Amorpheae

Amorpha L.

Shrubs with long slender grey-brown to brown twigs. **Buds:** only lateral buds, small and appressed to twig. **Fruits** mostly persistent in winter, many small legumes in long compound racemes.

Key to *Amorpha*

1　Shrub to 1 m high without accessory buds . . . . ***Amorpha canescens***
1*　Shrub over 1.5 m, with descending accessory buds . . . . . ***Amorpha fruticosa***

Amorpha fruticosa L., bastard indigo

Buds flat, ovoid, 2–3 mm long, with 1 to 2 smaller accessory buds lower down. Buds very appressed to twig and slightly sunk into it. **Twigs** matte grey-brown, with fine longitudinal stripes. **Fruits:** brown c. 8 mm long legumes in 15–20 cm long compound racemes. Frequent, shrub to 3 m high. Origin: southern North America, sometimes naturalised in central and southern Europe.

Amorpha canescens PURSH, lead plant

Buds grey-brown, just over 1 mm thick, globose, slightly flattened and appressed to twig. Two outermost prophyll scales cover buds about halfway, appressed basally and spreading towards the tip, so the next bud scale becomes visible. First year's growth **twigs** c. 3 mm thick, pale grey to pale ochre-brown, very finely hairy, slightly furrowed, with many small faint warty lenticels. **Fruits** c. 4 mm long, dark greyish-brown legumes, many in 10–15 cm long compound racemes. Sub-shrub to 90 cm high. Origin: eastern North America, in central and southern Europe, naturalised in places.

TRIBE Indigofereae

Indigofera heterantha WALL. ex BRANDIS, Himalayan indigo
[*Indigofera gerardiana* GRAHAM]

Buds alternate, at the shoot end almost opposite, c. 3 mm long and 2 mm broad, triangular seen from above, flat against the twig. With two outer bud scales, these grey-brown, ± tightly appressed-hairy, long-acute and longer than the bud. **Twigs** finely angular, grey-brown to dirty ochre, violet-brown towards the tip, especially towards the shoot tip ± densely and appressed fine white-hairy. **Leaf scars** on small cushions, dark brown, with 3 paler traces, the lateral ones occasionally indistinct and, due to this, apparently 1-traced. Frequent, 1–2 m high erect shrub. Origin: Himalaya.

TRIBE Robinieae

Robinia L., false acacia

No terminal buds, spiral lateral buds with accessory buds ± hidden under leaf scars. Next to the leaf scars often two stipule spines, mostly at base flattened lengthwise. The spinyness depends on age and position, with older plants often unarmed and young plants often with strong spines.

Four species from North America. Apart from the commonly planted and locally naturalised white-flowering *Robinia pseudoacacia*, there are a few species that are close to one another, with red flowers and often shrub-like. There are some hybrids, fairly common in cultivation, between these red-flowering species and the common false acacia; they are not easily identifiable in winter, e.g. *Robinia* ×*margarettae* ASHE (*R. hispida* × *R. pseudoacacia*) or *Robinia* ×*ambigua* POIR. (*R. pseudoacacia* × *R. viscosa*).

Key to *Robinia*

1 Twigs with glandular or bristly indument
. . . . 2
1* Trees; twigs without indument and mostly sturdy stipule spines . . . .
. . . . *Robinia pseudoacacia*
2 Twigs with many red-brown bristly hairs, over 3 mm long . . . . . . *Robinia hispida*
2* Twigs with shorter glandular hairs 3

3 Twigs long, glandular (up to 3 mm), hairy underneath; stipule spines mostly sturdy . . . .*Robinia neomexicana*
3* Twigs short-glandular, otherwise glabrous; stipule spines absent or weak . . . .
. . . .*Robinia viscosa*

Robinia pseudoacacia L., false acacia, black locust

Buds concealed below the leaf scars, without terminal buds. Lateral buds with accessory buds. **Leaf scars** 3-parted with three traces of vascular bundles (one per scar-segment), slightly raised and the segments internally pale-hairy. On both sides of the leaf scar there are mostly large orange-brown to dark violet-brown stipule spines but these can, depending on species or age, also be absent. **Twigs** olive-green to dark brown, with 5 descending ridges and small lenticels. Often planted and naturalised tree. Origin: eastern North America.

Robinia hispida L., rose acacia

Buds not entirely surrounded by leaf scars, without scales, dark brown hairy. Above the bud and leaf scar pale ochre-brown and appressed-hairy, ending in a small, grey-white hairy protuberance. **Twigs** grey-brown and round, occasionally with narrow, grey-black fissures. Bristly hairs long (4–5 mm), thin, the shorter ones partly with small glandular heads. **Leaf scar** two- to three-parted and stipule spines small. Low shrub spreading by root stolons or small tree, when grafted on *Robinia pseudoacacia*. Origin: south-eastern North America.

Robinia neomexicana A. GRAY, New Mexican locust

Buds surrounded by internally densely pale-hairy three-parted leaf scar; buds without scales, densely shaggy dark brown hairy. **Twigs** pale grey to grey-olive, spreading glandular hairy, underneath with 3 mm long appressed fine glandular hairs. Lenticels many, ochre-brown like the twig. Stipule spines sturdy, c. 10 mm long. Small tree to 10 m high. Origin: southern North America.

Indigofera heterantha
Part of twig

Indigofera heterantha
Lateral bud surrounded by prophylls

Robinia pseudoacacia
Part of twig at the node with flat stipule spines

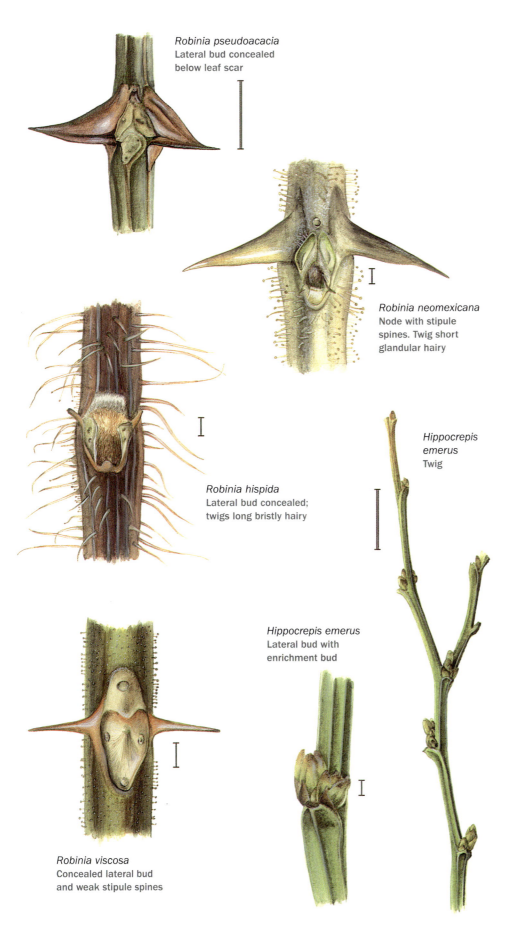

Robinia pseudoacacia
Lateral bud concealed
below leaf scar

Robinia neomexicana
Node with stipule
spines. Twig short
glandular hairy

Robinia hispida
Lateral bud concealed;
twigs long bristly hairy

*Hippocrepis
emerus*
Twig

Hippocrepis emerus
Lateral bud with
enrichment bud

Robinia viscosa
Concealed lateral bud
and weak stipule spines

Robinia viscosa Vent., clammy locust

Buds concealed below the merged leaf scar. **Twigs** olive-green to dirty red brown, slightly angular, only with short, c. 1 mm long glandular hairs. **Lenticels** dot-shaped, orange-ochre. **Leaf scars** with 3 traces, slightly raised at the centre where the three parts of the scar meet. Stipule-thorns small and slender, 3–5 mm long or absent. Tree to 12 m high, rarely planted. Origin: north-eastern and south-eastern USA.

TRIBE Loteae

Hipocrepis emerus (L.) Lassen, scorpion senna
[*Coronilla emerus* L.]

Buds: only lateral buds, mostly with some enrichment buds, c. 3–4 mm high and just as wide with 3–5 bud scales. Scales green, brown and hairy towards the tip; lower ones distinctly with adnate stipules and leaf scar. **Twigs** green, glabrous and strongly angular; later tearing open longitudinally, corky. **Leaf-scars** with 1 indistinct trace, on raised leaf cushions; leaf stipules persistent. Erect shrub, much-branched, 1.5–2 m high. Origin: Central and Southern Europe.

TRIBE Millettieae

Wisteria Nutt.

Lianas, difficult to tell apart in winter. One indicator is the direction of winding: this can be determined by thinking of the direction of winding as in a spiral staircase, from the bottom upwards.

Key to *Wisteria*

1 Winding to the left, anti-clockwise . . . . .
. . . .*Wisteria sinensis*
1* Winding to the right, clockwise . . . .
. . . . *Wisteria floribunda*

Wisteria sinensis (Sims) Sweet, Chinese wisteria

Exceptionally **lateral buds** only, appressed to the long-shoot, 6–10 mm long and 3–4 mm wide at the base, triangular, long-acute; surrounded by 2 outer bud scales, of which the first surrounds and overtops the second (equitant). Bud scales matte red-brown to dark brown, towards the tip often paler ochre-brown, finely longitudinally furrowed. On short-shoots the buds are slightly spreading and the bud scales slightly gaping apart: the subsequent leaves long and densely appressed-hairy. **Twigs** ± angular and strongly furrowed to almost round: pale ochre with green hues, pale grey when older, finely fissuring in rhomboid patterns, dark brown in the rhomboids, within the centre often inconspicuous ochre-coloured **lenticels**: small and many. **Leaf scar** on a cushion, with slightly conically raised traces of vascular bundles. Next to the leaf scar, particularly towards the tip of the twig, often with two spur-like appendages. **Pith** narrow and white. **Wood** pale greenish to ochre. Frequently planted, left-winding liana, to 10 m high. Origin: central China.

Wisteria floribunda (WILLD.) DC., Japanese wisteria

Similar to *Wisteria sinensis*. Can be told apart by the direction of winding. Frequently planted liana to 8 m high, right-winding. Origin: Japan.

TRIBE Hedysareae

The genera have in common that remnants of the leaf and its stipules persist in winter: in *Calophaca* these are fused with the leaf-base, in *Hedysarum* fused together and in *Caragana* and *Halimodendron* the leaf rachis often becomes spiny.

Key to Hedysareae

1 Lateral bud surrounded by persistent leaf at most for half its length 2
1* Lateral buds covered almost completely by persistent remnants of the leaf . . . . ***Calophaca wolgarica***
2 Stipules fused on the side away from the leaf scar . . . . . . ***Hedysarum multijugum***
2* Stipules free, sometimes turning into spines, as does the leaf rachis in some species . ***Caragana*** and ***Halimodendron***

Hedysarum multijugum MAXIM.

Younger twigs distinctly zig-zag, becoming ever more slender towards the tip, without terminal bud. **Lateral buds** ovoid, 2–2.5 mm long, standing away from the twig, loosely enclosed at base by a membranous scale, open at the top. From this scale emerges the densely white hairy tip of the bud. **Twigs** initially pale grey to greenish, densely white hairy. Older twigs brownish and finely longitudinally fissured. **Leaf scar** dark, 3-traced and membranous stipules, fused on the other side of the twig; these leave a ring-shaped scar. Open shrub to 1.5 m high. Origin: Mongolia.

Calophaca wolgarica FISCH.

The spirally arranged leaves leave behind the leaf base with the fused, wing-like enlarged stipules which half enclose the twig. They partly hide the lateral buds and merge towards the tip into the scales of the c. 6 mm long terminal bud. **Bud scales** matte red-brown to dark grey-brown, sparsely hairy, those of the lateral buds acute and with a keel. **Twigs** grey to ochre-brown, finely hairy towards the tip with glandular hairs and slightly striped longitudinally, without visible lenticels. Pith solid. **Leaf scar** with one trace. At the base of the twig remnants of the infructescence. Very rarely planted, shrub to 1 m high. Origin: south-east Europe.

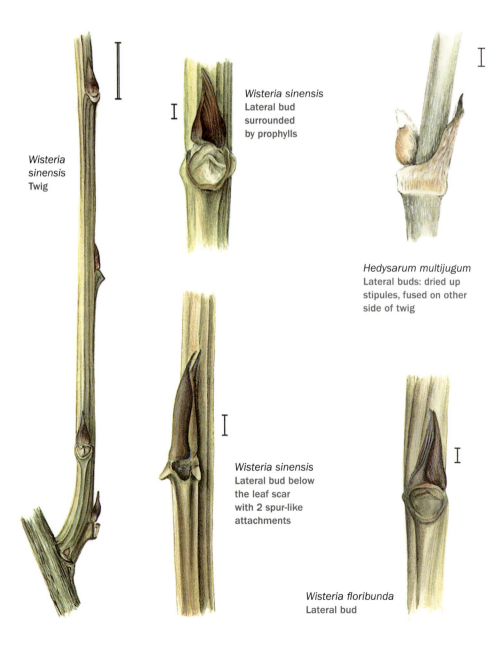

Wisteria sinensis
Twig

Wisteria sinensis
Lateral bud
surrounded
by prophylls

Hedysarum multijugum
Lateral buds: dried up
stipules, fused on other
side of twig

Wisteria sinensis
Lateral bud below
the leaf scar
with 2 spur-like
attachments

Wisteria floribunda
Lateral bud

Hedysarum multijugum
Lateral buds: dried up
stipules, fused on other
side of twig

*Caragana
arborescens*
Tip of twig

Caragana arborescens
Twig

Caragana arborescens
Tip of twig

Caragana spinosa
Lateral bud in the axil
of a leaf turned spiny;
the bud leaves also have
a small spiny rachis

Caragana spinosa
Twig with long leaf
rachis spines

Caragana FABR., peashrub, pea tree

Small to large shrubs, more rarely smaller trees. Terminal and lateral buds present. The leaf base and bud leaves (scales) persist with the stipules, so too in some species the rachides. They dry out and in some species turn spiny. The branching system has both long- and short-shoots. **Fruit** thin, cylindrical, to 5 cm long legume which opens early; each half twists, on opening, like a corkscrew. Distributed from Asia to eastern Europe.

Key to *Caragana* and *Halimodendron*

1 Plants ± strongly spiny 3
1* Hardly spiny. 2
2 Lateral buds 5–7 mm long . . . .
. . . . ***Caragana arborescens***
2* Lateral buds 3–4 mm long . . . .
. . . . *Caragana frutex*
3 Spiny leaf rachis less than 1 cm long . . .
. . . . *Caragana pygmaea*
3* Spiny leaf rachis 3–4 cm long. 4
4 Stipules basally flattened and parchment-like . . . . *Caragana spinosa*
4* Stipules thin and entirely turned spiny. .
. . . . *Halimodendron halodendron*

Caragana arborescens LAM., Siberian peashrub

Buds compact to narrowly ovoid. Terminal buds 7–10 mm long, lateral buds slightly smaller. Bud scales ochre-brown and red-brown to grey-brown, basally and towards the tip occasionally hairy, with membranous stipule lobes. **Twigs** below the terminal buds covered by a grey-brown epidermis, lower on the branch fissuring longitudinally and with 5 narrow angles, in-between glossy green. **Lenticels** few, ochre-brown, roundish, often wider than high. **Leaf scar** with one trace. Frequently planted, erect shrub to 6 m high. Origin: Siberia and Manchuria.

Caragana spinosa (L.) VAHL ex HORNEM.

Buds ovoid, 4–5 m long, grey-brown to pale ochre-brown. The bud leaves with acute-spreading stipules turning spiny and with up to 8 mm long leaf rachides. **Leaves** with 3–4 cm long spiny, persistent leaf rachis and membranous stipules with a spiny tip. **Twigs** slightly 5-angular, matte grey-brown to red-brown, with longitudinally fissuring bark, below green and smooth. Rarely planted, open shrub 0.7–1.5 m high. Origin: Siberia.

Caragana frutex (L.) K. KOCH, Russian peashrub

Buds 3–4 mm long, ovoid, ochre-brown to red-brown, towards the margin slightly membranous and grey. **Twigs** unarmed, at one year old slender, matte, pale grey-brown with ochre-brown longitudinal stripes; when older, darker, glossy olive-brown with slightly prominent longitudinal stripes that peter out, and few roundish to obliquely oblong lenticels. **Leaf scars** with one trace. Rarely planted, 1–2 m high shrub producing runners. Origin: south-east Europe.

Caragana pygmaea (L.) DC.

Buds 2–3 mm long, almost completely covered by persistent leaf base with its stipules. **Twigs** slender, at one year old pale grey with relatively broad, violet-brown ridges. **Leaf rachides** to 1 cm long, becoming spiny. Rarely planted shrub to 1 m high, often prostrate. Origin: Siberia and north-west China.

Halimodendron halodendron (PALL.) DRUCE, salt tree

Buds relatively small, ovoid, to 3 mm long, grey. **Twigs** initially pale grey to ochre-brown and finely hairy, the hairs falling off. **Bark** soon fissuring: green between narrow strips of the grey-brown first layer of bark. Of the **leaves** the spiny, 2–4 cm long leaf rachises persist, as well as the spiny stipules. **Fruit** a legume, to 3 cm long, looking slightly inflated with 4–5 mm long kidney-shaped seeds. Shrub to 2 m high. Origin: eastern Europe to western Asia. The single species of this [monotypic] genus is very similar to the *Caragana* species.

TRIBE Galegeae

Colutea arborescens L., common bladder senna

Buds 3–4 mm long, often developed only at the base of the shoot. Bud scales ± distinctly differentiated into leaf base and stipules, ochre-brown and green at the base initially, at the tip and margin densely white-hairy. **Twigs** finely appressed-hairy, grey-green to

Caragana pygmaea
Lateral bud concealed by leaf base

Caragana frutex
Lateral bud

Caragana frutex
Tip of twig

Caragana pygmaea
Twig

Caragana pygmaea
Short leaf rachis turned into spine

Halimodendron halodendron
Part of twig: spiny leaf rachides and stipules

Colutea arborescens
Fruit: bladder-like
enlarged 6–8 cm
long legume

Colutea arborescens
Lateral bud

Colutea arborescens
Lateral bud at the base
of twig

Lespedeza bicolor
Lateral buds
developed from
the prophyl-axillas of
the fallen lateral twig

grey-brown, the top layers of bark fissuring longitudinally. **Leaf scars** indistinct with (1 or) 3 traces, on leaf cushions, flanked by the dry, persistent stipules. In the axils of the upper leaves there are bladder-like, 6–8 cm long, grey-brown, dry legumes which mostly last throughout the winter; these make the shrub instantly recognisable. Very frequently planted shrub, 2–4 m high. Origin: southern central Europe and southern Europe to Africa.

TRIBE Desmodieae

Lespedeza bicolor TURCZ.,
shrubby bushclover

Buds in pairs (from prophyll axils), placed over the right and left side of the leaf scar, often with in-between a mostly dead shoot remnant; small, 2.5–3 mm long, ovoid, with many brown, finely ciliate bud scales. **Twigs** initially angular, soon ± roundish, glabrous, only near the lateral buds slightly white-hairy, with pale ochre grey-brown, fissuring bark, below the fissuring layer red-brown. In winter often considerable die-back. **Infructescence** in axils, 3–6 cm long. **Leaf scar** roundish with an indistinct trace and persistent long stipules. A frequently planted erect shrub to 1.5 m high. Origin: eastern Asia.

Also frequently planted is **Lespedeza thunbergii** (DC.) NAKAI, the Thunberg lespedeza, a shrub to 2 m high with branches hanging down. The inflorescences, 8–20 cm long axillary racemes, are united in 60–80 cm long panicles.

TRIBE Trifolieae

Ononis fruticosa L., shrubby restharrow

No terminal buds, **lateral buds** invisible, concealed below the leaf remains. **Twigs** grey, appearing dead, with spirally inserted remains of dry leaves. A remnant of the lower leaf persists, clasping and almost surrounding the twig, with winged stipules and with leaflet scars. At the base of the twig with tapering short-shoots. A rare sub-shrub to 1 m high. Origin: south-west Europe to Algeria.

The native restharrow/spiny restharrow, **Ononis spinosa** L., has distinctive thorns.

ORDER Fagales

The families of the Fagales differ in the structure of their buds. The stipules frequently are part of the bud protection (Nothofagaceae, Fagaceae p.p., Betulaceae), in the Juglandaceae the stipules are completely missing and here, as well as in the Myricaceae and the Fagaceae genus *Castanea*, the leaves or the leaf bases take on the role of bud protection.

Nothofagaceae
roble family

Nothofagus BLUME, southern beech

The genus occupies, with mainly evergreen species, the temperate and subtropical zones of the Southern Hemisphere: in South America and Australia to New Zealand and Indonesia. Many small lenticels are typical, and for the deciduous species, so are the dry stipules which persist long after leaf fall. Usually only lateral buds are formed. The distichous lateral buds are protected by pairs of stipules, also distichous (in 2 rows) so that the bud scales align in 4 rows. As these scales are not completely closed they are often sticky with strong bud glue. Sometimes we find amid the first pair of stipules a further scale (the leaf base). On the side away from the twig (abaxial), the buds are often slightly hollow.

Key to *Nothofagus*

1 Twigs hairy. 2
1* Twigs glabrous, buds 3–5 mm long . . . .
. . . . *Nothofagus obliqua*
2 Buds small, hardly exceeding 2 mm in
length. . . . . *Nothofagus antarctica*
2* Buds 8–15 mm long . *Nothofagus alpina*

Nothofagus antarctica (G. FORST.) OERST., Antarctic beech

Buds alternate, 1.5–2 mm long, broadly ovoid, red to orange-brown. **Twigs** grey-brown, very slender, less than 1 mm thick, with short internodes to 10 mm long, fine- and densely hairy. Later smooth, with distinct pale brown lenticels which extend, on older twigs, to oblique stripes (as in cherry or birch). **Leaf scars** very small, black-grey, without recognisable traces of vascular bundles; laterally with relatively large, dark brown dry stipules. Tree to 6 m high, but larger where native. Origin: Chile and Patagonia, to the tree line. The most frequently planted species of the genus.

The following two species are rarely encountered:

Nothofagus alpina (POEPP. & ENDL.) OERST., rauli [*Nothofagus procera* (POEPP. & ENDL.) OERST.], **lateral buds** 8–15 mm long, fusiform, ± spreading from the twig, with sticky, greenish-brown bud scales. Scales glabrous, only slightly ciliate. **Twigs** dark brown to olive-green, finely spreading-hairy, with many small pale lenticels. Tree to 20 m high. Origin: Chile.

Nothofagus obliqua (MIRB.) OERST., **buds** appressed to twig, 3–5 mm long, with ochre-brown, white-ciliate bud scales. **Twigs** glabrous, olive-brown with many lenticels. Tree to 30 m high. Origin: southern Argentina to Chile.

Nothofagus alpina
Lateral bud

Nothofagus antarctica
Lateral bud, flanked by
the persistent stipules
of the deciduous leaf

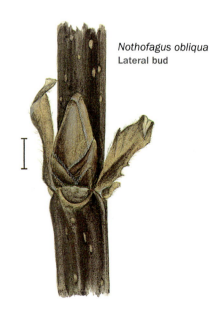

Nothofagus obliqua
Lateral bud

Fagaceae
beech family

Almost globally distributed, deciduous and evergreen woody plants. Mostly trees, more rarely shrubs, c. 900 species in 10 genera, of these five with only one or two species. The species of this family have in common the shape of infructescence: the fruits are solitary or a few united in a fruit cup, the cupule. The buds are alternate. In the beech (*Fagus*) and the oak (*Quercus*) the buds have many scales, which come from converted stipules. There are true terminal buds and the lateral buds are homologous to the terminal bud. The lateral buds of the chestnut (*Castanea*) are enclosed almost exclusively by 2 simple bud scales, formed by prophylls. The tip of the shoot dies back and there are no terminal buds.

Key to Fagaceae

1 Buds long-fusiform, bud scales in 4 rows
. . . . *Fagus*

1* Buds ovoid, bud scales in 2 or 5 rows . 2

2 With terminal buds, buds with many bud scales . . . .*Quercus*

2* Without terminal buds, buds with only few bud scales . . . .*Castanea*

Fagus L., beech

In winter beeches can be recognised with confidence by the smooth, grey bark and the long, fusiform buds. The buds are protected by many stipular scales: the paired stipules follow the distichous leaves (in two rows) and form four vertical rows. **Fruits** 3-angular nuts, in pairs (rarely three) in an initially closed fruit cupule, which later opens by four valves.

The species are difficult to distinguish from one another. However, species other than the native beech *Fagus sylvatica* are rarely planted. Characters for distinguishing the very similar species are the bud shape and colour, as well as the often-found fruit cupules. A great help too are the dry leaves which often persist on the twig for a long time.

Key to *Fagus*

1 Buds widest in the middle and narrower at the base 2

1* Buds very wide at the base . . . .
. . . . *Fagus grandifolia*

2. Buds 15–20 mm long, fruit cupules bristly . . . .*Fagus sylvatica*

2* Buds less than 15 mm long, fruit cupules at base with spatulate bracts . . . .
. . . .*Fagus crenata*

Fagus sylvatica L., common beech

Buds fusiform: long and slender, acute, thickest in the middle. **Bud scales** orange-brown to red-brown, the tip often darker and silver-white hairy. **Twigs** slender, slightly zig-zag, with grey-brown few small pale oblong lenticels and with a few remaining hairs. **Leaves** to 10 cm long, margin entire, on each side with up to ten lateral veins. **Fruit cupule** with ± straight spreading or slightly curved bristles. Forest and park tree to 30 m high. Distributed from central Europe to the Caucasus. Some cultivars differing in growth form are: **'Dawyck'** — slim and columnar; **'Pendula'** — with curved down twigs; and **'Tortuosa'** with zig-zag twigs. The subspecies distributed from the Caucasus to Asia Minor and northern Persia is hardly distinct: *Fagus sylvatica* subsp. *orientalis* (LIPSKY) GREUTER & BURDET, Oriental beech [*Fagus orientalis* LIPSKY]. It has probably slightly more matte, more ochre-brown buds. Leaves often longer than 10 cm with more than 10 lateral pairs of nerves. A tree to 40 m high with conical, ascending crown.

Fagus sylvatica
Twig

Fagus sylvatica
Fruit cupule and nut

Fagus sylvatica
subsp. *orientalis*
Tip of twig

Fagus grandifolia EHRH., American beech
[*Fagus americana* SWEET, *Fagus ferruginea* AITON]

Buds long, conical, broad-based on the twig and hardly thickening from the base to the middle. **Bud scales** dark red-brown mainly the upper ones silvery hairy at the tips. Twigs when young strongly violet-brown to red-brown with pale lenticels. **Leaves** with distinctly toothed margin. **Fruit cupules** with thin, straight or curved bristles. Rare, tree 20–30 m high. Origin: eastern U.S.A.

Fagus crenata BLUME, Japanese beech

Buds fusiform, in the middle distinctly thicker than at the base. **Bud scales** ochre-brown and slightly darker towards the margin, at the tips with small pale-hairy spots. **Twigs** slender, grey-brown with many paler lenticels; older twigs darker grey-brown. **Fruit cupule** at base with narrow, fusiform bracts, which change towards the tip into long bristles. Tree to 30 m high, very uncommon. Origin: Japan.

Castanea sativa MILL., sweet chestnut

Buds distichous, ± obliquely above the leaf scar, rarely (only on vertical twigs) also with spiral lateral buds, placed straight above the leaf scar and spreading from the twig. Buds to c. 6 mm long, ovoid to conical; surrounded for $^2/_3$ by two basally greenish to reddish prophylls with darker margins. The next scales are stipules of later developing foliage leaves. **Twigs** sturdy, olive-green to brown, glossy, occasionally covered in grey by the dying epidermis, with distinct, often slightly paler ochre-brown ridges and many pale, dot-shaped lenticels. **Leaf scars** dark with several indistinct traces. **Trunk** often twisted with deep, long-fissured bark. Tree to 30 m high, broad. Origin: frequent in southern central Europe to the Caucasus and Asia Minor. Long naturalised in central Europe.

Quercus L., oak

Over 400 deciduous and evergreen tree species, from the temperate to subtropical latitudes of the Northern Hemisphere, more rarely shrubs. Buds with stipular scales: the paired stipules of the leaves in five rows overlap laterally so that they

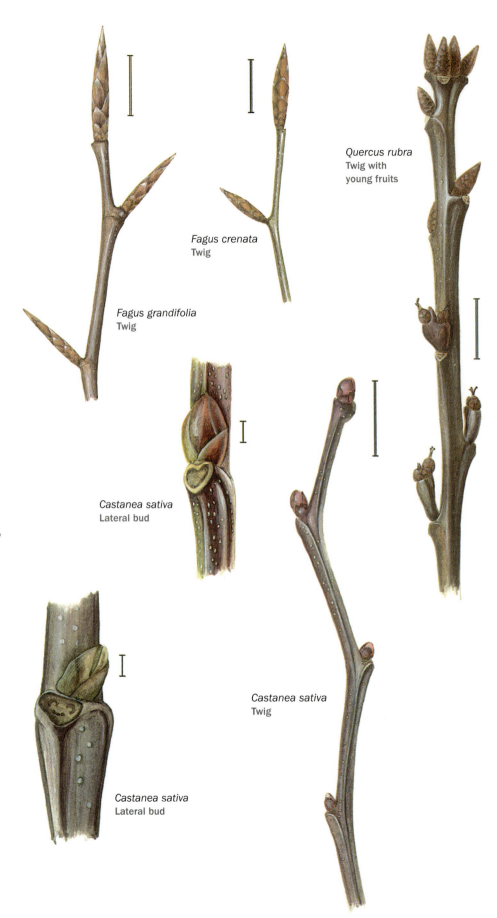

Quercus rubra
Twig with young fruits

Fagus crenata
Twig

Fagus grandifolia
Twig

Castanea sativa
Lateral bud

Castanea sativa
Twig

Castanea sativa
Lateral bud

Quercus rubra
Lateral bud

Quercus palustris
Tip of twig

Quercus imbricaria
Tip of twig

Quercus coccinea
Tip of twig

form 5 rows as well (in the gaps between the leaf insertions); the number of verticle stipule rows is not doubled, as in *Fagus*. Lateral buds additionally with simple prophyll scales. **Fruit**: a single nut in each fruit cupule. Important characters for recognising oak species are bud shape and size, the indument, the bark and often, for unequivocal attribution, the stage of ripening and the shape of fruits and their cupule.

Key to the groups

1 Fruits ripening in the second year, young fruit with short, thick stalk, buds with long thread-like stipules or bark remaining smooth for a long time, trunk bark not scaly 2
1* Fruits ripening in the first year, soon falling. Bark with deep furrows or scaly . . . .**Section Quercus, white oaks**
2 Bark smooth when young, cupule with blunt, appressed scales . . . .**Section Lobatae, red oaks**
2* Bark thick and furrowed, cupule with fimbriate-oblong and ± spreading scales . . . .**Section Cerris**

Section Lobatae, red oaks

Tall trees with smooth bark which turns rough only with age. **Fruits** ripening in the second year, with felty endocarp. Cupule with imbricate thin, blunt and appressed scales. Origin: North America.

Key to red oaks

1 Buds ± glabrous, red-brown 2
1* Bud scales with persistent hairs, ochre to brown . . . . ***Quercus coccinea***
2 Buds red-brown, relatively large: 5–7 mm long . . . . ***Quercus rubra*** and ***Quercus imbricaria***
2* Buds brown and smaller: 3–4 mm long . . . . ***Quercus palustris***

Quercus rubra L., red oak

Buds narrowly ovoid, slightly acute, 5–7 mm long, 3–4 mm thick, with many red-brown ciliate, but otherwise (apart from right at the tip of the bud) glabrous, bud scales. At the tip of the twig usually with groups of many buds; hardly any size difference between lateral and terminal buds. Lateral buds strongly spreading from the twig. **Twigs** glabrous, at one year old olive-green to dark brown, with many small, dot-like lenticels; at two years old twigs grey. **Bark** remaining smooth for a long time and only much later roughening. Ripening of fruit after two years, therefore on first year's growth frequently with young acorns: mostly two together on short thick stalks up to 10 mm long, globose, c. 3 mm thick, with remnants of style. Tree to 25 m high with broad crown. The most frequently planted species of this subgenus.

Very similar is ***Quercus imbricaria*** Michx., shingle oak. Also with almost glabrous, glossy bud scales and glabrous red-brown to violet-brown glossy twigs that are partly grey, due to dead epidermis. If it is not possible to distinguish this species with certainty from *Quercus rubra* by its winter characters, it is useful to look for the often persistent dry leaves, typically elliptical with entire margin. Old **bark** irregular flat and fissured. Tree to 25 m high.

Quercus palustris Münchh., pin oak

Buds ovoid, relatively small, mostly only 3(–4) mm long and hardly 2 mm thick. Bud scales dark red-brown, entirely glabrous or slightly ciliate. **Twigs** relatively slender, glabrous, towards the tip ± angular, glossy olive-brown to red-brown, with many very small, pale lenticels. **Stem** straight with pale longitudinally striped bark which stays smooth for a long time.

Quercus coccinea Münchh., scarlet oak

Buds 3–6 mm long, ovoid, with ochre-brown, slightly darker margin on bud-scales: the upper scale with closely appressed indument, the lower one becoming glabrous. **Twigs** initially hairy, but soon glabrous, orange to red-brown, may be olive green in shadow. **Stem** not straight, with finely fissured, dark grey bark. Tree to 25 m high.

Section Cerris

Bark mostly thick and furrowed, in some species corky. Cupule scales oblong, at least the upper ones spreading and reflexed. The fruits ripen in the second year and in winter of the first year young fruits become visible.

Key to Section Cerris

1 Terminal buds more than 7 mm long, bud scales oblong . . . . *Quercus castaneifolia*

1* Buds smaller, bud scales triangular to rounded . . . . *Quercus cerris*

Quercus cerris L., Turkey oak

Buds ovoid, 4–6 mm long, lowest bud scales grey-brown, densely fine-hairy, with long thread-like tip; subsequent scales red-brown, less hairy and finely ciliate. **Twigs** 2–4 mm thick, olive-green to brown, well appressed hairy, when older grey-brown to dark grey; with faint small lenticels. Twigs with deeply furrowed longitudinally fissured, dark grey bark. **Fruits** sessile, 2.5–3 cm long, half-covered by the cupule with oblong scales. Frequently planted tree, to 30 m high. Origin: from Asia Minor to southern-central and southern Europe.

Very difficult to distinguish in winter from *Quercus cerris* is **Quercus libani** OLIVIER, the Lebanon oak. **Buds** ovoid, 3–5 mm long, red-brown, relatively densely fine-hairy, basally with long thread-like scales. **Twigs** grey-brown to olive-brown, finely hairy. **Bark**, smooth for a long time, later flat, furrowed. Small tree to 10 m high. Origin: from Syria to Asia Minor.

Quercus castaneifolia C. A. MEY., chestnut-leaved oak

Buds narrowly ovoid, acute, to c. 10 mm long. **Bud scales** long, the outer ones thread-like and spreading but only a little longer than the bud, orange-ochre and red-brown to grey-brown, finely appressed-hairy particularly towards the tip. **Twigs** grey-brown to olive-green, at the tip sparsely finely hairy with pale, warty lenticels. Occasionally planted, tree to 25 m high, with a broad crown. Origin: from the Caucasus to Iran.

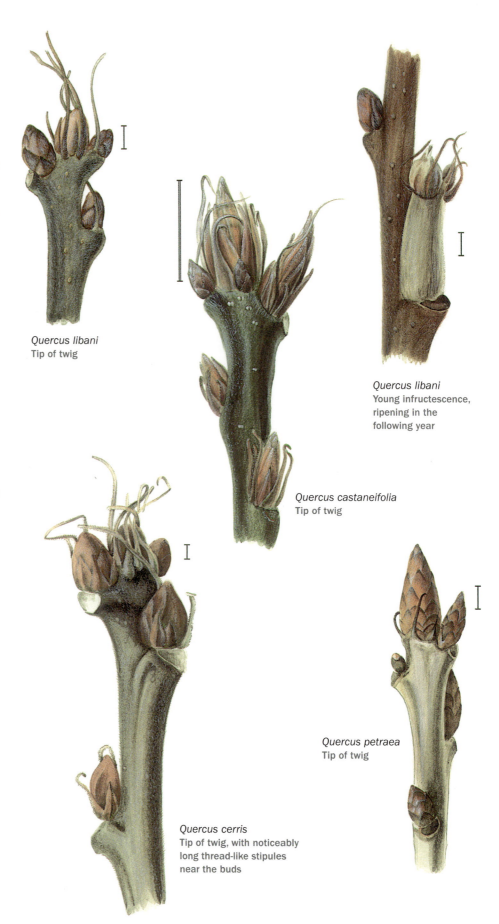

Quercus libani
Tip of twig

Quercus libani
Young infructescence,
ripening in the
following year

Quercus castaneifolia
Tip of twig

Quercus cerris
Tip of twig, with noticeably
long thread-like stipules
near the buds

Quercus petraea
Tip of twig

Quercus robur
Tip of twig

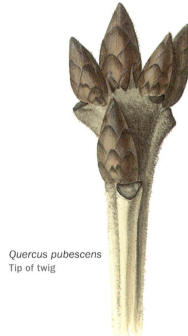

Quercus pubescens
Tip of twig

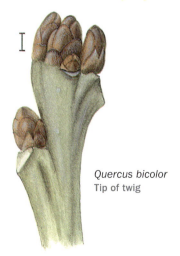

Quercus bicolor
Tip of twig

Section Quercus, white oaks

Bark (deeply) furrowed or flaking in scales. Scales of cupules thick and often dorsally convex or (in the species sometimes separated as section *Mesobalanus*, see below) cupule scales oblong and spreading.

Key to white oaks

1 Only 2–3 bud scales above one another 2

1* At least 3–4 bud scales above one another, bark mostly deeply fissured or with longitudinal furrows 4

2 Terminal buds more than 8 mm long, bark mostly deeply fissured or with longitudinal furrows . . *Quercus dentata*

2* Terminal buds shorter, bark peeling off in scales 3

3 Buds hairy in patches. . *Quercus bicolor*

3* Bud scales at most ciliate, twigs glabrous . . . . *Quercus alba*

4 Terminal buds more than 13 mm long. 11

4* Terminal buds to 12 mm long 5

5 Buds blunt, rounded, stipules not thread-like 6

5* Buds acute with thread-like stipules . . . . *Quercus macrocarpa*

6 Indument on bud scales not continuous 7

6* At least uppermost bud scales densely hairy. 9

7 Twigs grey-green with some oblong lenticels 8

7* Twigs brown with many round lenticels . . . . *Quercus montana*

8 Buds globose ovoid, transverse plane almost round . . . . *Quercus robur*

8* Buds conical-ovoid, transverse plane slightly 5-angular . . . . *Quercus petraea*

9 Twigs grey-brown, lenticels few . . . . *Quercus pubescens*

9* Twigs greenish, lenticels dense 10

10 Buds globose-ovoid, indument grey-brown. . . . . *Quercus frainetto*

10* Buds ovoid, indument pale ochre-brown . . . . *Quercus pyrenaica*

11 Bud scales ochre-green with dark margins, shrubs . . . . . . *Quercus pontica*

11* Bud scales red-brown, trees . . . . *Quercus macranthera*

The following 3 European species are closely related and connected by transitional forms / hybrids. **Q. petraea**, sessile oak; **Q. robur**, common oak and **Q. pubescens**, downy oak.

Quercus petraea (MATT.) LIEBL., sessile oak

Buds very variable in size: 3–10 mm long, pyramidal-ovoid, slightly 5-angular, more than 1.5 times as long as wide; with many ochre-brown bud scales with dark margins, and ciliate to hairy. **Twigs** glabrous, grey-brown to ochre-brown, slightly angular or slightly furrowed. Cupule: scales not fused, almost sessile, without peduncle. **Bark** dark, thick and furrowed, quite variable. Frequent tree 20–30 m high, with mostly straight stem. Origin: Europe to Asia Minor.

Quercus robur L., common oak

Buds mostly more compact than in the sessile oak, short, globose-ovoid, hardly angular to c. 1.5 times as long as wide. **Twigs** glabrous. **Cupules** with fused scales, on ± long peduncle. **Bark** dark and thick, initially flat, later deeply furrowed. Frequent tree, 30–40 m high, mostly with a broad crown, formed of sturdy branches, standing free, stem not straight. Origin: Europe to the Caucasus.

Quercus pubescens WILLD., downy oak

Buds ovoid, similar to the sessile oak (*Q. petraea*), 4–8 mm long, hairy on the bud scales, particularly towards the tip. **Twigs** grey-brown, slightly 5-angular, indument grey-brown, scattered with small, roundish lenticels. Large, densely branched shrub or tree to 20 m high. Distributed in west, central and southern Europe to the Caucasus

Quercus bicolor WILLD., swamp white oak

Buds short ovoid to almost globose, c. 4 mm long. Terminal buds hardly larger than lateral buds. Bud scales ochre-brown to red-brown, with dark margins and moderately hairy. **Twigs** grey-brown to olive-brown, with roundish to oblong pale lenticels. **Branches** (3–7 years old) often with bark which comes off in broad thin horizontally peeling strips. **Trunk** with scaly peeling bark, later with evenly deep longitudinally fissured bark. More frequently planted than *Q. alba*, tree to 20 m high. Origin: eastern North America.

Quercus alba L., white oak

Buds short, ovoid to almost globose, glabrous or bud scales slightly ciliate. **Twigs** glabrous, grey, with dried-up old leaves. **Bark** characteristic: grey, irregularly narrowly and longitudinally furrowed, peeling in scales, on older stems pieces of bark with the lower end bent outwards. Rarely encountered, medium-sized trees with a broad crown. Origin: eastern North America.

Quercus macrocarpa MICHX., bur (or burr) oak

Buds compact-ovoid, acute, appressed to twig, to 8 mm long and 5 mm broad, with grey-brown, appressed white hairy bud scales. Terminal bud at the base with some thread-like spreading stipules. **Twigs** 4–5 mm thick, olive green to pale grey-brown, ± densely hairy particularly towards the tip, partly with stellate hairs. Lenticels few, small, ochre-brown. **Trunk** with grey, longitudinally fissured bark, coming off as scales. Frequently planted, broad tree to 25 m high. Origin: North America.

Quercus montana WILLD., chestnut oak
[*Quercus prinus* L. nom. rej.]

Buds blunt-ovoid to 4–5 mm long, lateral buds slightly smaller than terminal buds; bud scales matte, ochre-brown to grey-brown, occasionally with a slightly hairy margin towards the tip of the bud. **Twigs** initially hairy, becoming glabrous, grey to dark brown, slightly glossy, with many warty pale lenticels. **Bark** with deep longitudinal furrows which are interrupted by fine cross-furrows. Tree to 20 m high. Origin: North America.

[Section Mesobalanus]

Bark deeply furrowed, scales of cupules oblong and spreading. Origin: Old World.

Quercus frainetto TEN., Hungarian oak

Buds blunt ovoid. Terminal buds 6–8 mm long and 3.5–5 mm thick, lateral buds distinctly smaller. **Bud scales** grey-brown long and densely hairy particularly towards the tip; basally with a few long thread-like bud scales, but not longer than the bud. **Twigs** olive-green to grey-brown, alternate, long-hairy with many inconspicuous lenticels. Tree to 30 m high, with a broad crown. Occasionally planted outside its native area. Origin: from south-eastern central Europe to southern Italy.

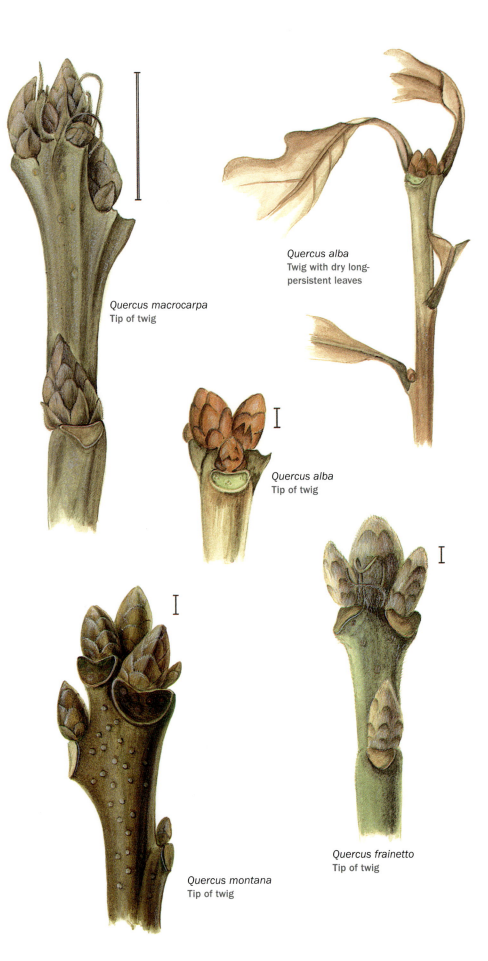

Quercus macrocarpa
Tip of twig

Quercus alba
Twig with dry long-persistent leaves

Quercus alba
Tip of twig

Quercus montana
Tip of twig

Quercus frainetto
Tip of twig

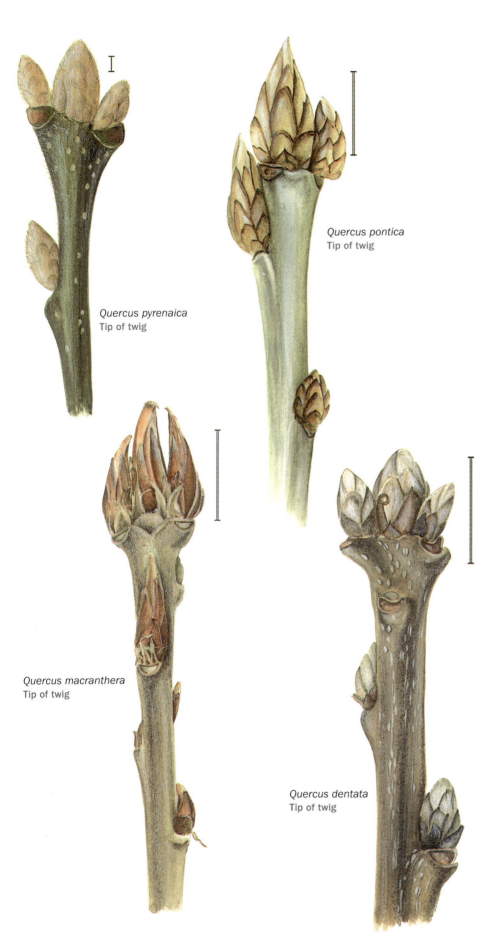

Quercus pyrenaica
Tip of twig

Quercus pontica
Tip of twig

Quercus macranthera
Tip of twig

Quercus dentata
Tip of twig

Quercus pyrenaica WILLD., Pyrenean oak

Buds ovoid, 4–7 mm long. Bud scales densely hairy, pale ochre-yellow to brownish, single-coloured or with distinctly darker margins. **Twigs** olive-green to grey-brown, with many lenticels, towards the tip with ± strong remains of felty indument. Tree to 15 m high. Origin: south-western Europe.

Quercus pontica K. KOCH, Armenian oak

Buds massed at the tip of the twig, short-ovoid, acute to c. 15 mm long, rounded 5-angular and slightly 5-furrowed. Lateral buds only a little smaller than terminal buds. **Bud scales** many, greenish to pale ochre with a broad brown margin and very finely appressed, hairy indument. **Twigs** thick, stiff and angular, olive-brown to grey-brown, mostly glabrous, with fine, pale ochre-brown lenticels; when older, smooth glossy silvery grey-brown, lenticels warty, hardly differing in colour from the twig. Shrub or small tree to 6 m high, with lax branching; rarely cultivated. Origin: Caucasus.

Quercus macranthera FISCH. & C. A. MEY., Caucasian oak

Buds oblong, blunt-conical, to 15(–20) mm long. **Terminal bud** a little longer than lateral buds, mostly flanked by one to two large lateral buds. Lateral buds appressed to twig. **Bud scales** many, lowest ones small and narrowly awl-shaped, ochre-brown hairy, the following bud scales almost glabrous, only the margins silvery-grey hairy and the uppermost scales densely ochre-orange-brown hairy. **Twigs** thick, slightly 5-angular, dark brown, towards the tip finely spreading hairy, below becoming glabrous and slightly glossy. Lenticels faint. **Leaf scars** dark grey-brown with several indistinct traces. **Bark**: thin and coming off in large plates. Unmistakable tree 20 m high, occasionally planted. Origin: northern Iran to the Caucasus.

Quercus dentata THUNB., Daimyo oak

Buds globose-ovoid, terminal buds c. 8–12 mm, lateral buds 6–8 mm long. Lateral buds slightly flattened, with spreading tips. Bud scales: lowest ones slightly brownish to grey, velvety dirty white hairy, the upper ones more densely so, and due to this, tip of bud paler. **Twigs** velvety hairy, thick and stiff, pale ochre to grey ochre, with many pale lenticels. **Leaf scars** distinctly red-brown, on relatively large leaf cushions. **Bark** black-grey, deeply furrowed. Seldom planted, tree to 25 m high, with open broad crown. Origin: eastern Asia.

Juglandaceae
walnut family

Mostly trees, more rarely shrubs with sturdy twigs. **Buds**: within the larger genera *Juglans*, *Carya* and *Pterocarya* some species lack bud scales, while others have bud scales (this, as well as the pinnate leaves and the distribution of taxa, are indicators of the tropical derivation of the family); always with terminal buds (except when the tip of the shoot is used up by the inflorescence = large scar), lateral buds alternate, often with descending accessory buds. The leaf scars are typical: large, round-triangular and notched on lateral margins, the top margin ± notched, shield-shaped. Traces of vascular bundles in three semi-circular to circular groups, united or clearly separate.

Key to Juglandaceae

1 Pith solid 3
1* Pith chambered 2
2 Terminal buds more than 15 mm long . . . .***Pterocarya** & Cyclocarya*
2* Terminal bud to 15 mm long . . . ***Juglans***
3 Lateral buds spreading from twig at a wide angle, both prophyll scales usually at least basally fused . . . . **Carya**
3 Lateral buds not strongly spreading from twig, first scales not fused. Nuts very small, in conical infructescences. . . . . *Platycarya*

Juglans L., walnut

Deciduous trees, often with branching main axis, seldom shrubby. Protection of the bud is quite different in the various sections: in the common walnut, *Juglans regia*, bud scales cover the buds, while buds of species from other sections have no bud scales and the outer bud leaves are distinctly pinnate. Here the shape of the outer leaves of the terminal buds differs distinctly: the bud leaves of the butternut *Juglans cinerea* and of the Asiatic species have widened wing-like margins, by which they surround the bud, while the bud leaves of the black walnut group (Section *Rhysocaryon*, from the Americas) are rounded in transverse section. Male flower catkins overwinter without bud scales: due to the densely arranged spiral bracts, they look like small cones. **Pith** dark, transversely chambered. Over 20 species occur from Southern Europe to Eastern Asia, in North and South America as well as in the West Indies. The wind-pollinated species of walnut hybridise easily with each other. Species which are rarely cultivated are often pollinated by the ubiquitous *Juglans regia*. As the latter differs by its bud scales from more rarely cultivated species, the hybrids, which have an intermediate position as to bud protection, can be recognised even in winter, although attribution of the parent species is not always possible. Fallen fruits (stones) found under the tree can be a help in winter.

Key to Juglans according to winter twigs

1 Buds without bud scales, ± densely hairy 2
1* Buds with bud scales, the outermost almost glabrous . . . . ***Juglans regia***
2 Above the leaf scar with finely hairy narrow bulge, bud leaves with wing-like margin 3
2* Without hairy bulge above leaf scar, bud leaves with round margins in transverse section 4
3 Terminal bud more than 8 mm long. . . . . ***Juglans nigra***
3* Terminal bud to 6 mm long . . . . *Juglans microcarpa*
4 Terminal buds lengthened, hardly acute. . . . .*Juglans cinerea*
4* Terminal buds short, globose-acute . . . . *Juglans mandshurica* and *J. ailantifolia*

Key to Juglans according to stone ("nut")

1 Stone with thin walls and relatively smooth, easy to crack *2*
1* Stone hard, thick walls or rough with sharp edges 3
2 Stone short-acute, more than 2 cm thick . . . . ***Juglans regia***
2* Stone long-acute, mostly less than 2 cm thick . . . . ***Juglans ailantifolia***
3 Stone with bulging edge. . . . . ***Juglans regia* — hybrids**
3* Stone with only thin edge 4

Juglans regia
Twig

Juglans regia
Lateral bud with accessory bud

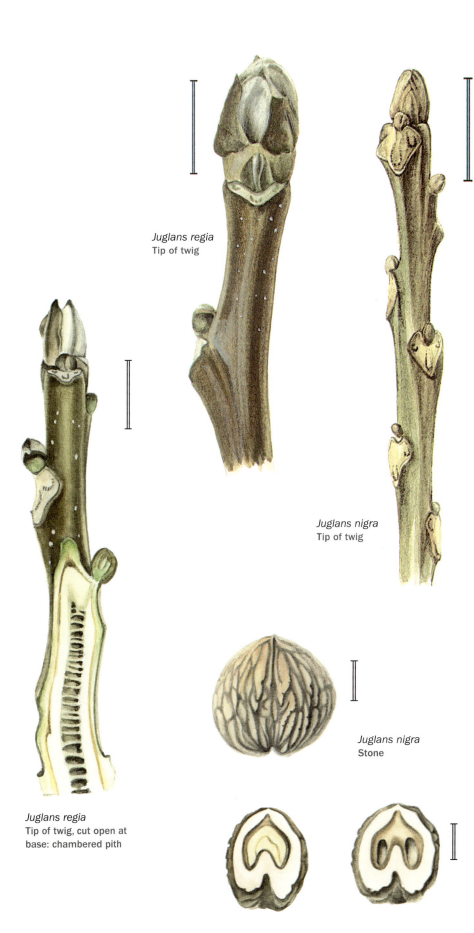

Juglans regia
Tip of twig

Juglans nigra
Tip of twig

Juglans nigra
Stone

Juglans regia
Tip of twig, cut open at
base: chambered pith

4 Stone globose or broader than high, with
   very thick walls 5
4* Stone oblong 6
5 Diameter of stone up to 1.5 cm . . . .
   . . . . *Juglans microcarpa*
5* Diameter of stone more than 2.5 cm . . .
   . . . . *Juglans nigra*
6 Stone sharp-angular . . . *Juglans cinerea*
6* Stone smooth . . . .*Juglans mandshurica*

Section Juglans

Juglans regia L., common walnut

Buds 5–8 mm long, with broad base, half-
globose to conical-ovoid, blunt to slightly
acute, particularly in larger buds. Lateral
buds on sturdy twigs towards the tip,
particularly when the terminal tip is absent
due to flower and fruit, with a clear stalk
and with descending accessory buds. **Bud
scales** thin, scaly and rounded at the tip,
grey-brown, finely hairy towards the tip.
The outer scales are differentiated from the
next ones: almost glabrous, initially greenish
with brown margin, later often brown and
deciduous. **Twigs** thick, slightly angular,
glabrous, olive-green and olive-brown
to red-brown with small, pale long and
thin lenticels. **Leaf scars** large, on robust
nodes as broad as the twig, shield-shaped
with 3 semi-circular groups of traces. **Bark**
pale grey to black-brown, long remaining
smooth, later fissured. Tree to 30 m high,
broad. Origin: eastern Europe to northern
Asia, in the rest of Europe naturalised and
planted since ancient times.

Section Rhysocaryon

Stones almost globose, with very thick walls.

Juglans nigra L., black walnut

Buds without bud scales, globose to ovoid.
Terminal buds 5–6 mm long, lateral buds
mostly smaller, pale white-grey hairy, at the
base merging with the colour of the twig.
Twigs thick: on first year's growth olive-
green to reddish-brown, c. 5 mm diameter,
with wiry stellate hairs and small, warty
lenticels; later glabrous and grey-brown to
grey. **Leaf scars** on distinct cushions, typical
for the family, triangular, shield- to heart-
shaped, so deeply notched on the upper
margin that the U-shaped central trace is
confluent at the top. Frequently planted, tree
to over 40 m high. Origin: North America.

Juglans ×intermedia Jacques: mostly vigorous forestry-managed F$_1$ hybrids with *Juglans regia*. Similar to that species, while the buds are longer and more acute.

Juglans microcarpa Berland. is rarely cultivated. This shrub or small tree is more dainty in all parts than *Juglans nigra*. Origin: southern USA and Mexico.

Section Cardiocaryon

Juglans mandshurica Maxim., Manchurian walnut

Buds without bud scales, the outer leaves clearly differentiated into leaf base with wing-like margins and small lamina with pinnate structure, velvety ochre-brown hairy. **Terminal buds** 7–10 mm long, conical-ovoid. Lateral bud globose-ovoid, 2–4 mm long. **Twigs** rust-brown, finely hairy, mixed with reddish-brown glandular hairs. **Lenticels** many, very pale. **Leaf scars** with 3 traces, at the upper margin with narrow, densely fine felty-hairy bulge. **Stone** acute-ovoid, smooth with sharp angles. Occasionally planted, broad tree to 20 m high. Origin: central Asia.

The Japanese walnut, *Juglans ailantifolia* Carrière, is sometimes planted. It hardly differs from the Manchurian walnut, but twigs do not have reddish glandular hairs, and are pale brown to darker brown. **Stone** globose, acute, smooth.

Section Trachycaryon

Juglans cinerea L., white walnut

Buds without bud scales, fine velvety, pale ochre-brown hairy. Terminal buds 10–15 mm long, oblong, hardly narrowing towards the tip. Lateral bud smaller, ovoid. **Twigs** thick, red-brown to olive, towards the tip with spreading hairs, becoming glabrous later and darker violet-brown, with areas of dying, grey epidermis. **Lenticels** many, small, pale, orange-brown to whitish. Above the **leaf scar** with a fine-downy bulge. **Stone** oblong, with sharp edges. Frequently planted, broad tree to 30 m high. Origin: North America.

Juglans microcarpa
Tip of twig

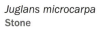

Juglans microcarpa
Stone

Juglans mandshurica
Tip of twig

Juglans mandshurica
Oblong stone

Juglans cinerea
Tip of twig

Juglans cinerea
Terminal bud

Juglans cinerea
Sharp-angled stone

  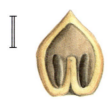

Juglans ailantifolia
Small stone

Cyclocarya paliurus
Tip of twig

Cyclocarya paliurus
Nut completely surrounded by wing

Pterocarya fraxinifolia
Tip of twig

Pterocarya fraxinifolia
Terminal bud and stalked catkin-buds

Pterocarya fraxinifolia
Short-winged nut

Pterocarya KUNTH and *Cyclocarya* ILJINSK., wingnuts

Terminal buds strikingly long. **Pith** of twigs chambered. **Fruits** winged nuts in long hanging spikes, which persist in winter. About six species of trees. Origin: Caucasus to eastern Asia.

Key to *Pterocarya* and *Cyclocarya*

1   Terminal buds with black-brown, deciduous bud scales, male catkins appear together with the female ones on the new shoots . . . . . *Pterocarya rhoifolia*

1* Buds without bud scales, catkins of male flowers arise on last year's wood and are naked in winter 2

2   Fruit distinctly winged on 2 sides. 3

2* Fruit completely surrounded by a wing. 3.5–4 cm diameter . . *Cyclocarya paliurus*

3   Wings of fruit broad, less than twice as long as broad . . .***Pterocarya fraxinifolia***

3* Wings of fruit oblong, more than twice as long as broad . . . . .*Pterocarya stenoptera*

Section Pterocarya

Buds without scales, ± velvety hairy. The lamina tips of the naked pinnate bud-leaves are spread away from the terminal bud. Similar to the species of this section is the rarely planted ***Cyclocarya paliurus*** (BATALIN) ILJINSK., a tree with nuts completely surrounded by a large wing. Their buds consist of a few, slim, oblong bud leaves. Origin: eastern China.

Pterocarya fraxinifolia (LAM. ex POIR.) SPACH, Caucasian wingnut

Buds without bud scales, with red-brown scaly indument which extends a little to the twig. Terminal buds to 20 mm long, surrounded by 3–4 bud leaves which are distinctly differentiated into stalk and pinnate blade, the rust-brown blades spreading at the tips. **Twigs** when young yellow-green to olive, slightly glossy, with pale round to oblong lenticels. Older twigs with dying, increasingly grey epidermis. **Leaf scars** broadly ovate to shield-shaped, with 3 groups of traces. **Bark** very dark, deeply furrowed. Male flower catkins persist without bud scales in winter, sessile to stalked, to 1.2 cm long and conical, coloured red-brown just as the buds. **Fruits** broad-winged one-seeded nuts, 1.5–2 cm broad,

many, in up to 40 cm long hanging spikes.
Very frequently planted tree 20–30 m high,
almost always with several stems. Origin:
Caucasus to northern Iran.

The rarely planted Chinese wingnut,
Pterocarya stenoptera C. DC., can be
distinguished in winter only by the longer
and narrower winged fruits. The hybrid
of both species, *Pterocarya ×rehderiana*
C. K. SCHNEID., is intermediate between
the parents.

Section Platyptera

Buds surrounded by (deciduous) bud
scales. Narrow scars of bud scales from the
previous year mark the annual growth of
each year on the twig.

Pterocarya rhoifolia SIEBOLD & ZUCC.,
Japanese wingnut

Buds surrounded by 2–3 dry, dark black-
brown bud scales, which fall off during
winter. Bud leaves lying beneath are very
close together, very pale grey-brown and
finely hairy. Terminal buds 15–25 mm long,
lateral buds smaller with a variously long
stalk. **Twigs** glabrous, grey-green with many
pale lenticels. **Leaf scars** shield-shaped with
3 groups of traces. Rarely planted, tree to
30 m high. Origin: Japan.

Carya NUTT., hickory

Twigs with solid pith. **Buds** as well as twigs
mostly hairy, often with thick peltate hairs.
Lateral buds mostly spreading away from
the twig, prophyll-scales basally fused. Male
inflorescences appear only after the leaves
appear and are therefore not visible in
winter.

Key to *Carya*

1 Terminal buds surrounded by bud scales
. . . . 2

1* Terminal buds ± naked (without scales),
bud leaves valvate, with distinctly
pinnate lamina 6

Section *Carya*

2 Terminal buds up to 12 mm long, bluntly
ovoid, twigs glabrous 3

2* Terminal buds over 12 mm long, ovoid
to conical, twigs and buds mostly with
indument 5

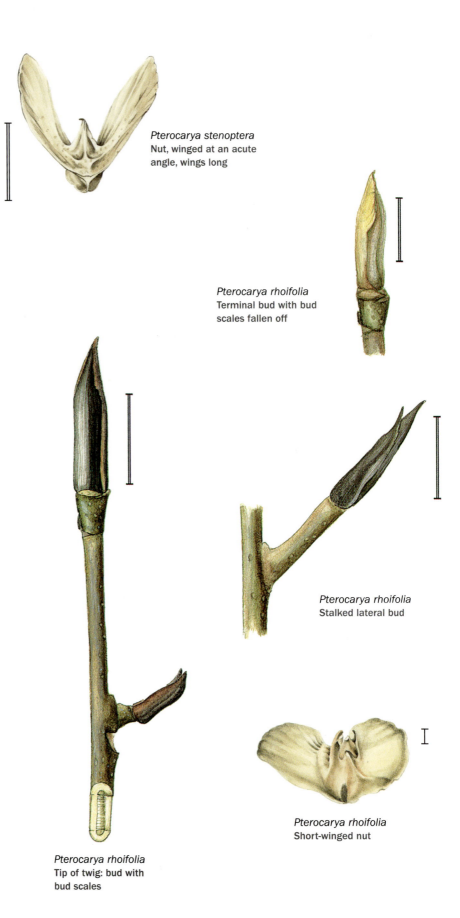

Pterocarya stenoptera
Nut, winged at an acute
angle, wings long

Pterocarya rhoifolia
Terminal bud with bud
scales fallen off

Pterocarya rhoifolia
Stalked lateral bud

Pterocarya rhoifolia
Short-winged nut

Pterocarya rhoifolia
Tip of twig: bud with
bud scales

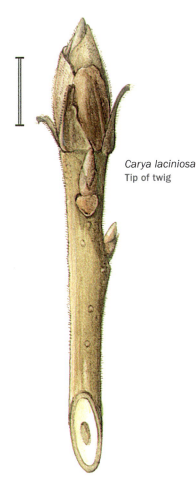

Carya laciniosa
Tip of twig

Carya ovata
Tip of twig

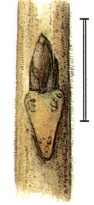

Carya laciniosa
Lateral bud

Carya ovata
Fruit and stone

3 Terminal bud with several scales, bud scales imbricate 4

3* Terminal bud with 2 scales, outer bud scales early deciduous. Tips of current year's twigs densely tufted-hairy . . . . *Carya tomentosa*

4 Stem ash-like, bark not flaking. . . . .*Carya glabra* var. *glabra*

4* Stem rough, with flaking bark . . . .*Carya glabra* var. *odorata*

5 Tips of current year's twigs leather-coloured to orange, angular, sturdy. Terminal buds more than 15–20 mm long . . . .*Carya laciniosa*

5* Tips of current year's twigs brown to grey, elliptic to roundish. Terminal buds smaller . . . . *Carya ovata*

Section *Apocarya*

6 Buds and ends of current year's twigs grey-brown with stellate hairs. . . . . *Carya illinoinensis*

6* Buds and ends of current year's twigs covered in yellow peltate hairs. . . . . **Carya cordiformis**

Section Carya

Buds with bud scales. When unfolding, these enlarge greatly through intercalary growth.

Carya laciniosa (F. Michx.) Loudon, shellbark hickory

Buds sparsely hairy and occasionally mixed with lower layer of peltate hairs. **Terminal buds** to 25 mm long, with some brown slightly keeled bud scales, the basal ones narrow, ± spreading and more densely hairy. **Lateral buds** long-ovoid, the prophyll scales only shortly fused. **Twigs** thick, ochre-brown to greenish-brown, softly hairy. **Bark** with unrolling, later flaking, sheets.

Similar, but in all parts smaller, is the shagbark hickory, ***Carya ovata*** (Mill.) K. Koch: **Terminal buds** 12–15 mm long. **Twigs** dark purple-brown, turning glabrous; branches early with flaking bark. Both species are planted only rarely; trees to 40 m high. Origin: North America

Carya glabra var. *odorata* (MARSHALL) LITTLE, pignut
[*Carya ovalis* (WANGENH.) SARG.]

Buds ovoid, scales covered in yellowish peltate hairs, yellowish-ochre to dirty ochre-brown hairy at the tips. **Terminal buds** larger with a few loosely appressed or slightly spreading bud scales. **Lateral buds** strongly spreading from the twig, the prophylls fused in the lower half. **Twigs** violet-red-brown to grey-whitish due to dying epidermis. **Lenticels** dot-shaped to oblong. **Leaf scars** triangular-ovoid, with many traces. **Bark** flimsy, later coming off in thin strips. Rarely planted tree to 30 m high. Origin: North America. Other rarely planted trees from North America are:

Carya tomentosa (LAM. ex POIR.) NUTT., mockernut hickory

A tree to 30 m high with flatly furrowed bark. Buds ochre-yellow to dirty yellowish-green, with simple hairs and occasionally with peltate hairs. Terminal buds ovoid, with appressed bud scales. Prophylls of lateral buds mostly fused more than half-way. Twigs yellowish-brown to violet red-brown, stiffly hairy, occasionally with whitish peltate scales. Bark smooth or finely wrinkled.

Carya glabra (MILL.) SWEET var. *glabra,* pignut

A tree 20 to 30 m high. **Buds** with few yellowish-brown bud scales, the outermost scales of terminal bud often deciduous. Lateral buds spreading from the twig with 2 basally fused scales. **Twigs** brown, only initially hairy, soon glabrous. **Bark** dark grey, finely fissured, not detaching.

Section Apocarya

Terminal buds with few valvate leaves with visible leaf blades (pinnate structure).

Carya cordiformis (WANGENH.) K. KOCH, bitternut hickory

Buds intensely dark yellow, due to yellow peltate hairs. **Terminal buds** naked, slim, c. 15 mm long, surrounded by few bud leaves with pinnate leaf lamina. **Lateral buds** 2–4 mm long, surrounded by adjacent but hardly fused prophyll scales, with descending

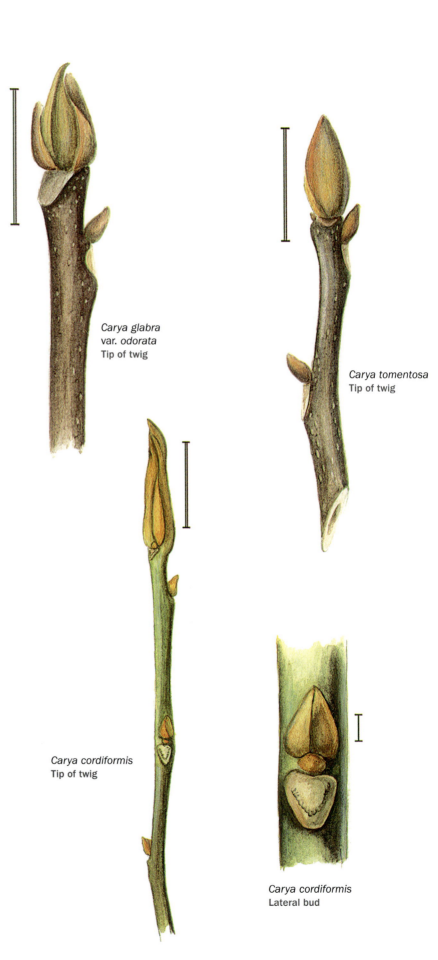

Carya glabra
var. *odorata*
Tip of twig

Carya tomentosa
Tip of twig

Carya cordiformis
Tip of twig

Carya cordiformis
Lateral bud

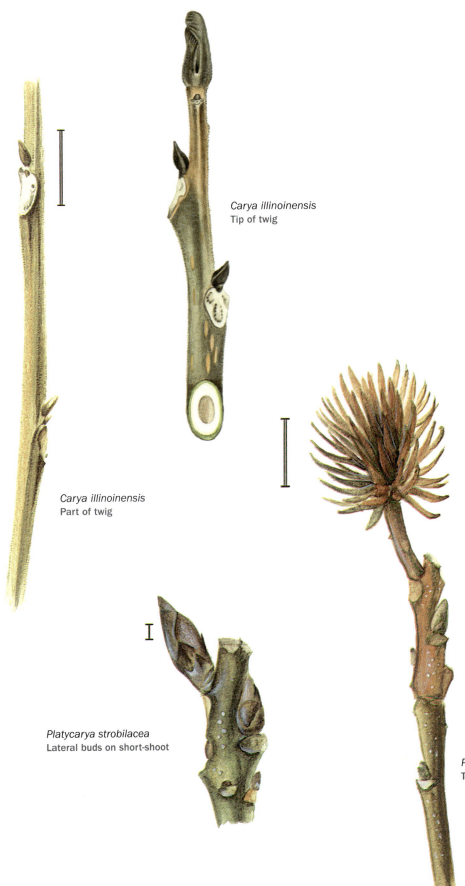

Carya illinoinensis
Tip of twig

Carya illinoinensis
Part of twig

Platycarya strobilacea
Lateral buds on short-shoot

Platycarya strobilacea
Twig with fruit cones

accessory buds. **Twigs** olive-green, at the tip with peltate hairs. **Pith** dark. **Stem** smooth, ribbed net-like and detaching in scales. Mostly sprouting before *Carya illinoinensis.* Occasionally planted, tree to 30 m high. Origin: North America.

Carya illinoinensis (WANGENH.) K. KOCH, pecan

Lateral buds 3–4 mm long, spreading from the twig, prophyll scales long-fused, with descending accessory buds. **Terminal buds** c. 12 mm long, surrounded by about three oblong leaves. Bud leaves dark red-brown, hairy, toward tip densely so, pale ochre with hardly noticeable yellow glands. **Twigs** ochre-brown to green-brown, round and narrowing distinctly towards tip, slightly angular, densely spreading fine-hairy. **Pith** dark brown. Trunk pale brown to grey, with bark-ribs forming a raised network. Rarely planted, tree to 30(–50) m high. Origin: North America.

Platycarya strobilacea SIEBOLD & ZUCC.

Buds to 7–9 mm long, ovoid with some slightly keeled bud scales, violet-brown or near the margin red-brown; the lowermost sparsely, the topmost densely white-hairy. **Twigs** towards the tip ochre-red-brown to dirty brown, sparsely appressed-hairy; older sections of twig grey-green to grey-brown, mostly glabrous. **Lenticels** many, pale and distinct from twig. **Leaf scars** shield-shaped, as is typical for the family. The fruit cones are distinctive, already developing on young individuals. Rarely planted shrub or small to 12 m high tree. Origin: China.

Myricaceae
bayberry family

A small family, consisting of 4 genera and more than 50 species. The largest genus is *Morella*, with predominantly evergreen species. Twigs and buds mostly with yellow, aromatic glands. Lateral buds spiral on the twig. In spite of the presence of stipules in *Comptonia*, buds always enclosed by scales of the leaf base, as in the sister family, walnuts.

Key to Myricaceae

1 Twigs with terminal bud, male inflorescences in globose-ovoid lateral buds, at least a few of the oblong leaves with ± entire margins persist. . . . . *Morella pensylvanica*

1* Twigs without terminal bud, tip of shoot dying off. Male inflorescence buds oblong 2

2 Next to the leaf scars stipule scars. At least a few of the narrow pinnate leaves persist when dry. . . *Comptonia peregrina*

2* Without stipule scars. The narrowly ovate to lanceolate leaves are deciduous . . . . *Myrica gale*

Myrica gale L., bog myrtle
[*Gale palustris* A. CHEV.]

Buds only lateral buds, differentiated into: **male inflorescence buds** at the ends of twigs following each other rather closely, spreading, cylindrical, 5–8 mm long and 2.5 mm thick with many brown, acute bud scales in 5 rows; **female inflorescence buds** 3 mm long, acute-ovoid; **leaf buds** basal on the twigs, below the region of inflorescence (buds), small, 1.5–2 mm long. **Twigs** slender, grey-brown to red-brown, fine whitish hairy and with yellow resin glands. **Lenticels** small, pale, more distinct towards twig base. **Leaf scars** with 3 traces, central trace large and distinct, the two on the margin smaller and relatively indistinct. Rare shrub 0.5 to 1.2 m high, dioecious, ascending. The aromatic twigs only sometimes with few dry leaves. Origin: North America and west to northern Europe.

Comptonia peregrina (L.) J. M. COULT., sweet fern

Buds: the tip of the shoot dies off, so there is no terminal bud. Below the tip of the shoot oblong cylindrical **male inflorescence buds**: 7 mm long and 3 mm thick with many (over 30) acute scales. **Female inflorescence buds** lower on the twig, c. 3–4 mm long with fewer than 30 scales. Lateral buds small, globose c. 2 mm diameter, surrounded by 4–6 scales (in spite of the presence of stipules, only simple scales). **Twigs** c. 1.5– 2.5 mm thick, glossy yellowish-green to red-brown, with long, spreading hairs and fine yellow glands. Lenticels dark, roundish. **Leaf scar** small, with 3 traces, with free stipule scars. Very rarely planted, monoecious shrub, to 1 m high, forming runners. Origin: north-east U.S.A.

Morella pensylvanica (MIRB.) KARTESZ, bayberry
[*Myrica pensylvanica* MIRB.]

Terminal bud terminates the shoot tip: oblong, c. 3 mm long with c. 8 bud scales. **Lateral buds** globose, spreading from twig, regularly distributed over the long-shoot, towards the base of the twig with the slightly larger male flower buds. **Twigs** brown, glossy, loosely covered by long spreading hairs and small yellow glands. Rarely planted shrub, c. 1.5 m high. Origin: eastern North America.

Myrica gale
Twig with male flower buds

Myrica gale
Male flower buds

Myrica gale
Female flower buds

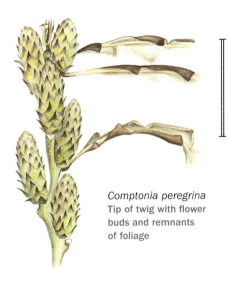

Comptonia peregrina
Tip of twig with flower
buds and remnants
of foliage

Comptonia peregrina
Lateral bud

Morella pensylvanica
Twig with terminal bud

Betulaceae
birch family

Buds, alternate (frequently distichous) lateral buds; terminal buds absent or present. Bud scales are stipules (exception: the first scale of the lateral bud in *Alnus* in subgenus *Alnaster*). **Leaf scars** with 3 traces. **Male flowers** mostly in naked oblong-cylindrical catkins which persist over winter (exception: *Carpinus)*. **Fruits** with bracts and bracteoles fused by their wing-margins, or forming an envelope. Deciduous shrubs or trees. Origin: Northern Hemisphere.

Key to Betulaceae

1 Long-shoots very rarely with terminal bud (only *Corylus* occasionally on few vertical shoots), lateral buds often distichous or in $^2/_5$ divergence. 2

Subfamily Betuloideae

1* All long-shoots with terminal bud, lateral buds spiral in $^1/_3$ divergence (pith and tips of twigs often 3-angular) rarely with more than 3 visible bud scales, infructescences woody cones which do not disintegrate. . . . . *Alnus*

2 Only short-shoots with terminal buds, buds with 3 pairs of stipule scales, fruits small, winged samaras in disintegrating mostly cylindrical spikes. . . . . *Betula*

2* Most twigs without terminal bud, fruits different 3

Subfamily Coryloideae

3 Buds with up to 10 visible bud scales, male inflorescences persist in winter without bud scales 4

3* Normally developed buds with more than 20 visible bud scales, the male inflorescence overwinters in bud . . . . *Carpinus*

4 Buds compact, blunt-ovoid, mostly with fewer than 10 bud scales, growth often shrubby; when tree-like, twigs cork pale grey-ochre . . . . *Corylus*

4* Growth mostly tree-like, buds narrowly ovoid, with c. 10 visible bud scales, infructescence hop-like with fruits enclosed by fused bracts and bracteoles forming an envelope . . . . *Ostrya*

SUBFAMILY Betuloideae

Mostly trees, more rarely small to medium-sized shrubs. Buds differentiated in terminal and lateral buds.

Alnus MILL., alder

Leaves on twigs tristichous, therefore twigs and pith often 3-angular. First leaf of the lateral bud, as also in other Betulaceae, turned towards the principal axis (adaxial). **Buds** of the tree species distinctly stalked. The male, and in the subgenus *Alnus*, also the **female inflorescences**, without bud scales, overwintering without a protective bud envelope. **Infructescences** 1–2.5 cm long cones becoming woody. **Leaf scars** very variable within the species; from distinctly 3-traced with a larger central trace (often of several parts) and 2 smaller traces in the corners to indistinctly V-shaped connected traces. Beside the leaf scars are two narrow stipule scars.

Key to *Alnus*

1 Buds ± distinctly stalked, mostly trees. 2

1* Buds sessile, shrub . . . ***Alnus alnobetula***

2 Stalk of bud less than half as long as the bud, twigs hairy or glabrous. 3

2* Stalk half the length of bud, twigs glabrous . . . . ***Alnus cordata***

3 Catkins red . . . . *Alnus rubra*

3* Catkins ± greenish. 4

4 Twigs hairy. 5

4* Twigs glabrous, old bark blackish and longitudinally fissured . . ***Alnus glutinosa***

5 Fruit cone small, sessile or very shortly stalked, bark pale grey or ± smooth . . . . ***Alnus incana***

5* Fruit cone large (to 25 mm long), distinctly stalked . . . . *Alnus japonica*

Subgenus Alnaster

Lateral bud sessile, the first bud scale is a simple scale (leaf base), after this stipular scales follow. Female inflorescences overwintering in buds. Growth often shrubby.

Alnus alnobetula (EHRH.) K. KOCH, green alder

[*Alnus viridis* (CHAIX) DC.]

Buds without stalks, 8–12 mm long with 3 outer bud scales: sticky, glossy, on the sun side red to violet-brown, on the shaded side yellowish-green. **Twigs** round, towards the tip also slightly 3-angular, ochre, red-brown to grey-brown with paler, small long-ovoid lenticels. **Leaf scars** ± 3-traced, central trace large and distinct, partly composed of several smaller ones, the lateral ones smaller and inconspicuous. Shrub with several stems 0.5–3 m high. Origin: mid-European mountains.

Subgenus Alnus

Lateral bud stalked, surrounded only by stipules of the outer leaves; these are formed like typical bud scales. As the linked foliage leaves develop mostly in spring, they are sometimes assessed as naked buds. (The transitions are fluid and similar buds of the genus *Hamamelis* are mostly called naked buds, due to the deciduous-ness of the stipules and their dense indument). The female inflorescences overwinter naked, without bud-scale protection, too. Mostly trees.

Alnus cordata (LOISEL.) DUBY, Italian alder

Buds 8–10 mm long, distinctly long-stalked: stalk about half the bud length. Actual bud ovoid, outer scale enveloping inner one. Bud scale matte wine-red, also greenish on the shaded side, without waxy layer. **Twigs** glabrous, round or slightly angular, smooth, glossy, in the shade olive green-brownish, on the sun side large areas covered by silver-grey epidermis, with fine pale lenticels. **Leaf scars** dark, semi-circular with 3 traces, the central trace consists, in turn, of 3 traces. Conical tree to 15 m high. Origin: Corsica and southern Italy.

Alnus alnobetula
Lateral bud

Alnus alnobetula
Twig with male catkins

Alnus cordata
Twig in March with male catkins

Alnus cordata
Lateral buds

Alnus rubra
Tip of twig

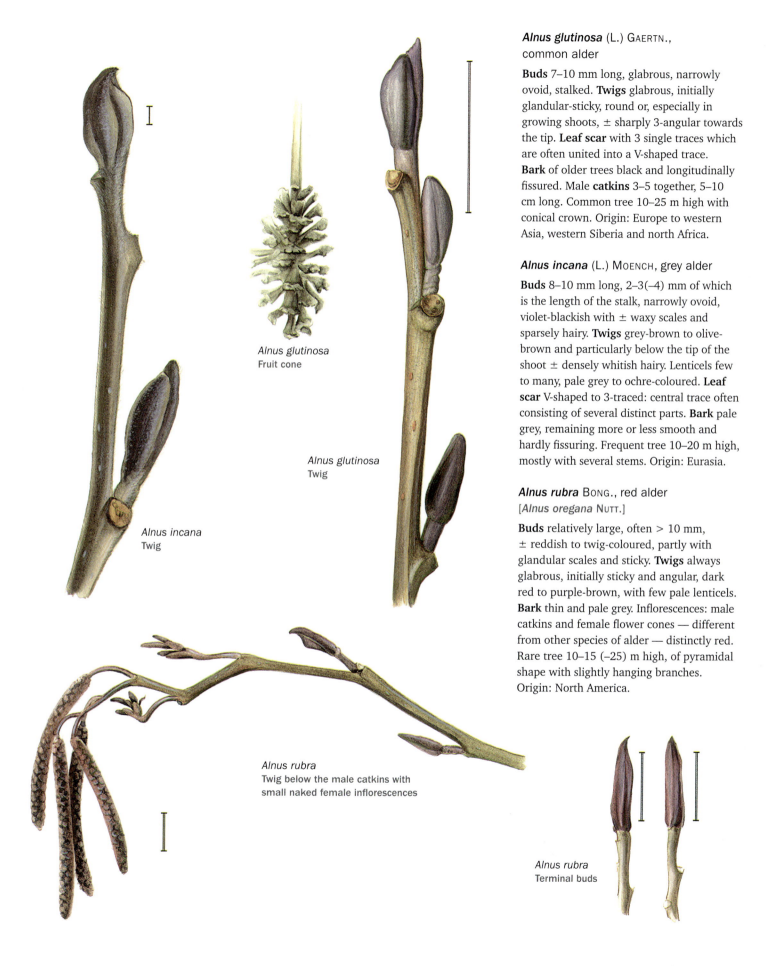

Alnus glutinosa
Fruit cone

Alnus glutinosa
Twig

Alnus incana
Twig

Alnus rubra
Twig below the male catkins with
small naked female inflorescences

Alnus rubra
Terminal buds

Alnus glutinosa (L.) Gaertn., common alder

Buds 7–10 mm long, glabrous, narrowly ovoid, stalked. **Twigs** glabrous, initially glandular-sticky, round or, especially in growing shoots, ± sharply 3-angular towards the tip. **Leaf scar** with 3 single traces which are often united into a V-shaped trace. **Bark** of older trees black and longitudinally fissured. Male **catkins** 3–5 together, 5–10 cm long. Common tree 10–25 m high with conical crown. Origin: Europe to western Asia, western Siberia and north Africa.

Alnus incana (L.) Moench, grey alder

Buds 8–10 mm long, 2–3(–4) mm of which is the length of the stalk, narrowly ovoid, violet-blackish with ± waxy scales and sparsely hairy. **Twigs** grey-brown to olive-brown and particularly below the tip of the shoot ± densely whitish hairy. Lenticels few to many, pale grey to ochre-coloured. **Leaf scar** V-shaped to 3-traced: central trace often consisting of several distinct parts. **Bark** pale grey, remaining more or less smooth and hardly fissuring. Frequent tree 10–20 m high, mostly with several stems. Origin: Eurasia.

Alnus rubra Bong., red alder
[*Alnus oregana* Nutt.]

Buds relatively large, often > 10 mm, ± reddish to twig-coloured, partly with glandular scales and sticky. **Twigs** always glabrous, initially sticky and angular, dark red to purple-brown, with few pale lenticels. **Bark** thin and pale grey. Inflorescences: male catkins and female flower cones — different from other species of alder — distinctly red. Rare tree 10–15 (–25) m high, of pyramidal shape with slightly hanging branches. Origin: North America.

Alnus japonica (THUNB.) STEUD., Japanese alder

Buds 8–10 mm long, distinctly stalked, the stalk about a third of the length of the entire bud. Bud scales: mostly 3 stipules visible, dark wine-red to violet-brown, whitish 'powdered' due to wax layering and sparsely hairy. **Twigs** glabrous or hairy in patches, becoming glabrous; matte or slightly glossy, olive-brown, round or, most of all on sturdy long-shoots, ± sharply 3-angular. Rare tree to 25 m high or large shrub. Origin: eastern Asia — Japan, Manchuria, Korea and the Ussuri area of Russia.

Betula L., birch

Widely distributed shrubs and trees of the Northern Hemisphere. A few shrubby species in the north reaching the tree line. Many taxa can hybridise with each other, dependent on the level of ploidy and how closely the species are related, and can produce fertile offspring in turn. Since the pollen are carried far by the wind, there may be only one white-barked diploid species in a region, such as *Betula pendula*. **Buds:** 5–8-scaly terminal buds on lateral short-shoots and 3–5-scaly lateral buds on the long-shoots. The male flowers overwinter in naked, oblong catkins without bud scales. The female ones appear only after the leaves unfold.

Leaf position spiral, in $^2/_5$ divergence, in which 5 leaves form a cycle and the 6th leaf appears after two turns above the first. Longer short-shoots appear distichous (cf. illustration "*Betula alleghaniensis* — short shoot"): because the two foliage leaves follow three leaves — as stipular scales of the terminal bud — there are 5 leaves each year, and in the next year the two new foliage leaves are in the same position as those of the previous year, and thus simulate distichy.

The most important characters to distinguish the various birch species are growth form, bark, size of buds as well as lenticels and twig indument. Trunks with very light, pure white to bronze-brown ring bark are common, the bark coming off in thin, paper-like patches from the periderm; less common are darker scaly or stripy barks.

Key to *Betula*

1   Trees, buds more than 5 mm long . . . 3
1*   Shrubs, buds to 4 mm long 2

Section *Apterocaryon*

2   Twigs densely hairy, male inflorescences concealed in the bud . . . . ***Betula nana***
2*   Twigs becoming glabrous, densely pale-warty, male catkins overwinter without bud scales . . . . *Betula humilis*

Section *Acuminatae*

3   Lateral buds more than 3 mm thick . . . .***Betula maximowicziana***
3*   Lateral buds thinner. 4
4   Bark on younger or older twigs/stems smooth and 'rolling off' in transverse strips, the rolls falling or persisting, white to grey-white, partly with red or yellow hues, occasionally basally with (in places) black bark. 6
4*   Bark brown to black-brown, relatively dark 5

Section *Lentae*

5   Buds to 8 mm long; bark aromatic, not 'rolling off' . . . . ***Betula lenta***
5*   Buds often more than 10 mm long, bark yellow- to grey-brown, rolling off at the border . . . . ***Betula alleghaniensis***

Section *Dahuricae*

6   Entire stem covered relatively closely with medium-sized remnants of persistent 'rolling off' bark, older bark blackish . . . . ***Betula nigra***
6*   Bark relatively smooth, if large patches of bark lift, then beside these larger, smoother patches of bark. 7
7   Twigs glabrous, often with warty lenticels 8
7*   Twigs hairy (at least remnants of indument retained) 9
8   Bark of older stems white to grey. . . 12

Section *Betula*, *Betula utilis* group

8*   Bark white only initially, soon tinged yellowish to pink . . . . . ***Betula ermanii***
9 (7)   Bark pure white to grey 10
9*   Bark orange-brown or yellow- to grey-brown, buds 5–8 mm long . . . .***Betula albosinensis***
10   Bark pure white, rolling off in large planes. 11
10*   Bark dirty-white to grey . . . . ***Betula pubescens***
11   Twigs with resin glands . . . .***Betula utilis*** var. ***jacquemontii***

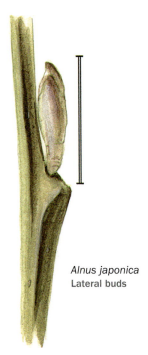

Alnus japonica
Lateral buds

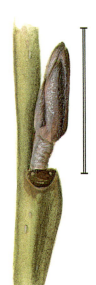

Alnus japonica
Lateral buds

Betula nana
Tip of twig

Betula maximowicziana
Twig

Betula humilis
Tip of twig

Betula maximowicziana
Lateral bud

Section *Betula*, *Betula pendula* group

11* Twigs with pale lenticels . . . .
. . . .***Betula papyrifera***

12 Bark hardly rolling off 13

12* Bark white, rolling off, basally often
with black bark . . . . . .***Betula pendula***

13 Bark floury, white . ***Betula platyphylla***

13* Bark grey, black-fissured when aged . .
. . . . ***Betula populifolia***

Section Apterocaryon, shrub birches

Betula nana L., dwarf birch

Buds globose to short-ovoid, 2–4 mm long, often larger towards tip of twig. Bud scales dark red-brown, glossy and sparsely long-hairy. **Twigs** dark grey-brown to black-brown, with fine felty indument; with large, pale lenticels. Leaf cushion distinct, but not changing the shape of the twig substantially. Male inflorescences concealed in the bud. Shrub 0.5–1 m high, with black-grey bark which hardly flakes off. Origin: northern Europe to Siberia.

Betula humilis SCHRANK

Buds small, 2–3 mm long, (narrowly) ovoid, with 3–5 grey-brown bud scales with ciliate margins. **Twigs** grey-brown to red-brown, mostly entirely glabrous and ± glossy, relatively slender, angular due to the proportionally large leaf cushions; with many small, pale warts. Uncommon, densely branched, 0.5–2 m high shrub. Origin: central Europe at the northern edge of the Alps and in north Europe and east to Kamchatka.

Section Acuminatae

Betula maximowicziana Regel, monarch birch

Buds 10–12 mm long and over 3 mm thick, at the tip of the twig very close to the twig, lower down the twig slightly spreading. Bud scales basally olive-yellowish-green to ochre-brown, the margin red-brown to dark brown. **Twigs** relatively thick, ochre-red-brown with remnants of fine grey hairy indument and distinct warty, pale grey lenticels. **Bark** initially orange-brown, later paler grey to orange and thin, rolling off. Occasionally planted, tree 20–30 m high. Origin: Japan.

Section Lentae

Betula alleghaniensis BRITTON, yellow birch
[*Betula lutea* MICHX., nom. illeg.]

Buds relatively large, often more than 10 mm long, narrowly ovoid, acute, with glossy dark red-brown bud scales with ciliate margin. **Twigs** initially finely hairy (lens!), pale ochre-brown to silvery-grey due to pale epidermis which later flakes off, older twigs glossy dark violet-brown. **Lenticels** many, dot-like, warty, pale ochre-brown to whitish-grey. **Fruits** often on the tree in winter: in erect 2–3 cm long and 2 cm thick catkins, the fruit scales 8–10 mm long and 3-lobed. **Bark** yellow-brown to grey-brown, rolling off in strips. Occasionally planted, 20–25 m high tree with a broad crown. Origin: North America.

Betula lenta L. sweet birch

Buds c. 8 mm long, mostly ± distinctly acute. Lateral buds slightly spreading from twig. Terminal buds at the tip of short-shoots, which often consist of several growth spurts: relatively thin and conical-acute. **Bud scales** with long hairs at the margin, at the base partly green, towards the tip violet-brown to dark brown or entirely brown. **Twigs** grey-green to grey-brown, long-shoots with long hairs to the tip, with pale, round lenticels. **Leaf scars** rounded, semi-circular with 3 traces. **Bark** dark, red-brown to blackish, cracking, but not rolling off, aromatic-sweet. Occasionally planted, tree to 25 m high with upright narrow crown. Origin: North America.

Section Dahuricae

Betula nigra L., river birch

Buds 8–10 mm long, narrowly ovoid, blunt, slightly spreading. **Bud scales** red-brown, with slightly ciliate margins. **Twigs** pale grey- to orange-brown, finely hairy or glabrous, with few small, pale lenticels. **Bark** red- to yellow-brown, crisply rolled up, not (!) falling; later dark brown to black, hard and rough. Frequently planted, 15(–20) m high tree with several stems. Origin: eastern USA. Closely related are *Betula dahurica* PALL. and *Betula raddeana* TRAUTV.

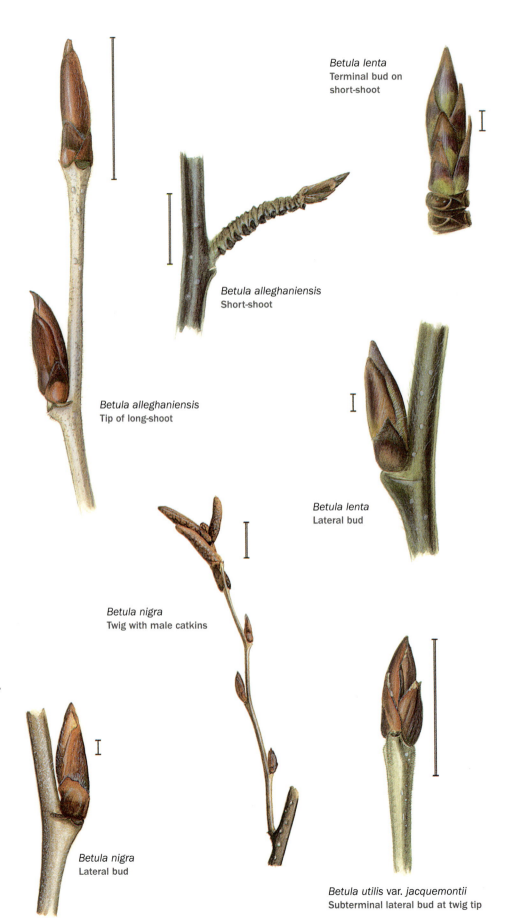

Betula lenta
Terminal bud on short-shoot

Betula alleghaniensis
Short-shoot

Betula alleghaniensis
Tip of long-shoot

Betula lenta
Lateral bud

Betula nigra
Twig with male catkins

Betula nigra
Lateral bud

Betula utilis var. *jacquemontii*
Subterminal lateral bud at twig tip

Betula utilis var.
jacquemontii
Lateral bud

Betula ermanii
Twig with male catkins
which overwinter without
bud scales

Betula ermanii
Lateral bud

Betula albosinensis
Lateral bud

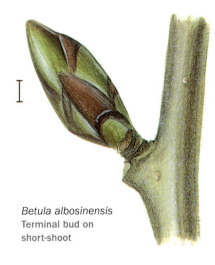

Betula albosinensis
Terminal bud on
short-shoot

Betula albosinensis
Lateral bud

Betula ermanii
Terminal bud on
short-shoot

Section Betula

Betula utilis D. Don aggr.

Due to the beautiful smooth bark, very attractive and a favourite in cultivation. Origin: from the Himalaya to Japan.

Betula utilis var. *jacquemontii* (Spach) H. J. P. Winkl., Himalayan birch
[*Betula jacquemontii* Spach]

Buds 7–9 mm long, narrowly ovoid, slightly acute or almost blunt, with red-brown, slightly resinous scales with a ciliate margin. **Twigs** initially grey and matte, very finely hairy with many small, pale, oblong lenticels. **Bark** smooth and pure white, peeling in fine rolls. Frequently planted, tree 15–20 m high. Origin: western Himalaya. The following two taxa are related:

Betula albosinensis Burkill, Chinese red birch

Buds narrowly ovoid, acute or blunt. **Lateral buds** 6–8 mm long, with 3–4 outer bud scales, the lowest 2 reaching about halfway of the bud. **Terminal buds** a little larger, mostly blunt with 5–7 outer scales. **Bud scales** green with a dark margin, sticky and glossy, the lowest ones slightly ciliate. **Twigs** grey-brown and finely hairy with few pale, dot-shaped lenticels. **Bark** orange to red-orange and initially bluish-white with waxy bloom, peeling off in very thin rolls. Frequent, tree 10–20 m high. Origin: China.

Betula ermanii Cham., gold birch, Erman's birch

Lateral buds 7–8 mm long with 3(–4) bud scales, the lowest 2 not reaching to halfway. **Terminal buds** to c. 9 mm long with 5–6 scales. **Bud scales** green, with broad brown margin, which is often so wide that the buds are almost brown. **Twigs** slender and glabrous, densely warty, glossy olive-brown to brown, violet-brown when older. **Bark** white when young, later yellowish-white to matte pink; peeling in broad thin transverse strips. Infructescences mostly persistent in winter: cylindrical-ovoid and 2–3 cm long. Frequent, large shrub or tree to 20 m high. Origin: eastern Asia.

Betula pubescens EHRH., common white birch, downy birch

Buds 6–8 mm long with some brownish, basally also ± greenish, ciliate bud-scales. **Twigs** brown when young and with dense fine downy hairs, later grey-brown with few lenticels. **Bark** initially grey-white, peeling off in thin rolls/strips, later slightly fissuring at the base of the stem. Crown upright and stiff: twigs not hanging down. Predominantly growing in moist habitats, tree 10–30 m high, often with several stems. Origin: Europe to Siberia and Asia Minor to the Caucasus.

Betula pendula ROTH, silver birch

Buds 6–8 mm long, narrowly ovoid with brown to greenish bud scales, often with ± glossy, waxy-resinous layer. **Twigs** slender, glabrous, grey-brown to brown; densely covered with small puberulous warts. **Bark** white, peeling off in transverse rolls and changing towards the stem base into black, deeply fissuring bark. Tree 10–25 m high with loose conical crown, the slender twigs pendulous. The most common species. Origin: throughout Europe to North Africa and northern Iran in the south and to the Caucasus and Siberia in the east.

The following 3 taxa are very similar to *Betula pendula* and easily hybridise with it. Because in older botanical collections seeds were often taken from cultivated trees, many old trees are difficult to attribute.

Betula papyrifera MARSHALL, paper birch

Buds blunt-ovoid, c. 6–7 mm long, with some dark brown to ochre-brown, ± hairy and resinous sticky bud scales. **Twigs** dark violet-brown, hairy to the tip and partly very densely grey-warty. Lenticels light ochre to orange-brown, round to oblong. **Bark** pure white and smooth for long, peeling in paper-like broad transverse strips and only in very old stems at the base with dark fissures. Frequently planted tree to 20–25 m high. Origin: North America.

Betula platyphylla SUKACZEV

Buds with deciduous resin/wax layer, lowest bud scales predominantly brown, upper ones partly green at base. Lateral buds 5–7 mm long with 3–4 outer bud scales of which the first two are to half as high as the

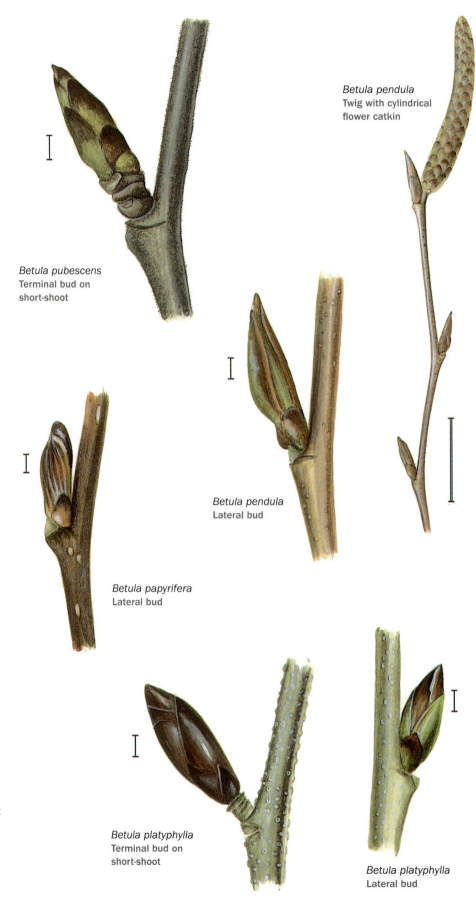

Betula pubescens
Terminal bud on short-shoot

Betula pendula
Twig with cylindrical flower catkin

Betula pendula
Lateral bud

Betula papyrifera
Lateral bud

Betula platyphylla
Terminal bud on short-shoot

Betula platyphylla
Lateral bud

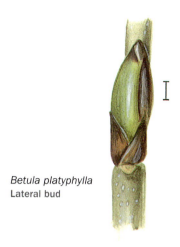

Betula platyphylla
Lateral bud

Betula populifolia
Terminal bud on
short-shoot

Corylus colurna
Twig in February: stigmas emerge
from the female flower bud

bud. Terminal buds of the short-shoots 6–7 mm long, with many scales, predominantly brown. **Twigs** mostly glabrous, matte grey-brown, when young 1–2 mm thick, densely covered with warty lenticels. **Bark** floury-white, not peeling/rolling off. Frequent, tree 10–20 m high. Origin: Japan, Manchuria.

Betula populifolia MARSHALL, grey birch

Buds 5–6 mm long, ovoid. Bud scales dark red-brown, near base also a little greenish, white-ciliate, sticky with a wax-like layer. **Twigs** dark violet-brown, glabrous, glossy with partly longitudinally fissuring wax layer and many warty lenticels. **Bark** white, not peeling, in old trees at the base black, cracking. Frequent, tree to 10 m high, often with several stems. Origin: North America.

SUBFAMILY **Coryloideae** hazels

Twigs mostly without terminal buds. Only the genus *Ostryopsis* DECNE. (not included here) always has terminal buds; and rarely the hazels, particularly the Turkish hazel *Corylus colurna*, on upright twigs with spiralling leaves, have such buds.

Corylus L., hazel

Leaves distichous, on vertical twigs also in spirals. **Flowering** very early before the leaves appear, in February to the beginning of March. **Male flowers** with four 2-parted anthers, without corolla, with two bracteoles; they are solitary in the axils of bract scales which are spirally arranged in catkins. **Female flowers** in pairs in axils of ± deciduous bract scales, with bracteoles which will later develop into a fruit envelope (husk). During flowering only the stigmas are visible, emerging from the buds of female flowers. **Buds** are surrounded by stipule scales of the outermost undeveloped leaves. In the shrubby species the fruits with their coverings, often present in early winter, are needed for certain identification.

Key to *Corylus*

1 Several-stemmed shrubs, twigs always without terminal bud 2

Subsection *Colurnae*, tree hazels

1* One stemmed tree, bark of first year's growth shoots very corky, ochre-pale grey, vertical shoots with spiral leaves and terminal buds........ *Corylus colurna*

2 Buds with more than 4 visible bud scales, husk at most 3 times as long as the nut 3

2* Buds with at most 4 visible bud scales, husk over the nut tubular and more than 3 times as long as the nut. 5

Subsection *Corylus*, shrub hazels

3 Husk fused all-round, narrowed above the nut .... *Corylus tubulosa*

3* Husk not narrowed above the nut, often in two parts or open on one side 4

4 Husk only little longer than the nut, in 2 parts............... *Corylus avellana*

4* Husk mostly distinctly longer than the fruit; fused all round or open to one side .... *Corylus ×maxima*

Subsection *Siphonochlamys*, beaked hazels

5 Lowest bud scales deciduous, the outer ones reaching the tip ... *Corylus cornuta*

5* Lowest bud scale not reaching tip, bud often reddish ...... *Corylus sieboldiana*

Subsection Colurnae, tree hazels

Corylus colurna L., Turkish hazel

Terminal buds rarely developed, 6–8 mm long with 2–3 loosely fitting pairs of scales, the outer one often deciduous. **Lateral buds** to 8 mm long, ovoid, with 3 visible pairs of scales. **Bud scales** grey-brown to pale brown, off-white to pale brown hairy, ciliate on margins towards the tip. **Twigs** initially densely (and partly glandular-) hairy, grey-brown to cinnamon brown, with few, initially slightly raised, soon lengthwise fissuring lenticels. **Bark** corky and rough. **Flower**: male catkins to 12 cm long. Female flowers with red stigmas. Frequently planted tree to 20 m high. Origin: south-eastern Europe.

Subsection Corylus, true hazels or shrub hazels

Corylus avellana L., hazel

Lateral buds with 7–8 visible scales, green, slightly reddish in the sun, with brown ciliate margin and otherwise almost glabrous. Buds slightly spreading from the twig, ovoid, usually slightly flattened: c. 5 mm long, 3.5–4 mm broad and 2.5 mm thick. **Twigs** grey-brown to olive-grey, matte, hairy especially towards the tip of the twig, with dense fine hairs with a lower layer of glands. **Lenticels** few, distinct on the lower part of young shoots: pale ochre, roundish to oblong. **Leaf scars** semi-circular on distinct leaf cushions, brown with 3 traces of vascular bundles. **Flowering** very early in February to March: 3–7 cm long male catkins with bright red stigmas protruding from female flowers. **Husk** split in two toothed lobes, hardly exceeding the nut. Frequently planted shrub to 6 m high. Origin: Europe to Asia Minor and the Caucasus.

Many **cultivars**: **'Aurea'** with yellowish catkins in winter, and **'Contorta'**, the corkscrew hazel with contorted twigs.

Also frequent, but difficult to tell apart in winter, is ***Corylus tubulosa*** Willd., the filibert. It is slightly more robust than *Corylus avellana*. Easiest to distinguish in fruit: husk more than double the size of the nut, completely closed and slightly narrowed above the nut. Origin: southeastern Europe to Asia Minor and the Caucasus.

In the frequently planted cultivar, *Corylus tubulosa* **'Purpurea'**, the purple-leaved filbert, even the bud scales are coloured red.

C. ×***maxima*** Mill., hazel hybrids

Between the filbert and the common hazel there are many hybrids which are difficult to tell apart without their fruits. These often have a husk slit on one side.

Corylus avellana 'Contorta'
Twig contorted

Corylus avellana
Twig in February with long flowering male catkins and female flower buds

Corylus tubulosa 'Purpurea'
Twig with reddish catkins and flower buds

Corylus cornuta
Twig with distinctly short
closed male catkins

Corylus cornuta
Twig with female
flowering buds

Corylus sieboldiana
Lateral bud

Corylus sieboldiana
Tip of twig

Subsection Siphonochlamys

In the beaked nut, the fused husk narrows above the nut to a 3–4 cm long beak-like tube. This envelope is covered externally with long, stiff hairs.

Corylus cornuta MARSHALL, beaked hazel

Buds 5–6 mm long, ovoid, with grey-brown to olive-brown bud scales which are ± densely hairy towards the tip. The outer ones mostly deciduous; due to this the persistent inner ones reaching to the tip. **Twigs** sparsely softly hairy, grey-brown to grey, towards the tip also ochre-brown to red-brown. Male catkins opened 2.5–3 cm long, with yellow anthers. Shrub to 3 m high. Origin: North America.

Corylus sieboldiana BLUME

Buds with only 4 visible bud scales, these deep wine-red, in shadow bright, in light darker violet wine-red, ciliate at the margin, the third scale sparsely and the fourth densely hairy. Shrub to 4 m high. Origin: Japan.

The east Asian *Corylus mandshurica* MAXIM., often treated as a variety of the above species, is sturdier and has distinctly more visible bud scales.

Carpinus L., hornbeam

No terminal bud is developed. In the lateral buds many stipules, following the distichous phyllotaxy, form four rows. Often 5 or 6 scales stand above one another so that, in all, more than 20 scales are visible. In contrast to the other genera of the family, this genus has male catkins that overwinter in the bud. These inflorescence buds are larger, in *Carpinus betulus* the scale-like flower bracts are visible in addition to the ordinary bud scales. The shape of buds and their size are mostly insufficient to identify a species with certainty: *Carpinus orientalis* has the smallest buds; the bud of *Carpinus caroliniana* are a touch fatter and darker, more glossy brown; and the buds of *Carpinus japonica* are pure green in shadow and reddish in the sun. To best identify the species, the infructescences, often long-persistent in winter, are required. The most diagnostic character is the shape of the large fruit bract (often with adnate bracteoles).

Key to *Carpinus*

1 Fruit not visible in the hop-like infructescence, hidden by turned down fruit bracts and additional small opposite scale . . . .*Carpinus japonica*

1* Fruits visible in the infructescence, only the bracts, no small opposite scales present 2

2 Buds to 3(–4) mm long, young twigs slender 3

2* Buds longer than 4 mm, fruit bract 3-lobed . . . . **Carpinus betulus**

3 Fruit bract at the base with 2 short lateral lobes, infructescences 5–10 cm long. . . . .**Carpinus caroliniana**

3* Fruit bract not lobed and dentate, in 3–6 cm long infructescence. . . . .*Carpinus orientalis*

Section Carpinus

Bark grey fissuring in oblong rhomboids, but remaining smooth.

Carpinus betulus L., common hornbeam

Lateral buds lying along the twig, fusiform-acute, 6–8 mm long. 5(4–6) bud scales stand in four lines above one another. Bud scales matte-glossy, almost glabrous but ciliate, upper ones more densely hairy; brown on the light side, in the shade green and only slightly brown at the margin. Male floral buds slightly longer than leaf buds, in the upper half the spiral scaly flower bracts are visible. **Twigs** initially finely hairy, becoming almost glabrous; first year's growth c. 2 mm thick, from violet brown to grey on the light side or olive-coloured in the shade, with many round, pale lenticels. **Leaf scars** ovate, on small cushions, with 3 traces, flanked by two oblong narrow stipule scars. **Bark** smooth, grey, with a narrowly rhombic net-like pattern. **Bracts** 3-lobed, 3–5 cm long, in 7–15 cm long hanging spikes. **Trunk** rarely round, usually twisted, with alternating raised and sunken parts. Medium-sized tree to 25 m high, often planted as a hedge. Origin: Europe.

Carpinus betulus
Twig

Carpinus betulus
Tip of twig

Carpinus betulus
Fruit with bract

Carpinus caroliniana
Fruit with bract

Carpinus caroliniana
Twig with subterminal lateral bud and dead shoot tip (right next to the bud base)

Carpinus betulus
Tip of twig: the upper bud is a leaf bud, the one below a male inflorescence bud

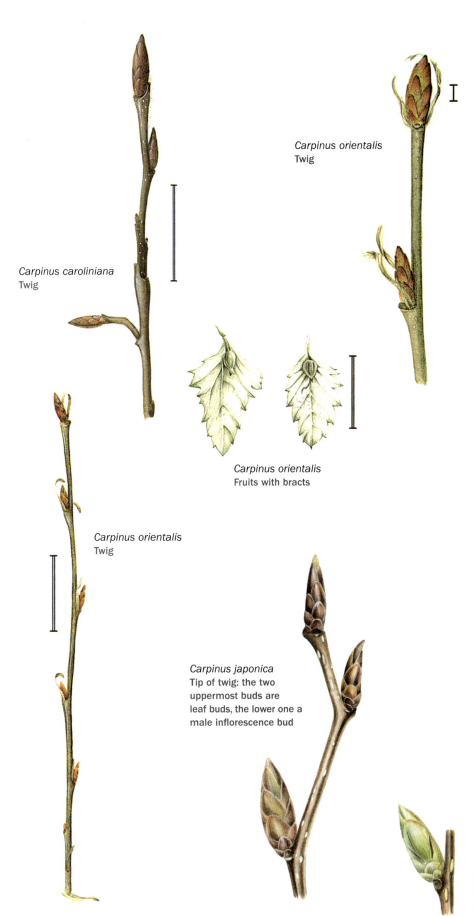

Carpinus caroliniana
Twig

Carpinus orientalis
Twig

Carpinus orientalis
Fruits with bracts

Carpinus orientalis
Twig

Carpinus japonica
Tip of twig: the two
uppermost buds are
leaf buds, the lower one a
male inflorescence bud

Carpinus japonica
Lateral bud, distinctly
green in shade

Carpinus caroliniana WALTER, American hornbeam

Buds blunt ovoid to fusiform, with many dark brown bud scales, c. 6 mm long, distinctly hairy towards the tip. **Twigs** slender when young, reddish, initially with pale hairs. **Bark** smooth, trunk fluted to below the branches. Tree to 10 m high, occasionally planted, with dense crown and daintily pendulous twigs. Origin: North America.

Carpinus orientalis MILL., Oriental hornbeam

Buds small, 3–5 mm long, acute-ovoid, ochre to orange-brown, bud scales darker at the margins and pale ciliate. **Twigs** grey-brown, first year's growth very slender, 2 mm thick, fine velvety hairy, below the buds with a few longer hairs. Phyllotaxy distichous. **Lenticels** small, slightly paler than the twig. **Leaf scars** small, raised, with three traces which are difficult to see. **Fruits** in 3–6 cm long infructescences. **Bracts** ovoid, dentate. Small tree to 5 m high. Origin: southern to eastern Europe, to Asia Minor.

Section Distegocarpus

Bark: furrowed, scaly.

Carpinus japonica BLUME, Japanese hornbeam

Buds oblong, acute, c. 8 mm long and 3 mm thick, pure green in shadow, reddish on the sun side, at the tip slightly hairy. Male inflorescence buds larger. **Twigs** hairy when young, first year's growth brownish to blackish. **Fruits** in 5–6 cm long infructescences; nutlets 4 mm long, surrounded by a dentate ovoid fruit bract, reflexed on one side and surrounding a second inner scale. Rarely planted, tree to 15 m high. Origin: Japan.

Carpinus cordata BLUME is similar. It has slightly longer buds and the inner scale under the reflexed fruit bract is absent. Origin: eastern Asia.

Ostrya Scop., hop hornbeam

Buds above the leaf scars often slightly oblique. **Bud scales** of stipules. **Leaf scars** with 3 traces with distinct stipule scars. **Fruits**: the individual nuts are enclosed in a sack-like envelope (connate bract and bracteoles) and united in loose, hop-like catkins. A few shrubs and trees. Origin: Northern Hemisphere.

Ostrya carpinifolia Scop., hop hornbeam

Buds 4–6 mm long, narrowly ovoid, slightly acute, with 6–9 greenish, sticky, often ± hairy bud scales with a brownish margin. **Twigs** round to slightly angular, brown to olive-brown, initially hairy, becoming glabrous later. **Lenticels** solitary, inconspicuous. **Bark** initially smooth or slightly transversely fissured, brown-grey, soon changing into ± scaly peeling bark. **Fruits** in hanging infructescences: 4–5 mm long nuts, hidden in sack-like structures with a tuft of hair at the tip. A tree to 20 m high. Origin: South-East Europe.

The following two, rarely planted, species are difficult to tell apart: the nuts of the American hop hornbeam *Ostrya virginiana* (Mill.) K. Koch are, in contrast to the common hop hornbeam, not hairy at the tip. In the east Asian *Ostrya japonica* Sarg., the Japanese hop hornbeam, the closed male flower catkins are, with a length of 15 mm, shorter than those of both other species.

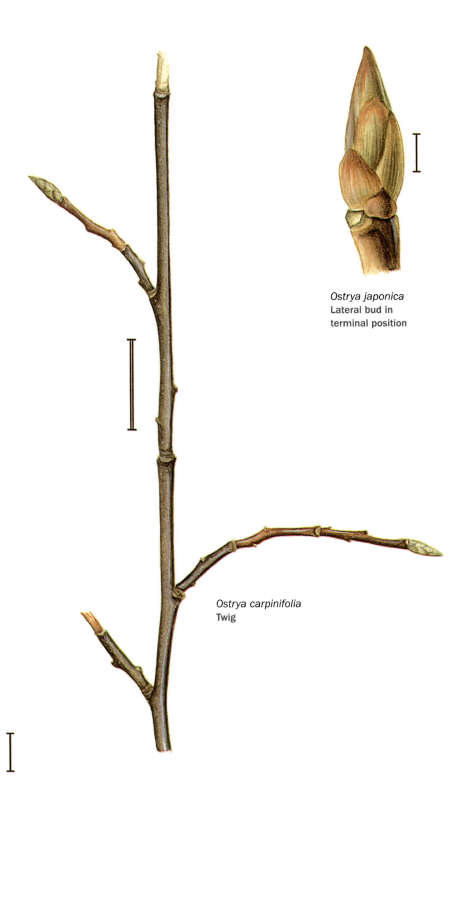

Ostrya japonica
Lateral bud in
terminal position

Ostrya carpinifolia
Twig

Ostrya carpinifolia
Lateral bud in
terminal position

ORDER Rosales

In the order Rosales belong, among others, the families Cannabaceae, Elaeagnaceae, Moraceae, Rhamnaceae, Rosaceae and Ulmaceae. These have in common that the bud scales are almost always formed from the leaf base, in the Rosaceae occasionally with part of some fused (adnate) stipules. While free stipules are often present, these rarely take part in the protection of the bud. Of the species mentioned here, fused stipules form the bud envelope only in *Ficus*. All other species have relatively few simple leaf base scales which follow the phyllotaxy.

Rosaceae
rose family

Many native species and many frequently planted fruit- and ornamental woody plants. Herbaceous species as well as shrubs and medium-sized to large trees. Buds mostly with simple bud scales, often with adnate stipules, more rarely naked (*Cotoneaster*). Lateral buds of some species with enrichment buds (*Prunus, Spiraea*) or with accessory buds (*Neillia*).

Key to Rosaceae

1　Twigs not armed. 3
1*　Twigs armed with prickles or thorns . . 2
2　Twigs with prickles or sharp bristles. . . .
　. . . . .**Rosoideae**
2*　Twigs with thorns. . . . . **Amygdaloideae**
3　Leaf base fused with the extended stipules and widened. Cup of the persistent fruits with 10 lobes . . . .*Dasiphora*
3*　All characters different. 4
4　Shrubs with remnants of multiple fruits. Either green- or red-twigged and leaf scar very narrow, enclosing the base of lateral buds or twigs thick with wide pith, matte grey-brown with persistent leaf base . . .
　. . . . . *Rosoideae*
4*　All characters different. .*Amygdaloideae*

SUBFAMILY Rosoideae

Many perennial herbs, and some shrubs, often armed with spines.

During the growth period most species have compound leaves, and in *Rubus,* leaf rachides occasionally persist in winter. Fruits variable: nutlets enclosed by hypanthium in the multiple fruits (hips) of the roses, deciduous multiple fruits in the genus *Rubus*, to the dry nutlets in open fruit cups of *Dasiphora*.

Key to Rosoideae

1　Leaf leaving a narrow scar, twigs in section mostly round, often with prickles or long, bristly hairs, fruits are rose hips
　. . . . . ***Rosa***
1*　Persistent parts of leaf base covering the lateral buds partly or entirely, at the leaf scar often an irregular break or persistent leaf or stipule remnants 2
2　Leaf stipules membranous and long, almost entirely covering the buds, twigs not armed, brown and hairy, dry multiple fruits in clusters with ten calyx lobes (5+5 epicalyx) . . . . ***Dasiphora fruticosa***
2*　Stipules not covering buds, fruits mostly deciduous, with 5 calyx lobes, twigs often with prickles. . . . . ***Rubus***

Rosa L., rose

The lateral and and terminal buds are protected by spiral scales from leaf bases and fused stipules. They are compact to narrowly ovoid, often with reddish bud scales, and offer few characters to distinguish between the species. To tell the species apart, the most useful characters are the fruits or infructescences, which often persist; prickles and hairs, bloom layers and growth form. Some species are so closely related, and therefore so similar, that to tell them apart in winter is not always possible.

Key to *Rosa*
(fruit remnants mostly required)

1　Twigs covered by large prickles and with bristles, or with bristles only . . 19
1*　Twigs with only prickles, or unarmed　2
2　Twigs reddish-violet, on the shaded side often greenish or entirely green, bark of twig smooth, when brown flaking off　4
2*　Twigs brownish, bark of twig flat warty, not flaking 3
3　Prickles flat, orange-brown . . . .
　. . . . . ***Rosa xanthina***
3*　Prickles not flattened, pale white-ochre to grey . . . .***Rosa foetida* f. *hugonis***
4　Climbing shrub. Fruits small, globose, 5–7 mm diameter, many in oblong panicles . . . . ***Rosa multiflora***
4*　Erect, partly overhanging shrubs 5
5　Bark of second year or older twigs flaking . . . . *Rosa roxburghii*
5*　Bark not flaking 6
6　Twigs violet, bluish-white with waxy bloom. 7
6*　Twigs without waxy bloom, often greenish 8
7　Sepals deciduous after fruit ripening . .
　. . . . . *Rosa tomentosa*
7*　Sepals persistent 11
8　Prickles straight (at most slightly curved), or plant unarmed. Sepals of fruit persistent 9
8*　Prickles distinctly curved, sepals often deciduous. 14
9　Fruits narrowly ovoid to bottle-shaped
　. 25
9*　Fruits globose to short-ovoid 10
10　Fruits to 3 cm diameter, twigs greenish
　. . . . . *Rosa villosa*
10*　Fruits slightly smaller, twigs ± reddish, partly with waxy bloom 11
11　Twigs greenish, at least on the shaded side . . . .*Rosa caesia*
11*　Twigs dark violet-brown to wine-red　12
12　Twigs with waxy bloom, prickly at least at the twig base 13
12*　Twigs neither armed nor with waxy bloom. . . . .*Rosa majalis*
13　Twigs prickly only at the base, at the tip of twigs mostly unarmed . **Rosa glauca**
13*　Twigs sparsely prickly up to the tip . . .
　. . . . .*Rosa mollis*

14 (8) Fruit sepals persistent or/and twigs
  with waxy bloom and reddish-violet. .15

14* Sepals deciduous, twigs mostly greenish
  17

15 Twigs without waxy bloom, wine-red,
  prickles on the nodes often in pairs . . .
  . . . .*Rosa majalis*

15* Twigs with waxy bloom, greenish when
  not with waxy bloom 16

16 Twigs wine-red violet, mostly with waxy
  bloom. . . . . *Rosa dumalis*

16* Very similar, twigs maybe a little
  greenish on the shaded side *Rosa caesia*

17 (14) Twigs slender (c. 3 mm), pendulous
  or climbing, prickles short (3–4 mm). .
  . . . . *Rosa arvensis*

17* Prickles longer and shrubs not climbing
  18

18 Fruit stalks glandular, buds red, twigs
  reddish on the sun side *Rosa micrantha*

18* Fruit stalks mostly glabrous . . . .
  . . . . **Rosa canina**

19(1) Fruits large, flattened-globose and to
  2.5 cm broad, twigs grey-brown . . . . .
  . . . . **Rosa rugosa**

19* Fruits smaller or deciduous, twigs
  when young green to wine-red (when
  brownish, see couplet 3) 20

20 Prickles flattened wing-like, straight . .
  . . . .**Rosa omeiensis** f. **pteracantha**

20* Prickles only slightly flattened, often
  curved 21

21 Twigs reddish at least on one side . . 23

21* Twigs green all round 22

22 Compact shrub to 1 m high, fruits
  globose to short-ovoid with glandular
  hairs. . . . . **Rosa gallica**

22* Shrub to 3 m high, fruits ovoid, often
  glabrous, only basally with some glands
  . . . . **Rosa rubiginosa**

23 Shrub to 2 m high, prickles and bristles
  sparse. . . . .*Rosa pendulina*

23* Shrubs to 1 m high, densely prickly
  bristly. 24

24 Fruits narrowly ovoid. . . . .**Rosa nitida**

24* Fruits globose. Stalk thickened, fleshy. .
  . . . .**Rosa spinosissima**

25 (9) Prickles paired at the nodes, twigs
  mostly wine-red all-round . . . .
  . . . .*Rosa moyesii*

25* Prickles scattered, twigs matte green on
  the shaded side . . . . . .*Rosa pendulina*

Subgenus Rosa

Section Gallicanae

Small upright shrubs, twigs mostly
with straight prickles or bristles; sepals
deciduous.

Rosa gallica L., French rose

Twigs sturdy, on the sun side deep green,
shaded side matte green. **Prickles** many,
slightly curved or almost straight or mixed:
merging via finer prickles to glandular
hairs. **Fruits** to 1.5 cm, globose to obovoid-
pyriform, bristly, with reflexed almost
pinnate sepals with densely glandular hairs,
as on the fruit stalks. Frequent, compact
shrub to 1 m high with wide-reaching
runners. In cultivation since ancient times.
Origin: south and central Europe.

One of the original garden roses. Other
cultivated roses belonging to the *Rosa
gallica* aggregate include: **Rosa centifolia**
L., Provence rose; **Rosa ×damascena** MILL.,
damask rose; and **Rosa ×alba** L., white rose
of York, an often only slightly armed shrub,
2–3 m high.

Section Pimpinellifoliae

Rosa spinosissima L., Scotch rose
[*Rosa pimpinellifolia* L.]

Twigs relatively slender, below the tip
to 2 mm; on the sun side dark wine red
to orange-red, greenish on the shaded
side. **Prickles** ± straight, narrow-based,
irregularly long and needle-like, the longest
ones on one-year-old twigs to c. 6 mm,
on small stems to 12 mm long mixed with
many needle-like bristles. **Fruits** on longer
stalks, glandular or lacking glands, globose,
1–1.5 cm thick, blackish, with persistent
narrow, acute sepals. Disc narrow, with wide
opening. Frequent, shrub 0.2–1 m high,
spreading by stolons. Origin: Europe from
the Mediterranean to western Asia.

Rosa gallica
Twig with fruits

Rosa gallica
**Lateral bud
surrounded
by prickles**

Rosa ×alba
Lateral bud

Rosa spinosissima
Part of twig with needle
prickles and lateral buds

Rosa spinosissima
Tip of twig

Rosa foetida
Part of twig

Rosa xanthina
f. *hugonis*
Part of twig

Rosa xanthina
f. *hugonis*
Lateral bud
and prickles

Rosa omeiensis
f. *pteracantha*
Tip of twig

Rosa omeiensis f. *pteracantha*
Prickles widened wing–like

Rosa foetida HERRM., Austrian copper rose

Buds globose, c. 2 mm thick with greenish to brown bud scales. **Twigs** initially greenish, soon red-brown, older twigs grey-brown to dirty grey. **Prickles** straight, with a broad base, white-ochre to pale grey., later darker, close to the ground mixed with bristles. **Fruits** dark red, flattened-globose. Occasionally encountered, shrub to 2 m, forming few stolons and with slightly hanging overhanging branches. Origin: central Asia.

Rosa xanthina LINDL. f. *hugonis* (HEMSL.) A. V. ROBERTS, Father Hugo's rose [*Rosa hugonis* HEMSL.]

Buds ovoid, 3–4 mm long. Lateral buds spreading from the twig. Lower bud scales dark wine-red to brown, upper ones paler. **Twigs** violet-brown to dark red-brown, often densely warty; with straight orange-brown to red-brown broad-based flat **prickles**, often, particularly at the base of the long-shoots, mixed with bristles. **Fruits** globose, c. 15 mm thick, dark red and mostly deciduous before winter. Occasionally planted, shrub 2–2.5 m high with branches hanging down. Origin: central China.

Rosa omeiensis ROLFE f. *pteracantha* (FRANCH.) REHDER & E. H. WILSON, winged thorn rose

Buds ovoid, spreading from the twig, with a few reddish bud scales. **Twigs** initially greenish, on the light side slightly reddish, later grey-brown with many short sharp bristles and particularly on the nodes of the twig with larger, longitudinally very flat and wing-like broadened prickles. **Fruits** with persistent upright sepals and fleshy thickened stalk. Occasionally planted, shrub 3–4 m high, unmistakable with its flat prickles.

Section Caninae
Subsection Caninae

Rosa canina L., dog rose

Buds half round to ovoid, reddish. **Twigs** glabrous, green, ± glossy with narrow, blackish leaf scars. **Prickles** many, all similar, large, slightly to strongly curved, rarely straight, with large base. **Fruits** orange-red to dark red, globose to narrowly ovoid, sepals deciduous. This is a species aggregate of many similar species which are difficult to identify even in the flowering season. The most common rose in central Europe; its area of distribution extends from north Africa and Asia Minor via Europe to central Asia.

Rosa caesia SM.
[*Rosa coriifolia* FR.]

Buds ovoid, c. 3 mm long, with few wine-red bud scales. **Twigs** violet-wine-red, on the shaded side also greenish ± with waxy bloom. **Spines** only slightly curved. **Fruits** persistent, globose to 2.5 cm diameter with longer persistent, upright sepals, like the fruit, ± densely beset with glandular hairs. Rarely planted, densely branching shrub to 1.5 m high. Origin: Europe and Asia Minor.

The closely related ***Rosa dumalis*** BECHST. is also a shrub, to 2 m high.

Subsection Rubrifolia

Rosa glauca POURR., red-leaved rose

Twigs dark wine red, bluish-white with waxy bloom. Unmistakable. **Prickles** straight to curved. **Fruits** globose, c. 1.5 cm thick, sepals deciduous. Frequently planted, erect shrub to 3 m high. Origin: mountains of the Pyrenees to south-east Europe.

Subsection Rubigineae

Rosa rubiginosa L., eglantine
[*Rosa eglanteria* L. nom. rej.]

Buds wide at base, together with some red bud scales which are darker towards the margin. **Twigs** green, slightly glossy, except in the buds and leaf scars without red. **Leaf scars** dark, broad and narrow. **Prickles** of two kinds: many large hook-like curved ochre-brown to pale grey-brown prickles with long, ovate base; and here and there

Rosa caesia
Tip of twig

Rosa canina
Twig with fruits

Rosa caesia
Tip of twig with fruit

Rosa glauca
Tip of twig

Rosa glauca
Part of twig

Rosa rubiginosa
Twig with fruits

Rosa rubiginosa
Part of twig

Rosa micrantha
Detail: lateral bud
and prickles

Rosa micrantha
Part of twig

Rosa villosa
Tip of twig

Rosa villosa
Twig with fruits

Rosa mollis
Tip of twig

Rosa villosa
Part of twig

many fine, slender, ± straight prickles, most of all on the stalks of fruit, here mixed with glandular bristles. **Fruits** red, ovoid, 1.5–2 cm long with upright, spreading, ± pinnate and glandular sepals. Dense shrub attaining a height of over 2 m; branches initially upright, later hanging down. Some forms planted more frequently. Origin: Europe to Caucasus and Asia Minor.

To the species aggregate of the eglantine (*Rosa rubiginosa* agg.) belongs ***Rosa micrantha*** BORRER ex SM. with slightly sturdier reddish twigs and similarly sturdy hook-like, curved prickles. **Fruits** globose to ovoid, 1–1.5 cm long, glandular; sepals deciduous, recurved, of various shapes: the outer ones with lanceolate pinnae, the inner ones ± with entire margin, glandular-ciliate otherwise without glands or the latter only at the base.

Subsection Vestitae

Rosa villosa L., apple rose

Twigs green to brownish, initially hairy, becoming glabrous. **Prickles** ± straight on rounded bases, all the same or towards the tip of the twig of two kinds: sturdier prickles mixed with sharp bristles. **Fruits** globose to ovoid, red-glandular to softly spiny, with persistent glandular bristly, erect sepals. Rare compact shrub over 2 m high.

Similar is the closely related ***Rosa mollis*** SM.: **twigs** dark wine red, on the shaded side greenish and with some waxy bloom. **Fruits** under 1 cm thick, sepals often without glands. Low shrub hardly exceeding 1.5 m high. Origin: northern and western Europe to central Russia.

Rosa tomentosa SM., harsh downy rose

Buds ovoid, dark wine-red. **Twigs** dark wine-red to violet-brown and whitish to pale blue with waxy bloom; ± zig-zag. **Prickles** straight to slightly curved, often in pairs by the lateral buds. **Fruits** globose 1–2 cm thick, covered in stalked glands, sepals deciduous before the fruit is ripe. Origin: Europe to Asia Minor and the Caucasus.

Section Rosa (Cinnamomeae including Section Carolinae)

Rosa nitida WILLD.

Twigs densely bristly. **Prickles** slender, 3–5 mm long. **Fruits** on glandular bristly stalks, globose, c. 1 cm thick, slightly bristly; sepals erect, glandular, bristly. Small erect shrub, 50–70 cm high. Origin: eastern North America.

Rosa virginiana MILL., American dwarf wild rose
[*Rosa lucida* EHRH.]

Twigs red-brown. **Prickles** initially bristly, later hook-shaped. **Fruits** on glandular-hairy stalks, flat globose to 1.5 cm broad, smooth and red. Shrub without runners, to c. 1.5 m high. Origin: eastern North America.

Rosa pendulina L., alpine rose

Buds with terminal bud, lateral buds ovoid and spreading. Twigs reddish on the sun side, on the shaded side greenish, occasionally with some waxy bloom. **Prickles** few on older twigs, slender to 6 mm long spines with small, round base, only basal part persistent. Fruits red, mostly hanging, 2–2.5 cm long, **narrowly** ovoid, towards the tip narrowed; sepals persisting, entire and erect. Shrub forming runners 0.5 m to over 2 m high. Origin: in the mountains of central and southern Europe.

Rosa majalis HERRM., cinnamon rose

Buds red, terminal buds ovoid, 3–4 mm long, lateral buds (particularly below the tip of the twig) just as long or much smaller. **Twigs** slender (c. 2 mm), brown-red. **Prickles** to 5 mm long, hook-like, pale grey to ochre; fruit shoots mostly unarmed. **Leaf scars** narrow, surrounding the twigs for c. $^2/_3$. Fruits few or solitary, globose or flattened, 1–1.5 cm thick, smooth and red. Slightly branching shrub to 1.5 m high, with runners. Main centre of distribution: northern and eastern Europe to western Asia; locally in central Europe.

Rosa nitida
Part of twig

Rosa pendulina
Short-shoot with fruit

Rosa majalis
Short-shoot with fruits

Rosa majalis
Tip of twig

Rosa pendulina
Tip of twig

Rosa majalis
Detail of twig: lateral bud and prickles

Rosa rugosa
Part of twig: lateral bud and prickles

Rosa rugosa
Fruit

Rosa moyesii
Short-shoot:
two prickles
at the base

Rosa moyesii
Lateral bud and prickles

Rosa moyesii
Tip of twig
with fruit

Rosa multiflora
Twig with many
pea-sized fruits

Rosa arvensis
Twig

Rosa arvensis
Part of twig with lateral
bud and prickles

Rosa rugosa THUNB., hedgerow rose, Japanese rose

Buds ovoid, spreading from the twig, with brown bud scales. **Twigs** grey-brown, with many unequal straight prickles. **Fruits** several in a group, short-stalked, large, flattened-globose with upright, persistent sepals which are leaf-like. Very frequently encountered, sturdy shrub, 1–1.5(–2) m high. Origin: eastern Asia.

Rosa moyesii HEMSL. & E. H. WILSON

Buds globose to ovoid, 2–4 mm long, with wine-red scales. **Twigs** matte orange wine-red, on the shaded side also yellowish green. **Prickles** straight, often in pairs at the nodes, with medium to large bases. **Fruits** persistent over winter, narrowly ovoid, 5–7 cm long with erect, pinnate sepals, these with stalked glandular hairs. Occasionally planted, lax shrub to 3 m high. Origin: China.

Section Systylae

Rosa multiflora THUNB., multiflora rose

Twigs relatively slender, green, on the light side only slightly reddish. **Prickles** few or sometimes entirely absent, slightly curved, with small base. **Fruits** red, small: c. 5 mm across, very many in large conical panicles. Very frequent, climbing shrub to 3(–5) m high. Origin: Eastern Asia.

The similar *Rosa wichuraiana* CRÉP. is a half evergreen prostrate shrub which does not climb; it has ovoid deeply red fruits to 1.5 cm long, united in short-conical groups of 6–10. This is a species that is also used in the breeding of climbing roses. Origin: eastern Asia.

Rosa arvensis HUDS., field rose

Buds small, short-ovoid, with reddish bud scales. **Twigs** slender: first year's growth to 3 mm diameter, green. **Prickles** on broad bases, 4 ×1.5 mm and 4–5 mm long, gently curved. Low creeping or 1–2 m high climbing shrub. Rare species, distributed from south-west to central Europe

Subgenus Platyrhodon

Rosa roxburghii TRATT., burr rose

Bark of twigs initially green, later grey-brown, flaking. **Prickles** mostly in pairs at the nodes. **Fruits** green, 3–4 cm thick, globose, densely spiny. Occasionally planted, shaggy shrub, to over 2.5 m high. Origin: Japan and China.

Rubus L., blackberry, raspberry

The leaf base with its stipules is mostly retained, occasionally, particularly in blackberries, even the rachis or the whole leaf. Towards the tip of the twig, the buds are more slightly developed. The shoots have a short life span, up to 2 years old: in the first year they shoot from the base of the shrub and in the second year they flower and bear fruit, after which they die.

Key to *Rubus*

1 Twigs grey to ochre-brown, unarmed or ± dense with slender, straight spines . . 2

1* Twigs green to dark wine-red, mostly with many, often curved spines 5

Raspberries

2 Bark tearing open, unarmed, buds mostly > 8 mm . . . . *Rubus odoratus*

2* Bark often furrowed, but not tearing open, buds mostly smaller 3

3 Twigs dense, long spreading, with glandular hairs . . . . *Rubus phoenicolasius*

3* Twigs glabrous or short and appressed, never with glandular hairs 4

4 Twigs pale grey-brown fine whitish hairy or with waxy bloom . . . . . . *Rubus idaeus*

4* Twigs orange-brown, not with waxy bloom . . . . *Rubus spectabilis*

Blackberries

5 (1) Twigs slender (<3 mm), round and with waxy bloom . . . .*Rubus caesius*

5* Twigs mostly sturdier, often angular and not with waxy bloom . . .*Rubus fruticosus*

Blackberries

Rubus caesius L., dewberry

Buds ovoid, 3–5 mm long, with a few brown bud scales in a spiral. **Twigs** slender: one year old 2–3 mm thick, round, with bluish-white waxy bloom, green to (mostly

Rosa roxburghii
Twig with flaking bark

Rubus caesius
Lateral bud

Rubus caesius
Twig

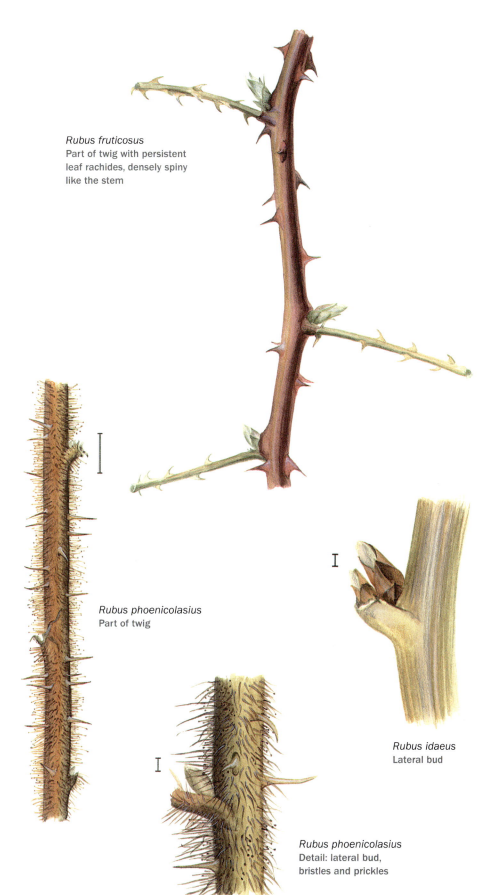

Rubus fruticosus
Part of twig with persistent
leaf rachides, densely spiny
like the stem

Rubus phoenicolasius
Part of twig

Rubus idaeus
Lateral bud

Rubus phoenicolasius
Detail: lateral bud,
bristles and prickles

on older parts of the twig) violet wine-red. **Prickles** small, c. 2 mm long, gently curved. Overhanging shoot tips which touch the ground take root. Creeping or climbing shrub, occasionally seen. Origin: Eurasia.

Rubus fruticosus L., blackberry

Species with many forms, a deciduous to evergreen shrub, erect, overhanging or clambering to creeping. Some authors see this as an aggregate with many agamospermous microspecies. **Buds** medium-sized, ovoid with some spiral fine-appressed-hairy bud scales. **Twigs** round to strongly 5-angular, particularly on the sun side often strongly wine-red to green, glabrous to hairy, also glandular; with equal or unequal straight or ± strongly curved **prickles**. In the deciduous blackberries, the spiny rachis often persists in winter (rambler). Frequent, native shrub c. 2 m high, and cultivated as fruit shrub. Origin: Europe.

Raspberries

Rubus phoenicolasius MAXIM., wineberry

Buds relatively small: 3–5 mm long, ovoid, densely grey-white-hairy; with 4–6 mm wide leaf base. **Twigs** thick and round; red-brown, in places ochre-yellow, to dark grey-brown and violet-brown; dense with 6–7 mm long hairs, partly glandular, and with scattered 6–8 mm long slender, straight or gently curved, **prickles**. Occasionally planted. Origin: east Asia.

Rubus idaeus L., raspberry

Buds 5–7 mm long, with brown, keeled bud scales, most of all the upper ones fine white-pale grey hairy, as if with waxy bloom; on sturdier basal sections of twigs often with small descending accessory bud. **Twigs** pale ochre-brown to grey-brown, slightly striped longitudinally, but not fissured, very fine white-hairy, as if with waxy bloom; with short straight prickles or unarmed. Leaf base large, ending in an indistinct fringed scar. Frequently planted fruit-bearing shrub, 1–2 m high, forming dense stands. Origin: Eurasia.

Rubus odoratus L., flowering raspberry

Buds ovoid, 8–10 mm long with 5–6 imbricate, slightly spreading, red-brown to grey bud scales, especially the upper ones densely appressed dark-grey silvery hairy. **Twigs** erect, unarmed, deciduous-hairy towards the tip; with pale cinnamon brown longitudinally fissured bark. Stipules joined in pairs and with the leaf base, basally enveloping the lateral buds. The leaf scar with indistinct traces of vascular bundles. Often, remnants of the many-fruited panicle persist in winter. Shrub, to 3 m high, often planted as an ornamental. Origin: North America.

Rubus spectabilis PURSH, salmonberry

Buds ovoid, 4–7 mm long with 6–8 bud scales, these acute, lowest one small, dark brown and glabrous, the upper one larger, orange-brown and finely appressed-hairy. **Twigs** ochre-brown, slightly longitudinally striped, glabrous and only at the base with slender, straight prickles. **Leaf scars** on distinct leaf cushions, pale, with 3 traces. Next to the leaf scar there are often two thread-like stipule remnants. Shrub 1–2 m high, occasionally planted. Origin: North America

Dasiphora RAF.

This used to be classified with the herbaceous species of *Potentilla*, but the few shrubs are now segregated in their own genus.

Dasiphora fruticosa (L.) RYDB.
[*Potentilla fruticosa* L.]

Lateral buds are surrounded by persistent leaf-base remnants; these consist of the leaf base and large, fused membranous lateral stipules with attenuate tips; adaxially there is also a small scar of the petiole. **Twigs** relatively slender, orange to red-brown to grey-brown, especially towards the shoot tips densely hairy. **Fruits** in terminal panicles, with the dry remnants of ten (each 5) brown incurved "sepals" of calyx and epicalyx. Decumbent and flat to erect, shrub to 1.5 m high. Spread over the entire Northern Hemisphere.

Similarly persistent leaf bases can be found in the genera *Caragana* and *Colutea*. A clear difference is the predominantly brown coloration without green, and the almost always present fruits.

Rubus odoratus
Tip of twig with remnants of fruit

Rubus odoratus
Lateral bud

Rubus odoratus
Tip of twig with remnants of fruit

Rubus spectabilis
Lateral bud

Rubus spectabilis
Lateral bud

Rubus spectabilis
Lateral bud

SUBFAMILY Amygdaloideae

Many European species and very many ornamental woody plants belong to this (new) large subfamily. The tribes Amygdaleae (*Prunus*) and Pyreae, which once had their own subfamilies, belong here next to the Spiraeas and other shrubby tribes.

Key to Amygdaloideae

1 Twigs not armed. 4
1* Twigs thorny 2
2 Twigs with chambered pith, leaf scars with one trace . . . . . **Prinsepia uniflora**
2* Twigs with solid pith, leaf scars with several traces 3
3 All twigs without terminal buds . **Prunus**
3* Long-shoots with terminal buds . . **Pyreae**
4 (1) Buds opposite. . **Rhodotypos scandens**
4* Buds alternate 5
5 Leaf scars with a single trace, twigs without terminal bud 6
5* Leaf scars with several traces or not clearly marked 7
6 Buds solitary or with enrichment buds . . . . **Spiraea**
6* Buds with descending accessory buds or itself an accessory bud below the lateral shoot . . . . **Neillia**
7 Twigs green all-round, without terminal bud. . . . . **Kerria japonica**
7* Not so. 8
8 No leaf scar: leaves dry and fall off late, together with the outermost layer of bark . . . . **Sibiraea**
8* Leaves leave distinct leaf scars 9
9 Twigs with distinctly warty lenticels. . . . . **Photinia**
9* Twigs with flat lenticels or with terminal buds 10

10 Shrubs, twigs thick with wide pith. . . . . **Sorbaria**
10* Shrubs with slender twigs or trees, pith not noticeably wide 11
11 Pith solid 12
11* Pith chambered . . . . *Oemleria*
12 Buds naked or with two ± spreading bud scales, prostrate to erect shrubs 13
12* Buds with distinct bud scales 14
13 Terminal buds surrounded by only loosely appressed 2-pinnate green leaves. Twigs and buds with stellate hairs and glandular sticky . . . . *Chamaebatiaria*
13* Buds brownish, leaves never pinnate . . . . **Cotoneaster**
14 Remnants of fruits long persistent, dry fruits 15
14* Fruits mostly deciduous, fleshy, berry-like multiple fruits or stone fruits. . . 18
15 Fruit with five large dry brown dentate sepals. *Neviusia*
15* Fruits not surrounded by distinctive large sepals 16
16 Multiple fruits with dry follicles in corymbs or panicles 17
16* Multiple fruits of five woody fruit chambers in oblong racemes . . . . **Exochorda**
17 Lateral buds surrounded by two bud scales . . . . *Holodiscus*
17* Lateral bud with more than two bud scales . . . . **Physocarpus**
18 Buds fusiform and/or deep red . **Pyreae**
18* Buds more compact to acute ovoid, at most reddish on one side 19
19 Either to 1.5 m high slender-twigged shrubs with enrichment buds or medium-sized to large trees with ringed bark. Lateral buds often spreading from twig, round in transverse section. . . . . **Prunus**
19* Shrubs without enrichment buds, or trees with mostly scaly bark. Lateral bud on long-shoots mostly appressed or distinctly flattened, i.e. distinctly broader than thick . . . . **Pyreae**

Dasiphora fruticosa
Twig with persistent dry fruit calyces

Dasiphora fruticosa
Lateral buds surrounded by persistent leaf base and the membranous fused stipules

TRIBE Spiraeeae

The genera differ in the size and shape of the buds, the number of traces of vascular bundles on the leaf scar, the presence of enrichment buds and, not least, by the fruits and inflorescences which tend to persist during winter.

Key to Spiraeeae

1 Leaf scars distinctly visible, twigs often without terminal bud. 2

1* Leaves deciduous with adjacent bark, and due to this no clear leaf scars, terminal buds often swollen . . . *Sibiraea altaiensis*

2 Leaf scars with 3 traces, terminal bud on long-shoots . . . . *Holodiscus discolor*

2* Leaf scars with one trace, without terminal buds . . . . **Spiraea**

Holodiscus discolour (PURSH) MAXIM., creambush

Buds 2–3 mm long, appressed to twig or, at the tip, slightly spreading, lateral buds surrounded by mostly 2(–3) bud scales. Terminal buds longer. Bud scales dark wine-red to violet-brown in places, in places also pale brown, and densely appressed white-hairy particularly towards the tip. **Twigs** slightly angular, basally sturdy becoming very slender towards the tip and hanging down, brown to grey-brown, when older longitudinally fissured, losing fibres, grey to violet-brown. **Leaf scar** on distinct leaf cushion, dark brown with 3 blackish vascular traces. **Fruit**: small multiple fruit with 5 nutlets, many in a large pendulous panicle. Frequent, a shrub 1–5 m high with erect twigs hanging down at the end. Origin: western North America.

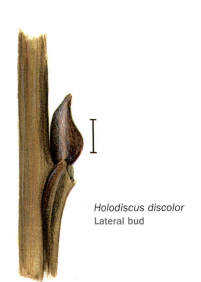

Holodiscus discolor
Lateral bud

Spiraea salicifolia
Lateral bud

Spiraea L., meadowsweet

Buds: only lateral buds, in spirals around the twig. **Twigs** round or ± 5-angled, always without lenticels. **Leaf scars** with one vascular trace. The mostly persistent infructescences offer good characters for identification. Differences in bud formation often occur (number of bud scales, enrichment buds) between very sturdy and slender twigs.

Key to Spiraea

1 Inflorescences terminal or axillary umbels or umbel-like corymbs, mostly broader than high 4
1* Inflorescences terminal racemes, higher than broad, twigs mostly angular 2
2 Buds almost appressed to twig, with ciliate bud scales . . . . **Spiraea salicifolia**
2* Buds spreading from twig, at least initially densely hairy 3
3 Twigs densely hairy . . **Spiraea douglasii**
3* Twigs becoming glabrous . . . .
. . . .**Spiraea douglasii** var. **menziesii**
4 (1) Inflorescences broad, terminal umbel-like corymbs 5
4* Inflorescences small axillary umbels or racemose corymbs, occasionally deciduous, buds distinctly spreading from twig 8
5 Buds appressed to twig, ± acute 6
5* Buds spreading from twig, ovoid . . . . . .
. . . . Spiraea betulifolia

6 The two outermost bud scales longer than bud . . . . **Spiraea henryi**
6* Buds with many bud scales, lowest ones relatively small 7
7 To 25 cm high shrub, decumbent with ascending tips, forming runners . . . . . .
. . . .**Spiraea decumbens**
7* To 1.5 m high (but in some forms much lower) erect shrub . . **Spiraea japonica**
8 (4) Twigs with fissuring and flaking bark, buds very small (c. 1 mm), globose . . 9
8* Bark not flaking, buds mostly larger . .10
9 Twigs densely hairy at least at the tip, slightly angular . . . **Spiraea thunbergii**
9* Twigs glabrous or becoming glabrous, round . . . . .Spiraea hypericifolia
10 Twigs angular, the outer two bud scales overtop the bud and cover it almost completely 12
10* Twigs round, buds with some bud scales, the outermost small 11
11 Buds small: 1(–1.5) mm long, globose-ovoid; twigs hairy . . **Spiraea prunifolia**
11* Buds 2–3 mm long, narrowly ovoid, mostly with enrichment buds; twigs glabrous . . . . **Spiraea ×vanhouttei**
12 Twigs pale ochre-grey . . . .
. . . .**Spiraea chamaedryfolia**
12* Twigs dark red-brown to grey-brown . .
. . . . **Spiraea nipponica**

Section Spiraea

Fruits in oblong panicles.

Spiraea salicifolia L., bridewort

Buds appressed to twig or slightly spreading, ovoid, 3–5 mm long, with a few scales. Bud scales orange-red-brown, darker brown towards the margin, ciliate, at the tip of the bud also hairy. **Twigs** red-brown, matte, angled, very finely and sparsely hairy. **Leaf scars** half circular, with a central trace. Shrub to 1.5 m high, erect, producing runners. Frequently planted and sometimes naturalised. Origin: south-eastern Europe to north-east Asia and Japan as well as in North America.

Spiraea douglasii HOOK., steeple bush

Buds to 45° spreading from the twig, 2–3 mm long, compact-ovoid with blunt tip, in sturdier shoots with enrichment buds. Bud scales ochre-brown, densely hairy, only basal scales more sparsely hairy. **Twigs** brown, slightly angled or in sturdier twigs round, longitudinally striped, densely fine hairy, later becoming glabrous. **Infructescences**: terminal conical panicles. **Fruits** glabrous, glossy, with erect style and recurving sepals. Frequently planted, shrub to 2 m high, forming runners. Origin: western North America.

The var. **menziesii** (HOOK.) PRESL [Spiraea menziesii HOOK.] differs in its twigs which soon become glabrous and by more sparsely hairy buds. Shrub to 1.5 m high.

Spiraea ×billardii hort. ex K. KOCH

A hybrid, frequently encountered, of Spiraea douglasii × Spiraea salicifolia. It can be distinguished from the parent species only with some difficulty: it has brown, initially densely hairy twigs and as the parent species do, forms runners.

Spiraea douglasii
Lateral bud

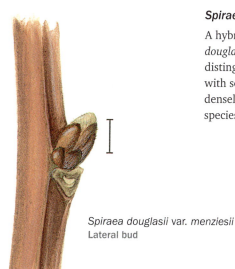

Spiraea douglasii var. *menziesii*
Lateral bud

Section Calospira

Fruits in terminal corymbs. Buds ± appressed to twig, many-scaled and acute.

Spiraea japonica L.f., Japanese spiraea

Buds 3–4 mm long, narrowly ovoid, appressed to twig and slightly spreading towards the tip; with ± oblong triangular, fibrous, ochre-brown to grey-brown bud scales. **Twigs** ochre-brown to bronze-brown, glabrous, relatively thick (at the base of the annual shoot c. 4–5 mm), ± round and only towards the tip slightly 5-angled. Bark of twig finely longitudinally furrowed, at the base splitting longitudinally. **Leaf scars** pale brown with a large central dark brown trace. **Fruits** in very broad (15–20 cm) flat corymbs. Shrub 1.5 m high, but cultivars that stay much smaller are frequently planted, for instance the up to 30 cm high compact cultivar **'Nana'**. Origin: Japan.

Spiraea betulifolia PALL., white meadowsweet

Buds globose to ovoid, 1.5–2.5 mm long, distinctly spreading from twig, on sturdy shoots with enrichment buds. Bud scales ochre-brown to red-brown, the margin slightly darker and toward the tip of the bud shaggy-ciliate or appressed-hairy. **Twigs** glabrous, red-brown, partly with dead epidermis and then grey-brown, slightly angled. **Fruits** in 3–6 cm broad corymbs. Rare, 0.5–1 m high shrub. Origin: eastern Asia.

Spiraea decumbens W. D. J. KOCH

Buds 2–3 mm long, flattened-ovoid, acute, almost appressed to twig. **Bud scales** brown, darker towards the tip. **Twigs** glabrous or hairy (in subsp. *tomentosa* (POECH) DOSTÁL), slender, round or slightly angled, ochre-brown. Frequently planted, small dense shrub, only c. 25 cm high, forming runners. Origin: south-eastern Alps.

Spiraea henryi HEMSL.

Buds to 2 mm long, ovoid, often abruptly acute, slightly spreading from twig only in upper part; with 2–4 outer loose, slightly ciliate, matte ochre-brown to dark brown bud scales. **Twigs** round, brown, initially hairy, mostly becoming glabrous. **Fruits**

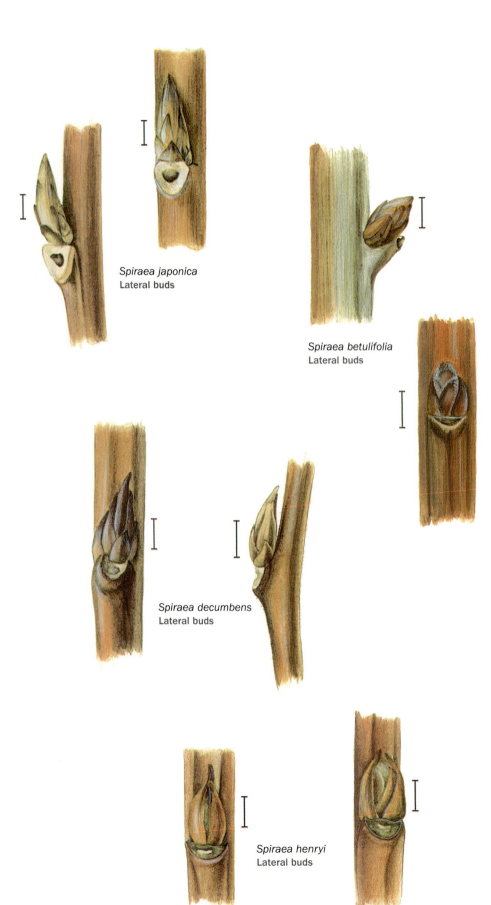

Spiraea japonica
Lateral buds

Spiraea betulifolia
Lateral buds

Spiraea decumbens
Lateral buds

Spiraea henryi
Lateral buds

Spiraea thunbergii
Part of twigs

Spiraea hypericifolia
Lateral bud in upper
part of twig

Spiraea hypericifolia
Lateral bud with
enrichment bud in
lower part of twig

Spiraea prunifolia
Lateral bud in upper
part of twig

Spiraea prunifolia
Lateral bud

in axillary corymbs, hairy and slightly spreading. Occasional, 2–3 m high shrub. Origin: China.

Section Chamaedryon

Fruits in umbels or racemose corymbs on axillary short-shoots.

Spiraea thunbergii Siebold ex Blume, Thunberg's meadowsweet

Buds very small: less than 1 mm, globose, with some ochre-brown bud scales, slightly sunken between leaf cushion and twig. **Twigs** angled, initially very slender, grey-brown, fine and densely hairy, later bark fissuring and flaking in shreds: below deep red-brown. Very frequently planted, dense shrub to 1.5 m high. Origin: China.

Similar and difficult to tell apart is the similarly frequent *Spiraea ×arguta* Zabel, a hybrid with *Spiraea ×multiflora* Zabel.

Spiraea hypericifolia L., Iberian meadowsweet

Buds small, c. 1(–2) mm long, globose to ovoid. Bud scales dark brown to ochre-brown, densely white-hairy particularly at the tip of the bud. **Twigs** finely hairy, ochre-brown to dark red-brown with dying and flaking grey epidermis. **Leaf cushion** clearly raised with a small, one-trace leaf scar. Rarely planted shrub to 1.5 m high. Origin: south-east Europe to Siberia and central Asia.

Spiraea prunifolia Siebold & Zucc., bridal wreath spiraea

Buds very small, less than 1–1.5 mm long, globose, slightly spreading from twig, surrounded by few ochre-coloured slightly hairy bud scales. **Twigs** curving down, slender, grey-green to dark red-brown, ± angled and hairy; often densely so at the tip, glabrous towards base of the twigs and only on the angles with persistent hairs. **Leaf scars** small, semi-circular to half-moon shaped, with one trace on weak, orange-ochre-coloured leaf cushion. Shrub to 2 m high. Origin: eastern Asia. In the frequently planted double-flowered species, fruits are hardly developed.

Spiraea nipponica MAXIM.

Buds spreading from the twig, c. 2(–3) mm long, protected by 2 ± long acute ochre-brown bud scales. **Twigs** furrowed and ridged, particularly distinct on growing long-shoots, dark red to violet-brown, partly dark grey due to dying epidermis. **Leaf scars** on developed leaf cushion with a prominent remnant of vascular bundle. Frequently planted, shrub to 2.5 m high, erect and then curving down. Origin: Japan.

Spiraea ×vanhouttei (BRIOT) CARRIÈRE
[*Spiraea cantoniensis* LOUR. × *Spiraea trilobata* L.]

Buds blunt conical, spreading; on weak branches c. 1 mm long, almost completely surrounded by 2 prophyll-bud scales, on thicker twigs multi-scaled, 2–3 mm long, mostly with enrichment buds. Bud scales brown, glabrous, strongly white ciliate. **Twigs** glabrous, red-brown to violet-brown, sturdier twigs a little paler, ochre-brown to brown, at the tip hanging down. **Fruits** in axillary corymbs, with spreading sepals and half-erect styles. Frequently planted, shrub 2–3 m high.

Spiraea media F. SCHMIDT

Buds 2–3 mm long, narrowly ovoid, spreading from twig, solitary or with enrichment buds; bud scales 6–8(–12), basal ones often slightly darker grey-brown, otherwise ochre-brown to brown, ± densely hairy towards the tip of the bud. **Twigs** round, only first year's growth indistinctly angled, pale ochre-brown to darker red-brown, when older grey-brown. Slender twigs zig-zag, the thicker ones vertical and straight, slightly longitudinally striped. **Leaf scars** on distinct cushions, with one trace, with fine ridges running down the twig from the margins of the cushion for some distance. Rarely planted, stiffly erect shrub 1.5 m high. Origin: eastern central Europe to north-east Asia.

Spiraea chamaedryfolia L.

Buds spreading from twig, ovoid, 3–4 mm long, with 2(–4) outer, very pale grey-brown, often ± long-acute bud scales which are slightly hairy towards the tip. **Twigs** glabrous, pale grey-brown, slightly zig-zag, strongly angular with 5 wing-like ridges.

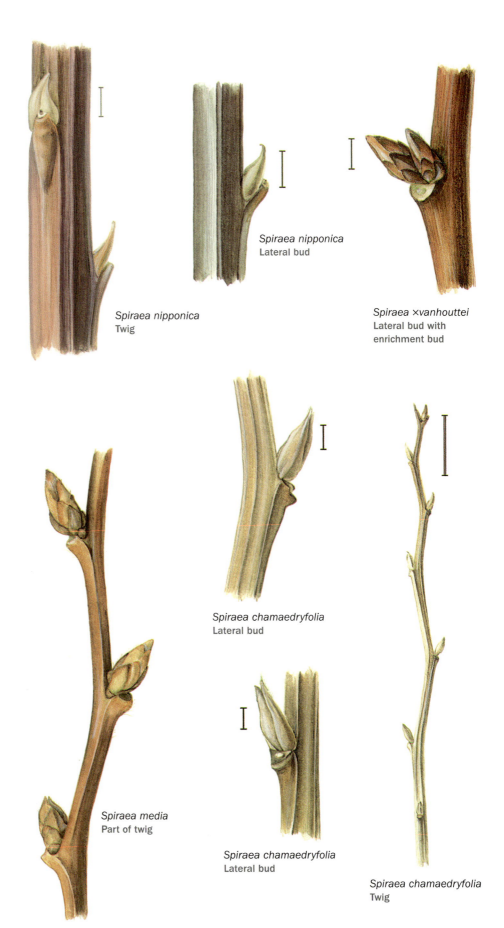

Spiraea nipponica
Lateral bud

Spiraea nipponica
Twig

Spiraea ×vanhouttei
Lateral bud with
enrichment bud

Spiraea chamaedryfolia
Lateral bud

Spiraea media
Part of twig

Spiraea chamaedryfolia
Lateral bud

Spiraea chamaedryfolia
Twig

Sibiraea altaiensis
Lateral bud

Sibiraea altaiensis
Twig distinctive: leaves
deciduous together with
outer bark

Chamaebatiaria millefolium
Short-shoot with terminal bud

Chamaebatiaria millefolium
Lateral bud

Infructescences: corymbs on lateral short-shoots. Frequently planted, erect shrub, 1.5 m high, forming runners. Origin: European mountains, Siberia and north-east Asia.

Sibiraea altaiensis (LAXM.) C. K. SCHNEID. [*Sibiraea laevigata* (L.) MAXIM.]

Lateral buds glabrous, 5–6 mm long, acute-ovoid; dry, brown bud scales loose, basally spreading, pointed, keeled and easily deciduous. **Terminal buds** mostly not closed and rather open, often with terminal infructescence. **Twigs** initially surrounded by old, dry leaves which are firmly attached to the outer bark; the outer layer of bark flakes off together with these old leaves, and the layer below is red-brown to dark wine-red or, in places, olive-green. **Leaf scars** not present because of this, only 3 small round scars of vascular bundles. **Fruits** c. 4 mm long with 5 follicles which are half hidden by the flower cup. Occasionally planted, shrub, unmistakable, to 1 m high. Origin: eastern Europe to Siberia and the Altai.

TRIBE Sorbarieae

The spirally placed leaves are mostly pinnate and form leaf scars with 3 traces.

Key to Sorbarieae

1 Buds without bud scales, bud leaves bipinnate, twigs resinous and sticky.... *Chamaebatiaria*
1* Buds with bud scales, twigs thick, not sticky . . . . **Sorbaria**

Chamaebatiaria millefolium (TORR.) MAXIM., fern bush

Although the species is described in literature as deciduous, it does lose its leaves along the shoots, but around the terminal buds many of the finely twice-pinnate leaves are retained. All parts with felty shaggy tufted hairs, resinous-aromatic (hairs in sticky resin). **Twigs** light ochre-brown, underneath the tufted hairs with many globose resin glands. Bark first tearing open, later flaking, then smooth and brown. **Pith** white. **Leaf scars** pale ochre with 3 traces, stipule scars with one central trace each. Very rarely planted, shrub to 1 m high. Origin: west and south-west USA.

Sorbaria (SER.) A. BRAUN

Coarse-branched shrubs with large buds and thick twigs from Asia and North America. The terminal fruit panicles, to 30 cm long, often persist, with multiple fruits 5(–6) mm long consisting of 5 follicles.

Key to Sorbaria

1   Bud scales, at least partly, green. . . . . . .
. . . .***Sorbaria sorbifolia***
1* Bud scales brown to grey-brown . . . . . .
. . . .*Sorbaria kirilowii*

Sorbaria sorbifolia (L.) A. BRAUN, false spiraea

Buds large, to 8(–10) mm long, ovoid, with some **bud scales** in spiral arrangement. The lowest scales small and dry brown, often with enrichment buds in the axils, the next enclosing the bud, green with dried brown-grey tips. **Twigs** pale ochre-brown; lenticels slightly raised initially, soon grey, tearing open longitudinally. Pith broad and pale brown. **Leaf scars** shield-shaped, grey-brown with three traces of vascular bundles. **Fruit panicles** often persistent in winter, terminal, erect, dense. Follicles 4.5–6 mm long, style terminal, recurving. Frequently planted, shrub 1–2 m high, spreading by runners, sparsely branching and coming into leaf early. Origin: north-east Asia.

Sorbaria kirilowii (REGEL & TILING) MAXIM. [Sorbaria arborea C. K. SCHNEID.]

Buds ovoid, 5–8(–10) mm long, globose to ovoid. **Bud scales** brown to grey-brown, slightly appressed, especially the upper ones ± densely hairy. **Twigs** pale grey-brown, sparsely hairy or glabrous, not as thick as in *Sorbaria sorbifolia*. **Leaf scars** with 3 traces, very light ochre-yellow. Fruit panicle loose, reflexed. Follicle 3–4.5 mm long, style fixed distictly below the apex. Rarely encountered, broad shrub 6 m high. Origin: eastern Asia.

TRIBE Neillieae

Key to Neillieae

1 Terminal buds present, lateral buds mostly solitary . . . .*Physocarpus*
1* Without terminal buds, lateral buds often with accessory buds . . . . *Neillia*

Neillia D. DON

Shrubs 1.5–2.5 m high, difficult to tell apart in winter. In addition to the typical winter characters, mostly with many dry leaves. **Fruits** in terminal inflorescences, either in panicles, with a single follicle, c. 2 mm long, in a shallow flower cup, or in racemes, the 1–2 follicles c. 5 mm long, enclosed by the tubular flower cup. Origin: eastern Asia.

Key to Neillia

1 Infructescences (remnants of) panicles . .2
1* Infructescences (remnants of) racemes . . .
. . . . *Neillia sinensis*
2 Buds mostly oblong and acute . . . .
. . . .***Neillia tanakae***
2* Buds often blunt, narrowly ovoid . . . . . .
. . . .***Neillia incisa***

Sorbaria sorbifolia
Lateral bud

Sorbaria sorbifolia
Lateral bud

Sorbaria kirilowii
Tip of twig

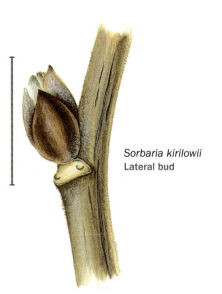

Sorbaria kirilowii
Lateral bud

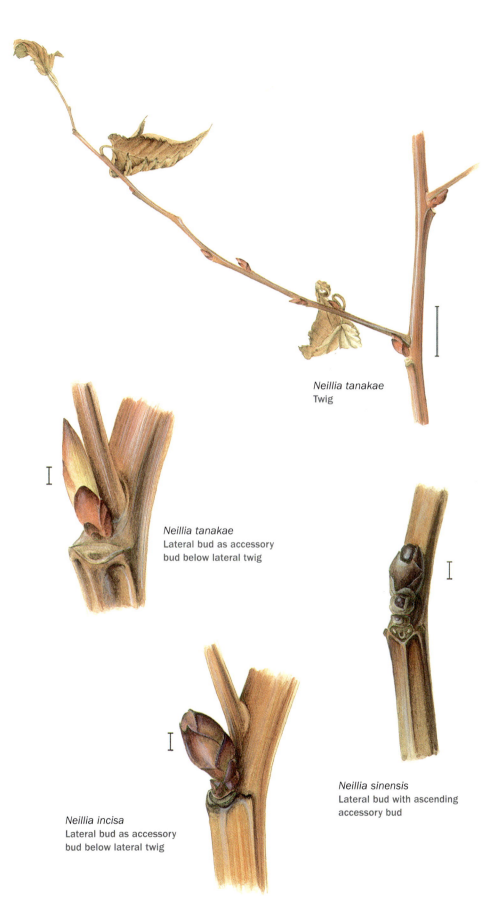

Neillia tanakae
Twig

Neillia tanakae
Lateral bud as accessory
bud below lateral twig

Neillia sinensis
Lateral bud with ascending
accessory bud

Neillia incisa
Lateral bud as accessory
bud below lateral twig

Neillia tanakae FRANCH. & SAV.
[*Stephanandra tanakae* FRANCH. & SAV.]

Buds only lateral buds, narrowly ovoid, to 10 mm long, mostly with descending accessory buds or themselves accessory buds below shoots formed in the last growth period. With few reddish bud scales. **Twigs** initially slightly angled, orange-brown to red-brown, zig-zag. **Leaf scar** with one trace. **Leaves** persistent dry on the twig, 3–8 cm long, double-dentate. Frequently planted, shrub to 2 m high. Origin: Japan.

Neillia incisa (THUNB.) S.H.OH
[*Stephanandra incisa* (THUNB.) ZABEL]

Buds 5–8 mm long, narrowly ovoid, blunt or slightly acute, wine-red; often with descending accessory buds or themselves accessory buds below lateral twigs. **Twigs** round, glabrous, orange-ochre to deep red-brown, zig-zag, older twigs grey-brown. **Leaf scars** with a large central trace, with faint traces of stipules. **Leaves** lobed, 2–6 mm long. Frequently planted, shrub 1.5–2.5 m high. Origin: Japan and Korea.

Neillia sinensis OLIV.

Buds: only lateral buds, globose to short-ovoid, 4–5 mm long, mostly with 1–2 descending accessory buds. Bud scales many, often loosely appressed, dark red-brown, slightly lighter at the margin and sparsely ciliate. **Twigs** round and glabrous, at first year's growth 1.5–4 mm thick, orange-brown to dark brown, slightly zig-zag. **Leaf scars** broad-triangular, with 3 distinct traces: the middle one U-shaped, the lateral ones oblique, broad-dot-shaped; next to the leaf scar with traces or remnants of stipules. **Fruits** 2 mm long, consisting of 1–2 follicles with 1–3(–5) seeds, enclosed by the flower tube. Rare; shrub to 2 m high. Origin: China.

Neillia thyrsifolia D. DON from the Himalaya is occasionally planted. It remains slightly lower and has angled twigs.

Physocarpus (CAMBESS.) RAF.

Small to medium-sized shrubs. The species can be recognised and told apart by their fruits, the shape of buds and the indument. The multiple **fruits** consist of 2–5 fruitlets: pods, capsules, opening on both sides in contrast to the one-sided bursting follicles which occur in many other genera of the subfamily. Origin: ten species from North America and one species from northern Asia; some species are more frequently planted and some of these have naturalised.

Key to *Physocarpus*

1 Buds compact-ovoid, twigs and fruits often with very fine indument . . . .
. . . . *Physocarpus amurensis*
1* Buds oblong, twigs and fruits glabrous. . . . . *Physocarpus opulifolius*

Physocarpus amurensis (MAXIM.) MAXIM.

Buds short-ovoid, 3–5 mm long, with some matte dark brown bud scales densely ciliate to the tip and at the tip ± yellowish-white hairy. **Twigs** glabrous or finely hairy, orange-brown to red-brown, angled by ridges running down from the leaf scar margins. **Leaf scars** rounded-triangular to broadly U-shaped, with 3 or 5 traces; on the edges with indistinct stipule scars. **Fruits** many in corymbs, with 3–4(–5) stellate hairy legumes. Frequently planted, shrub 1–3 m high. Origin: Amur region, Manchuria and Korea.

Physocarpus opulifolius (L.) MAXIM., Atlantic ninebark

Buds narrowly ovoid, 4–9 mm long, with olive-brown to matte violet-brown, towards the tip finely appressed-hairy bud scales. **Twigs** angled, glabrous, orange-brown to red-brown, on older twigs bark flaking-fissured. **Leaf scars** with 3(–5) traces. **Fruits** in c. 5 cm broad, flat-hemispherical corymbs, with 3–5 glabrous legumes. Frequently planted, 2–3 m high shrub. Origin: central and eastern North America.

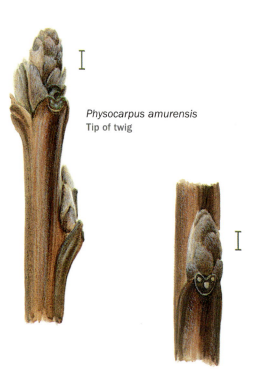

Physocarpus amurensis
Tip of twig

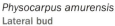

Physocarpus amurensis
Lateral bud

Physocarpus opulifolius
Fruit

Physocarpus opulifolius
Lateral

Rhodotypos scandens
Twig with fruit

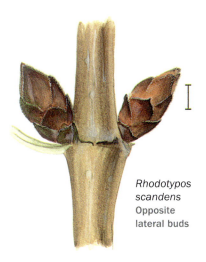

Rhodotypos scandens
Opposite lateral buds

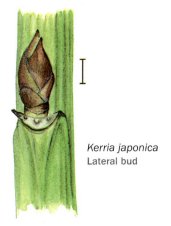

Kerria japonica
Lateral bud

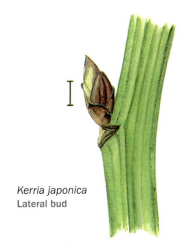

Kerria japonica
Lateral bud

TRIBE Kerrieae

Shaggy shrubs, twigs without terminal buds.

Key to Kerrieae

1 Buds/leaf scars opposite, older twigs brown. . . . . ***Rhodotypos scandens***

1* Buds, leaf scars alternate 2

2 Twigs green . . . . ***Kerria japonica***

2* Twigs brown . . . . . . *Neviusia alabamensis*

Rhodotypos scandens (THUNB.) MAKINO, jet bead

In contrast to most Rosaceae, **buds** opposite: 3–4 mm long, ovoid, spreading from the twig, with 8–10 outer slightly hairy or only ciliate bud scales; on the light side dark violet-brown to red-brown, on the shaded side more green and only margins of bud scales brown. In the axils of prophylls there are, occasionally, enrichment buds. **Twigs** light grey-brown to yellow-brown in the shade, on the light side darker red- to olive-brown, mostly ± glossy. **Lenticels** few, small and indistinct. **Leaf scars** with 3 traces, the opposite scars connected by a fine line; next to these small scars or dried up remnants of stipules. **Fruits**: 4(3–5) hard dry, glossy purple-black fruitlets, overtopping the dry sepals. Frequently planted, shrub to 2 m high. Origin: Japan and central China.

Kerria japonica (L.) DC., kerria

Buds slightly spreading, 2–3 mm long, acute-ovoid, with 6–8 spiral red-brown bud scales (basally partly also greenish), at the margin fine white-ciliate. **Twigs** green all-round (also on the light side), slightly long-furrowed and slightly zig-zag. Ends of shoots drying out, often with remnants of flowers or fruit. **Leaf scars** brown-grey, hardly proud of the twig's surface, with 3 vascular traces. Without lenticels. Very frequently planted, shrub to over 2 m high. Origin: China; long cultivated in Japanese gardens.

Neviusia alabamensis A. GRAY, snow wreath

Buds: only ovoid, 2–3 mm long lateral buds, with several red-brown ciliate bud scales. Frequently with enrichment buds. **Twigs** become more slender towards the tip, slightly zig-zag, red-brown, deeper lighter grey-brown or light grey, finely longitudinally striped, slightly angled to round. **Leaf scar** with 3 traces with fused stipule remnants. **Pith** solid, white. The long persistent dry infructescences are helpful in winter: each fruit surrounded by 5 spreading c. 8–12 mm long dry sepals, dentate towards the tip. Up to 4 nutlets inside, surrounded by the remnants of many filaments. Very rarely planted, shrub 1.5 m high. Origin: USA.

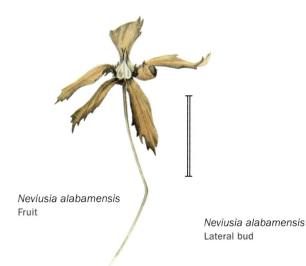

Neviusia alabamensis
Fruit

Neviusia alabamensis
Lateral bud

TRIBE Osmaronieae

Twigs often with terminal bud (absent in *Prinsepia*), pith occasionally chambered.

Key to Osmaronieae

1 Pith of the twigs solid. Fruits of 5 woody 8–12 mm high legumes . . . . ***Exochorda racemosa***
1* Pith of the twigs chambered, fruits deciduous stone fruits 2
2 Twigs unarmed. . . . *Oemleria cerasiformis*
2* Twigs with axillary thorns . . . .*Prinsepia uniflora*

Oemleria cerasiformis (TORR. & A. GRAY) J. W. LANDON, oso berry

[*Osmaronia cerasiformis* (TORR. & A. GRAY) GREENE]

Buds glabrous and glossy, pure green to strong wine-red. **Terminal buds** c. 8 mm long, narrowly ovoid to fusiform, as flower buds bulging-globose. **Lateral buds** mainly below the tip of the shoot, often tiny, towards base of twig larger, but distinctly smaller than terminal buds. **Twigs** glabrous, glossy olive-brown, with pale ochre-brown lenticels. **Pith** chambered. **Leaf scars** narrow with 3 traces. Rarely used, shrub over 2 m tall. Origin: North America.

Prinsepia uniflora BATALIN

Buds spirally placed, very small, without bud scales and densely grey-brown hairy, ± hidden behind the persistent stipule tips. **Twigs** slender: c. 1–2 mm diameter, at one year old silvery grey with 12 mm long axillary thorns, which are absent only at the base and at the shoot tip. The accessory buds below the spines do not differ from the lateral buds in leaf axils. **Twigs** with many, very small, ochre-brown to blackish dots, partly with tufts of hairs. The two-year-old red-brown twigs are in strong contrast to the first year's silvery grey ones. **Leaf scars** slightly prominent, edged ochre-brown with an, often pale, scar of a central vascular bundle. Shrub to 1.5 m high, occasionally planted. Often coming into leaf as early as February. Origin: north-west China.

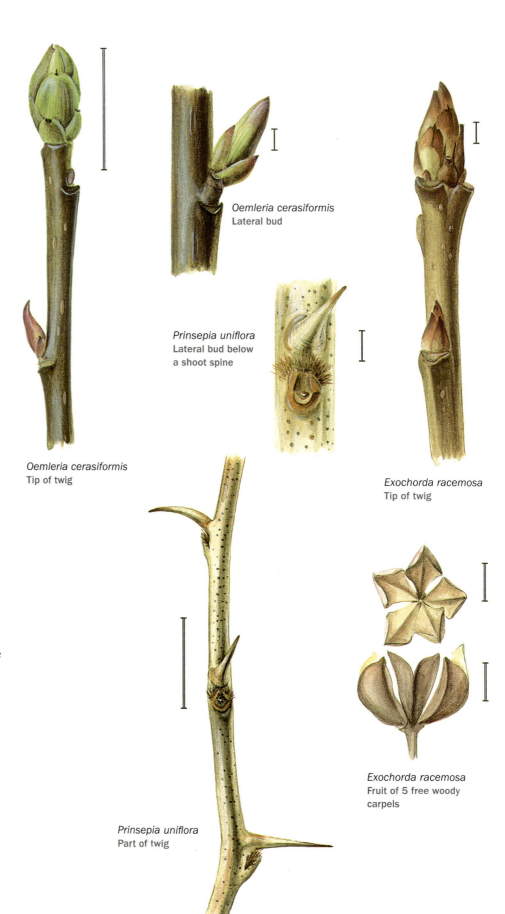

Oemleria cerasiformis
Lateral bud

Prinsepia uniflora
Lateral bud below a shoot spine

Oemleria cerasiformis
Tip of twig

Exochorda racemosa
Tip of twig

Prinsepia uniflora
Part of twig

Exochorda racemosa
Fruit of 5 free woody carpels

Exochorda racemosa (LINDL.) REHDER, pearlbush

Buds narrowly ovoid, differentiated in terminal and lateral buds. Terminal buds 3–4 mm long, slightly larger than lateral buds. Bud scales spirally arranged, deep brown, basally paler orange-brown and partly greenish. Basal bud scales partly with oblong stipule lobes. **Twigs** slender, olive-brown to grey-brown, glabrous or with fine remnants of indument. **Fruits** 1–1.5 cm high, consisting of five 2-valvate woody single carpels connected at the centre. Shrubs to 4 m high. Origin: Eastern China to Central Asia. Frequent ornamental shrubs, unmistakable particularly due to fruit persisting throughout the winter. Sometimes 4 or 5 species are recognised: *E. giraldii* HESSE, *E. korolkowii* LAVALLÉE, *E. racemosa*, *E. serratifolia* S. MOORE, *E. tianschanica* GONTSCH.) but the characters are so variable that even during the growth period no definite separation is possible.

Exochorda racemosa
Fruit cluster

Only the large genus *Prunus* belongs to this tribe.

Prunus L.: cherry, plum, apricot, almond and peach

Small shrubs to large trees. Many ornamental plants belong to the genus, for instance, the frequently planted Japanese flowering cherries, as do many fruit shrubs. Major groups are distinct even without leaves, and these are sometimes treated as separate genera. Within these groups an exact classification can be problematic. The deciduous species are native to the temperate to boreal regions of the Northern Hemisphere.

Key to *Prunus*

Fruit plants <u>underlined</u>

1 Twigs with terminal buds 2
1* Twigs without terminal bud 26
2 Twigs brown to grey 4
2* Twigs red and green 3
3 Buds hairy. Trees . . . .
. . . . . ***Prunus persica*** and ***Prunus dulcis***
3* Buds glabrous, twigs slender. Low shrub
. . . . ***Prunus glandulosa***
4 Stipules persistent or at least not entirely deciduous. Shrub with slender twigs to 1.5 m high 5
4* Stipules deciduous with the leaves or shrub > 2 m high or tree 7
5 Glabrous or finely hairy, stipules thread-like, digitate, ciliate . . . . **Prunus triloba**
5* Stipules not thread-like 6
6 Twigs hairy, lateral buds with many enrichment buds (mostly >2) . . . .
. . . .*Prunus tomentosa*
6* Twigs glabrous, lateral buds mostly with 2 enrichment buds . . . . *Prunus prostrata*
7 (4) Twigs distinctly hairy 8
7* Twigs glabrous (occasionally with a few hairs) 11
8 Buds small, c. 3 mm long, ovoid, spreading . . . .***Prunus mahaleb***
8 Buds larger, narrowly ovoid to conical . 9
9 Bark very smooth, glossy, coming off in rolls. Buds conical, with 8–11 visible bud scales 24
9* Bark dull. Buds acute-ovoid, with c. 7–8 visible bud scales 10

10 Small to medium sized shrubs, always with enrichment buds 21
10* Mostly trees 22
11 (7) Lateral buds with enrichment buds or buds in groups at short-shoots 17
11* Lateral buds always solitary 12
12 Bark, when scratched, very aromatic (of bitter almond) 13
12* Bark not noticeably scented 14
13 Buds ovoid, to 6 mm long . . . .
. . . . ***Prunus serotina***
13* Buds narrowly conical, mostly longer than 6 mm . . . .
. . **Prunus padus** and *Prunus virginiana*
14 Bark distinctly glossy, coming off in rolls; buds narrowly conical 24
14* Bark different, buds (acute-)ovoid . . . 15
15 Small shrub, rarely above 1 m high, buds narrowly ovoid, c. 4 mm long . . .
. . . .*Prunus fruticosa*
15* Larger shrubs or small trees 16
16 Internodes relatively short, mostly less than 2 cm long, buds globose-ovoid, c. 3 mm long . . . .***Prunus mahaleb***
16* Internodes mostly longer, buds acute-ovoid, mostly > 4 mm long 25
17 (11) Buds c. 2–3 mm long. Small slender-twigged shrubs 18
17* Buds longer. Large shrubs or trees . . 19
18 Buds globose-ovoid, twigs sturdy and angled, brown as the buds, partly decumbent to erect shrub . . . .
. . . . *Prunus pumila*
18* Buds narrowly ovoid, twigs slender, partly with white-grey epidermis. Erect shrub . . . . **Prunus tenella**
19 Twigs grey-ochre, from the start with large lenticels tearing open 20
19* Twigs olive-green to grey-brown. Fruit shrub or forest tree . . . .***Prunus avium***
20 Buds more than 8 mm long, with more than 10 bud scales. Tree. Ornamental plant . . . . **Prunus serrulata**
20* Buds smaller, mostly with fewer bud scales. Shrubs 21
21 Twigs more than 3 mm thick and/or buds 5 mm long or longer . . . .
. . . .***Prunus nipponica***
21* Twigs to 2 mm thick and buds 2–4 mm long . . . . *Prunus incisa*
22 Lateral buds dark brown, appressed to twig, twigs pale grey densely hairy towards tip, often hanging down . . . 23

22* Lateral buds red-brown distinctly spreading from twig, tips of twigs sparsely hairy. . . . .*Prunus ×subhirtella* 'Autumnalis'

23 Buds to 5 mm long. . .*Prunus pendula*

23* Buds over 5 mm long. . . . . *Prunus ×yedoensis*

24 (14) Bark glossy mahogany brown . . . . *Prunus serrula*

24* Bark glossy orange-brown . . . . *Prunus maackii*

25 (16) Buds 4–6 mm long . . . . <u>*Prunus cerasus*</u>

25* Buds larger. . . . . **Prunus sargentii**

26 (1) Twigs glabrous, ± glossy wine-red to green 27

26* Twigs red-brown, olive-green to grey, ± matte 29

27 Buds dark, bud scales ciliate, twigs sturdy. . . . . <u>*Prunus armeniaca*</u>

27* Buds pale brown, slightly hairy, twigs slender. 28

28 Twigs, at least in shade, predominantly green . . . . *Prunus cerasifera*

28* Twigs all-round dark wine-red. . . . . *Prunus cerasifera* 'Pissardii'

29 (26) Buds small and globose, not much longer than thick, twigs velvety hairy and becoming thorny. .*Prunus spinosa*

29* Buds conical-ovoid, distinctly longer than thick. 30

30 Twigs glabrous. . . . . <u>*Prunus domestica*</u>

30* Twigs velvety hairy . . . . <u>*Prunus domestica*</u> subsp. <u>*insititia*</u>

Subgenus Padus, bird cherries

Rubbed bark smelling distinctly of bitter almonds.

Prunus padus L., bird cherry
[*Cerasus padus* (L.) DC., *Padus avium* MILL.]

Buds 6–8(–10) mm long and c. 2 mm, thick, fusiform and acute, appressed or slightly spreading from the twig. About 8 externally visible bud scales: dark brown at the base, with paler upper margin, especially the lowest bud scales keeled, almost glabrous, but often with single hairs. **Twigs** mostly glabrous or becoming glabrous, only at the upper margin of the leaf scar often with a small tuft of hair. Initially dark to red-brown, smooth, matte to faintly glossy, with few small, pale lenticels, later grey-brown to violet-grey. **Leaf scars** indistinctly 3-traced. Frequent, tree to 15 m high. Origin: Europe to northern Asia, Japan and Korea. Very similar, and not easily told apart in winter, is the Virginian bird cherry, *Prunus virginiana* L., from central and eastern North America, with entirely glabrous twigs and buds.

Prunus serotina EHRH., black cherry
[*Padus serotina* (EHRH.) BORKH.]

Buds ovoid, c. 3–5 mm long and 2–3 mm broad. Bud scales basally greenish, towards the margin via olive-brown to red-brown to dark-brown, faintly glossy. Bud scales of terminal buds spirally inserted, in the lateral buds only changing to spiral position after the ± opposite prophyll scales. **Twigs** glabrous, red-brown to olive, glossy, with many distinct light ochre to orange lenticels. To 30 m high tree, but often smaller and shrubby. Origin: North America, occasionally naturalised in Europe.

Subgenus Cerasus, cherries

Buds ± ovoid, brown, shrub species with enrichment buds, tree species have these more rarely. **Twigs** olive-green to grey-brown, or pale grey due to dying epidermis. Sweet or wild, sour and steppe cherries are relatively closely related and are linked by hybrids. They have in common that there are buds on the base of long-shoots, in the axils of the leaf scars, in relatively close sequence. As a rule, enrichment buds are absent. Only with a good food supply and under expert pruning do some sour cherries form enrichment buds.

Prunus avium (L.) L., sweet cherry, wild cherry
[*Cerasus avium* (L.) MOENCH]

Buds 6–8 mm long, glabrous, ovoid to conical, ± acute, red-brown, glossy. Flower buds globose, ovoid, hardly acute. Massing on short-shoots: also called spurs. **Twigs** relatively thick, glabrous, red-brown to olive-green, often pale grey due to dead epidermal layer. Lenticels few, older twigs uniting into distinct transverse bands. Long-shoots become horizontal, with short-shoots bearing many flower buds. **Bark:** ringed, flaking in transverse strips. Upright growing tree to 20(–30) m high, never with

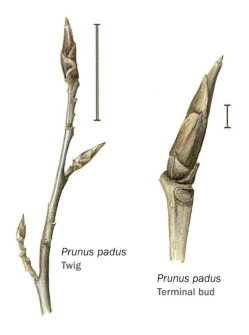

Prunus padus
Twig

Prunus padus
Terminal bud

Prunus serotina
Tip of twig

Prunus avium
Tip of long-shoot

Prunus avium
Flower buds grouped on the short-shoot (spur)

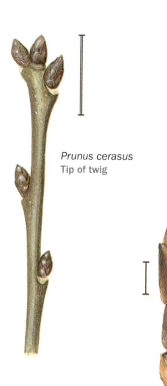

Prunus cerasus
Tip of twig

Prunus fruticosa
Tip of long-shoot

Prunus fruticosa
**Short-shoot with
terminal bud**

Prunus fruticosa
**Lateral bud
(dorsal view)**

Prunus mahaleb
Tip of twig

Prunus mahaleb
Long-shoot

overhanging branches. In cultivation since ancient times as a fruit tree; the cultivated forms differ little in vegetative characters. The wild form, var. *avium*, has slightly weaker twigs and smaller buds. Origin: Europe to western Asia, Asia Minor and north Africa.

Prunus cerasus L., sour cherry
[*Cerasus vulgaris* MILL.]

Smaller than the sweet cherry in all parts, and growing more weakly. **Buds** 4–6 mm long and only rarely in a few forms grouped at the short-shoots. Large shrub generating runners or small tree to 5(–8) m high. In its vegetative characters quite variable. Origin: Asia Minor to southern Europe, widely cultivated.

The two subspecies, *Prunus cerasus* subsp. *acida* (DUMORT) DOSTÁL, mostly a shrub or small tree with pendulous twigs, and *Prunus cerasus* subsp. *cerasus*, a tree with upright twigs, differ mainly in growth form.

The steppe cherry, *Prunus fruticosa* PALL., is close to the acid cherry. The shrub is smaller in all parts, and rarely reaches 1 m high: **buds** c. 4 mm long, narrowly ovoid and acute; lateral buds are spreading from the twig, bud scales glabrous, red-brown, darker towards the margin. **Twigs** slender, initially brown, due to the dying epidermis soon grey-brown; in the second year entirely grey and fissuring lengthwise. **Lenticels** few, small and round. **Leaf scars** black-grey, on distinct leaf cushions with 3 tiny traces of vascular bundles. Origin: in the west of the original area of dispersal from continental central Europe to Siberia, there is much hybridisation with the sour cherry, *Prunus cerasus*, so that here the hybrid *Prunus ×eminens* BECK is dominant. The improved, mostly tall-stemmed globose cultivar 'Umbraculifera' (*Prunus fruticosa* 'Globosa') also belongs to the hybrid *Prunus ×eminens*.

Prunus mahaleb L., St Lucie cherry
[*Cerasus mahaleb* (L.) MILL.]

Buds small, to 3 mm long, ovoid, glabrous and hairy only at the tip, with some red-brown bud scales. Buds are about the same along the whole length of the shoots. **Twigs** with grey outer skin, underneath olive-green; densely, bristly-felty hairy with some ochre-brown lenticels. **Leaf scars** small, dark, vertical on weak leaf cushions. **Bark** aromatic, initially smooth, later longitudinally fissured, changing into black-brown bark. Frequently planted, but also used as a source for sour cherry graftings. Shrub or small to 10 m high tree with broad, finely branching, overhanging crown. Origin: Europe to Asia Minor.

Prunus maackii Rupr., Manchurian cherry

Buds oblong, blunt, surrounded by a few rather loose bud scales. Terminal buds c. 5 mm long, lateral buds slightly smaller, appressed to the twig or gently spreading. **Twigs** brown, slightly hairy when young, becoming glabrous; with grey lenticels. **Bark** distinctly yellow-brown to bronze-brown, very smooth, glossy and coming off in thin strips — similar to many birches. Frequently planted, small to 10 m high tree with broad, conical crown. Origin: eastern Asia, Manchuria to Korea.

Prunus serrula Franch., Tibetan cherry

Buds narrowly conical, only slightly bulging, 5–7 mm long, bud scales slightly hairy, lowest ones dark brown and keeled, upper ones red-brown. **Twigs** grey-brown when young and ± densely hairy, becoming glabrous and later glossy dark violet-brown, sturdier twigs with mirror-smooth, mahogany-coloured bark which rolls off in narrow strips. **Lenticels** few, ochre-brown, warty. **Leaf scars** ochre-brown, 3-traced. Tree to 7(–12) m high. Origin: western China.

Japanese blossom cherries

These are native to Eastern Asia and are highly esteemed, particularly in Japan. Blossom cherries are difficult to tell apart, as they are closely related and have been interbred for centuries. They have large orange-red lenticels which merge on the stems in transverse bands. The difference from *Prunus avium* with more matte lenticels is particularly distinct in blossom cherries grafted on this species.

Prunus serrulata Lindl., Japanese cherry

Buds relatively large, 8–10 mm long, acute-ovoid, glabrous, with many red-brown and basally darker violet-brown bud scales. Flower-buds mostly (like *Prunus avium*) grouped on short-shoots. **Twigs** of the most frequent cultivar 'Kanzan' of first year's growth distinctly orange-brown with ovate, longitudinally tearing lenticels. **Leaf scars** with 3 traces on distinct cushions. **Bark** dark-brown. Ornamental cherry from Japan, planted frequently, with many cultivars. 'Kiku-Shidare-Sakura' is distinctive in winter with downward growing twigs or the columnar 'Amanogawa'. The cultivated forms grow into more than 10 m high trees.

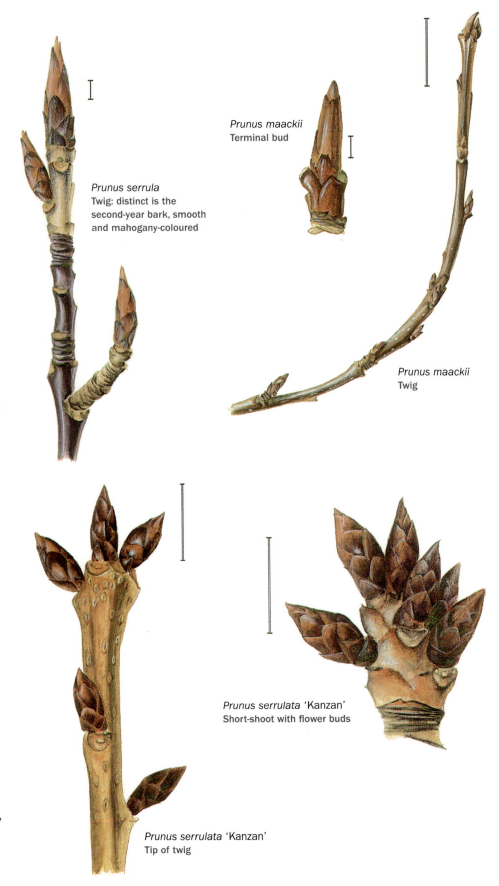

Prunus serrula
Twig: distinct is the second-year bark, smooth and mahogany-coloured

Prunus maackii
Terminal bud

Prunus maackii
Twig

Prunus serrulata 'Kanzan'
Short-shoot with flower buds

Prunus serrulata 'Kanzan'
Tip of twig

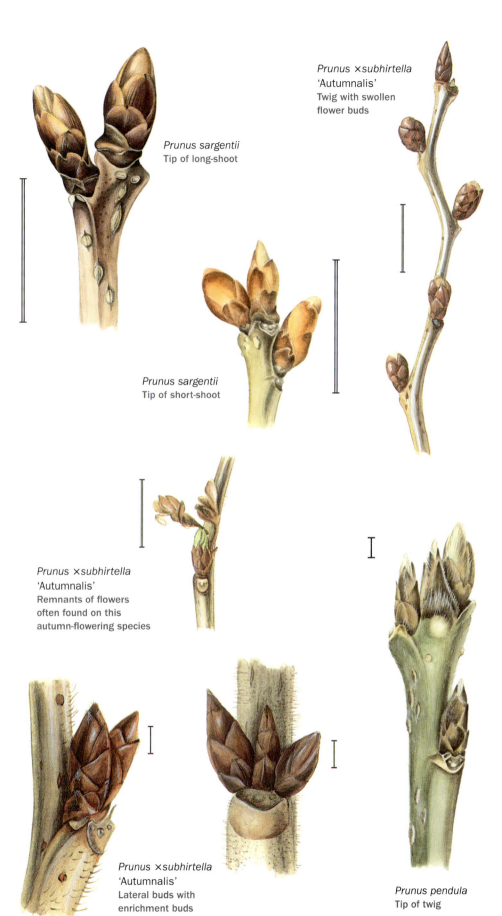

Prunus sargentii
Tip of long-shoot

Prunus sargentii
Tip of short-shoot

Prunus ×subhirtella
'Autumnalis'
Twig with swollen
flower buds

Prunus ×subhirtella
'Autumnalis'
Remnants of flowers
often found on this
autumn-flowering species

Prunus ×subhirtella
'Autumnalis'
Lateral buds with
enrichment buds

Prunus pendula
Tip of twig

Prunus sargentii REHDER., Sargent's cherry

[*Prunus serrulata* var. *sachalinensis* (F. SCHMIDT) E. H. WILSON]

Just like *Prunus serrulata* a tree and, in winter, difficult to tell apart. **Terminal buds** 8–12 mm long, bud scales glossy red-brown to orange-brown, glabrous, lowest ones almost black-brown. **Lateral buds** distinctly spreading from twig. Tips of buds slightly rounded, not as distinctly acute as in *P. serrulata*. Bark relatively smooth horse-chestnut-brown. Tree reaching 15–18 m. Hardly distinct in winter and, maybe, con-specific is *Prunus jamasakura* SIEB. ex KOIDZ. [*Prunus serrulata* var. *spontanea* WILS.].

Prunus ×subhirtella 'Autumnalis', rosebud cherry

Buds 4–5 mm long, narrowly ovoid with red-brown, sparsely hairy bud scales. Flower buds often starting very early and globose-ovoid. In the prophyll axils there often are one or two enrichment buds. **Twigs** finely hairy, light ochre to red-brown, with many, initially small, later distinctly wart-like ochre to dark brown lenticels. **Flowers** shortly before the leaves appear. Tree to 8 m high, frequently planted, which often flowers in autumn and, in periods with mild weather, also in winter. Origin: Japan.

Prunus pendula MAXIM., weeping cherry

The frequently planted typical form has slender, pendulous twigs. Particularly in the upper half of the long-shoot the **lateral buds** are relatively closely appressed to the twig (a distinct difference with the other species of the subgenus). Bud scales dark to ochre-brown, the margins shaggy spreading-hairy. **Twigs** olive-ochre-grey, slightly glossy, hairy at the tip, lower down glabrous and tearing open in rhomboids, within the rhomboids greenish. Lenticels brown, lower on the twig relatively large. When flowering, this differs from the other ornamental cherries by the funnel-shaped or cylindrical hypanthium and glabrous sepals, by its urn-shaped (globose-bulging) corolla and by the hairy style.

The characters of *Prunus pendula* are found in a weaker form in the hybrids such as 'Accolade' (*Prunus pendula × Prunus sargentii*) or in **Prunus ×yedoensis** MATSUM., the Yoshino Cherry, a hybrid that was bred in the 19th century in Japan with an unknown species (perhaps *Prunus speciosa*). Its **lateral buds** are oblong and larger, 7–9 mm long, ± acute and appressed to the shoot. As flower buds they are broadly ovoid and ± blunt-tipped. Bud scales 9–11, red-brown to violet-brown, the lowest ones also grey-brown, towards the margin densely pale hairy. **Twigs** ochre-brown to light grey-brown and slightly hairy, with ovate, longitudinally tearing warty lenticels. **Leaf scar** dark brown, with 3 traces. Broad tree 12–15 m high. Encountered occasionally. Origin: Japan.

Prunus nipponica MATSUM., Kuril cherry

Buds c. 5–6 mm long, narrowly ovoid, ± glabrous. Lateral buds with enrichment buds, which in early spring give rise to many flowers. **Twigs** medium thick, in winter mostly completely glabrous with ochre-brown lenticels tearing open longitudinally. **Leaf scars** narrow, with 3 traces. Frequently planted, 1–2 m high shrub. Origin: Eastern Asia: Japan and Kurile Islands.

The cultivar 'Brilliant', mostly ascribed to var. *kurilensis* (MIYABE) E. H. WILSON (*Prunus kurilensis* MIYABE) has many enrichment buds and stocky twigs with shorter internodes, causing a large number of slightly smaller flowering buds.

Prunus incisa THUNB., Fuji cherry

Buds 2–4(5) mm long. Bud scales matte brown, ochre to dark brown (darker areas predominate), glabrous. Flower buds mostly as enrichment buds beside the regular lateral buds, partly grouped on shorter shoots. **Twigs** at the tip c. 2–3 mm diameter, pale grey with a trace of pale brown, hardly ochre, glabrous. Lenticels relatively large (1 mm). Elegant tree to 5 (10) m high. Origin: Japan. The dwarf form 'Kojo-no-mai' is frequently sold as a pot plant.

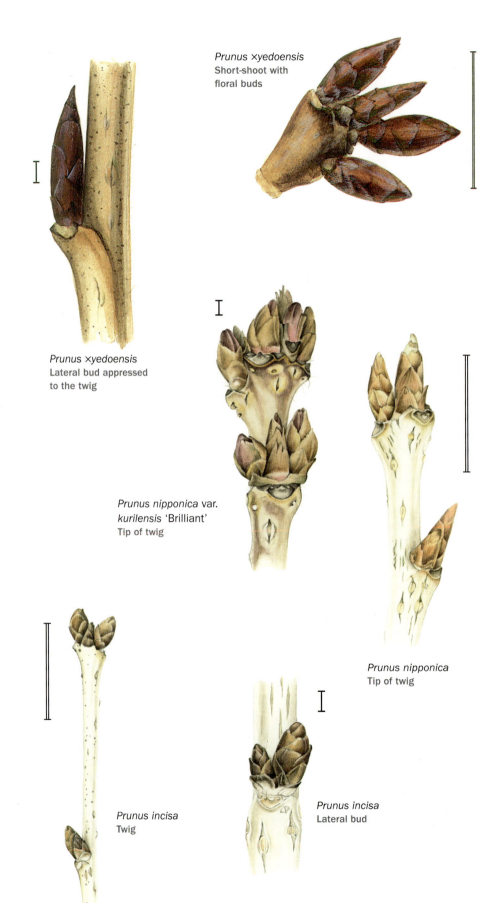

Prunus ×yedoensis
Short-shoot with floral buds

Prunus ×yedoensis
Lateral bud appressed to the twig

Prunus nipponica var. *kurilensis* 'Brilliant'
Tip of twig

Prunus nipponica
Tip of twig

Prunus incisa
Twig

Prunus incisa
Lateral bud

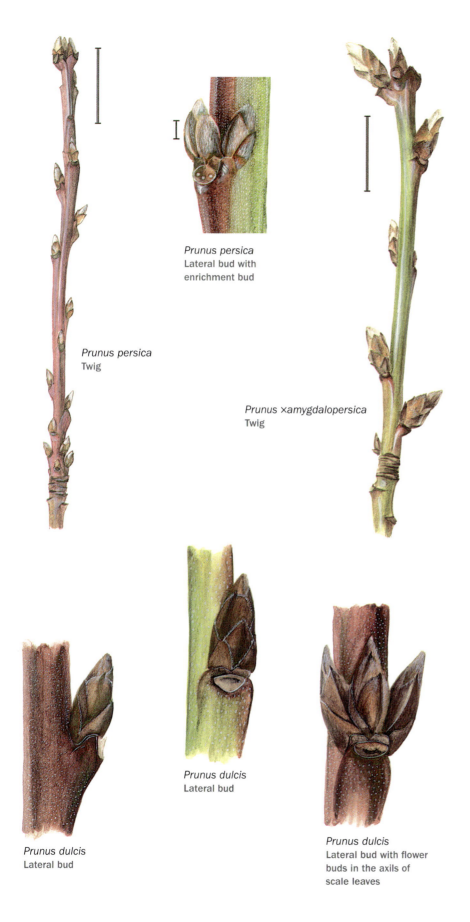

Prunus persica
Lateral bud with
enrichment bud

Prunus persica
Twig

Prunus ×amygdalopersica
Twig

Prunus dulcis
Lateral bud

Prunus dulcis
Lateral bud

Prunus dulcis
Lateral bud with flower
buds in the axils of
scale leaves

Subgenus Amygdalus, almond and peach

Buds: terminal buds present; lateral buds frequently with lateral enrichment buds. Above the leaf scar mostly persistent stipules or their remnants. **Leaf scars** on distinct leaf cushions, broadly rounded to rounded-triangular, with 3 distinct traces (vascular bundles). The sessile flowers appear before, more rarely with, the leaves.

Prunus persica (L.) BATSCH, peach
[*Persica vulgaris* MILL., *Amygdalus persica* L.]

Buds narrowly ovoid, 3–6 mm long, grey-brown at the base and towards the tip lighter ochre-brown, ± densely white hairy. Lateral buds frequently with enrichment buds. **Twigs** glabrous, smooth, red on the sun side, in the shade green very dense and regularly beset with small, white dots. **Flowers**, according to form, pink to red, mostly before the leaves appear, in March to April. Tree or large shrub to 8 m high, cultivated for centuries in many forms for fruit and as an ornamental. Origin: China.

The small tree *Prunus ×amygdalopersica* REHDER [*Prunus dulcis × Prunus persica*], a rare ornamental is similar. It is sturdier than the peach. **Buds** blunt-ovoid, to 8(–10) mm long, with many brown-grey, hairy bud scales, which are densely felty towards the tip. **Twigs** glabrous, smooth, greenish and on the sun side reddish, especially underneath the leaf scar; becoming grey-brown in the second year, in patches with pale grey epidermis.

Prunus dulcis (MILL.) D. A. WEBB, almond [*Amygdalus communis* L., *Prunus amygdalus* BATSCH]

Buds 5–7 mm long, ovoid, with some dark brown bud scales, hairy only towards the tip. Particularly on sturdy twigs the lateral buds often with one to two lateral enrichment buds in prophyll axils. **Twigs** glabrous, on the light side dark violet-red, in the shade green, finely pale-dotted. Flowers before the leaves, 1–2 and almost sessile, white to pale pink. Upright broad shrub or tree, to 10 m high. Origin: Syria to North Africa. In cultivation since ancient times, heat-loving. Planted particularly in wine-growing areas and occasionally naturalised in the Mediterranean region.

Prunus triloba LINDL., flowering almond

Buds 3–4 mm long, red-brown to dark brown with bud scales that are darker towards the margin and white-ciliate. Lateral buds small, conical, in the axils of prophylls mostly with 2 slightly larger, ovoid and acute flower buds. **Twigs** dark brown and velvety hairy, becoming glabrous. Epidermis fissuring narrowly rhomboid, underneath this olive to brown. **Lenticels** initially small and indistinct, later ochre-brown, round and corky-warty. **Stipules** persistent, ending in thread-like lobes. Densely branched shrub or, when bred as a standard tree, to 3 m high. Frequently planted ornamental shrub. The pink double flowers (to 3.5 cm diameter) appear in March to April. *Prunus triloba* is known only in cultivation. Origin: China.

Prunus tenella BATSCH, dwarf Russian almond
[*Amygdalus nana* L.]

Buds 2–3 mm long, red to dark brown with glabrous bud scales with slightly ciliate margins. Terminal bud ± conical, lateral buds ± conical, often with two acute-ovoid floral enrichment buds. **Twigs** glabrous, slightly curving, spotted with pale grey patches of epidermis and an orange olive-brown glossy layer below. **Lenticels** very many, initially small and grey, then warty brown. **Leaf scars** round, 3-traced, relatively small on a distinct leaf cushion. **Flowers** together with leaves. Frequently planted, erect shrub 0.5–1.5 m high, spreading by runners. Origin: eastern central Europe to Siberia.

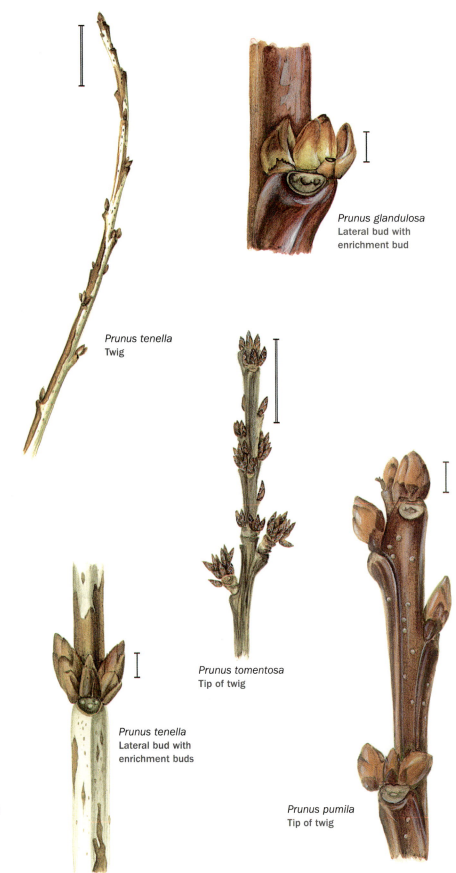

Prunus glandulosa
Lateral bud with enrichment bud

Prunus tenella
Twig

Prunus tomentosa
Tip of twig

Prunus tenella
Lateral bud with enrichment buds

Prunus pumila
Tip of twig

Prunus triloba
Distinctive thread-like stipules at the leaf base cover the lateral bud and its enrichment bud

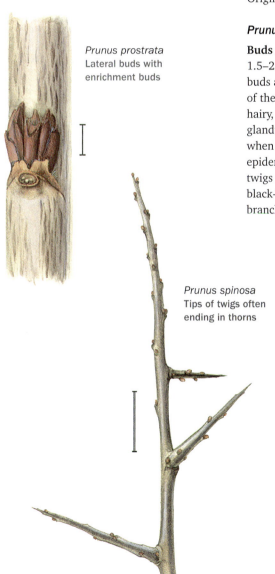

Prunus tomentosa
Out of the axils of
the lateral bud
emerge many
enrichment buds

Prunus prostrata
Lateral buds with
enrichment buds

Prunus spinosa
Tips of twigs often
ending in thorns

Subgenus Lithocerasus, bush cherries

Small shrubs with slender twigs and small buds.

Prunus glandulosa THUNB., Chinese bush cherry

Buds blunt-ovoid, dark brown on the shaded side, on the light side pale ochre-brown to greenish, glabrous. Lateral buds with mostly smaller enrichment buds, which frequently emerge 2(3–4) from the axils of the bud scales of the primary lateral bud. **Twigs** green to dark wine-red, glabrous, with few lenticels and, in places, flaking grey-white epidermis. **Leaf scars** grey-brown, above small glandular stipular remnants. Frequent, shrub to c. 1.5 m high. Origin: mid- and north China and Japan.

Prunus pumila L., sand cherry

Buds red-brown, small, globose-ovoid, 1.5–2.5 mm long, often with two flowering buds as enrichment buds from the prophylls of the lateral buds. Bud scales very slightly hairy, at the margins (hardly noticeable) glandular-ciliate. **Twigs** glossy red-brown when young, distinctly furrowed, later epidermis fissuring longitudinally, older twigs matte grey-brown. **Leaf scars** dark black-brown. **Lenticels** many, initially branch-coloured and small, later large,

ochre-coloured, transverse cork warts. Rare, to 1 m high shrub with ascending twigs. Origin: north-east North America. The variety ***Prunus pumila*** var. ***depressa*** (PURSH) BEAN is slightly smaller in all parts, is of prostrate growth, and hardly reaches 15 cm height.

Prunus tomentosa THUNB., downy cherry

Buds 2.5–4 mm long, narrowly ovoid and acute, in dense groups in the leaf axils and on short-shoots. A single lateral bud can have two, four or more enrichment buds. Bud scales ochre-brown to brown, towards the margin dark violet-brown, white-ciliate on the margins. **Twigs** dark black-brown, like the buds slightly lighter in the shade, densely-felty hairy. Twigs of several years' growth with longitudinally rhomboid-fissuring epidermis. Densely branched, rarely planted shrub to 1.5 m high. Origin: Japan, northern and western China to the Himalaya.

Prunus prostrata LABILL.

Buds small, to 2 mm long, glabrous or slightly ciliate; lateral buds with two enrichment buds. **Twigs** brown with grey, ± rhombic-fissuring epidermis. **Lenticels** few to many, small and pale. **Leaf scars** very small with three dot-shaped traces. Small shrub to 1 m high. Origin: Mediterranean region to western Asia.

Subgenus Prunus, plums

Without terminal buds, growing sympodially. Vernation (ptyxis) involute: leaves mostly rolled inside the bud. Flowers mostly before the leaves. Occasionally thorny, particularly stool- and coppice-shoots.

Prunus spinosa L., blackthorn, sloe

Buds very small (1–2 mm), lateral buds dark brown, hemispherical and densely hairy, mostly with slightly larger globose enrichment-(flower-)buds, light to dark brown-red, and are only sparsely hairy. **Twigs** with thorny short-shoots, grey-brown when young, felty-hairy. **Flowers** before the leaves, 1–2 together, very many. Frequent, shrub to 4 m high. Origin: Europe, Asia Minor and Africa.

Prunus spinosa
Globose flower bud as
enrichment bud next
to small lateral bud

Prunus cerasifera EHRH., cherry plum
[*Prunus myrobalana* LOISEL.]

Buds 3(–5) mm long, ochre-brown, the fine indument often only visible at the margin of the bud scale. Conical, ovoid-acute leaf buds, often with 1–2 larger globose flower buds in the axils of prophylls. **Twigs** greenish, often reddish in the sun, lenticels not visible initially. Older twigs grey-brown to blackish, finely fissured and with distinct ripples of bark. **Flowering** March to April. Frequently planted, tree-like shrub to 8 m high. Origin: Asia Minor and Caucasus.

Prunus cerasifera 'Pissardii', 'Nigra' and other red forms: the purple-leaved plums differ from the typical species by a deep and dark wine-red bark on the twigs, and by pink flowers. Specimens with some terminal buds are *P.* ×*cistena* (N. E. HANSEN) KOEHNE, the purple-leaf sand cherry, a hybrid with *P. pumila*.

Prunus domestica L., common plum

Buds globose and acute with dark violet-brown bud scales, ciliate at the margin, spreading, on large leaf cushions. **Twigs** mostly glabrous, olive-green to brown, rarely with thorns. **Lenticels** initially not noticeable, later more distinct. **Leaf scars** ± vertical with 3 traces. **Bark** smooth when young and black-grey, later brown-grey and slightly splitting. **Flowers** in groups of 1–3 in April. Mostly a small tree to 8 m high. In Eurasia frequently cultivated for centuries, origin unknown. The species is very rich in forms. In winter one variety stands out: *Prunus domestica* var. *insititia* (L.) C. K. SCHNEID, damson or bullace, one-year-old twigs are densely hairy and sometimes thorny.

Subgenus Armeniaca, apricots

No terminal buds. Vernation (ptyxis) involute: leaves rolled in the bud. Close to plums.

Prunus armeniaca L., apricot
[*Armeniaca vulgaris* LAM.]

Buds: only lateral buds, with distinct leaf and flower buds. Leaf bud flattish-conical: bud scales black grey with silver-white epidermis flaking off. Flower-buds from the prophyll axils, narrowly ovoid and larger, with red-brown, at the margin dark brown bud scales. **Twigs** wine-red to red-brown, glabrous and glossy. Lenticels initially few,

Prunus cerasifera
Tip of twig

Prunus cerasifera 'Pissardii'
Lateral bud with global
enrichment flower buds

Prunus domestica
subsp. *insititia*
Lateral bud

Prunus domestica
Twig with short-shoots

Prunus armeniaca
Tip of twig

Prunus armeniaca
Lateral bud with large
enrichment flower bud

Prunus brigantina
Tip of twig

and faint: later many and more distinct, pale ochre-grey. **Leaf scars** grey with 3 traces on distinct leaf cushions. **Flowers** mostly single, in April, white to pink. Frequent, small tree or large shrub, 5–10 m high. In cultivation since ancient times. Origin: northern China.

Related, but smaller in all parts is the Briançon apricot, **Prunus brigantina** VILL. It originated in south-east France and it has been cultivated since ancient times. **Buds** globose to short-ovoid, slightly acute, small and c. 2 mm long with some ochre-brown, basally slightly keeled bud scales. Often with small enrichment buds in the axil of one or both prophylls. **Twigs** glabrous, slender: first year's growth 2–4 mm thick, mostly wine-red, on the shaded side yellowish-green, regularly dotted with small white spots. Leaf scars small, on small cushions with indistinct traces, also with the remnants of stipules and some hairs in between. Shrub or small tree, 3–6 m high.

TRIBE Pyreae

The apple fruit, typical of this tribe, is similar to a berry or stone fruit with an inferior ovary, in which the axial tissue envelops the carpels with the seeds. The actual carpels can be like parchment or woody. In many species fruits are present at the start of winter, and offer valuable help for identification. Small to large trees, more rarely small shrubs. Fruit trees such as pear, quince and apple are included in this tribe, as well as frequently planted ornamentals such as cotoneaster and juneberry.

Key to Pyreae

1 Twigs thorny. 2
1* Without thorns 6
2 Thorns sharply acute 3
2* Thorns often like short-shoots, without sharp tip 5
3 Buds inconspicuous, at the base of thorns, buds matte . . . . *Chaenomeles*
3* Buds clearly visible. 4
4 Buds globose-ovoid, glossy . . . *Crataegus*
4* Buds acute . . . . *Pyrus*
5 Fruits long-persistent, 2–3 cm thick, brownish, with large erect sepals . . . . . .
. . . . *Crataegus germanica*

5* Fruits deciduous, or long-stalked, often less than 2 cm thick, small apples with inconspicuous small and occasionally deciduous sepals . . . . *Malus*
6 (1) Shrubs without terminal buds, lateral buds with 2 spreading bud scales or without bud scales . . . . *Cotoneaster*
6* Bud scales of lateral buds appressed, often with terminal bud, or tree. 7
7 Leaf scars with 5 traces . . . .
. . . . **rowans**, **whitebeams** etc.
7* Leaf scars with 3 traces 8
8 Terminal buds large, mostly more than 8 mm long, green or reddish: salmon pink, wine red to violet-brown. 9
8* Terminal buds mostly smaller or absent, brown. 11
9 Leaf cushions different in colour from the twig, buds green, partly reddish, bud scales often with brown margin, twigs relatively thick (3–5 mm diameter) . . . **rowans**, **whitebeams** etc.
9* Leaf cushions of the same colour as the twig, terminal buds almost fusiform, acute, often predominantly red to violet-brown, twigs slender to medium thick (2–4 mm diameter) 10
10 Terminal buds with 3–5 all-red bud scales . . . . *Aronia*
10* Terminal buds with 6 or more bud scales . . . . *Amelanchier*
11 (8) Young twigs densely beset with relatively large ochre-brown, warty lenticels . . . . *Pourthiaea*
11* Lenticels occasionally many, mostly pale and dot-shaped 12
12 Basal bud scales of terminal buds acute and keeled, short-shoots mostly with terminal buds, long-shoots often felty hairy towards the tip, leaf scars blackish . . . . *Pyrus*
12* Terminal bud absent or basal bud scales of terminal bud hardly keeled, ± blunt, leaf scars brownish. 13
13 Fruits persistent, 2–3 cm thick, brownish, with large, upright sepals . .
. . . . *Crataegus germanica*
13* Fruits mostly deciduous, less than 2 cm or more than 4 cm thick, sepals small .
. . . . 14
14 Long-shoots with terminal bud . *Malus*
14* Long-shoots without terminal bud . . . .
. . . . *Cydonia*

Chaenomeles LINDL., flowering quince

Mostly shrubs armed with thorns; 3–4 species from eastern Asia. **Flowers** before or together with the leaves, on older wood from clustered buds on short-shoots; mostly ± deep red, in some of the many garden forms also white or double flowers. **Fruits** often long-persistent.

Key to *Chaenomeles*

1 Twigs warty 2
1* Twigs smooth and glabrous, shrub to over 2 m high . . . . *Chaenomeles speciosa*
2 Shrub to 1 m high with very warty twigs . . . . *Chaenomeles japonica*
2* Shrub with slightly warty twigs, intermediate between the parents . . . . *Chaenomeles ×superba*

Chaenomeles japonica (THUNB.) LINDL., Japanese flowering quince

Buds very small, 1–2 mm, globose to ovoid, with few (3–4) bud scales. Flower buds in groups on short axillary shoots, the whole flower bud group c. 4 mm in diameter. **Twigs** red-brown to grey-brown, rough-warty, relatively slender and densely branching. In the leaf axils there are usually sharp thorns with small leaf buds at the base. **Flowers** from March to April, orange to brick-red, 2.5–3 cm in diameter. **Fruits** flattened and longitudinally furrowed, to 4 cm diameter. Frequently planted, small shrub to 1 m high. Origin: Japan.

Chaenomeles speciosa (SWEET) NAKAI, Chinese flowering quince

Buds small, c. 2 mm, short-ovoid, with few brown, basally lighter bud scales which have a dark margin. **Twigs** red-brown to dark violet-brown, smooth: without warts, with fine silvery, partly detaching, epidermis. **Flowers** in March to April, pink to dark red, never orange. **Fruits** oblong, to 7 cm long. Frequent, shrub to 2 m high. Origin: China.

Chaenomeles ×superba (FRAHM) REHDER, hybrid flowering quince [*Chaenomeles japonica* × *Chaenomeles speciosa*] is very common, with many forms in which the characters of the parents are combined in various ways.

Chaenomeles japonica
Flower buds at the base of a short-shoot thorn

Chaenomeles japonica
Twig with short thorns

Chaenomeles speciosa
Flowering branch

Chaenomeles speciosa
Short-shoot thorn

Chaenomeles speciosa
Long-shoot with thorns

Chaenomeles speciosa
Lateral buds

Cydonia oblonga
Tip of twig

Cydonia oblonga
Twig

Cydonia oblonga
Lateral bud

Cotoneaster multiflorus
Lateral bud at the base
of the short-shoot

Cotoneaster multiflorus
Lateral bud

Cotoneaster multiflorus
Lateral bud

Cotoneaster multiflorus
Twig with short-shoots

Cydonia oblonga MILL., quince

Buds: only lateral buds, red-brown, at the base of the twig ± glabrous, towards the tip densely ciliate and felty hairy, on slender long-shoots c. 3 mm long and appressed to twig, surrounded by 2 bud scales; shorter on fruiting short-shoots, c. 5 mm long, ovoid and with several scales. **Twigs** glossy olive-brown to violet-brown, towards the tip ± felty hairy. Long-shoots mostly slender, but fruiting short-shoots thicker. **Leaf scars** dark, 3-traced. **Lenticels** many, small and round, ochre-brown and slightly raised. Shrub to 6 m high, or small tree. Origin: Transcaucasia, Persia, Turkestan to southern Arabia. Very widespread as fruit tree, naturalised in central and southern Europe.

Cotoneaster MEDIK., cotoneaster

Buds without bud scales and densely hairy, or lateral buds surrounded by two outer bud scales, becoming entirely glabrous in some species. Many of the species are difficult to tell apart in winter, and are often in cultivation. There are differences in the height of the plant, degree of hairiness and branching. Deciduous or evergreen, small to large shrubs, more rarely small trees. About 50 species. Origin: Europe to China.

Key to *Cotoneaster*

1. Branching distinctly dense, distichous and in one plane (herringbone pattern) . . . 5
1* Branching irregular 2
2 Prostrate to erect shrubs to 1 m high . . 4
2* Erect, more than 1 m high shrubs 3
3 Buds surrounded by 2 distinct bud scales which become ± glabrous . . . . *Cotoneaster multiflorus* and *Cotoneaster integerrimus*
3* Buds without bud scales, densely hairy . .6
4 To 25 cm high, prostrate shrub with very short internodes . . . . *Cotoneaster adpressus*
4* To 80 cm high shrub . . . . *Cotoneaster adpressus* var. *praecox*
5 (1) Low, often prostrate shrub, mostly less than 1 m high. *Cotoneaster horizontalis*
5* Erect shrub, usually more than 2 m high . . . . *Cotoneaster divaricatus*
6 Fruits red . . . . . . *Cotoneaster dielsianus* and *C. bullatus*
6* Fruits black . . . *Cotoneaster moupinensis*

Cotoneaster multiflorus BUNGE, many-flowered cotoneaster

Buds with 2 outer brown bud scales, 4–5 mm long, flat-appressed to the long-shoot, or basal to the short-shoot and 3–7 mm long. Scales on short-shoot buds spreading, derived from leaf base and, often slightly spreading, stipules: long, hairy inner bud leaves visible; on long-shoots bud scales simpler and covering the bud almost entirely. **Twigs** differentiated into glabrous, smooth glossy olive brown to orange-brown long-shoots and into mostly darker sparsely hairy short-shoots, a few cm long. Two-year-old long-shoots violet-brown, glossy partly with silvery-grey, dead epidermis. **Lenticels** small, round and paler than the twig. Frequent, shrub 3–4 m high. Origin: north-west China.

Cotoneaster adpressus BOIS, creeping cotoneaster

Buds appressed to twig or slightly spreading, small, to 2 mm long, ovoid and without bud scales. Bud leaves with dense grey-green indument, the lowest ones with dark scars of lamina. **Twigs** glabrous, very slender, first year's growth 1–1.5 mm diameter, internodes less than 1 cm long; on the light side dark wine-red with grey-white lifting epidermis, on the shaded side paler coloured reddish-ochre, epidermis dying later and lifting off. Surface of twig dense with very fine tuberculate points. **Leaf scars** faint, relatively narrow. Frequently planted, flat shrub, reaching only 25 cm in height. Origin: western China.

Cotoneaster adpressus var. praecox BOIS & P. BERTHAULT

[*Cotoneaster praecox* (BOIS & P. BERTHAULT) M. VILM.]

Buds small: 2–3 mm, densely hairy, with oblong leaf stipules. **Twigs** branching irregularly, glossy red-brown and violet-brown, on the shaded side also slightly yellowish-green and (most of all towards the tip of the branch) densely hairy. Frequently planted, shrub to 50(–80) cm high with prostrate or ascending branches curving down at the tip. Origin: western China.

Cotoneaster horizontalis DECNE., wall cotoneaster

Buds distichous, appressed to twig or slightly spreading, 2–3 mm long, outer bud scales slightly spreading, dark wine-red and sparsely hairy, inner ones densely grey-green hairy. Of the small almost round foliage leaves some persist. **Twigs** brown to grey-brown and finely hairy, older dark-brown upper layer of bark tearing in a rhomboid pattern revealing ochre-brown underneath. The typical lateral branching with the appearance of herringbone marks this frequent shrub which attains 1 m in height. Origin: western China.

Cotoneaster adpressus
Twig

Cotoneaster adpressus
Lateral bud

Cotoneaster adpressus
var. *praecox*
Lateral buds

Cotoneaster adpressus
var. *praecox*
Lateral buds

Cotoneaster horizontalis
Typical distichous branching

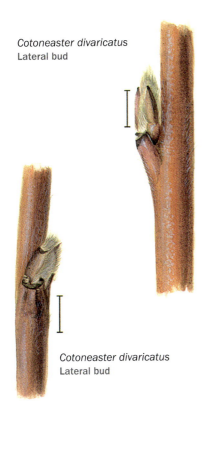

Cotoneaster divaricatus
Lateral bud

Cotoneaster divaricatus
Lateral bud

Cotoneaster dielsianus
Short-shoot

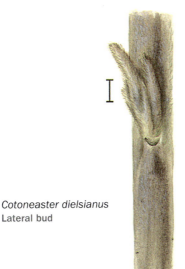

Cotoneaster dielsianus
Lateral bud

Cotoneaster integerrimus
Lateral buds

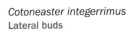

Cotoneaster tomentosus
Short-shoot

Cotoneaster divaricatus REHDER & E. H. WILSON, spreading cotoneaster

Buds densely long-hairy. Lateral buds on long-shoots c. 2 mm long; on short- and long-shoots longer. **Twigs** distichous in one plane (fan-shaped), on the underside young twigs smooth orange-brown to dark red-brown, on the upper side dark red-brown to violet-brown with lifting grey epidermis, towards the tip hairy, otherwise mostly glabrous. Older twigs dark grey to violet-brown. **Lenticels** few, roundish-square, ochre-brown. Frequently planted, shrub to 2 m high with spreading twigs. Origin: China.

Cotoneaster dielsianus E. PRITZ, Diels' cotoneaster

Buds 4–5 mm long, ± without bud scales, with 2(–3) sometimes spreading oblong, dense dirty grey-green to grey-brown hairy leaves: the outer ones with small stipules. **Twigs** grey-brown when young, occasionally also red-brown to dark grey, hairy; later turning glabrous, dark grey and finely fissuring. Lateral twigs branching irregularly, and not in one plane. **Fruits** red and 6 mm thick; with 3–5 seeds. **Lenticels** few, especially on older twigs obliquely oblong with cork warts and ochre-brown. Frequent, shrub to 2 m high. Origin: China.

Cotoneaster integerrimus MEDIK., common cotoneaster

Buds to 4–5 mm long and mostly as broad, enveloped by two often spreading bud scales, almost glabrous and dark red-brown, with in between a few densely pale grey-brown hairy, compact leaves. **Twigs** at the tip densely felty ochre-grey hairy, later (on the base of long-shoots in the first year) becoming glabrous and orange-brown to red-brown; in the second year dark violet-brown with flaking epidermis. Lateral twigs irregular. **Fruits** red, 6 mm long and roundish with 2 seeds. Rarely planted, 1.5 m high shrub, variable in shape. Origin: south to central Europe.

Similar, but with denser indument is ***Cotoneaster tomentosus*** (AIT.) LINDL. A lax shrub to 2 m high. Origin: southern Europe.

Cotoneaster bullatus BOIS, hollyberry cotoneaster

Buds without bud scales, with dense grey-green indument; 4–5 mm long with prophylls spreading wide. **Twigs** initially with hair, dark black-grey, red-brown towards the tip. **Fruits** pale red, globose, 7–8 mm thick, with 4–5 seeds. Broad and lax shrub, to 3 m high. Frequently planted. Origin: western China.

Cotoneaster moupinensis FRANCH.

Buds 6–8 mm long, formed from densely ochre-green hairy leaves with ± glabrous, violet-brown and acute stipules. **Twigs** often curving down, ochre-brown to red-brown, initially densely hairy, turning glabrous. Older twigs dark black-brown, slightly glossy. **Fruits** black, 6–8 mm thick, almost globose with 4–5 seeds. Squarrose branching; occasionally planted, shrub 2–3 (–5) m high. Origin: China.

Amelanchier MEDIK., juneberry, serviceberry

Buds reddish, also greenish on the shaded side or dark violet-brown, ± acute, narrowly ovoid to fusiform. **Twigs** relatively slender, first year's growth hardly more than 2 mm. **Leaf scars** dark, narrow and faintly 3-traced. One species, *Amelanchier ovalis*, is native to Europe and other species, difficult to tell apart, are fairly frequently planted and sometimes naturalised.

Key to *Amelanchier*

1 Buds salmon pink to brown-red or reddish, on the shaded side also green 2
1* Buds dark wine-red to violet-brown all round 3
2 Buds slim-fusiform, slightly curved, scales long ciliate . . . . ***Amelanchier spicata***
2* Buds mostly thicker, hardly curved. . . . 3
3. Twigs slender, when young distinctly less than 2 mm thick. 4
3* Twigs thicker, about 2 mm thick when young 5
4 Twigs reddish-brown, buds slightly bulging, leaf scar encircling the bud . . . . *Amelanchier ovalis*
4* Twigs olive-brown, buds fusiform. . . . . *Amelanchier laevis*

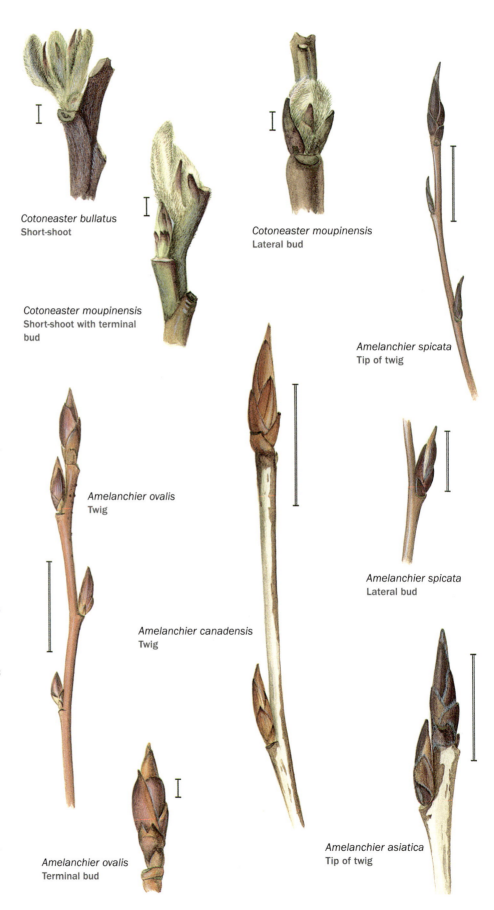

Cotoneaster bullatus
Short-shoot

Cotoneaster moupinensis
Short-shoot with terminal bud

Cotoneaster moupinensis
Lateral bud

Amelanchier spicata
Tip of twig

Amelanchier ovalis
Twig

Amelanchier spicata
Lateral bud

Amelanchier canadensis
Twig

Amelanchier ovalis
Terminal bud

Amelanchier asiatica
Tip of twig

Amelanchier lamarckii
Terminal bud

Amelanchier lamarckii
Twig

Amelanchier lamarckii
Lateral bud

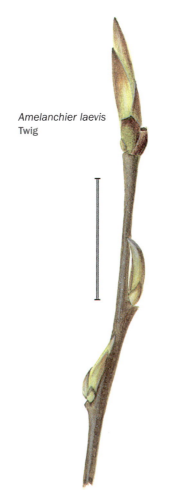

Amelanchier laevis
Twig

5 Twigs pale brown to cinnamon-brown with traces of lifting epidermis. *Amelanchier canadensis*

5* Twigs violet-brown to greenish. . . . . *Amelanchier lamarckii*

Amelanchier ovalis MEDIK., common juneberry

Buds acute, narrowly ovoid: terminal buds 8–10 mm long and up to 4 mm thick; lateral buds smaller, almost appressed to twig or slightly spreading; with wine-red to brown-red basally and at the margins also yellowish-brown bud scales, these slightly hairy, often becoming glabrous and at the margin short-ciliate. **Twigs** orange-red-brown, initially woolly-felty-hairy, mostly glabrous in winter, on the weather side with many grey-white patches of epidermis; older twigs dark violet-brown. **Lenticels** small and dark. Irregular 2–3(–6) m high shrub with stems which remain relatively slender. Origin: central and southern Europe to Asia Minor.

Very similar is the rare *A. canadensis* (L.) MEDIK., Canadian serviceberry, from North America with slightly more fusiform buds.

Amelanchier spicata (LAM.) K. KOCH, low juneberry

Buds narrowly fusiform, terminal buds 10–12(–15) mm; lateral buds 8–10 mm long, appressed to twig and with the tip slightly curved around the twig. Bud scales dark violet-brown, at the margin slightly paler, the uppermost long white-ciliate. **Twigs** dark red-brown to violet-brown with small patches of grey, dead epidermis. **Lenticels** initially few, later many: small and round. Shrub 2–5(–7) m high, growing stiffly upright. Origin: North America, sometimes naturalised.

Similar is the rarely planted *Amelanchier asiatica* (SIEBOLD & ZUCC.) ENDL. from Japan and Korea. **Buds** slightly more stubby and lateral buds not curved around the twig.

Amelanchier lamarckii F. G. SCHROED., snowy mespilus

Buds fusiform, slightly acute, to 11 mm long and 3 mm thick, salmon-coloured to wine-red, in the shade also greenish. Lowest bud scales often long-ciliate. **Twigs** relatively sturdy: first year's growth c. 2 mm thick, matte grey-brown to olive-brown, in the shade also greenish, slightly hairy towards the tip but soon turning glabrous, and later, on thicker twigs, characteristically rhomboid-fissured. **Lenticels** few, small, brown and warty. Broad shrub or small tree to 10 m high. Stems originate from a common root collar, without runners. Very frequently planted tree. Origin: Eastern North America, in Europe sometimes naturalised.

Amelanchier laevis WIEGAND, Allegheny serviceberry

Buds slim, fusiform, to c. 8 mm long and 2 mm thick, on the shaded side yellowish-green and in the sun with a pink-salmon-coloured hue. **Twigs** glabrous, very slender, 1–1.5 mm thick, matte red-brown to ochre-brown with small, faint, pale lenticels. Flowers in April together with leaves expanding, c. 2 weeks before *A. lamarckii*. **Leaf scar** narrow, dark with 3 traces. Frequently planted large shrub or tree to 12 m high. Origin: North America.

Malus Mill., apple

Large shrubs or small trees. Some species with thorny short-shoots. The thorns mostly resemble short-shoots and are rarely sharply acute. Found mainly in young trees, they are often lost entirely in older specimens. Apart from a few closely related species, many hybrids are in cultivation. Even the wild species are very difficult to identify without flowers and fruits. In winter, with few exceptions, only identifiable when fruits are present.

Key to Malus

1　Fruits with persistent sepals 2
1*　Sepals deciduous, leaving a circular scar
　. 6
2　Older twigs with smooth, pale grey bark
　. . . . . **Malus coronaria**
2*　Older twigs with rough, mostly brownish
　bark 3
3　Twigs sturdy, fruits over 3 cm across . . 4
3*　Twigs medium sturdy, fruits to 2 cm
　across 5
4　Twigs hairy at least to the tip . . . .
　. . . . . **Malus domestica**
4*　Twigs and buds glabrous . . . .
　. . . . . Malus sylvestris
5　Wood of twigs tinged wine-red . . . .
　. . . . .**purple crab apples**
　　　　for example **Malus ×purpurea**
5*　Wood of the twigs without red tinge . . .
　. . . . . **Malus ×spectabilis** and
　　　　　　　　　Malus ×prunifolia
6　Terminal bud less than 10 mm long . . . .
　. . .series **Baccatae** and series **Kansuenses**
6*　Terminal bud more than 12 mm long . . .
　. . . . . .Malus yunnanensis

Subgenus Malus

Series Malus

Fruits often relatively large, mostly with only a short stalk, sepals persistent.

Malus domestica BORKH. nom. cons. prop., cultivated apple

[Malus pumila MILL. nom. rej. prop.]

Buds mostly persistently hairy, the uppermost hairs on the twig grey-white and felty, bud scale base at the twig base becoming glabrous. **Twigs** relatively thick at shoot tips, densely felty-hairy and towards the base becoming ± glabrous, reddish-brown to dark violet-brown. **Lenticels** few to many, pale cream-white. Characters are variable and often distinctive for the cultivar: the common 'Golden Delicious' is more slightly hairy and has many distinct pale lenticels. A very common medium-height fruit tree, in cultivation since ancient times.

Malus sylvestris MILL., crab apple

Buds almost glabrous, only slightly ciliate; only terminal bud and top lateral buds of long-shoots a bit more hairy. Terminal buds 2–5 mm long; lateral buds 1.5–3 mm, towards tip of twig often slightly flattened. There are sometimes distinctive differences in the colouration between the sunny and the shaded side. **Twigs** glabrous, glossy, olive-green to wine-red and only towards the tip slightly hairy, with grey-white longitudinally striped epidermis. short-shoots often becoming thorny. **Lenticels** small, pale oblong to round. Rare, tree to 7 m high. Origin: Europe.

Malus domestica
'Golden Delicious'
Almost glabrous tip
of twig

Malus domestica
'Winter-Gold'
Densely hairy tip
of twig

Malus sylvestris
Lateral bud

Malus sylvestris
Twig

Malus ×spectabilis
Lateral bud

Malus ×spectabilis
Part of twig with
short-shoots

Malus ×prunifolia
Lateral bud

Malus ×purpurea
Short-shoot with remnants
of inflorescence

Malus ×purpurea
Lateral bud

Malus ×purpurea
Tip of twig, wood
reddish where cut

Malus sieversii var.
niedzwetzkyana
Tip of short-shoot

Malus ×spectabilis (SOL.) BORKH., Asiatic apple

Buds strongly ciliate and partly hairy; on long-shoots triangular 3 mm high lateral buds with a few scales, and also small terminal buds with a few scales. short-shoots with c. 6 mm long fusiform terminal buds with 6–8 visible dark violet-wine-red white-ciliate bud scales. **Twigs** when young wine-red to brownish grey, towards the tip loosely appressed-hairy with long white hairs and with small, pale dot-shaped lenticels. Two-year-old twigs grey-brown with reddish hues, ± glossy. Tall shrub or small tree to 8 m, in its youth conical, later a broad tree. Origin: China, only known from cultivation.

Malus ×prunifolia (WILLD.) BORKH.

Buds 4–6 mm long, relatively dense white-hairy; lateral buds with 3–5 violet-brown bud scales. **Twigs**: long-shoots olive green to wine-red-brown, towards the tip with short indument, partly felty, ± angular, lower down turning glabrous, with grey-white patches of epidermis, two-year-old twigs grey-brown. Many oblong grey-white to grey-brown lenticels. Frequent, small 5 (–10) m high tree. Origin: north-east Asia, only known from cultivation.

Malus ×purpurea (EUG. BARBIER) REHDER [*Malus atrosanguinea* × *Malus sieversii* var. *niedzwetzkyana*]

Buds very dark, almost glabrous, especially the lowest bud scales ciliate. Terminal buds c. 6 mm, lateral buds to 5 mm long. **Twigs** dark violet-brown to red-brown, glabrous and glossy, partly with lifting grey epidermis. Wood of twigs in transverse section red. **Fruits** with long stalks, ± deciduous, globose, 1.5–3 cm across, dark wine-red with persistent calyx. Tree 6–8 m high. One of the very frequently planted purple crab apples. Other red-flowering hybrids, such as *Malus ×moerlandsii* J. DOOR., have been bred with *Malus ×purpurea* or its parents; in winter they cannot be distinguished.

Malus sieversii (LESEB.) M. ROEM. var. **niedzwetzkyana** (DIECK) LIKHONOS [*Malus niedzwetzkyana* DIECK], a large shrub or small tree from Turkestan is the most important parent species of the dark-red flowering apples. Twigs sturdy. Fruits 5–6 cm across.

Series Baccatae
(incl. series Sieboldianae)

Fruits mostly rather small, longer stalked and calyx deciduous (a circular scar remaining).

Malus baccata (L.) BORKH., Siberian crab apple

Buds relatively long, about twice as long as broad: 5–6 mm long and 2.5–3 mm thick. Lateral buds appressed to twig with brown, partly slightly wine-red 3–5 outer bud scales with ciliate margin. Terminal buds of short-shoots almost fusiform, c. 7 mm long and 2 mm thick. **Twigs** ± glabrous, wine-red to brown, on the shaded side also greenish, with silvery-grey epidermis. Two-year-old twigs olive-green to grey-green or grey-brown. **Lenticels** few and brown. **Fruits** to 4 cm long, stalked, globose, 1 cm across, yellow or reddish. Frequently planted, tree to 5 m high, with dense ovoid crown. Origin: north eastern Asia to western China.

Less common is the similar and hard to differentiate in winter *Malus hupehensis* (PAMP.) REHDER, Hupeh crab apple, its twigs are sparsely hairy at the tip. **Fruits** to 3 cm long, stalked, globose, 1 cm across, greenish-yellow, frequently with red cheeks. Rarely planted, tree 5–7 m high with spreading growth. Origin: China, Northern India.

Malus floribunda SIEBOLD ex VAN HOUTTE, showy crab apple

Buds c. 4 mm long with some red-brown to dark-brown hairy scales, ciliate at the margin and towards the tip of the bud. Lateral buds on short-shoots ovoid, on long-shoots flat. **Twigs** wine-red-violet-brown, glossy, towards the tip sparsely hairy, becoming glabrous, on the weather side with dying grey epidermis. Frequently planted, 4–6 m high ornamental tree with wide spreading branches, forming a wide crown. Origin: Japan.

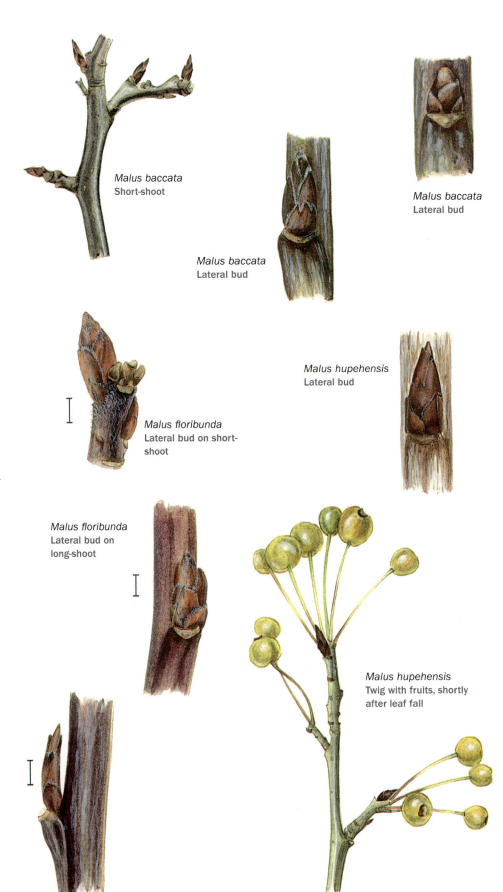

Malus baccata
Short-shoot

Malus baccata
Lateral bud

Malus baccata
Lateral bud

Malus floribunda
Lateral bud on short-shoot

Malus floribunda
Lateral bud on long-shoot

Malus hupehensis
Lateral bud

Malus hupehensis
Twig with fruits, shortly after leaf fall

Malus floribunda
Lateral bud on long-shoot

Malus sargentii
Tip of twig

Malus sargentii
Lateral bud
(lateral view)

Malus sargentii
Lateral bud
(dorsal view)

Malus toringo
Fruits on short-shoots
persistent long into winter

Malus toringo
Lateral buds

Malus toringoides
Lateral bud

Malus toringoides
Short-shoot with fruits

Malus sargentii REHDER
[*Malus toringo* var. *sargentii* (REHDER) PONOMAR.]

Buds: lateral buds appressed to long-shoots, c. 3 mm long with 4–5 outer bud scales, basally ± densely long-hairy. Terminal buds slightly larger, mostly densely hairy. **Twigs**: long-shoots at the tip appressed ± densely hairy and c. 2 mm thick, lower down becoming glabrous and 4–5 mm in diameter; wine-red to violet-brown, glossy, with oblong, pale lenticels. Two-year-old long-shoots olive-brown with ± round lenticels. **Bark** of twigs distinctly smooth, epidermis hardly dying. Short-shoots densely long-hairy; during the winter with small fruits in groups, long-stalked, 8 mm across, without calyx. Planted occasionally, extensively branching shrub c. 2 m high. Origin: Japan.

Malus toringo (SIEBOLD) SIEBOLD ex de VRIESE, Toringo crab apple
[*Malus sieboldii* (REGEL) REHDER]

Buds on relatively pronounced leaf cushions, lateral buds especially below the tip of the long-shoot, strongly flattened, with 3–5 bud scales, ciliate at their margin, 2.5–3 mm long and 1.5–2 mm broad. **Twigs** slightly hairy only at the shoot tip, otherwise glabrous and glossy, violet-wine-red. Long-shoots partly slightly furrowed. short-shoots only slightly short-hairy. **Lenticels** roundish, pale and few. Rare shrub, 4–5 m high. Origin: Japan.

Series Kansuenses

Malus toringoides (REHDER) HUGHES
[*Malus bhutanica* (W. W. SM.) J. B. PHIPPS nom. rej. prop.]

Buds remain densely hairy for relatively long, lateral buds c. 3 mm long, terminal buds of the short-shoots small as well, of irregular shape. **Twigs**: long-shoots matte wine-red to brownish, on the shaded side also greenish, with one-sided denser cover of longitudinally fissured silver-grey epidermis. Densely hairy towards the tip, but not continuous. Lateral shoots ending thorny. **Lenticels** few, small and brownish. Shrub to 8 m high, or tree. Origin: western China.

Subgenus Sorbomalus

Series Yunnanensis

Malus yunnanensis (FRANCH.) C. K. SCHNEID.

Buds: largest among the apples: terminal buds c. 15 mm long with 6–7 lateral buds, each c. 12 mm long with 4–5 bud scales; in shape reminiscent of *Sorbus aucuparia*. Bud scales wine-red glossy white ciliate at the margin, towards the tip of the bud tufted shaggy-hairy. **Twigs** sturdy, glossy, olive green and particularly below the leaf scars wine-red, with silver-grey bits of the epidermis. **Lenticels** few to many, ochre-brown. Rare, tree 10 m high. Origin: China.

Subgenus Chloromeles

Section Chloromeles

Malus coronaria (L.) MILL.,
sweet crabapple

Buds: lateral buds on long-shoots flattened triangular, appressed, mostly slightly narrower than the leaf scars. Terminal buds broadest at the base, cylindrical-conical and the general shape slightly irregular. Bud scales wine-red, brown-violet near the upper white-ciliate margin. **Twigs**: long-shoots dark wine-red to violet, with grey-white epidermis, glabrous, towards the tip wine-red, on the shaded side olive-green and with slightly sparsely hairy patches; several year-old twigs pure grey. Lenticels few, ochre-brown. Tree to 7 m high. Origin: eastern North America.

Section Docyniopsis

Malus tschonoskii (MAXIM.) C. K. SCHNEID.,
Chonosuki crab apple

Buds with wine-red-violet glossy, white-ciliate bud scales. Lateral buds of long-shoots at the tip of the twig 3–4 mm long, flat-appressed to twig and ± felty-hairy; lower on the shoot to 5 mm long, glabrous, roundish-ovoid and spreading. **Terminal buds** of short-shoot ovoid, slightly larger, 5–6 mm long; lowest bud scale of long-shoot terminal bud consisting mostly of leaf base with distinct stipules, ± felty. **Twigs** densely felty-hairy towards the tip, later just like the base of the twig also turning partly glabrous. Strongly glossy below the

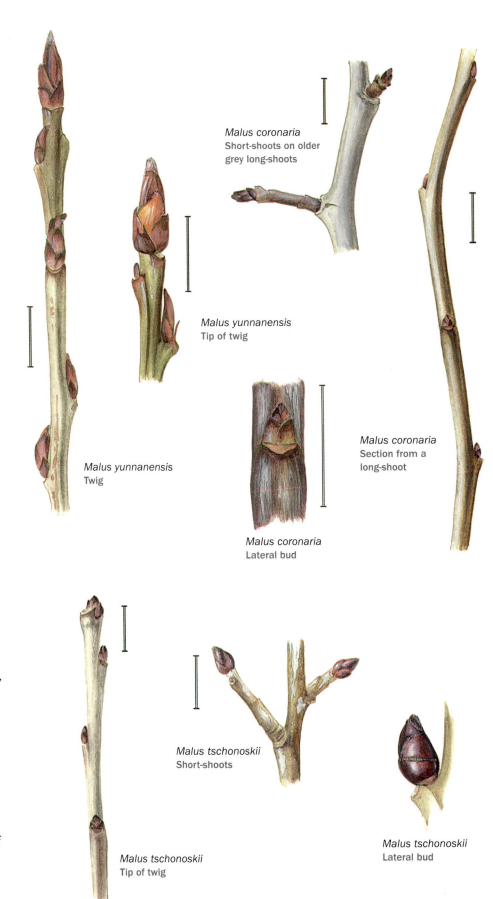

Malus coronaria
Short-shoots on older grey long-shoots

Malus yunnanensis
Tip of twig

Malus yunnanensis
Twig

Malus coronaria
Section from a long-shoot

Malus coronaria
Lateral bud

Malus tschonoskii
Short-shoots

Malus tschonoskii
Tip of twig

Malus tschonoskii
Lateral bud

indument, olive-brown to greenish. Two-year-old twigs grey-green to grey-brown, slightly glossy with few to many, grey-brown round lenticels. Upper epidermis silvery white, lifting off. Rarely planted, tree to 12 m high. Origin: Japan.

Pyrus L., pear

Genus distributed from Europe to eastern Asia and North Africa. About 30 species of deciduous trees and shrubs. Shoots in some of the species becoming thorny. Several species have been involved in the breeding of cultivated taxa, which are therefore quite variable. In most species the terminal buds of the long-shoots are different from those of the many short-shoots. The terminal buds of the long-shoot have mostly fewer bud scales and are more densely hairy; they are often also slightly smaller and stouter than the terminal buds of the short-shoots.

Pyrus communis
var. *sativa*
Terminal bud

Pyrus communis var. *sativa*
Tip of twig

Key to *Pyrus*

1 First year's growth twigs relatively slender (c. 3 mm), often thorny 2
1* First year's growth twigs thicker (rarely less than 4 mm), mostly unarmed 7
2 Twigs mostly completely glabrous, mostly strongly thorny. 3
2* Twigs with distinct, mostly felty indument 4
3 Twigs red-brown to dark brown . . . . *Pyrus pyraster*
3* Twigs olive-brown to grey . *Pyrus spinosa*
4 First year's growth long-shoots almost completely white-felty . **Pyrus salicifolia**
4* First year's growth shoots predominantly glabrous in the lower half 5
5 Terminal buds conical, dark grey-brown . . . . *Pyrus elaeagrifolia*
5* Terminal buds light ochre-brown or ovoid 6
6 Terminal buds of short-shoots red to ochre-brown, globose conical-ovoid. . . . .*Pyrus calleryana*
6* Terminal buds of short-shoots dirty-brown, ovoid . . . . *Pyrus ussuriensis*
7 (1) Long-shoots red-brown, bud conical; rarely thorny . . . .**Pyrus communis**
7* Long-shoots dark violet-brown, buds globose-ovoid; always unarmed . . . .*Pyrus nivalis*

Pyrus communis L. var. *sativa* DC., common pear
[*Pyrus domestica* MEDIK.]

Buds compact-conical to ovoid. Terminal buds 5–8 mm, lateral buds to 6 mm long. Bud scales red-brown to grey-brown, often grey through dead epidermis. **Twigs** mostly glabrous, first year's growth long-shoots c. 4–5 mm thick, red-brown to dark-brown with many pale lenticels. Mostly unarmed, thorny in those specimens which reverted to the wild. Frequent, broad tree to 15 m high. In cultivation since ancient times. Origin: Europe and Asia Minor.

Pyrus pyraster (L.) BURGSD., wild pear

Buds slim, conical, terminal buds 4–5 mm and lateral buds to 3 mm long. Bud scales brown and in patches silvery grey due to dead epidermis. **Twigs** first year's growth to 3 mm thick, red-brown to ochre or olive, glabrous. **Lenticels** few to many, pale. Mostly with many shoot thorns, lacking only

Pyrus pyraster
Twig with thorns

Pyrus pyraster
Tip of twig

in some very old specimens. Rare, shrub or tree to 20 m high. Origin: central to south-western Europe.

Pyrus elaeagrifolia PALL.

Buds conical-acute, with slightly keeled and fusiform red to dark brown bud scales, partly grey due to dead epidermis. Terminal buds of long-shoots 5–6 mm long with 6–8 scales. Lateral buds 4–5 mm long, slightly spreading, only towards the tip of the twig almost appressed. **Twigs** relatively slender: first year's growth long-shoots c. 3 mm thick, olive-brown to dark-brown, with remnants of felty indument. **Lenticels** few. Lateral branching often ending in thorns. Rare, small tree. Origin: Asia Minor.

Pyrus nivalis JACQ., snow pear

Buds with very dark, towards the tip red-brown appressed-hairy bud scales. Terminal buds globose to short-ovoid, to 7 mm long, with 8–11 scales. Lateral buds smaller, conical, to 4(–5) m long, only spreading slightly from the twig, with 4–7 bud scales. **Twigs** thick: c. 5 mm, slightly angular, dark brown, initially short felty hairy, soon becoming glabrous, the indument persistent only at the tip of the twig; with many round, ochre-brown lenticels. **Leaf scars** dark with 3 indistinct traces. Rarely planted, small, unarmed tree to 10 m high. Origin: south-eastern Europe.

Pyrus salicifolia PALL., willow-leaved pear

Buds: lateral buds of long-shoots ovoid to triangular, 3–5 mm long with 4–7 scales. Flowering bud at the end of the short-shoot broadly ovoid, 6–8 mm long, with more than 10 scales. Bud scales red-brown, ± very hairy and ciliate. **Twigs** grey-brown to violet-brown, especially the long-shoots densely felty hairy towards the tip. Older twigs sparsely short-hairy, often thorny. **Lenticels** many, on older and more sparsely hairy parts of twig distinct, round-warty, ochre-brown. **Leaf scars** narrow, dark, with 3 traces. Frequently planted, tree to 9 m high. Origin: Iran, Asia Minor to Caucasus.

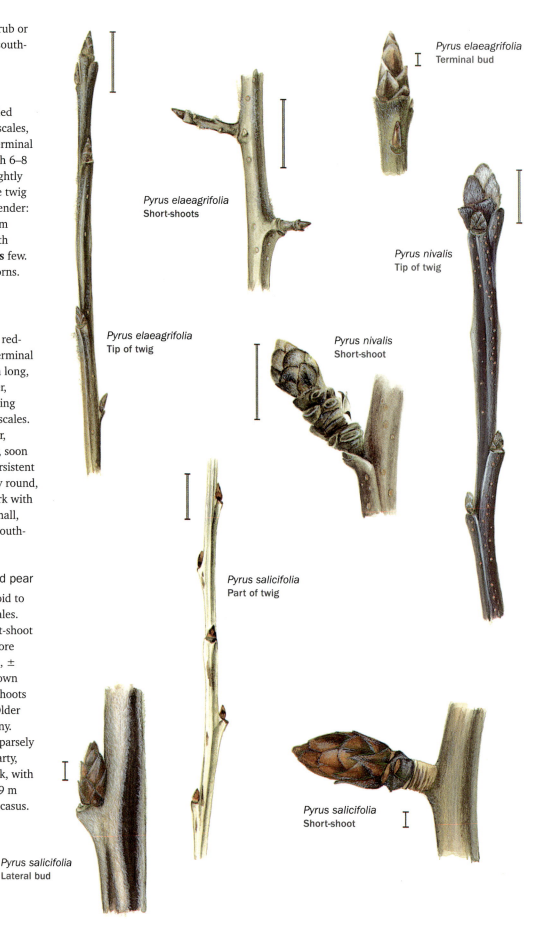

Pyrus elaeagrifolia
Terminal bud

Pyrus elaeagrifolia
Short-shoots

Pyrus nivalis
Tip of twig

Pyrus elaeagrifolia
Tip of twig

Pyrus nivalis
Short-shoot

Pyrus salicifolia
Part of twig

Pyrus salicifolia
Short-shoot

Pyrus salicifolia
Lateral bud

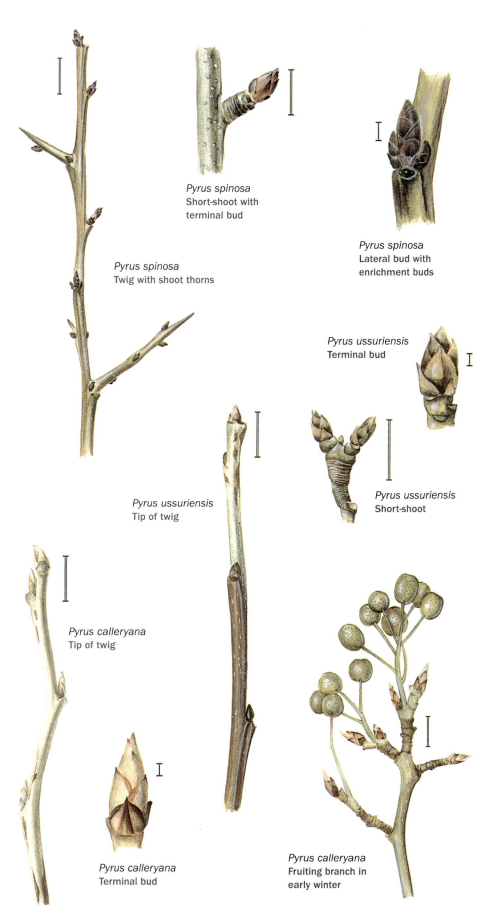

Pyrus spinosa
Short-shoot with
terminal bud

Pyrus spinosa
Twig with shoot thorns

Pyrus spinosa
Lateral bud with
enrichment buds

Pyrus ussuriensis
Terminal bud

Pyrus ussuriensis
Tip of twig

Pyrus ussuriensis
Short-shoot

Pyrus calleryana
Tip of twig

Pyrus calleryana
Terminal bud

Pyrus calleryana
Fruiting branch in
early winter

Pyrus spinosa FORSSK.
[*Pyrus amygdaliformis* VILL.]

Buds with dark red-brown, keeled, ciliate to hairy bud scales. Lateral buds spreading from the twig, ovoid, to 3(–4) mm long, occasionally with enrichment buds. Terminal buds to 6 mm long, compact: globose-ovoid with c. 10 bud scales. **Twigs** relatively slender, when young c. 3 mm thick, ochre-olive to grey (weather side) with many fine whitish lenticels. Hairy during the growth period, towards winter mostly glabrous. Lateral short-shoots mostly thorny. **Leaf scar** dark, 3-traced. Rare, shrub to 6 m high or small tree. Origin: southern Europe to Asia Minor.

Pyrus ussuriensis MAXIM., Ussurian pear

Buds: Lateral buds appressed, c. 3 mm long; glabrous and with felty indument at the tip of the shoot. Terminal buds of short-shoots 6–8 mm long, those of long-shoots shorter, blunt-conical and a little felty. **Bud scales** dirty brown, the lower ones ± keeled, the upper ones darker and towards the tip with few hairs. **Twigs**: long-shoots near the tip almost covered with felty indument, becoming glabrous lower down and with only sparse remnants of hair; on the sun side matte wine-red, to olive-green in the shade. **Leaf scar** narrow, blackish. short-shoots grey to olive-brown, only slightly hairy. **Lenticels** few only, very narrow, oblong to round, light ochre-coloured. Rarely planted, tree to 15 m high. Origin: north-east-Asia.

Pyrus calleryana DECNE., Callery pear

Buds with keeled and long-acute bud scales. Terminal buds on long-shoots 5–8 mm long, light ochre and felty-hairy, with c. 6 bud scales; on short-shoots 7–9 mm long with c. 10 bud scales: the lower ones with sparse hairs and red-brown, the upper ones paler and more densely hairy. Lateral buds smaller and slightly spreading from twig. **Twigs**: long-shoots robust, 3–4 mm in diameter, towards the tip with a felty cover, lower down slowly turning glabrous. Glabrous sections of twig glossy red-brown to olive-green with oblong to roundish, almost white lenticels. short-shoots not glossy, irregularly hairy; occasionally thorny. **Leaf scar** pale grey, narrow, with 3 traces. Occasionally used as street tree. Origin: China.

Pourthiaea villosa (THUNB.) DECNE., Oriental photinia
[*Photinia villosa* (THUNB.) DC.]

Buds narrowly ovoid, acute, 3–4 mm long with few bud scales: red to ochre-brown, partly ciliate. **Twigs** slender, olive-grey to grey-brown, initially long-hairy, later becoming glabrous; with many, warty lenticels. **Leaf scars** with 3 traces. **Fruits** in racemose corymbs, glossy red, c. 8 mm long and ovoid with strongly warty stalks. Shrub to 5 m high, or small tree. Origin: east Asia.

Aronia MEDIK., chokeberry

Buds: red, long-acute terminal bud. **Leaf scars** with 3 traces. The fruits are grouped in upright corymbs and are, depending on the species, deciduous or persistent on the shrub. Differences between the species are found in the indument, the size of fruits and the thickness of twigs. Forms and hybrids of this North American genus are used as fruit plants. Species of the genus were earlier attributed to other genera of the tribe: *Pyrus*, *Sorbus* and *Crataegus*.

Key to *Aronia*

1 Twigs glabrous . . . ***Aronia melanocarpa***
1* Twigs hairy. 2
2 Basal bud scales hairy . . . .
. . . . ***Aronia arbutifolia***
2* Buds glabrous. . . . . . . ***Aronia prunifolia***

Aronia arbutifolia (L.) PERS., red chokeberry

Buds wine red, in the shade also slightly greenish, slightly glossy and basally slightly hairy. Terminal buds fusiform, slightly flattened on one side, 6–8 mm long with 3–4 bud scales. **Twigs** grey to grey-green or red-brown, relatively slender; densely appressed pale-hairy. **Fruits** 4–7 mm across, dark wine-red and persistent to December. Frequent, shrub to 2 m high.

Aronia prunifolia (MARSHALL) REHDER, purple chokeberry [*Aronia arbutifolia* var. *atropurpurea* (BRITTON) C. K. SCHNEID.]
is similar to the red chokeberry, with glabrous buds and felty-hairy twigs. **Fruits** 8–10 mm across, dark black-purple, persistent to December.

Pourthiaea villosa
Lateral bud

Pourthiaea villosa
Twig with fruits

Pourthiaea villosa
Lateral bud

Aronia arbutifolia
Short-shoot with terminal bud

Aronia arbutifolia
Twig with long persistent fruits

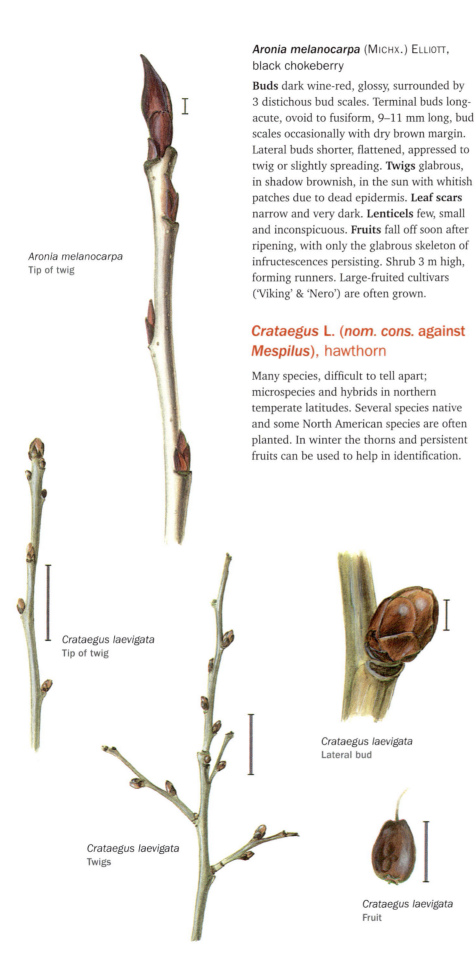

Aronia melanocarpa
Tip of twig

Crataegus laevigata
Tip of twig

Crataegus laevigata
Twigs

Crataegus laevigata
Lateral bud

Crataegus laevigata
Fruit

Aronia melanocarpa (MICHX.) ELLIOTT,
black chokeberry

Buds dark wine-red, glossy, surrounded by
3 distichous bud scales. Terminal buds long-
acute, ovoid to fusiform, 9–11 mm long, bud
scales occasionally with dry brown margin.
Lateral buds shorter, flattened, appressed to
twig or slightly spreading. **Twigs** glabrous,
in shadow brownish, in the sun with whitish
patches due to dead epidermis. **Leaf scars**
narrow and very dark. **Lenticels** few, small
and inconspicuous. **Fruits** fall off soon after
ripening, with only the glabrous skeleton of
infructescences persisting. Shrub 3 m high,
forming runners. Large-fruited cultivars
('Viking' & 'Nero') are often grown.

Crataegus L. (*nom. cons.* against *Mespilus*), hawthorn

Many species, difficult to tell apart;
microspecies and hybrids in northern
temperate latitudes. Several species native
and some North American species are often
planted. In winter the thorns and persistent
fruits can be used to help in identification.

Key to *Crataegus*

1 Twigs distinctly hairy, fruits 1.5–5 cm in
   diameter. 5
1* Twigs mostly glabrous, fruits smaller . . 2
2 Thorns more than 3 cm long; if unarmed,
   terminal bud > 4 mm long. 6
2* Thorns, if present, mostly less than 3 cm
   long, terminal buds < 4 mm long 3

European species (can be identified only by
their fruits, but apart fom the wild species
there are natural hybrids which are sometimes
more common than the parent species)

3 (2) Fruits with one style. 4
3* Fruits with 2 styles and 2 stones. . . . . . .
   . . . . *Crataegus laevigata*
4 Short-shoots rarely end in thorns, sepals
   longer than broad . . . .
   . . . . *Crataegus rhipidophylla*
4* Short-shoots mostly end in thorns, sepals
   not much longer than broad . . . .
   . . . . *Crataegus monogyna*

Medlars

5 (1) Tips of twigs with dense woolly hairs.
   Fruits with distinctly long sepals . . . . . .
   . . . . *Crataegus germanica*
5* Twigs with sparse hairs . . . .
   . . . .*Crataegus azarolus*

North American species

6 (2) Longest thorns over 6 cm long. . . . . .
   . . . . *Crataegus crus-galli*
6* Thorns less than 6 cm long. 7
7 Buds ochre-brown to brown. . . . .
   . . . .*Crataegus coccinea*
7* Buds wine-red to violet-brown . . . .
   . . . . *Crataegus punctata*
   and *Crataegus persimilis*

European species

Crataegus laevigata (POIR.) DC.,
midland hawthorn

Buds ovoid, with brown bud scales: red to
orange-brown, darker at base, lighter near
tip. Lateral buds globose to ovoid, 3(–4)
mm. Terminal buds to 4 mm long and
slightly thicker. **Twigs** relatively pale, ochre-
brown to pale grey, slightly darker only
below the tip of the twig; slightly thorny to
unarmed. **Fruits** initially pale red, 8–12 mm
long, with reflexed sepals hardly or very
little longer than broad as well as with two
styles and stones. Frequent, to 6 m high,
shrub or small tree.

Crataegus monogyna JACQ., common hawthorn

Thorns mostly many, 2–2.5 cm long, only old specimens sometimes unarmed. **Fruits** with one seed, dark red, 8–10 mm long, sepals reflexed, short and broad. Tree to 10 m high, but mostly smaller and shrubby. Very common species in Europe to North Africa, Asia Minor and western Asia.

Crataegus rhipidophylla GAND.

Twigs mostly unarmed: short-shoots rarely thorny. **Fruits** relatively large: 9–15 mm long, red, ovoid with reflexed [subsp. *rhipidophylla*] or erect [subsp. *lindmanii* (HRABETOVA) BYATT] narrow oblong sepals. Shrub, occasional. Origin: central to south-eastern Europe.

Medlars (species of southern Europe)

Crataegus germanica (L.) KUNTZE, common medlar
[*Mespilus germanica* L.]

Buds acute-ovoid, 3–4 mm long, with a few red-brown to dark-brown white-ciliate bud scales. **Twigs** occasionally thorny, brown to grey-brown, especially towards the tip ± densely long-hairy; with few to many small round, pale lenticels. Older twigs violet-brown to grey. **Leaf scars** small, dark, indistinctly 3-traced, also frequently round scars of fruit stalks with a paler, round central trace. A 3–6 m high shrub or small tree. Origin: south-east Europe, naturalised in central and western Europe.

Crataegus azarolus L., azarole

Buds almost glabrous, glossy wine-red to dark violet-brown. Terminal buds ovoid, to 4 mm long, lateral buds globose to ovoid, c. 3 mm long. **Twigs** wine red-brown to ochre-brown with some pale lenticels, hairy especially towards the tip of the twig, but not continuous. **Thorns** lacking or relatively short, to 1.5 cm long. Shrub-like or to 10 m high. **Fruits** c. 2 cm thick roundish, yellowish to orange-red, with 1–2 stones. Shrubby or to 10 m high and tree-like. Origin: north Africa to western Asia, naturalised in southern Europe.

Crataegus monogyna
Tip of twig

Crataegus monogyna
Fruit

Crataegus germanica
Short twig with lateral buds and scars of fruit stalks

Crataegus rhipidophylla
Twig with shoot thorns

Crataegus germanica
Terminal bud

Crataegus germanica
Lateral buds on short-shoot and scars of fruit stalk

Crataegus azarolus
Terminal bud

Crataegus azarolus
Tip of twig with shoot thorns

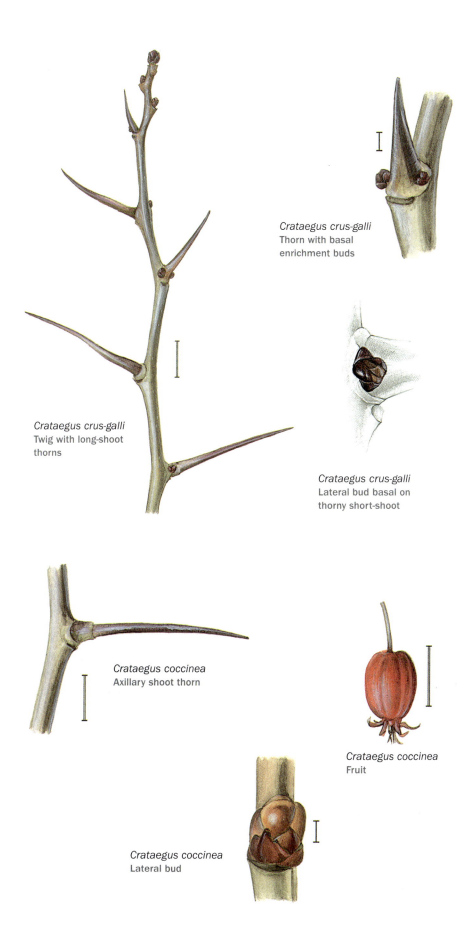

Crataegus crus-galli
Thorn with basal enrichment buds

Crataegus crus-galli
Twig with long-shoot thorns

Crataegus crus-galli
Lateral bud basal on thorny short-shoot

Crataegus coccinea
Axillary shoot thorn

Crataegus coccinea
Fruit

Crataegus coccinea
Lateral bud

North American species

Apart from the native species, some North American species are planted frequently. They differ by longer thorns (as long as these are present), by mostly slightly larger buds or by a rather thick terminal bud compared to its slender twig.

Crataegus crus-galli L., cockspur thorn

Buds compact, terminal buds globose, c. 4 mm long, regular lateral buds globose-ovoid, 2–4 mm, buds at the base of thorns 1–2(–3) mm, with wine-red to dark-violet-brown glabrous bud scales. **Twigs** glabrous, grey-brown to violet-brown, partly with peeling off grey epidermis. **Lenticels** inconspicuous, few, small and round. **Thorns** the colour of the twig, to 8 mm long, straight and sharp-acute. Frequently planted, tree to 12 m high with broad, round crown. The common hybrid cockspur thorn *Crataegus* ×*lavallei* HERINCQ ex LAVALLE [*Crataegus crus-gallii* × *Crataegus mexicana* DC.] is similar, with 5 cm long thorns and initially hairy twigs. Fruits persistent over winter, 1.5–2 cm across, initially brick to orange-red, with 2–3 stones.

Other species from North America are planted frequently but are difficult to tell apart:

Crataegus punctata JACQ.: **buds** wine-red to brown-violet. **Thorns** 3–6 cm long or absent. **Twigs** stiffly spreading horizontally, initially hairy, towards winter mostly glabrous. Fruits deciduous. Tree to 10 m high.

Scarlet haw, *Crataegus coccinea* L.: **buds** glossy ochre-brown to red-brown. **Thorns** 3–5 cm long, straight or slightly curved. **Twigs** slender. Fruits pear-shaped, c. 1 cm long, with large, erect sepals and 4–5 stones. Tree to 7 m high with round crown.

Crataegus persimilis SARG., broad-leaved cockspur thorn: **thorns** to 4 cm long, slightly curved. **Fruits** deciduous. Shrub or tree to 7 m high, with branches at right angles to the trunk.

Rowans, whitebeams, etc.

Some species are native, other species from Asia and North America are planted quite frequently. Rowan, whitebeam, wild and true service tree are quite distinct from each other, and in spite of many hybrids, their respective genera are not closely related. In winter genera can be told apart without too much trouble, but it is not always easy to distinguish individual species.

The hybrid complexes between the genera can, in winter, be roughly classified as so-called 'collective species'. For the apomictic varieties that are described as 'microspecies', classification is very difficult.

Key to species once united under *Sorbus*

1 Buds wine-red, brown or violet-brown, without any green 2

1* Buds predominantly green to reddish-brown. 3

2 Buds very large (> 10 mm), more than twice as long as broad, leaf scars with 5 traces 5

2* Buds smaller or more compact. . . . . [*Aria* × *Sorbus*] ***hybrida***

3 (1) Buds almost globose. . . . . ***Torminalis clusii***

3* Buds ovoid, ± acute. 4

4 Leaf scars with 5 traces . . . . **Cormus domestica**

4* Leaf scars with 3 traces 8

5 (2) Buds to 15 mm long, fruits white. . 6

5* Buds more than 15 mm long, fruits red 7

6 Buds dark violet-brown . . . . **Sorbus koehneana**

6* Buds wine-red . . . . **Sorbus cashmiriana**

7 Upper bud scales partly paler, ± densely felty hairy. . . . .**Sorbus aucuparia**

7* Upper bud scales only towards the tip more densely hairy, only a little paler . . . .*Sorbus decora*

8 Terminal buds often more than 10 mm long, acute, lateral buds much smaller; shrubby . . . . . . *Chamaemespilus alpina*

8* Terminal buds to c. 8 mm long, mostly tree-like 9

9 Buds and twigs densely hairy. 10

9* Buds and twigs only sparsely hairy, becoming glabrous. 11

10 Terminal buds narrowly ovoid (more than twice as long as wide) . . . . [*Sorbus* × *Torminalis* × *Aria*] ***intermedia***

10* Terminal buds ovoid (up to twice as long as wide) . . . .***Aria nivea***

11 Buds relatively deep reddish . . . . [*Sorbus* × *Aria*] ***mougeotii***

11* Buds mostly predominantly green . . . . ***Aria nivea***

Sorbus L., rowan, mountain ash

Buds mostly dark wine-red, violet-brown to grey-brown, oblong-ovoid, glabrous to ± densely brown- to grey-hairy. **Fruits** present at the beginning of winter: white, yellowish to red. **Leaf scars** with 5 traces.

Sorbus aucuparia L., common rowan

Buds with dark brown to grey-black bud scales which are, particularly toward the apex, ± densely whitish to yellowish-brown-hairy. Terminal buds irregular cylindrical-conical: large, 10–20 mm long: lateral buds smaller, appressed to twig. **Twigs** pale brown to grey brown, initially hairy to the tip, later ± glabrous. **Bark** on older twigs and stems smooth with transverse bands of lenticels. **Leaf scars** on distinct cushions with narrow leaf scar incorporating 5 traces. Frequently planted, shrub or tree to 15 m high. Origin: Europe to Asia Minor and the Caucasus as well as western Siberia.

Some rarely planted species such as *Sorbus decora* (SARG.) C. K. SCHNEID. are similar. **Buds** dark wine-red-violet, narrowly ovoid. Terminal buds to 20 mm long and 6–8 mm thick. Bud scales slightly sticky and brown hairy at the tip. **Twigs** thick, matte brown, later grey with silver-grey epidermis. **Lenticels** few to many. **Leaf scars** with 5 traces, initially dark, on cushions with violet-brown margin, later pale grey. Rarely planted, tree to 10 m high. Origin: North America.

Sorbus koehneana C. K. SCHNEID., Koehne's rowan

Buds with dark, black-brown bud scales which are whitish-hairy at the margins and toward the apex. Terminal bud conical-ovoid, c. 10 mm long. Lateral buds smaller. **Twigs** mostly glabrous, red-brown, on the weather side with grey patches of epidermis. **Lenticels** many, pale, oblong. **Fruits** long persistent: to 7 mm thick, globose, white. Frequent, shrub 2–3 m high. Origin: central China. *Sorbus prattii* KOEHNE cannot be distinguished from this in winter.

Sorbus aucuparia
Twig with infructescence

Sorbus aucuparia
Terminal bud

Sorbus decora
Terminal bud

Sorbus aucuparia
Terminal bud

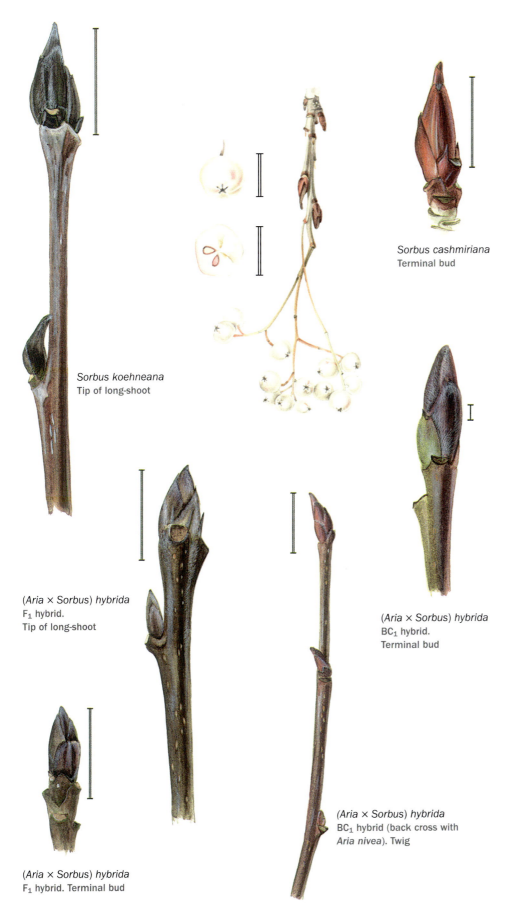

Sorbus koehneana
Tip of long-shoot

Sorbus cashmiriana
Terminal bud

(Aria × Sorbus) hybrida
F$_1$ hybrid.
Tip of long-shoot

(Aria × Sorbus) hybrida
BC$_1$ hybrid.
Terminal bud

(Aria × Sorbus) hybrida
F$_1$ hybrid. Terminal bud

(Aria × Sorbus) hybrida
BC$_1$ hybrid (back cross with
Aria nivea). Twig

Similar too is **Sorbus vilmorinii** C. K. Schneid., Vilmorin's rowan, planted occasionally. Buds and twigs rust-red hairy. **Fruits** in hairy corymbs, to 8 mm thick, reddish, later pale pink. Large shrub or small tree to 6 m high. Origin: western China.

Sorbus cashmiriana Hedl., Kashmir rowan

Buds narrowly ovoid, to 15 mm long and 4–6 mm thick, with bud scales which are purple-wine-red, basally slightly pale yellow-red, glabrous, brown-hairy only at the tip of the bud and the margins. Lowest bud scales of lateral buds ± keeled. **Twigs** brick-brown with patches of grey-white, dead epidermis, later darker and grey. **Lenticels** few, grey-brown, few. **Leaf scars** with 5 traces. Rarely planted tree, 5–8 m high. Origin: Himalaya, Kashmir and Afghanistan.

Hybrids with whitebeams and other rowans

Aria nivea × *Sorbus aucuparia* [Syn. *Sorbus* ×*hybrida* L., *Sorbus* ×*pinnatifida* (Sm.) Düll., *Sorbus* ×*thuringiaca* (Nyman) Schönach, *Sorbus* ×*mougeotii* Soy.-Will. & Godr.]: various hybrids, distinct from the rowan in slightly smaller and compact buds, and from the whitebeam in the dark colour of the buds.

F$_1$ hybrids [(***Aria*** × ***Sorbus***) ***hybrida***], which have basally free pinnule leaflets in their growth period, are in winter closer to the rowan: **buds** with dark-violet bud scales which are densely hairy at the tip. Terminal buds ovoid, 8–10 mm long, lateral buds to 6 mm long, slim and acute, ± appressed to twig. **Twigs** dark wine-red to violet-brown, on long-shoots felty hairy in patches towards the tip. **Lenticels** many, oblong, whitish to ochre.

 'Secondary' BC$_1$ hybrids [(***Aria*** × ***Sorbus***) ***mougeotii***] without free pinnule leaflets, and on predominantly deep red-coloured buds also with green hues on the shaded side; the glossy brown twigs turn glabrous more quickly. Large shrubs or trees to 20 m high. In various regions of Europe with many "microspecies".

Cormus domestica (L.) SPACH., service tree
[*Sorbus domestica* L.]

Buds with slightly sticky, glossy green bud scales with reddish-brown margin. Terminal buds large: to c. 15 mm long, cylindrical, individual bud scales thick and especially convex towards the apex of the bud. Lateral bud considerably smaller, flat. **Twigs** olive-green to brownish, glossy especially towards the tip, with remnants of long, felty hair. **Lenticels** distinct: small, oblong, pale ochre. Older bark of twig grey-brown with warty lenticels. **Leaf scars** on small cushions with green margins, red to dark brown, with 5 traces. Tree 10–20 m high. Plant cultivated since ancient times, with the fruits used to prepare a kind of cider. Origin: southern central and southern Europe to North Africa and Asia Minor.

Aria nivea HOST., common whitebeam
[*Sorbus aria* (L.) CRANTZ]

Buds with yellow-green to green bud scales; these can be reddish in the sun and have a dark-brown margin with loose felty hair. Terminal buds c. 10 mm long, ovoid, acute; lateral buds smaller, to c. 7 mm long. **Twigs** slightly hairy at the tip below the buds, becoming glabrous; on the shaded side olive, on the light side reddish-brown; on older twigs grey-brown to brown, due to dying epidermis. **Lenticels** many, small, round to elliptic. **Leaf scars** with 3 traces. Frequently planted, tree to 15 m high. Origin: Europe, the Mediterranean area to North Africa Asia Minor and the Caucasus.

This can be easily confused with, especially, generic hybrids with *Torminalis* (wild service tree) and with *Sorbus* s.str. (rowan):

The Swedish whitebeam is a hybrid of three species: *Sorbus aucuparia*, *Torminalis clusii* and *Aria nivea* (syn. *Sorbus intermedia* (EHRH.) PERS.): **buds** oblong to ovoid, terminal buds 10–12 mm long; slightly hairy only at the tip. Bud scales 3–4, basally often glabrous and slightly sticky, glossy greenish to dark reddish-brown. **Twigs** of first year dark wine-red and dark brown in the shade, on the weather side soon grey due to dead epidermis. Frequently planted, to 15 m high tree. Origin: northern Europe.

Also frequent is the broad-leaved whitebeam, *Torminalis clusii* × *Aria nivea* (syn. *Sorbus latifolia* L.): **buds** ovoid with ± glabrous, glossy green bud scales which are not or hardly reddish, with a dry, dark-brown, shaggy-ciliate margin. Terminal buds 8–14 mm long, lateral buds distinctly smaller. **Twigs** (almost) glabrous, glossy olive-green to dark grey-green. Two-year-old twigs grey-brown with dying grey-white epidermis. Tree to more than 15 m high. Origin: southern to central Europe.

Chamaemespilus alpina (MILL.) K. R. ROBERTSON & J. B. PHIPPS, dwarf whitebeam
[*Sorbus chamaemespilus* (L.) CRANTZ]

Terminal buds very large, ovoid, acute, to 18 mm long. Lateral buds distinctly smaller, ± appressed to twig, to 10 mm long. **Bud scales** olive-green to dirty green in the light with a reddish tinge, with dry, dark brown, slightly ciliate margin. **Twigs** olive-brown, glabrous and glossy, with a few pale lenticels. **Fruits** rarely present in winter; fruit stalks persistent, deep red-brown with many lenticels. Terminal buds often on short-shoots. Rarely planted, shrub to 2 m high. Origin: mountains of central and southern Europe.

Torminalis clusii (M. ROEM.) K. R. ROBERTSON & J. B. PHIPPS, wild service tree
[*Sorbus torminalis* (L.) CRANTZ]

Buds green and glossy. Bud scales occasionally ciliate at the margins, more rarely slightly felty hairy. Terminal buds short ovoid-globose, 5–6 mm long; lateral buds smaller. **Twigs** olive-brown to dark brown, ± woolly hairy mainly towards the tip, but hairs hardly continuous; with many small lenticels. **Leaf scars** dark brown, 3 traces. Tree with a round crown, 10–15 m high, occasional. Origin: Europe to Asia Minor and north Africa.

Cormus domestica
Terminal bud

Cormus domestica
Twig

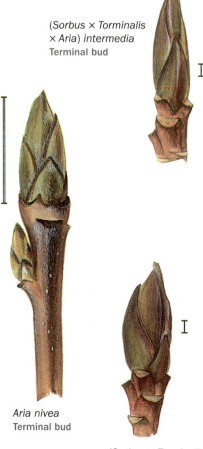

(*Sorbus* × *Torminalis* × *Aria*) *intermedia*
Terminal bud

Aria nivea
Terminal bud

(*Sorbus* × *Torminalis* × *Aria*) *intermedia*
Terminal bud

Aria nivea
Lateral bud

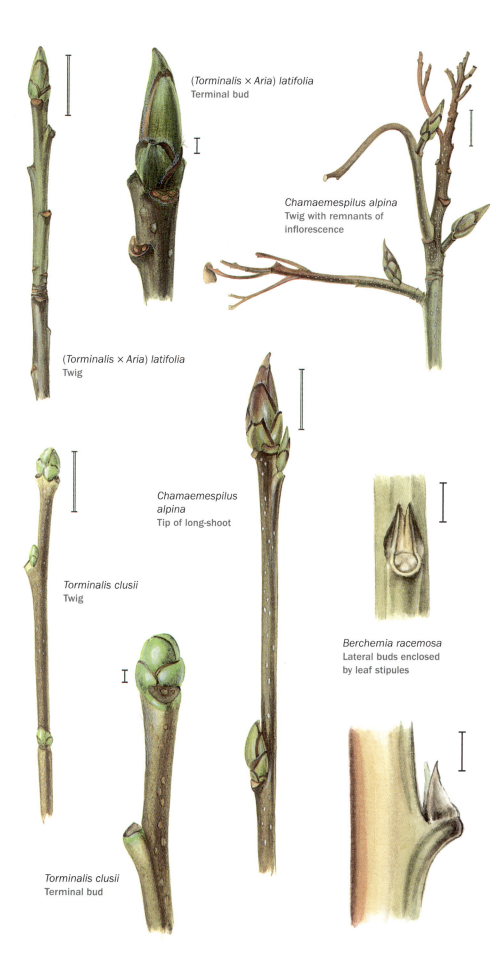

(Torminalis × Aria) latifolia
Terminal bud

(Torminalis × Aria) latifolia
Twig

Chamaemespilus alpina
Twig with remnants of inflorescence

Chamaemespilus alpina
Tip of long-shoot

Torminalis clusii
Twig

Torminalis clusii
Terminal bud

Berchemia racemosa
Lateral buds enclosed by leaf stipules

Rhamnaceae
buckthorn family

Buds alternate or sub-opposite, with bud scales or without. Shrubs and trees, often armed with thorns or spines, more rarely lianas.

Key to Rhamnaceae

1 Plants with thorns or spines 2
1* Plants unarmed 4
2 Two unequal (stipular) spines at every node, buds alternate. 3
2* Shoot tips sharply thorny; buds obliquely opposite . . . . ***Rhamnus***
3 Spines distinctly longer than 1 cm . . . . *Ziziphus jujuba*
3* Spines short, rarely more than 1 cm long . . . . *Paliurus spina-christi*
4 Tree-shaped, twigs sturdy, buds with bud scales and accessory buds. *Hovenia dulcis*
4* Shrubs or lianas, often with slender twigs, when tree-shaped, buds without bud scales. 5
5 Winding liana, only with lateral buds, these enclosed by two leaf stipules . . . . *Berchemia*
5* Erect shrubs or (small) trees 6
6 Living twigs greenish to reddish, fruits or their remnants many in corymb . . . . ***Ceanothus***
6* Twigs grey-brown, fruits (mostly deciduous) solitary or few in racemes . . . . ***Rhamnus***

Berchemia racemosa SIEBOLD & ZUCC.

Buds: lateral buds, spirally inserted, surrounded by leaf stipules. The covering of the bud consists of the leaf base with an oblique-ovate scar of petiole with 3 traces and two persistent acute and distinctly keeled stipules. Their densely ciliate margins meet on the inside and cover the axillary bud. The hidden bud is c. 2 mm long and its outer leaves are green, scale-like, but relatively soft. **Twigs** glabrous, round, first year's growth to 3 mm thick; on the shaded side matte grey-green, on the light side violet-brown without distinct lenticels. Liana to 4 m high, winding to the left. Origin: Japan and Taiwan. Another species, ***Berchemia scandens*** (HILL) K. KOCH. is rare in cultivation. Origin: eastern North America.

Ceanothus americanus L., New Jersey tea

Buds: mostly lateral buds, 3–4 mm long, flattened-ovoid, appressed to twig or slightly spreading, with few bud scales from leaf bases and stipules. **Twigs** initially hairy, often becoming glabrous, yellowish-green to olive-green in the shade and dark wine-red in the sun; partly, and especially the fruit shoots dead and dry. **Infructescences** terminal and below the shoot tip, corymb on axillary shoots with many 3–4 mm large fruits. **Fruit:** three-parted stone fruit surrounded by the dark-brown, long persistent calyx; when the fruit disintegrates, the base of the calyx remains throughout the winter. Occasionally planted, small erect shrub to 1 m high Origin: North America. There are also hybrids with other species and some cultivars of these are: the 5 m high deciduous or evergreen ***Ceanothus ×delilianus*** SPACH [*Ceanothus americanus × Ceanothus caeruleus* LAG.] and ***Ceanothus ×pallidus*** LINDL. [*Ceanothus ×delilianus × Ceanothus ovatus* DESF.], a deciduous shrub to 1.5 m high with many shoots.

Hovenia dulcis THUNB., raisin tree

Buds spreading from the twig, to 5 mm long, ± blunt, globose to conical-ovoid; with few, mostly 2, outer, relatively densely hairy bud scales. Lateral buds with 1–2 descending accessory buds ± shifted across the leaf scar in one direction. **Twigs** unarmed; glossy dark brown, finely longitudinally fissured; initially hairy, in winter mostly glabrous; with many pale, dot-like lenticels. **Leaf scars** distinctly 3-traced, partly at a slight angle to the twig, often encircling the accessory buds; on both sides with narrow stipule scars. Rarely planted, tree to 10 m high. Origin: eastern Asia.

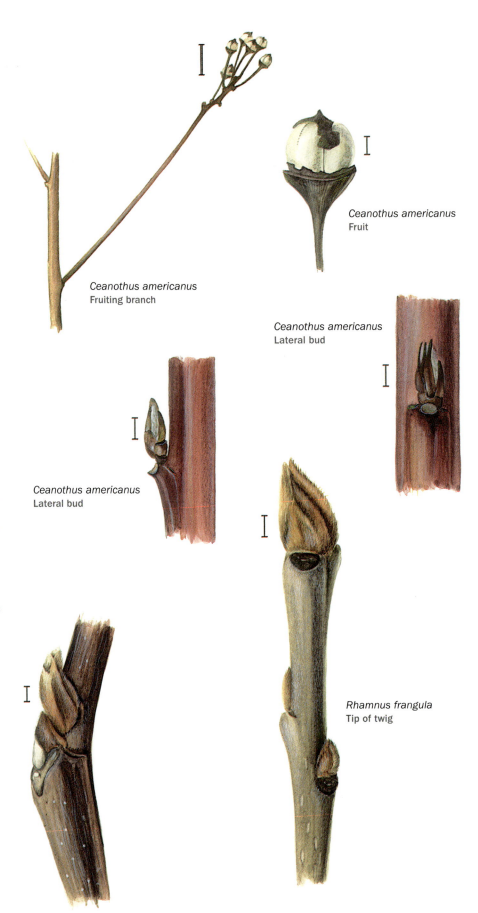

Ceanothus americanus
Fruiting branch

Ceanothus americanus
Fruit

Ceanothus americanus
Lateral bud

Ceanothus americanus
Lateral bud

Rhamnus frangula
Tip of twig

Hovenia dulcis
Lateral bud with descending
accessory bud

Rhamnus cathartica
Lateral bud below the thorny tip of twig

Rhamnus L., buckthorn

Buds with bud scales (subgenus *Rhamnus*) or without bud scales and alternate (subgenus *Frangula*).

Key to *Rhamnus*

1 Buds with bud scales 2
1* Buds without bud scales . . . .
. . . .*Rhamnus frangula*
2 Buds alternate 4
2* Buds (often sub-) opposite, twigs thorny
. . . . 3
3 Buds 2–4 mm long . . . *Rhamnus saxatilis*
3* Buds 5–9 mm long . . . .
. . . .*Rhamnus cathartica*
4 Dwarf shrub to 30 cm high; buds less than 6 mm long . . . . . . *Rhamnus pumila*
4* Erect shrub 2–3 m high; buds more than 6 mm long . . . .*Rhamnus alpina*

Subgenus Frangula

Buds alternate, without bud scales, unarmed shrubs.

Rhamnus frangula L., alder buckthorn

Buds densely hairy, ochre-brown to grey-brown. Terminal buds ovoid, 4–5(–6) mm long, with 2–4 visible outer leaves. Lateral buds smaller, appressed to twig. **Twigs** matte wine-red-brown to violet-brown, towards the tip finely white-hairy; basal and older parts of twigs glabrous, smooth, dark violet-brown, glossy; mainly towards the tip with large flakes of grey-white, matte, dead epidermis. **Lenticels** very pale, dot-to-dash-shaped, more distinct on older twigs. **Leaf scars** with 3 traces on distinct leaf cushions. In the axils of leaf scars there are often some round scars of fruit stalks. Frequent shrub to 3 m high or small tree to 7 m high. Origin: From Europe to central Asia and around the Mediterranean area.

Subgenus Rhamnus

Buds with bud scales, lateral buds alternate or sub-opposite.

Section Rhamnus, buckthorn

Buds sub-opposite, twigs mostly thorny.

Rhamnus cathartica L., common or purging buckthorn

Buds c. 5–9 mm long, oblong, acute, with 8–10 dark brown, glabrous, bud scales which are finely ciliate at the margin. Lateral buds sub-opposite, slightly smaller than the terminal buds. The latter ones at the end of short-shoots, but often missing at the end of long-shoots: in their place the shoot ends in a thorn. **Twigs** mostly glabrous, grey-brown to olive-brown, later grey, with warty lenticels. **Leaf scars** narrow to semi-circular, on small but distinct leaf cushions, 3-traced. Frequent, to 3 m high shrub with the branches at right angles to the stems. Origin: Europe to central Asia as well as north Africa and Asia Minor.

Rhamnus saxatilis JACQ., Avignon berry

Buds: mostly without terminal buds due to shoot tips ending in thorns. Lateral buds opposite, ovoid, 2–4 mm long, appressed to twig, dark red-brown with few bud

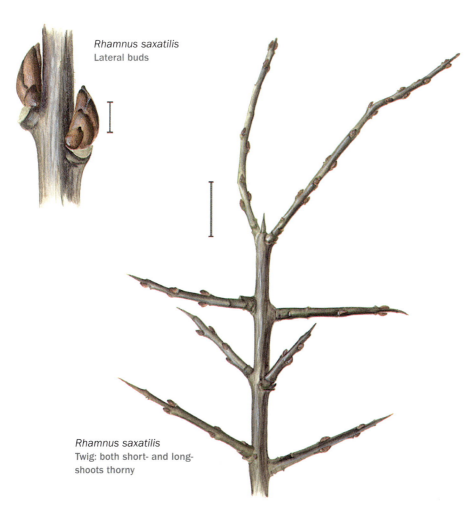

Rhamnus saxatilis
Lateral buds

Rhamnus saxatilis
Twig: both short- and long-shoots thorny

scales. **Twigs** almost glabrous or sparsely and inconspicuously hairy, olive-brown to violet-brown, with silvery-grey epidermis. Very thorny due to many short lateral shoots ending in thorns. **Leaf scar** grey-brown, 3-traced. At the leaf scars 2 long persistent stipules. Rare, shrub 0.5–2 m high with branches at right angles to stems. Origin: south and southern central Europe

Section Oreoherzogia

Buds alternate. Shrubs. Always unarmed.

Rhamnus alpina L., Alpine buckthorn

Buds: only lateral buds: narrowly ovoid, 6–9 mm long with some dark-brown bud scales, slightly lighter ochre-red-brown towards the margin. **Twigs** glabrous, red-brown with grey areas of dead epidermis, especially near the base of the first year shoot, two-year-old twigs entirely grey. **Lenticels** initially few, small and grey; later more, rhombic, grey-brown but inconspicuously merging into the fine-fissured structure of bark. **Leaf scars** grey-brown with 3 indistinct traces, and with narrowly oblong stipule scars with several traces; and basally on the first-year shoot in the leaf axils with round infructescence scars. Rare, shrub 2–3 m high. Origin: southern European mountains.

Rhamnus pumila TURRA, dwarf buckthorn

Buds acute-ovoid, to 5 mm long, with few matte dark ochre- to red-brown bud scales. **Twigs** grey-brown to olive-brown, short-hairy, with pale, warty lenticels. **Leaf scars** on distinct cushions. Rare, shrub c. 20 cm high. Origin: mountains of central- and southern Europe.

Ziziphus jujuba MILL., jujube

Buds c. 2 mm long, hemispherical and densely hairy. **Bud scales** 3-acute due to remnants of stipules and the leaf base; tips of scales dark brown and ± glabrous between areas of lighter hair. **Twigs** zig-zag, initially brown-violet to wine-red, pale grey when older with many small whitish lenticels. On the nodes with two unequal stipular spines: one straight and slender, to 3 cm long, the other shorter and more squat.

Rhamnus alpina
Unarmed twig

Rhamnus alpina
Tip of twig

Rhamnus pumila
Twig

Ziziphus jujuba
Lateral bud flanked
by stipular spines

Ziziphus jujuba
Twig

Leaf scars small, brown with an oblique oblong trace, which appears occasionally to be composed of several traces. Above the bud frequently a scar of an infructescence with a round trace. Shrubs to 9 m high, or small tree. Origin: south-eastern Europe to eastern Asia.

Paliurus spina-christi MILL., Jerusalem thorn

Similar to *Ziziphus jujuba*. **Buds** distichous at the base of first year's growth twigs, c. 2 mm thick, compact, densely grey-brown hairy with some bud scales at the base. **Twigs** finely hairy or glabrous, in the first year clearly distichous in one plane, slender, towards the tip red-brown to ochre-brown, often darker at the base. In the second year violet-brown and partly covered by dead grey epidermis. On many nodes two branch-coloured spines of unequal size, to 8(–12) mm long. At the tips of one-year–old shoots remnants or scars of the infructescences. Shrub to 6 m high, or small tree. Origin: the Mediterranean area, across the Balkans and Transcaucasia to China.

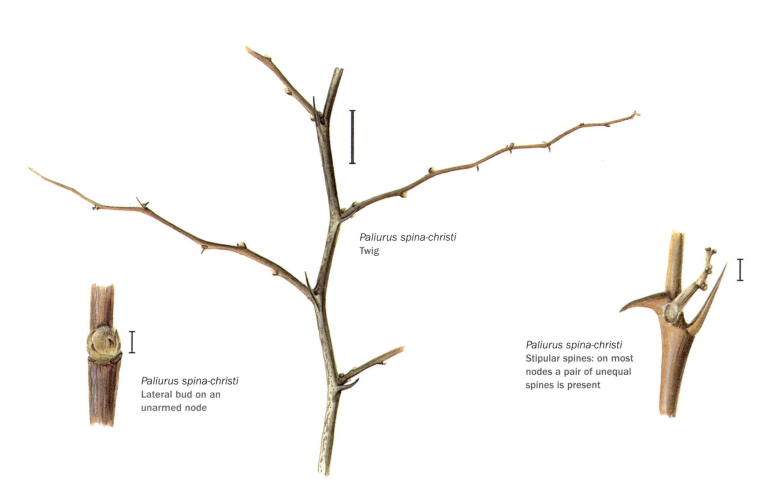

Paliurus spina-christi
Twig

Paliurus spina-christi
Lateral bud on an unarmed node

Paliurus spina-christi
Stipular spines: on most nodes a pair of unequal spines is present

Elaeagnaceae
oleaster family

The species of the oleaster family have a special kind of hair, a scaly trichome or 'scurfy scale': several-celled, disc-shaped hairs with a short stalk, which give a metallic silvery- or bronze sheen to buds and twigs. Leaf scars with one trace.

Key to Elaeagnaceae

1 Buds alternate 2
1* Buds opposite. . . . . *Shepherdia*
2 Terminal bud mostly present . *Elaeagnus*
2* Without terminal buds, end of shoot mostly thorny 3
3 Twigs sturdy, with dense brown scurfy scales . . . . *Hippophae rhamnoides*
3* Twigs slender, mostly with white scurfy scales . . . . *Elaeagnus multiflora*

Hippophae rhamnoides L., sea buckthorn

Dioecious. **Buds**: only lateral buds; bronze-brown glossy, compact, c. 2 × 2 mm. **Flower buds** of male shrubs relatively large, 6 × 4 mm, in close succession on the shoot; those on the female plant smaller, c. 4 × 3 mm. The outer bud leaves relatively thick and, due to this, the bud appears slightly swollen. **Twigs** 2–3 mm thick, sometimes furrowed to round, with grey to bronze-brown scurfy scales. The ends of shoot ends and lateral shoots mostly dry out. The **flowers** in March to April are inconspicuous. The orange-coloured **fruits** on female plants persist until winter. Common, shrub or small tree 4–6 m high. Origin: Europe and Asia, with several forms.

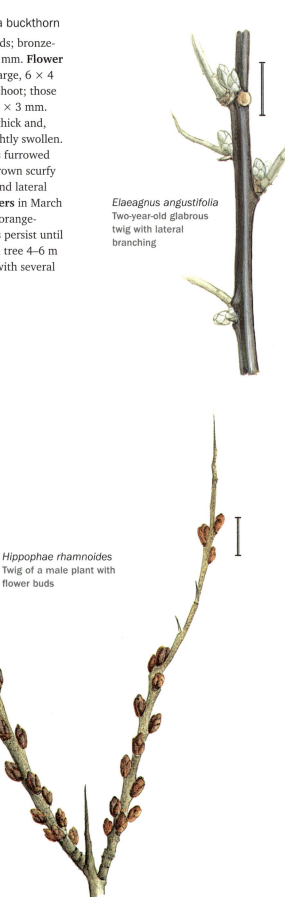

Elaeagnus angustifolia
Two-year-old glabrous twig with lateral branching

Hippophae rhamnoides
Twig of a male plant with flower buds

Hippophae rhamnoides
Twig of a female plant with fruits

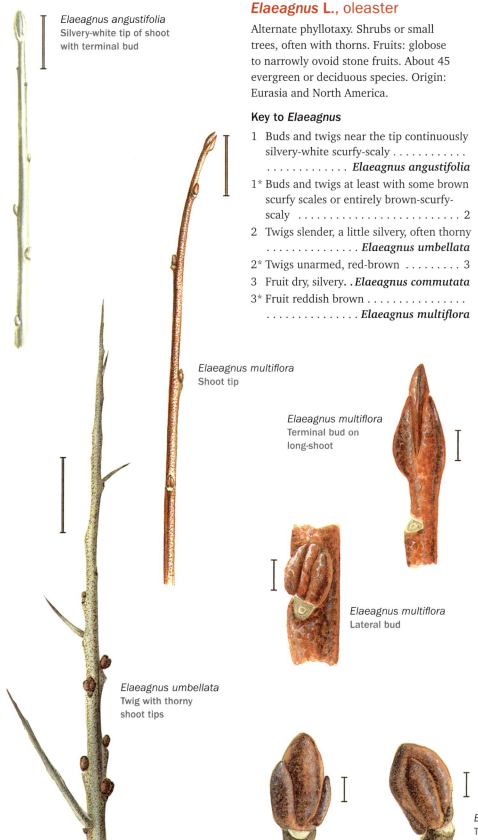

Elaeagnus angustifolia
Silvery-white tip of shoot
with terminal bud

Elaeagnus multiflora
Shoot tip

Elaeagnus multiflora
Terminal bud on
long-shoot

Elaeagnus multiflora
Lateral bud

Elaeagnus umbellata
Twig with thorny
shoot tips

Elaeagnus multiflora
Terminal buds on
short-shoots

Elaeagnus L., oleaster

Alternate phyllotaxy. Shrubs or small trees, often with thorns. Fruits: globose to narrowly ovoid stone fruits. About 45 evergreen or deciduous species. Origin: Eurasia and North America.

Key to *Elaeagnus*

1 Buds and twigs near the tip continuously silvery-white scurfy-scaly . . . . *Elaeagnus angustifolia*

1* Buds and twigs at least with some brown scurfy scales or entirely brown-scurfy-scaly 2

2 Twigs slender, a little silvery, often thorny . . . . *Elaeagnus umbellata*

2* Twigs unarmed, red-brown 3

3 Fruit dry, silvery. . *Elaeagnus commutata*

3* Fruit reddish brown . . . . *Elaeagnus multiflora*

Elaeagnus angustifolia L., common oleaster

Buds silvery-white scurfy-scaly, globose to ovoid, with several bud scales. Lateral buds on long-shoots c. 2 mm, at the base of ± thorny lateral shoots, to 4–6 mm. **Twigs** densely white scurfy-scaly, olive-green below, older twigs dark brown to brown-violet, glabrous and glossy. **Bark** of stronger twigs coming off in longitudinal strips. **Fruit** whitish to greyish-yellow. Frequently planted, shrub or tree to 7 m high. Origin: Eurasia.

Elaeagnus multiflora THUNB., cherry oleaster

Buds brown-scurfy-scaly, without bud scales, shining like bronze. Lateral buds compact, c. 2 × 2 mm, the terminal buds longer, 3–4 mm, with 2–3 outer leaves which stand ± apart, with the leaf base rolled upwards. **Twigs** straight and slender: 2 mm thick, without thorns and glossy brown-scurfy-scaly. **Leaf scars** on a small cushion with a round central trace. **Fruit** reddish-brown. Frequently planted, shrub to 3 m high. Origin: China and Japan. Also unarmed is the frequently planted *Elaeagnus commutata* BERNH. ex RYDB., silver berry, a shrub 1.5–2.5 m high with runners. **Twigs** and **buds** with silvery and brown scurfy scales. **Fruit** dry, silvery.

Elaeagnus umbellata THUNB., autumn oleaster

Buds usually only lateral buds, because shoots often end in thorns: buds 2–3 mm long, bronze-brown, surrounded by two laterally convex prophylls and one to two subsequent bud scales. **Twigs** scurfy-scaly, on the shaded side olive-brown to yellowish-brown, silvery on the light side, mostly ending in long thorns and with many axillary thorns. **Fruit** reddish. Frequently planted, shrub to 4 m high, with runners. Origin: China, Japan and Korea.

Shepherdia NUTT., buffalo berry

Rarely planted shrubs with opposite phyllotaxy. Distinctive and unmistakable due to the many small flower buds without bud scales. Origin: North America.

Key to *Shepherdia*

1 Buds and twigs silvery-white . . . . *Shepherdia argentea*
1* Buds and twigs brown . . . . *Shepherdia canadensis*

Shepherdia argentea (PURSH) NUTT., silver buffalo berry

Buds with silvery-white scurfy scales: at the centre of each scale a reddish or brownish dot. **Leaf buds** oblong, terminal bud to c. 6 mm, lateral buds 4–5 mm long and to 2 mm broad. **Flower buds** in groups on short-shoots, globose to compact-ovoid, to 2 mm in diameter, stalked. Initially **twigs** as scaly as the buds, later becoming glabrous and darker, occasionally thorny. Large shrub or tree to 6 m high.

Shepherdia canadensis (L.) NUTT., Canadian buffalo berry

Similar to the silver buffalo berry, but twigs and buds with red-brown scurfy scales so that the **buds** and shoot tips appear bronze-brown. Terminal buds slightly flattened, c. 8 mm long and short-stalked. **Flower buds** 1.5 mm long, oblong-globose. **Twigs** later covered by remnants of grey scurfy scales. Unarmed shrub to 2.5 m high, the branches at right angles to the stem.

Shepherdia canadensis
Tip of twig: below the terminal bud there are flower buds on short-shoots

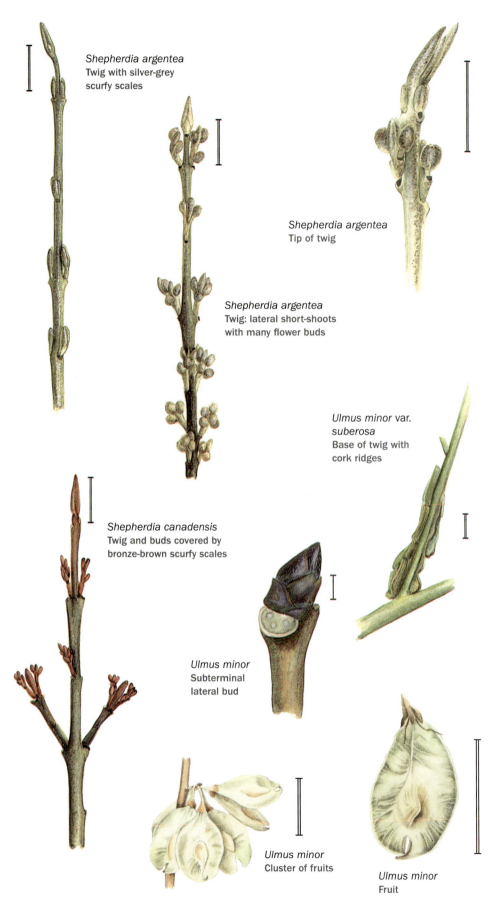

Shepherdia argentea
Twig with silver-grey scurfy scales

Shepherdia argentea
Tip of twig

Shepherdia argentea
Twig: lateral short-shoots with many flower buds

Ulmus minor var. suberosa
Base of twig with cork ridges

Shepherdia canadensis
Twig and buds covered by bronze-brown scurfy scales

Ulmus minor
Subterminal lateral bud

Ulmus minor
Cluster of fruits

Ulmus minor
Fruit

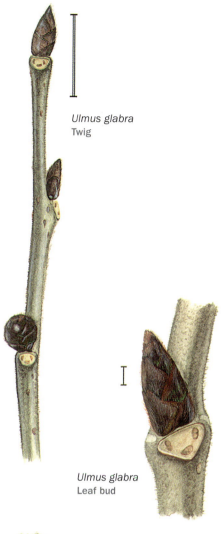

Ulmus glabra
Twig

Ulmus glabra
Leaf bud

Ulmus glabra
Globose flower buds

Ulmus glabra
Fruit

Ulmaceae
elm family

Without terminal buds, continual growth from a terminal lateral bud. Phyllotaxy mostly distichous, including that of the simple bud scales.

Key to Ulmaceae

1 With thorny short-shoots . . . . *Hemiptelea*
1* Unarmed 2
2 Buds mostly less than 2 mm long, with small enrichment buds . . . . ***Zelkova***
2* Buds more than 2 mm long, slightly oblique above the leaf scar, enrichment buds mostly larger, globose flower buds . . . . ***Ulmus***

Ulmus L., elm

The buds on the twigs as well as the simple bud scales of the buds are distichous. The buds stand mostly at a slight angle above the leaf scar. As there are only lateral buds, elms continue the growth of their tips from the uppermost lateral bud. Flower buds ± globose and distinctly different from the leaf buds, sometimes in groups. Flowering very early, from February, and when the leaves appear, already with young fruits. Origin: temperate latitudes of the Northern Hemisphere.

Key to *Ulmus*

1 Twigs persistently hairy 2
1* Twigs glabrous or becoming glabrous, with only sparse remnants of hair 3
2 Buds blunt-ovoid . . . . ***Ulmus glabra***
2* Buds narrowly ovoid, acute. Bud scales with dark margin . . . . ***Ulmus laevis***
3 Leaf buds very small, to 2 mm, with 4 visible bud scales . . . . *Ulmus pumila*
3* Buds larger, with more visible bud scales 4
4 Twigs without cork ridges . . . .*Ulmus minor*
4* Twigs with cork ridges . . . .*Ulmus minor* var. ***suberosa***

Ulmus minor MILL., field elm
[*Ulmus campestris* L. nom. rej., *Ulmus carpinifolia* GLED.]

Buds relatively small, compact, 2–4(–6) mm; dark brown and slightly white-ciliate (lens!). **Twigs** slender, mostly glabrous, brown, longitudinally fine-fissured, with some lighter small lenticels. Trunks often with water shoots and without buttresses. With sucker shoots and, particularly following injury such as coppicing, with stool shoots from stumps. **Flowers** almost sessile in dense clusters, with 4–5 stamens and white scars. Very variable tree, rich in forms, 20(–30) m high with erect to hanging branches. Origin: almost the whole of Europe to Asia Minor, the Caucasus and north Africa. In the quite common *Ulmus minor* var. *suberosa* (MOENCH) REHDER the **twigs** have strong corky ridges.

Closely related is ***Ulmus procera*** SALISB., the English elm. It has persistently hairy and relatively thick, often corky shoots. Tree to 30 m high, with a few stiff branches in the lower part and an irregular, dense crown. Origin: southern and western Europe

Ulmus glabra HUDS., wych elm

Buds 5–7(–9) mm long, rust-red to brown hairy, without cork ridges. Partly becoming glabrous, then only bud scales ciliate on the margin. **Twigs** densely red-brown hairy, without cork ridges. **Trunk** mostly straight, with bark which is smooth for a long time, rarely with buttresses and almost without water shoots or sucker shoots. **Flowers** very shortly stalked, in dense clusters, with 5–6 stamens and reddish stigmas, appearing February to March. Frequent, tree 30–40 m high. Origin: Europe to Asia Minor and western Asia.

Here and there ***Ulmus ×hollandica*** MILL., Dutch elm, is more frequent than its parents: this is a natural hybrid between field elm and wych elm [*Ulmus glabra* × *Ulmus minor*]. It is intermediate between the parent species and, due to repeated back-crosses with them, is very rich in forms.

Ulmus pumila L., Siberian elm

Buds: leaf buds very small, dark brown, 1.5–2 mm long; flower buds red-brown, globose, 3–4 mm and mostly in small groups: enrichment buds from the axils of the leaf bud. Bud scales densely shaggy-white ciliate at the margin. Young **twigs** slender, only initially hairy and soon becoming glabrous. Bark rough. **Flowers** short-stalked with 4–5 violet stamens. Rarely planted, small tree 3–6(–10) m high. Origin: eastern Asia.

Ulmus laevis PALL., European white elm [*Ulmus effusa* WILLD.]

Buds 4–6(–7) mm long, acute, ovoid to conical, bud scales with red-brown to dark brown margins. **Twigs** hairy, olive-to ochre-brown, lenticels distinct. Trunk frequently with water shoots and buttresses. **Flowers** long-stalked, in clusters, with 6–8 stamens and white stigmas. Frequent, tree 10–30 m high, with hanging branches. Origin: western central Europe to Asia Minor and the Caucasus.

Ulmus americana L., the American elm,

is similar: **buds** 3–4(–5) mm long, slightly spreading, short-ovoid and ± blunt. Bud scales brown with a distinct darker margin. **Twigs** turning glabrous, grey-brown to rust-brown, with many small, ochre-coloured lenticels, later epidemis fissuring rhombically, particularly on the weather side. Rarely planted, tree 20–40 m high with spreading and hanging branches. Origin: North America.

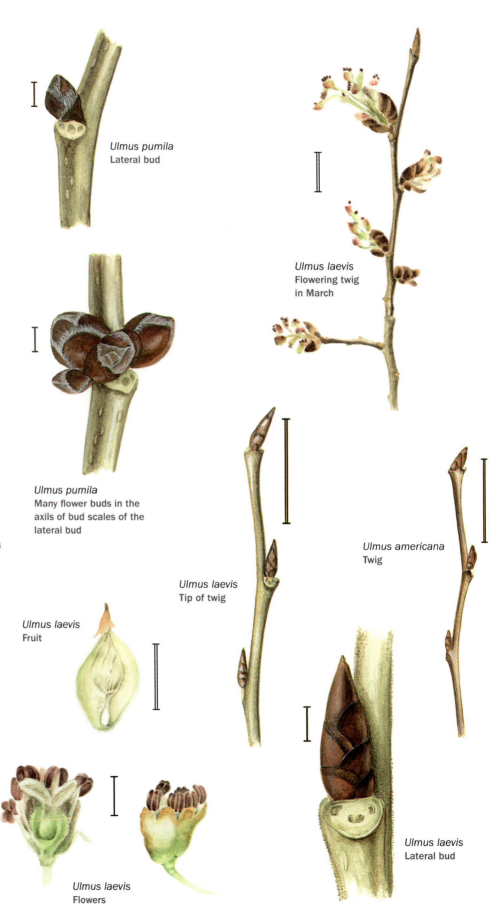

Ulmus pumila
Lateral bud

Ulmus laevis
Flowering twig
in March

Ulmus pumila
Many flower buds in the
axils of bud scales of the
lateral bud

Ulmus laevis
Tip of twig

Ulmus americana
Twig

Ulmus laevis
Lateral bud

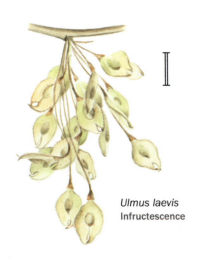

Ulmus laevis
Infructescence

Ulmus laevis
Fruit

Ulmus laevis
Flowers

Ulmus americana
Lateral bud

Zelkova carpinifolia
Twig

Zelkova Spach

Leaf scars apparently with 1 trace, the 3 vascular bundle traces are occasionally grouped centrally and difficult to tell apart. Bark long remaining smooth and grey, similar to the beech. Large trees, often with several stems. Origin: Eastern Mediterranean to Eastern Asia.

Key to *Zelkova*

1 Twigs densely hairy, buds with wide base and frequently with a lateral enrichment bud. Leaf scar broad, obliquely rhomboid . . . . *Zelkova carpinifolia*
1* Twigs slightly hairy or glabrous, buds clearly globose to ovoid. Leaf scars oblique-ovate . . . . *Zelkova serrata*

Zelkova carpinifolia (Pall.) K. Koch, Caucasian elm

Buds short-conical, c. 2 mm. Bud scales sparsely hairy and ciliate, lower ones a little more densely so, dull brown, upper ones red-brown. Lateral buds with mostly one, more rarely with several axillary enrichment buds in the prophyll axils. **Twigs** 1.5–2 mm thick, strongly zig-zag, ochre-brown to dirty brown, sparsely hairy, more densely hairy at the tip. Bark of twig splitting longitudinally and slightly flaking, below initially greenish. **Lenticels** few, ochre-brown, slightly raised. **Leaf scars** broad, rhomboid, at the sides narrowed into a large, broad trace with 3 faint traces of vascular bundles. Frequently planted, tree to 25 m high with erect main branches. Origin: Caucasus.

Zelkova serrata (Thunb.) Makino, Japanese zelkova, keaki

Buds very small: 1–2 mm, globose-ovoid, spreading from the twig. Bud scales ochre- to orange-red-brown, mostly glabrous and sparsely ciliate. **Twigs** very slender, youngest ones mostly less than 1 mm thick, red-brown and violet-brown to olive, with a few roundish, very pale ochre-coloured lenticels; older twigs glossy violet-brown and olive-brown to grey-brown, with pale grey lenticels. **Leaf scars** oblique-ovate with a clearly demarcated central area: in this area 3 faint traces. **Bark** remaining smooth for a long time, grey. Crown roundish and broad. Tree to 30 m high, often with several stems. Origin: Japan, China and Korea.

Hemiptelea davidii (Hance) Planch.

Buds small, less than 2 mm high, relatively broad due to many axillary enrichment buds. Bud scales dirty ochre-brown, margins ciliate and partly slightly fimbriate. **Twigs** hairy, brownish, violet-brown to olive-brown, with small round orange-brown lenticels. Epidermis tearing early and then grey. Mostly with many lateral thorns, often several on each shoot node, emerging from enrichment buds. **Leaf scar** on distinct leaf cushion, 3-traced. Shrub or small tree, rare in cultivation. Origin: northern China and Manchuria.

Zelkova carpinifolia
Lateral bud with enrichment bud

Hemiptelea davidii
Twig with lateral stem thorns

Zelkova serrata
Lateral bud

Hemiptelea davidii
Lateral bud with axillary enrichment buds

Cannabaceae
hemp family

No terminal buds, just as in the Ulmaceae. The lateral buds are very small and seldom grow to more than 3 mm long. They are appressed to the slender twigs and have, in contrast to some Ulmaceae, no enrichment buds.

Key to Cannabaceae

1 Largest buds with up to 5 visible red-brown bud scales . . . . *Celtis*

1* Largest buds with more than 5 visible black-brown bud scales . . . *Aphananthe*

Aphananthe aspera (THUNB.) PLANCH.

Buds: only lateral buds close to twig, narrowly ovoid, to c. 4 mm long, with 5–6 distichous bud-scales. At the base these are often slightly green, higher up relatively dark, black-brown, dorsally white-hairy. **Twigs** very slender (uppermost internode c. 1 mm thick), grey-brown and hairy. Bark of two-year-old twigs slightly obliquely striped, partly grey, scaly, with many ochre-brown lenticels. Pith white with slightly darker margin towards the xylem. Leaf scars very dark, indistinctly 3-traced. Very rarely planted, tree to 20 m high. Origin: eastern Asia.

Celtis L., hackberry, nettle tree

Buds small: 2–4 mm, without enrichment buds, lying close to the twig, with 3–5 bud scales.

Key to *Celtis*

1 Bark smooth, twigs with inconspicuous remnants of long indument. . . . .
. . . . *Celtis australis*

1* Bark irregularly bulging, twigs densely fine-hairy. . . . .*Celtis occidentalis*

Celtis occidentalis L., common hackberry

Buds 2–3 mm long, conical-ovoid, with 3–5 ciliate bud scales: the upper ones occasionally more hairy and pale ochre-brown, the lower ones basally glabrous, glossy dark brown to dark green on the shaded side. **Twigs** dark brown to olive green, sparsely fimbriate-hairy with many slightly raised ochre-brown lenticels. **Bark** irregularly bulging, with rhombic ridges, older stems also furrowed and with flat scales. **Leaf scars** round-triangular, with 3 traces. **Fruits**: stone-fruits with up to 2 cm long stalk, persistent on the tree to autumn or early winter, 7–10 mm long, ovoid, orange-brown. Frequently planted, broad-crowned tree to 25 m high. Origin: North America.

Celtis australis L., European nettle tree

Buds c. 3 mm long, oblong-triangular from above, lying close to twig and relatively flat, or slightly spreading. Bud scales 2–5, red-brown, hairy or at least ciliate. **Twigs** olive-brown to dark grey-green, ± densely fine hairy, with pale, rounded lenticels. **Bark** grey and remaining smooth. **Fruits** 10–12 mm thick, round, the stalk to 3 mm long. Frequently planted, tree to 25 m high, slightly sensitive to frost when young, later with a spreading crown. Origin: southern Europe, Mediterranean region to North Africa and Asia Minor.

Celtis occidentalis
Tip of twig

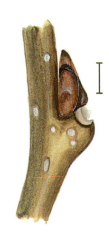

Celtis occidentalis
Lateral buds

Aphananthe aspera
Lateral bud

Celtis australis
Lateral buds

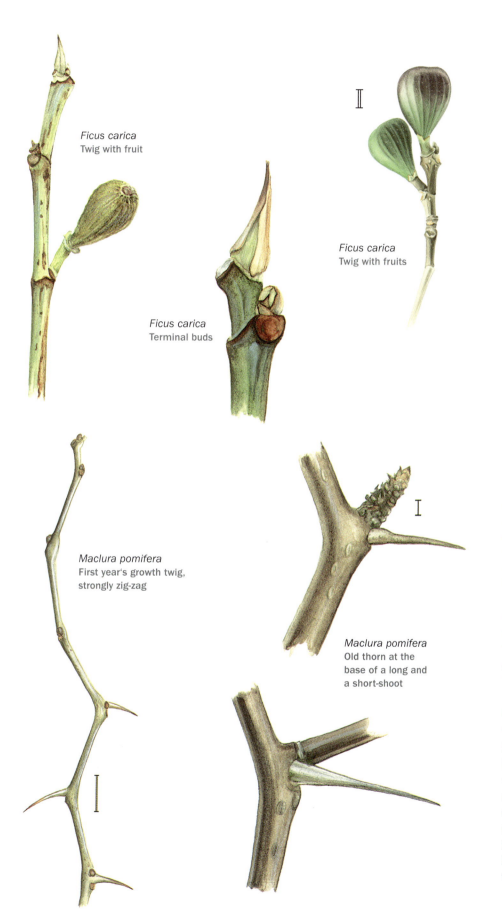

Ficus carica
Twig with fruit

Ficus carica
Twig with fruits

Ficus carica
Terminal buds

Maclura pomifera
First year's growth twig,
strongly zig-zag

Maclura pomifera
Old thorn at the
base of a long and
a short-shoot

Moraceae
mulberry family

Family widely distributed in the tropics and subtropics. The Moraceae have over a thousand, predominantly evergreen species, with only a few deciduous; most species containing latex.

Key to Moraceae

1 Twigs unarmed. 2
1* Twigs with thorns. . . . . *Maclura*
2 Only lateral buds present, with several visible bud scales 3
2* Terminal buds present, these acute-conical, completely surrounded by the outermost bud scale. . . . . . .***Ficus carica***
3 Twigs relatively thick, with wide pith, densely long-hairy towards tip . . . . *Broussonetia papyrifera*
3* Twigs with narrow pith, glabrous or short-hairy . . . . *Morus*

Ficus carica L., common fig

Buds green to yellowish-brown or darker brown, differentiated in terminal and lateral buds: **terminal buds** long acute, conical. **Lateral buds** spiral, smaller and mostly not acute. The bud scales consist of fused stipules of a leaf. These structures, also called median-stipels, completely cover the tips of shoots and buds, or split apart on the side opposite the fused side. **Twigs** thick, greenish to olive-brown. **Leaf scars** red-brown, at each node with a narrow stipule scar encircling the twig. Shrub or small tree. As a plant cultivated since ancient times, widely spread beyond its usual area. Naturalised in western Europe and north Africa. Origin: south-eastern Europe to Asia Minor.

Maclura pomifera (RAF.) C. K. SCHNEID., Osage orange

Buds c. 2 mm long and 1 mm broad. Lateral buds hemispherical, almost contiguous with the twig, often at the base of a thorn, surrounded by c. 4 outer bud scales. **Bud scales**: the lowest ones twig-coloured, the uppermost red-brown. **Twigs** differentiated into long- and short-shoots. **Long-shoots** light grey-green, strongly zig-zag, without terminal buds, with sharp shoot thorns, slightly bent, basally twig-coloured, towards the tip dark red-brown, when older uniformly grey-brown. **Short-shoots** on older long-shoots many (in every leaf axil), short, covered densely with leaf scars and dry remnants of stipules; only with terminal bud. Occasionally planted, tree to 10 m high. Origin: North America.

Maclura tricuspidata CARRIÈRE (*Cudrania tricuspidata* (CARRIÈRE) BUREAU ex LAVALLÉE), a shrub or small tree from Eastern Asia is similar. **Leaf scar** with a trace of vascular bundle (of 3 combined traces). Above the leaf scar a hairy scale, the fused leaf stipules. With enrichment buds next to the short thorns.

Broussonetia papyrifera (L.) L'HÉR. ex VENT., paper mulberry

Buds: without terminal buds. Lateral buds partly alternate, partly opposite, with 2–3 outer bud scales, red-brown to greenish. **Twigs** especially at the tip densely soft and long-hairy, grey-green, with a few ochre-coloured lenticels. Above the leaf scars there are also occasionally some larger scars of inflorescence axes. Rarely planted, mostly a several-stemmed shrub to 5 m high, in Asia a tree to 15 m high. Origin: south-east Asia, naturalised in southern Europe.

Morus L., mulberry

No terminal buds. Lateral buds distichous. Medium-sized trees with latex. Twelve deciduous species in the subtropics and the temperate latitudes of the Northern Hemisphere; two species are often planted and naturalised in southern Europe. As *Morus alba* and *Morus nigra* were grown in the same areas for centuries, there are many hybrid plants which are not easily classifiable. Traditionally the colour of their fruit led to these being assigned to one of the parent species. As the fruit is not present in winter, this is not possible here.

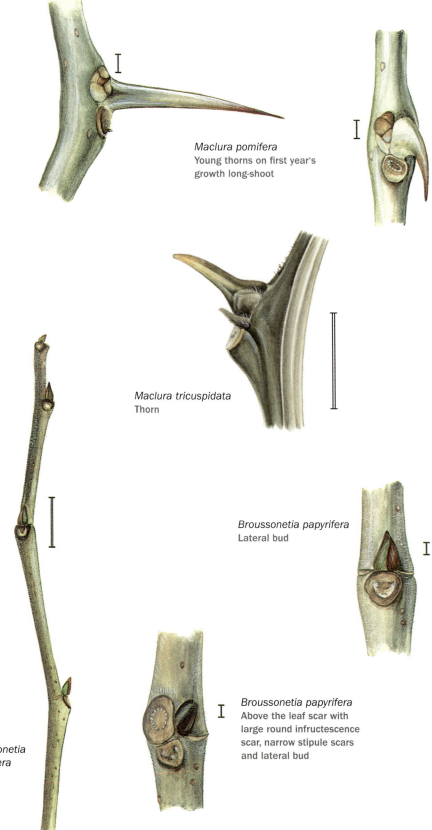

Maclura pomifera
Young thorns on first year's growth long-shoot

Maclura tricuspidata
Thorn

Broussonetia papyrifera
Lateral bud

Broussonetia papyrifera
Above the leaf scar with large round infructescence scar, narrow stipule scars and lateral bud

Broussonetia papyrifera
Twig

Morus alba
Lateral buds

Morus nigra
Lateral bud

Morus nigra
Twig

Punica granatum
Twig, with tips of twig
becoming thorny

Punica granatum
Part of twig: thorny lateral shoots,
at the base with lateral buds

Morus alba L., white mulberry

Bud 4–5 mm long, triangular-conical to globose, with c. 5 red-brown ± fimbriate bud scales with darker margin. **Twigs** mostly glabrous grey-brown to grey-green to red-brown, partly, due to pale grey epidermis, pale grey to grey-yellow. Older twigs with finely fissured bark, yellow-brown to grey-brown. **Lenticels** many, roundish to mostly oblong, pale ochre. **Leaf scars** with many distinct traces. Frequent, tree to 15 m high with round crown, the branches at right angles to the trunk. Origin: China, naturalised in southern Europe.

Morus nigra L., black mulberry

Buds mostly slightly larger, ovoid-acute, ± appressed to twig, with 4–5 outer, ochre to red-brown bud scales with darker margin. **Twigs** grey-brown, with inconspicuous fine hairs, relatively thick with many pale, predominantly oblong **lenticels**, later, on older twigs ochre-brown corky. **Leaf scars** on large leaf cushions, with many indistinct traces. Often shrubby, or tree to 10 m high with dense round crown. In culture since ancient times. Origin: Orient.

Lythraceae
pomegranate family

Punica granatum L., pomegranate

Buds sub-opposite, 2.5–3 mm long, spreading from the twig with a few red-brown, acute bud scales. **Twigs** slender, ± square, often ending in thorns. **Bark** longitudinally fine-fissured, coming off in strands, grey-brown to yellow-brown. **Leaf scars** small, dark, with a central paler vascular bundle. Small fruit tree, 5(–10) m high. Frequently planted; for container planting, the cultivar '**Nana**' is used, as it remains smaller and has finer twigs. Origin: south-eastern Europe to the Himalayas, introduced from Asia Minor to the entire Mediterranean area and in cultivation since ancient times.

Onagraceae
fuchsia family

The family is distributed world-wide, with mostly annual and perennial herbs and also a few woody species.

Fuchsia magellanica Lam., hardy fuschia

Buds opposite or in whorls of 3, small, ovoid with few dry ± hairy scales and, occasionally, with ascending accessory buds. **Twigs** pale grey-brown, initially fine spreading-hairy, mostly dying to the ground. **Leaf scars** on raised cushions, roundish, with one trace. Between the leaf scars of a whorl there are frequently remnants of the partly fused stipules (interfoliar stipules). Frequently planted, sub-shrub to 2 m high and dying down to the ground. Origin: South America.

Staphyleaceae
bladdernut family

Staphylea L., bladdernut

In winter the species are difficult to tell apart. They have opposite buds, surrounded by 2 or 4 visible bud scales, often only two prophylls, fused just basally or over their entire length. True terminal buds are rarely present; usually one of the terminal lateral buds moves into the place of the end of the shoot that has died off. Distinct stipule scars and the bark, which soon becomes longitudinally fissured, are characteristic of the genus. The fruits which persist throughout the winter are helpful for identification: inflated, dry-skinned, pale brown capsules.

Key to *Staphylea*

1 Only two bud scales visible (sometimes fused into one) 2
1* Buds with 4 visible bud scales . . . . *Staphylea trifolia*
2 Buds short-acute, fruits distinctly longer than wide . . . . **Staphylea colchica**
2* Buds acute, fruits rarely longer than wide . . . . **Staphylea pinnata**

Staphylea pinnata L., common bladdernut

Buds glabrous, keeled on two sides, completely surrounded by the two first, mostly fused, bud scales. **Twigs** green to brown-red, glossy, lower down on the twig with many whitish lenticels which later combine in streaks, giving rise to a finely-fissured bark. **Pith** wide and white. **Leaf scars** grey to grey-brown, usually with 5–7 traces of vascular bundles and with two distinct stipule scars between the leaf scars of a node. **Fruits** in hanging panicles, ± globose, c. 3 cm diameter. **Seeds** c. 10 mm diameter. Often planted, shrub 2–5 m high or small tree. Origin: central and southern Europe to Asia Minor.

Staphylea colchica Steven, ivory-flowered bladdernut

Buds short, ovoid, to 10 mm long, surrounded by a single bud scale, which originates from the two fused prophylls, acute and with ± keeled margin. **Twigs** slightly longitudinally furrowed, wine-red to greenish. **Lenticels** initially faint, then roundish to oblong, pale grey and forming a longitudinally fissured bark. **Leaf scars** dark brown to pale grey-brown, usually with (3–)5 traces of vascular bundles uniting in a semi-circle, as well as distinctive pale grey stipule scars. **Fruits** in upright, slightly overhanging panicle, two to three-acute; 5–8 cm long, distinctly longer than broad, with a conical base. Seeds to 7 mm long. Occasionally planted, erect shrub to 4 m high. Origin: Caucasus.

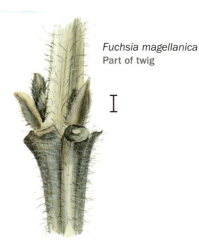

Fuchsia magellanica
Part of twig

Staphylea colchica
2-lobed fruit capsule

Staphylea colchica
**Subterminal lateral bud:
at the shoot tip with
infructescence scar**

Staphylea pinnata
**Pair of lateral buds
at the shoot tip**

Staphylea pinnata
**Lateral bud at shoot tip:
bud scales fused into one**

Staphylea pinnata
Seeds

Staphylea colchica
Twig with fruit

Staphylea trifolia
Lateral buds with
several lateral buds

Staphylea trifolia
Tip of twig

Staphylea trifolia L., American bladdernut

Buds narrowly ovoid to globose, with 4 outer brown-red bud scales, the first two reaching to about half the height of the bud. **Twigs** glossy olive-green or ± brownish, with many lenticels. **Bark**, as in the other bladdernut species, developing a typical longitudinally-fissured pattern. **Fruits** in short overhanging infructescences, mostly 3-lobed, 3–4 cm long. Seeds oblong, c. 6 mm long. Rarely planted, erect shrub to 5 m high. Origin: eastern North America.

The eastern Asian **Staphylea bumalda** DC. is rarely planted. **Buds** surrounded by a single bud scale, the two fused prophylls. It has relatively slender twigs and its flat 2-lobed fruit capsules are persistent in winter, and relatively small: only a little over 2 cm long. The seeds are c. 4–5 mm long,

Staphylea colchica
3-lobed fruit
capsule

Staphylea trifolia
Seed

Staphylea bumalda
Fruiting branch

Staphylea colchica
Seed

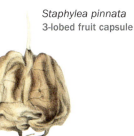

Staphylea pinnata
3-lobed fruit capsule

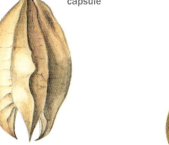

Staphylea trifolia
3-lobed fruit
capsule

Staphylea bumalda
Fruit

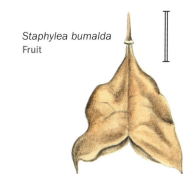

Staphylea bumalda
Seed

Stachyuraceae
spiketail family

Stachyurus praecox Siebold & Zucc., early spiketail or stachyurus

Terminal buds mostly present, c. 3 mm long. **Lateral buds** alternate, c. 2 mm long, along the twig, ovoid when seen from above, flattened in side view, with 2–3 outer bud scales. **Twigs** slender, glabrous, glossy, red- to olive-brown long and slightly curving down. **Leaf scar** small, dark with very faint traces, one larger central one and a smaller one on each side. Flowering: early March to April: small, to 8 mm long, yellowish flowers in 5–8 cm long, hanging spikes. Carpels glabrous, not projecting from flower. Occasionally encountered, shrub 1–2 m high. Origin: Japan.

Less common in cultivation is **Stachyurus himalaicus** Hook. f & Thomson (incl. *Stachyurus chinensis* Franch.) from China. Twigs mostly without terminal bud. Stigma emerging from flower and with finely hairy ovary.

Thymelaeaceae
mezereon family

About 50 genera with some 900 species. The centre of distribution is in the Southern Hemisphere. Predominantly trees and shrubs, more rarely herbs and lianas. In the genus *Daphne* a few deciduous shrubs are native to Europe, while other species are cultivated; other genera are rarely in cultivation. Mostly unpleasantly scented, strongly poisonous plants.

Key to Thymelaeaceae

1 Leaf scars with a trace 2
1* Leaf scars surrounding the lateral buds, with several traces . . . . . . *Dirca palustris*
2 Buds yellowish-green, fine-hairy, terminal bud oblong, buds of inflorescence c. 15 mm diameter, cup-shaped . . . .
. . . . *Edgeworthia chrysantha*
2* Terminal bud ovoid, when green ± glabrous, buds of inflorescence not different . . . . ***Daphne***

Daphne L., mezereon, daphne

Many evergreen and some deciduous species. The buds have laminar bud scales, the leaf scars always have a single trace. Small, 0.3–1.5(–2) m high, shrubs which are poisonous in all parts.

Key to *Daphne*

1 Terminal bud with few scales, often replaced by terminal inflorescence 2
1* Terminal buds with many scales, flower buds axillary, ± many below the terminal bud . . . . ***Daphne mezereum***
2 Bud scales turned glabrous. Twigs slender, red-brown . . . .
. . . .*Daphne ×burkwoodii*
2* Bud scales densely hairy. Twigs sturdy, dark violet-brown . . . . . . ***Daphne alpina***

Section Mezereum

Flowers before the leaves, lateral; continuing the main axis.

Daphne mezereum L., mezereon

Buds differentiated in vegetative terminal buds, floral and vegetative lateral buds. **Terminal buds** 7–9 mm long, acute-conical, with some spiral, dark brown, finely ciliate and acute bud scales. **Flower buds** in groups towards tip of twig, 6–8 mm long, narrowly ovoid, with mostly green bud scales, only the basal ones entirely dark brown, and the central ones with brown margins. Vegetative lateral buds small, globose and dark brown. **Twigs** grey-brown, sparsely hairy, becoming glabrous, when older with many corky lenticels. **Flowers** very early, February to March: 2–3 from one flower bud, with 5–7 mm long pink to purple-lilac-coloured clayx, in some cultivars white. Frequently planted, open shrub to 1 m high. Origin: Siberia and Asia Minor to Europe.

Stachyurus praecox
Twig with flower spike

Stachyurus praecox
Single flower

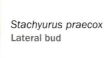

Stachyurus praecox
Lateral bud

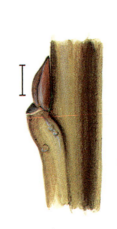

Stachyurus praecox
Lateral bud

Daphne mezereum
Tip of twig with
many flower buds

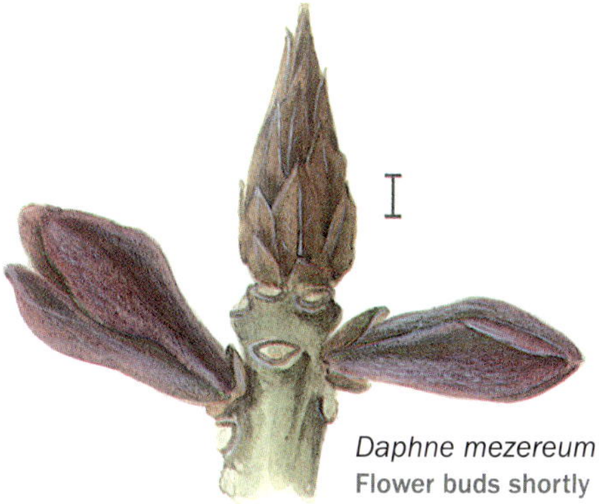

Daphne mezereum
Flower buds shortly
before opening: bud
scales already fallen

Section Daphnanthes

Flowers after the leaves, in terminal inflorescences. One or several lateral axes overtop the end of shoot occupied by last year's inflorescence, therefore without continuous main axis.

Daphne alpina L., Alpine daphne
Buds dark grey-brown. Bud scales at the margin partly wine-reddish, closely appressed-hairy. Terminal buds ovoid, 5–7 mm long. Lateral buds globose, spreading from the twig, c. 3 mm thick. **Twigs** thick and rigid, violet-brown, glossy, sparsely hairy. Prostrate and ascending, compact shrub to 50 cm high. Origin: mountains of southern and central Europe.

Daphne ×burkwoodii TURRILL, Burkwood daphne
[*Daphne caucasica × Daphne cneorum*]

Buds: lateral buds 1–2 mm long, surrounded by two prophylls; terminal buds often absent, 4–5 mm long, ovoid with few grey-brown to olive-green ± glabrous bud scales. **Twigs** orange-brown to red-brown, initially finely appressed white-hairy: older twigs grey-brown. The remnants of the ± densely hairy stalks of the tufted inflorescences are often found at the shoot ends and at the forks of branching of last year's shoots. Occasionally planted, shrub 0.5–0.7 m high.

Daphne alpina
Twig

Daphne alpina
Tip of twig with
terminal bud

Daphne ×burkwoodii
Terminal bud

Daphne ×burkwoodii
Lateral bud

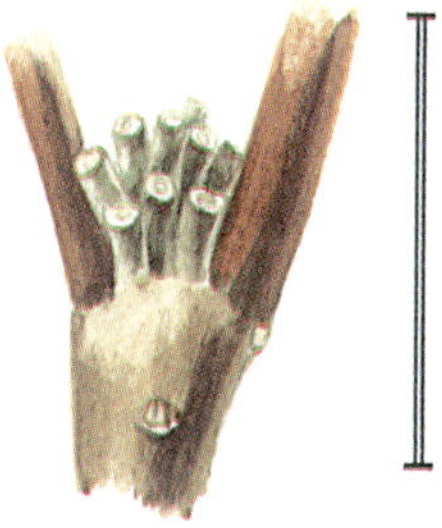

Daphne ×burkwoodii
Branching: detail

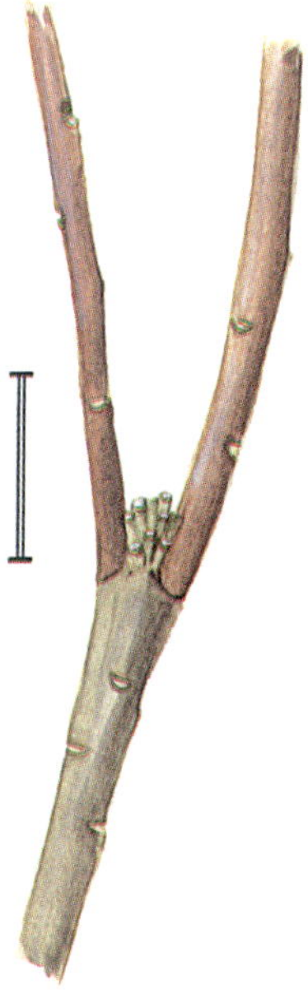

Daphne ×burkwoodii
Branching starts below tip
of twig replaced by last
year's inflorescence

Edgeworthia chrysantha LINDL., paperbush
[*Edgeworthia papyrifera* SIEBOLD & ZUCC.]

Buds: alternate, small and inconspicuous, densely hairy lateral buds and oblong, also densely hairy terminal bud, which is enfolded by leaves. During winter the axillary inflorescence buds below the tip are visible: many single flower buds are grouped in a 1.5 cm broad cup, which is surrounded by some deciduous involucral bracts (similar to *Cornus nuttallii*, but pure green). **Twigs** dark matte green at the tip, lower down deeper brownish and with variably dense, long appressed hairs. When older, grey-brown and flaking. **Leaf scar** semicircular, hardly enclosing lateral bud, with a trace at the top margin. Shrub to 2.5 m high. Origin: eastern Asia.

Dirca palustris L., leatherwood

Without terminal buds. **Lateral buds** spiral, hemispherical, naked, partly two-acute with 4 dark-silvery hairy leaves. **Twigs** very flexible, grey to red-brown, glabrous, with rather a lot of pale, almost white lenticels. Very gnarled due to leaf cushions above the twig surface. **Leaf scar** very broad, encircling the bud, with 5 traces. **Pith** solid but rather open. Shrub c. 5 m high, rarely planted. Origin: eastern North America.

Malvaceae
mallow family

Key to Malvaceae

1 Shrubs 2
1* Buds glabrous or with stellate hairs, stipules deciduous, trees. . . . .*Tilia*
2 Twigs grey-brown, buds visible, closely hairy, stipules persisting . . . .*Grewia*
2* Twigs pale ochre-brown, buds concealed below leaf cushions, twigs with fruit capsules at their tip . . . . **Hibiscus**

SUBFAMILY Grewioideae

Grewia biloba D. DON, bilobed grewia or starbush

Buds always distichous, usually obliquely above the leaf scar lateral buds, 1–3 mm long, globose-ovoid to oblong-acute, incompletely surrounded by 2 outer dark brown short-hairy scales, underneath these densely and long pale-ochre hairy. **Twigs** slender, brown to grey-brown, initially with stellate hairs. **Leaf scars** almost circular to semicircular, with one trace. Next to the leaf scar the thread-like stipules are often persistent. These are particularly distinctive at the twig tips, where the internodes are short and buds are dense. Rarely planted, warmth-loving shrub to 2.5 m high. Origin: eastern China.

SUBFAMILY Tilioideae

Mostly woody plants, more rarely herbaceous. End of shoot often dying, lateral buds naked or covered in bud scales, mostly distichous. Origin: world-wide.

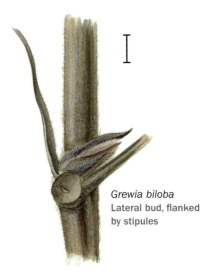

Grewia biloba
Lateral bud, flanked by stipules

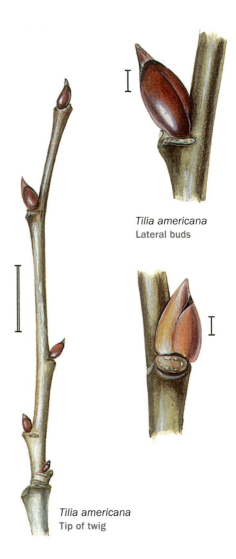

Tilia americana
Lateral buds

Tilia americana
Tip of twig

Edgeworthia chrysantha
Tip of twig with inflorescence buds

Tilia platyphyllos
Hairy tip of twig

Tilia cordata
Glabrous tip of twig

Tilia L., lime

Buds distichous on the twig, mostly ± obliquely above the leaf scar. The end of the shoot dies off at the end of the growth period; growth continues in the next season with the topmost lateral bud. With few scales: two externally visible simple prophyll scales and mostly invisible internal stipule scales. There is frequently an infructescence beside a bud, or, if this has fallen, a scar with a single trace. The bud is then axillary to the lowest bract of the infructescence. This bract resembles a scale, just as the bud scales, and does not fall with the infructescence. Therefore the bud has one more scale than one borne in the axil of a foliage leaf. The buds axillary to the foliage leaf are used in identification.

Leaf scars 3- (more rarely 5- or multi-) traced, with narrow to ovate stipule scars, the latter often with several small, close together dot-shaped traces. Origin: warm-temperate latitudes. A few species native, and several frequently planted. Also many hybrids which are difficult to identify.

Key to *Tilia*

1 Twigs densely felty with stellate hairs . 6
1* Twigs glabrous or with simple hairs . . . 2
2 Buds and twigs predominantly green . . 3
2* Buds and twigs predominantly reddish to red-brown. 4
3 Twigs hairy. . . . . *Tilia platyphyllos*
3* Twigs glabrous . . . . *Tilia ×euchlora*
4 Buds ovoid with broadly rounded tip . . 5
4* Buds a little acute, ± triangular when seen from above . . . . . . *Tilia americana*
5 Buds compact-ovoid, with 2 scales, the lowest one to over half the height of the bud; twigs glabrous, infructescences persistent, fruits above the second wing-like bract . . . .*Tilia cordata*
5* Buds narrowly ovoid, often acute, with 3 outer scales: the lowest of these less than half the bud height; twigs hairy, infructescences deciduous . . . .
. . . . *Tilia platyphyllos*
6 (1) Branches growing strongly upwards .
. . . . *Tilia tomentosa*
6* Branches curving down . . . .
. . . . *Tilia tomentosa* 'Petiolaris'

Tilia americana L., American basswood

Buds spreading from the twig, ovoid-acute, c. 5–6 mm high and 3 mm broad, with 2 **outer bud scales**: the first envelops the inner one almost entirely and is more than half of the bud's height. On the light side strongly wine-red, on the shaded side ochre yellow to greenish-yellow, glossy, with a narrow, dark, slightly ciliate margin. **Twigs** initially brown, with grey epidermis and some faint lenticels; at two years old predominantly pale grey to brown, with a whitish cover of dead epidermis, with dark lenticels tearing open longitudinally. **Leaf scar** ovate with some unequal traces, in between 2 narrow stipule scars with up to 5 dot-shaped traces close together. Occasional, tree to 40 m high. Origin: North America.

Tilia platyphyllos SCOP., large-leaved lime

Buds 6–8 mm long, ovoid, with (2–)3 outer green to orange-brown bud scales which are red only in the sun. The lowermost bud scale rarely reaches the middle of the bud. **Twigs** green to olive-green, in the sun slightly reddish, sparsely spreading-hairy. **Lenticels** few, small, ochre-brown. **Leaf scars** brown, 3-traced, with stipule scars. **Infructescences** pendulous, with up to 5, five-angular, more than 8 mm long fruits, mostly early deciduous. **Bark** long-fissured, densely ribbed bark. Frequent, tree to 40 m high. Origin: Central- and Eastern Europe to Asia Minor.

Tilia cordata MILL., small-leaved lime

Buds 4–6 mm long, short-ovoid, with 2(–3) reddish to wine-red bud scales which are greenish only in the shade. The lowest bud scale more than half the height of the bud. **Twigs** glabrous, brown-red and greenish in the shade, with many lenticels. **Leaf scars** 3(–5)-traced with small stipule scars. **Bark** longitudinally fissured and deeply ribbed. **Infructescences** persistent during winter, with 5–7 slightly angular fruitlets above the second wing-like bract. Frequent tree to 40 m high with a broad crown. Origin: Europe to Asia Minor and western Siberia.

Tilia ×*europaea* L., common lime
[*Tilia* ×*vulgaris* HAYNE]
[*Tilia cordata* × *T. platyphyllos*]

The hybrid between the two previous species is very common. It is intermediate between the parent species. The cultivar 'Pallida', "Kaiserlinde" is frequently planted, and can be recognised by the straight stem, which is unusual for lime trees.

Tilia ×*euchlora* K. KOCH, Caucasian lime
[*Tilia cordata* × *Tilia dasystyla*]

Buds ovoid, 5–6 mm long, glabrous, with 2 (–3) relatively light olive green to yellowish-green bud scales, which are hardly reddish even on the light side. **Twigs** glabrous, initially also mostly strong green and not reddish. **Lenticels** few, small, ± round, later tearing open longitudinally, corky, connecting in a net formation. Frequently planted, tree to 20 m high.

Tilia tomentosa MOENCH, silver lime

Buds 4–6 mm long, ovoid, dark green to olive-brown, densely covered in stellate hairs. **Twigs** dark olive-green to ochre-green, at least initially ± densely felty with stellate hairs. **Lenticels** initially faint, later tearing open longitudinally. **Leaf scars** rounded semicircular, with 3 traces and 2 stipule scars. Often planted, tree to 30 m high. Origin South East Europe. The cultivar '**Petiolaris**', weeping silver lime, distinguished by pendulous twigs, is sometimes regarded as a species in its own right [*Tilia petiolaris* DC.].

SUBFAMILY Malvoideae

Hibiscus syriacus L.

Buds very small and inconspicuous, ± concealed below the leaf scar and the remnants of the two prophylls, which consist of the leaf base with distinct scar and thread-like stipules. **Twigs** hairy, becoming glabrous pale brown to grey brown; with small, pale, warty lenticels. **Leaf scars** on distinct cushions, 3(–5)-traced. Next to the leaf scars short thread-like stipules or small stipule scars. From the margins of the leaf scars two thin ridges descend. **Fruits** grouped at shoot tips, dry 5-valvate axillary capsules, 2–3 cm long and yellow-ochre-

Tilia ×*europaea*
Twig

Tilia ×*europaea*
Tip of twig

Tilia tomentosa
Lateral bud

Tilia tomentosa
Tip of twig

Tilia ×*euchlora*
Twig distinctly yellowish green. Next to top-most bud with infructescence remnant

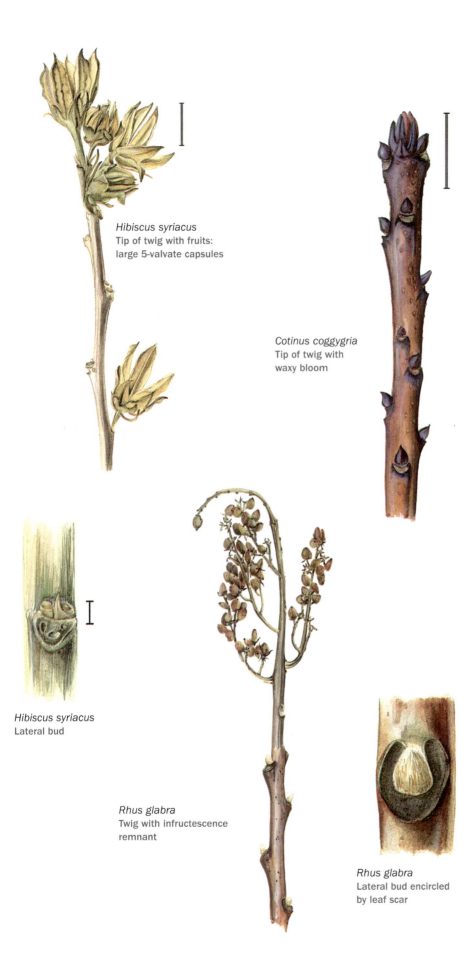

Hibiscus syriacus
Tip of twig with fruits:
large 5-valvate capsules

Cotinus coggygria
Tip of twig with
waxy bloom

Hibiscus syriacus
Lateral bud

Rhus glabra
Twig with infructescence
remnant

Rhus glabra
Lateral bud encircled
by leaf scar

brown to grey-green. They fall off late and then leave large, round horizontal scars. Frequently planted, 2–3 m high shrub. Origin: southern and eastern Asia, partly naturalised in southern Europe.

Anacardiaceae
cashew family

Trees and shrubs, often aromatic, occasionally poisonous. Growth monopodial from a terminal bud or with several axes from lateral buds: this notably in *Rhus typhina*.

Key to Anacardiaceae

1 Buds with bud scales, twigs with waxy bloom, wood yellow. . *Cotinus coggygria*
1* Buds without bud scales. 2
2 Twigs with terminal buds. . *Toxicodendron*
2* Without terminal buds . . . . **Rhus**

Cotinus coggygria Scop., smoke tree

Buds blueish-white with waxy bloom. Terminal buds with many spirally-imbricate bud scales. Lateral buds alternate, protected by two valvate prophylls. **Twigs** orange to dark brown, particularly below the buds, also with waxy bloom. **Lenticels** few, the colour of the twig. **Wood** strongly yellow. In winter often with remnants of the infructescence: large, loose panicles with long spreading-hairy fruit stalks, the actual fruits inconspicuous and small. Frequently planted, broad shrub to 8 m high. Origin: south eastern Europe to central China.

Rhus L., sumac

Large shrubs, more rarely trees. **Leaf scars** alternate, surrounding the small, naked lateral buds; terminal buds absent. **Fruits** in compact terminal panicles.

Key to *Rhus*

1 Twigs glabrous . . . . *Rhus glabra*
1* Twigs hairy. . . . . **Rhus typhina**

Rhus glabra L., smooth sumac

Lateral buds without bud scales, densely pale ochre-brown, short-appressed-hairy, blunt-conical. **Leaf scar** broad, dark grey to blackish, widely encircling the bud, with three indistinct vascular traces. **Twigs** glabrous, orange-brown to grey, blueish-white with waxy bloom. **Lenticels** few, small and dark. **Pith** very wide, pale yellow. **Infructescences** long persistent: hairy panicles, with many, initially scarlet red, later matte red to olive-brown pubescent fruits. Rare, erect shrub 3–5 m high. Origin: North America.

Rhus typhina L., stag's horn sumac

Lateral buds without bud scales, pale orange-brown to ochre-brown, with slightly longer hairs than *Rhus glabra*, blunt-conical to hemispherical. **Leaf scar** relatively small, widely encircling the bud. **Twigs** orange- to red-brown, hairy. **Lenticels** small, branch-coloured and faint. **Pith** ochre-yellow. **Infructescences**: very dense, compact panicles persistent through and after the winter. Fruits dark red, densely hairy. Very frequently planted, erect shrub or small tree to 10 m high. Origin: eastern North America.

Toxicodendron MILL.

Flowers appear in axillary panicles long after the leaves. These very rarely planted and very **poisonous**(!) species belong to the genus: *Toxicodendron vernicifluum*, *T. radicans* (L.) KUNTZE, poison ivy, a shrub climbing by adhesive rootlets and *T. pubescens*.

Key to Toxicodendron

1 Erect tree . . . *Toxicodendron vernicifluum*
1* Smaller or climbing shrubs. 2
2 Shrub climbing with adhesive rootlets . .
. . . . *Toxicodendron radicans*
2* Creeping to upright shrub . . . .
. . . . *Toxicodendron pubescens*

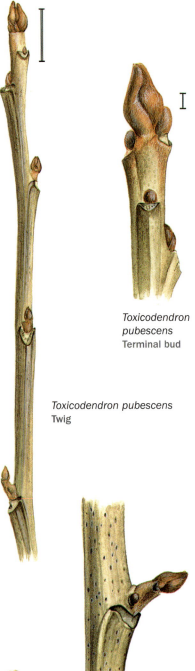

Toxicodendron pubescens
Terminal bud

Toxicodendron pubescens
Twig

Rhus typhina
Lateral bud surrounded by leaf scar

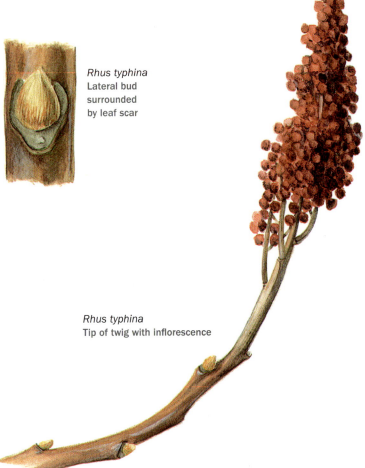

Rhus typhina
Tip of twig with inflorescence

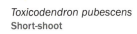

Toxicodendron pubescens
Short-shoot

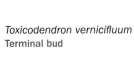

Toxicodendron vernicifluum
Terminal bud

Toxicodendron pubescens MILL.,
Atlantic poison oak
[*Rhus toxicodendron* L., *Toxocodendron quercifolium* (MICHX.) GREENE]

Buds without bud scales, densely ochre-brown hairy. Terminal buds 6–8 mm long, onion-shaped, enclosed by 3(–4) bud leaves. Lateral buds 3–5 mm, often stalked (at the end of short-shoots), with stalk 8(–10) mm long. **Twigs** grey-brown, glabrous, only below the terminal buds ochre-brown hairy, like the buds; slightly angular, grooved; with many very small lenticels. **Leaf scars** large dark grey to pale brown, with many vascular traces (c. 9). Low shrub, to 50 cm high, spreading by runners. Origin: USA.

Toxicodendron vernicifluum (STOKES) F. A. BARKLEY, Chinese lacquer tree
[*Rhus verniciflua* STOKES]

Terminal bud to c. 10 mm long and 6 mm thick, irregularly blunt-conical, without bud scales, dense hairy ochre-brown hairy. **Twigs** thick; c. 8 mm diameter and rigid, pale grey to olive-grey, with many ochre-brown lenticels. **Leaf scars** large, ovate to shield-shaped, with many irregularly arranged traces. Rarely planted tree, 10–20 m high. Origin: Japan and central China.

Sapindaceae
maple family

Only a few species of this family, which is largely distributed in the tropics and subtropics, are able to survive winter conditions. The genera *Acer* (maple) and *Aesculus* (horse chestnut) belong here, although they were formerly placed in their own families.

Key to Sapindaceae

1   Leaf scars and lateral buds opposite. . . 2
1* Leaf scars and lateral buds alternate . . 5
2   Buds without bud scales. . . . . *Dipteronia*
2* Buds with bud scales 3
3   Buds less than 1.5 cm long. 4
3* Terminal buds more than 1.5 cm long . .
. . . . ***Aesculus***
4   Buds with white wax, on the shoot tips persistent infructescence rachides, large shrub becoming very broad . . . .
. . . . ***Aesculus parviflora***
4* Characters different . . . .*Acer*
5   Tree without terminal bud . *Koelreuteria*
5* Shrub, twigs without terminal bud. . . . .
. . . . *Xanthoceras*

Xanthoceras sorbifolium BUNGE, shinyleaf yellowhorn

Terminal buds enclosed by c. 8 scales which reach to the tip of the bud. **Lateral buds** surrounded almost completely by the two prophylls. Scales brown, glossy, at the base also greenish, turning glabrous. Strong first year's growth **twigs** to c. 5 mm thick, bark grey-brown, initially fine whitish-hairy, tearing longitudinally in rhomboid shapes, with many ochre-brown lenticels. **Pith** compact, relatively wide, ochre-coloured. **Leaf scars**, particularly at the tip of twig, on large leaf cushions, rounded-triangular, with narrow, ochre-coloured border, with three traces or groups of traces. Rarely planted, shrub to 7 m high or small tree. Origin: northern China.

Koelreuteria paniculata LAXM., pride of India

Buds: only alternate lateral buds, with a fat base, 4–5 mm long and just as wide, surrounded by 2 ± acute bud scales. Bud scales dark grey-brown, sparsely hairy, not deciduous after unfolding and present at the

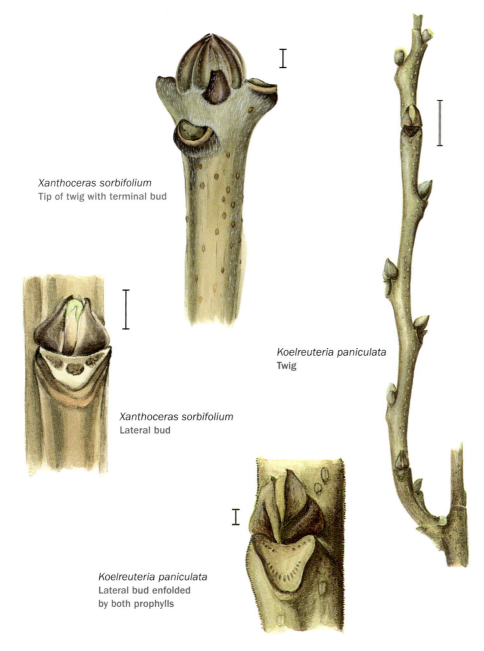

Xanthoceras sorbifolium
Tip of twig with terminal bud

Xanthoceras sorbifolium
Lateral bud

Koelreuteria paniculata
Twig

Koelreuteria paniculata
Lateral bud enfolded by both prophylls

shoot base of older twigs. **Twigs** sparsely short-hairy, brownish to pale grey, with some roundish, warty lenticels. **Leaf scar** dark, rounded-triangular with many traces. **Fruits** many in large terminal panicles, persistent during the winter, 3–4 cm long, 3-angular, bladder-like, thin-walled capsules with 3 globose seeds. Often planted, tree 8–15 m high. Origin: northern China and Korea.

Aesculus L., horse chestnut, buckeye

Buds mostly rather large, with simple bud scales. Lateral buds opposite. **Twigs** sturdy. Young growth monopodial from terminal buds. When flowering age is reached, the inflorescences often use up the shoot tip and there remain two lateral buds which continue the growth next season. **Leaf scars** with several traces (3–9); opposite leaf scars connected with one another by a line. Trees and shrubs. Origin: temperate northern latitudes, mostly in mountains.

The tree species are unmistakable through bud-shape and size. The shrub by *Aesculus parviflora*, on the other hand, can easily be confused with species of maple. It differs from other species mainly by its growth form.

Key to *Aesculus*

1 Buds slightly to strongly sticky, red-brown to dark green-brown. 2
1* Buds not sticky, pale brown to yellowish-brown 3
2 Buds very strongly sticky and glossy red-brown. . . . . . . *Aesculus hippocastanum*
2* Buds dark olive-green-brown, slightly sticky . . . .*Aesculus ×carnea*
3 Bud scales smooth and tightly clinging to bud. 4
3* Tips of the sharply keeled bud scales slightly spreading . . . . . . *Aesculus glabra*
4 Mostly shrubs 5
4* Trees, bud scales partly with pink margin . . . .*Aesculus flava*
5 Buds blunt, broad-growing shrubs . . . .*Aesculus parviflora*
5* Buds acute, shrubs to 4 m high, rarely trees . . . . *Aesculus pavia*

Section Aesculus

Aesculus hippocastanum L., common horse chestnut

Buds red- to dark brown and strongly sticky. Terminal buds c. 2 cm long, acute-ovoid. Flower buds thicker than leaf buds. Lateral buds smaller, slightly spreading from twig and occasionally slightly stalked. **Twigs** thick, grey-brown. Young twigs ascending, older branches in lower parts of the crown often hanging down with the tips curving upwards. **Leaf scars** shield-shaped with (5–) 7–9 horse-shoe-shaped traces of vascular bundles; the number of traces depends on the number of leaflets. **Bark** small-scaled, flaking. Tree to 25 m high. Origin: Balkan; introduced to central Europe in 1576 and in Britain in the 17th Century; these days widely distributed.

Aesculus ×carnea HAYNE, red horse chestnut
[*Aesculus hippocastanum* × *Aesculus pavia*]

Intermediate between the parents. **Buds** only slightly sticky, green-olive-dark brown. Terminal buds 1.5–2 cm long. Tree to 20 m high. Mostly grafted on *Aesculus hippocastanum*, but growing more slowly than that species. In older specimens mostly with strong differences in diameter. The most often planted chestnut besides the horse chestnut.

Section Pavia

Aesculus pavia L., red buckeye

Buds with ochre-brown bud scales with darker margin; similar to *Aesculus flava*, but bud scales without pink border. Terminal buds 1.2–1.5 cm long. **Twigs** glabrous (in var. *discolor* (PURSH) TORR. & A. GRAY shoots softly hairy), pale grey-brown with few lenticels. **Bark** smooth, dark brown. Occasional, shrub 1–4 m high or, only when grafted on other species, a tree to 12 m high. Origin: North America.

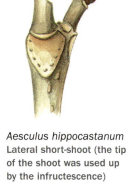

Aesculus hippocastanum
Lateral short-shoot (the tip of the shoot was used up by the infructescence)

Aesculus hippocastanum
Tip of twig with terminal bud

Aesculus hippocastanum
Unfolding of the bud in spring

Aesculus ×carnea
Twig

Aesculus ×carnea
Terminal bud

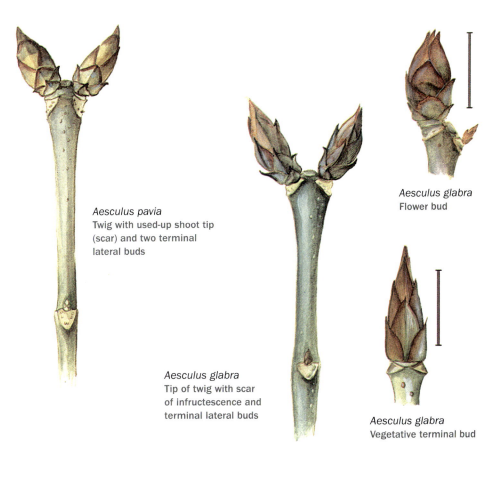

Aesculus pavia
Twig with used-up shoot tip
(scar) and two terminal
lateral buds

Aesculus glabra
Flower bud

Aesculus glabra
Tip of twig with scar
of infructescence and
terminal lateral buds

Aesculus glabra
Vegetative terminal bud

Aesculus glabra WILLD., Ohio buckeye

Buds grey-brown, the bud scales externally keeled, acute, spreading at the tip and ciliate at their margins. Terminal buds acute-ovoid, c. 1.5 cm long. Flower buds shorter, compact-ovoid. Buds and other parts, when crushed, emit an unpleasant odour. **Twigs** round, glossy olive-grey, on the sun side ± tanned, partly with pale blueish grey epidermis. **Bark** on stem, branches and twigs fissured and rough, corky; very pale, whitish in var. *leucodermis*. Small, rarely planted tree 10(–20) m high. Origin: eastern USA, where it may reach 30 m high.

Aesculus flava SOL., sweet buckeye
[Aesculus octandra MARSHALL]

Buds cinnamon- to ochre-brown and brown. Terminal buds 1.2–1.5 cm long. Bud scales appressed, some with a pink membranous margin. **Twigs** pale brown, grey-with waxy bloom, many small lenticels and six-angular pith. In older trees twigs pendulous. **Bark** deep brown and smooth. The most common yellow-flowering species. Tree 25(–30) m high. Origin: North America. The natural hybrid *Aesculus ×hybrida* DC. with *Aesculus pavia*, cannot be distinguished in winter.

Section Macrothyrsus

Aesculus parviflora WALTER, bottlebrush buckeye

Buds to 1 cm long and 5 mm thick, blunt (not acute), pale brown with white-scaly cover, but not sticky. **Twigs** grey to light brown, with many lenticels. **Leaf scars** often only 3-traced. Shrub growing very broadly: twigs basally decumbent and ascending to a height of 4 m. A single specimen can grow to 10(–30) m diameter, through the outer twigs taking root and through runners. Commonly planted species. Origin: North America.

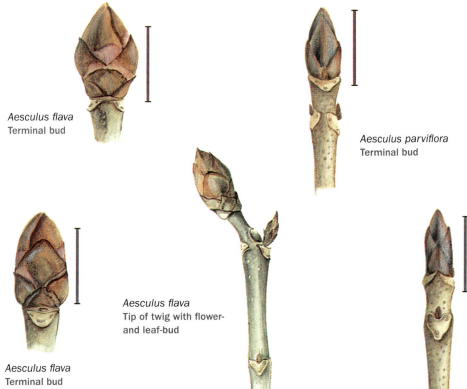

Aesculus flava
Terminal bud

Aesculus parviflora
Terminal bud

Aesculus flava
Tip of twig with flower-
and leaf-bud

Aesculus flava
Terminal bud

Aesculus parviflora
Terminal bud

Dipteronia sinensis OLIV.

Buds without bud scales, fine appressed-hairy, grey-brown to orange brown, partly also red-brown and with greenish hues. **Terminal bud** to 8 mm long, outer leaves with pinnate primordia. **Lateral buds** smaller and appressed to twig. **Twigs** straight, rigid and stiff, first year's growth 3–4 mm thick, olive-green, smooth, glossy, minimally hairy on the top-most internodes, otherwise glabrous. Lenticels few, pale ochre, few, distinct. Bark with thin longitudinal fissures, more distinctly so on two year-old twigs; fissures rhomboid, paler than the matte grey-brown bark. **Pith** wide, pale ochre-brown. **Leaf scars** shield-shaped, opposite, ± connected, with 5 traces. **Fruit** as in *Acer* a bipartite schizocarp, the partial fruits with a wing all around. Rarely planted, tree to 10 m high. Origin: central China.

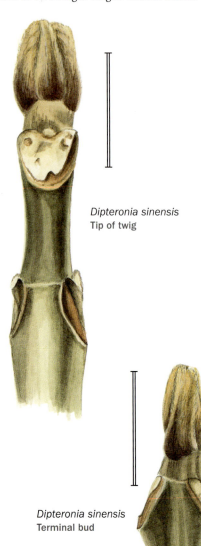

Dipteronia sinensis
Tip of twig

Dipteronia sinensis
Terminal bud

Acer L., maple

Aproximately 100–150 species in the temperate zone, particularly frequent in the mountains. Some species native and many species planted due to their showiness: e.g. snake-bark maples with their interesting bark and Japanese maples due to the attractive shape of their leaves and their autumn colours. All species have opposite leaves (rarely in threes), in which the leaf scars at a node abut or are connected by a line. Useful characters to tell the species apart are: the number of bud scales, bud size, the presence of terminal buds, colour, indument and the waxy nature of bud and twigs; lenticels; leaf cushions with presence of petiole remnants; colour and structure of the bark; growth form and the hanging fruits and infructescences that often remain on branches far into winter, as well as flowers and inflorescences in early flowering (before the leaves).

The shrubby species can be confused with *Viburnum*, with *Aesculus parviflora* and with genera of the Caprifoliaceae.

Key to *Acer*

1 Lateral buds enclosed at base by petiole remnants, these internally densely hairy, often without terminal bud, twigs without visible lenticels when young . . . . Section **Palmata**

1* Leaf scars distinctly visible 2

2 Buds with two valvate outer bud scales 3

2* Buds with more than two outer bud scales 7

3 Buds ± stalked, glabrous 4

3* Bud scales ± hairy 6

4 Twigs and trunk with distinctly striped bark . . . . Section **Macrantha**

4* Bark not like this 5

5 Twigs red . . . . *Acer glabrum*

5* Twigs green to brown, with white waxy bloom . . . . **Acer negundo**

6 Next to the terminal bud usually with two large lateral buds. . . *Acer cissifolium*

6* Usually without terminal buds . . . .*Acer spicatum*

7 Buds with few bud scales: 2–4(–5) pairs 8

7* Buds with many bud scales: 5 or more pairs 12

8 Buds very small (2–4 mm), on large leaf cushions; shrubs . . . . . Section **Ginnala**

8* Buds rarely so small or on slightly developed leaf cushions 9

9 Flower buds many, grouped in axils of lateral buds, flowering long before leaves appear; trees . . . . Section **Rubra**

9* Lateral buds without axillary accessory buds 10

10 Upper pair of lateral buds large and on same level as terminal bud, lateral buds often slightly stalked, with a pair of bud scales which often open. .*Acer negundo*

10* Buds with several bud scales 11

11 Bud scales with latex, lateral buds usually appressed. . . . . Section **Platanoidea**

11* Bud scales without latex, lateral buds spreading . . . . Section **Acer**, series **Acer**

12 (7) Mostly several-stemmed small trees or shrubs 13

12* Large trees with a single trunk. . . . . *Acer saccharum*

13 Leaves without lobes, dried up, long remaining on the tree; buds green to red . . . . *Acer carpinifolium*

13* Buds brown, often very dark 14

14 Shoots glabrous, buds brown. . . . . Section **Acer**, series **Monspessulana**

14* Shoots hairy and/or bark flaking, buds very dark . . . . Section **Trifoliata**

Section Acer

Series Acer

Buds medium-size and slightly acute in the upper half; coloured green and grey-brown to wine-red-violet. Externally visible bud scales (4–)6–8(–10). Flowers with or after the leaves, in axillary panicles, the resulting infructescences often persistent well into winter. *Acer pseudoplatanus* is the type of the genus and thus of the name *Acer* at all levels. In contrast to all other species presented in the section *Acer*, *A. pseudoplatanus*, which is the only species with green bud scales in the section, probably does not belong in the section itself.

Key to series *Acer*

1 Buds green, bud scales with narrow brown margin. . . . **Acer pseudoplatanus**

1* Buds wine-red. . . . . **Acer heldreichii**

Acer pseudoplatanus
Twig

Acer pseudoplatanus
Tip of twig

Acer opalus
Tip of twig

Acer heldreichii
Tip of twig

Acer monspessulanum
Twig

Acer pseudoplatanus L., sycamore

Buds ovoid, with green bud scales, particularly the lower ones with dry, grey-brown margins. Terminal buds c. 8–15 mm long. Lateral buds opposite, slightly spreading from twig, 6–8 mm long. **Twigs** grey-brown, with few to many paler lenticels. **Leaf scars** with three traces. Bark of older trees irregularly scaly-flaking, similar to the bark of plane trees (*Platanus)*. Frequently planted, tree to 35 m high. Origin: Central Europe to the Caucasus and Asia Minor.

Acer heldreichii ORPH. ex BOISS.

Buds large with dark wine-red to chestnut brown glossy bud scales with pale ciliate margins. **Twigs** red to chestnut brown with many small, pale lenticels. Older twigs often very dark, glossy. Rare, tree 20–25 m high from the mountains of the Balkans.

Series Monspessulana

Buds brown, conical to narrowly ovoid with many: (4–)5–6(–7) bud scale pairs. Shrubs or small trees.

Key to series *Monspessulana*

1 Terminal buds to 10 mm long. Bud scales densely hairy at margins . . . . *Acer monspessulanum*
1* Terminal buds more than 10 mm long. Margins of bud scales becoming glabrous . . . .*Acer opalus*

Acer opalus MILL., Italian maple

Buds narrowly ovoid with many (5–7) pairs of bud scales; lowest pairs slightly pale hairy at the centre, otherwise glabrous and brown. Successive bud scales, except for the narrow brown margin, densely and short ochre-silver-grey hairy. Terminal buds 10–12 mm long, lateral buds mostly smaller and with fewer bud scales. **Twigs** olive-ochre at the tips, then via red-brown and violet-brown basally changing to grey-brown colour. **Lenticels** paler, initially roundish, later tearing open longitudinally. **Bark** with oblong pale brown stripes on dark grey ground. Occasionally planted, large shrub or tree to 15 m high. Origin: south-western Europe.

Acer monspessulanum L., Montpelier maple

Buds conical, narrowly ovoid, 6–8 mm long, terminal buds also slightly longer, acute, dark-brown to red-brown, with 10–12 bud scales. Scales darker toward the margin, lower ones slightly hairy, upper ones more densely hairy at the margins and paler. **Twigs** initially dark red-brown, later grey, with many twig-coloured lenticels. **Leaf scars** pale ochre-coloured, 3-traced. Frequently planted, shrub or tree to 15 m high. Origin: south-western Europe.

Series Saccharodendron

Buds similar to those of series *Monspessulana*, but large trees.

Acer saccharum MARSHALL., sugar maple

Buds basally shaggy hairy, conical to narrowly ovoid. Bud scales slightly acute, dark brown, velvety, whitish hairy towards tip of the bud, with glabrous, darker margin. Terminal buds with 5–6 lateral buds with 3–4 pairs of bud scales. **Twigs** towards the tip cinnamon-brown to olive-brown, basally pale grey. **Lenticels** pale whitish, round to oblong. **Bark**: dark grey, flat-furrowed. Rarely planted, tree to 40 m high. Origin: North America.

Section Indivisa

Acer carpinifolium SIEBOLD & ZUCC., hornbeam maple

Buds conical to ovoid, with many scales: terminal buds with 5–6 and lateral buds with 2–5 pairs of bud scales. Bud scales acute, reddish on the light side, on the shaded side greenish, paler towards the margin. **Twigs** glabrous, initially olive-cinnamon-brown, on the sun side red-brown; later dark red-brown to violet-brown, finally grey. Foliage dries and persists: dentate leaves without lobes. At first sight similar to the hornbeam, but with the opposite leaves characteristic for all maples. Large shrub or tree to 10 m high. Origin: Japan.

Section Ginnala, step maples

Acer tataricum L., Tatar maple

Buds small: terminal buds to 4 mm, with 6–8 visible, dark violet wine-red to brown bud scales or reddish ones with a darker, densely pale ciliate margin. **Twigs** slender, glabrous, carmine to red-brown with large leaf cushions. **Bark** smooth, dark grey, brown-striped; blackish when older. Often planted species. Mostly a dense, finely branched shrub, more rarely a tree to 12 m high with a deep and broad crown. Origin: south-eastern Europe to central Asia.

The often encountered **Acer tataricum** subsp. **ginnala** (MAXIM.) WESM., Amur maple [*Acer ginnala* MAXIM.] from Asia is

Acer saccharum
Tip of twig

Acer saccharum
Terminal bud

Acer tataricum
subsp. *ginnala*
Tip of twig

Acer carpinifolium
Terminal bud

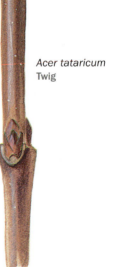

Acer tataricum
Twig

Acer maximowiczianum
Terminal bud

very similar. **Buds** pale- to red-brown and slightly ciliate. **Bark** dark brown, smooth. A tall shrub or small tree to 7 m high. Origin: China, Manchuria and Japan.

Section Trifoliata

Rarely seen large shrubs or small trees to 12 m high. **Buds** conical with many dark bud scales. **Lenticels** small and very many. **Leaf scars** relatively narrow, encasing lateral buds, 3-traced. Origin: eastern Asia.

Key to section *Trifoliata*

1 Buds to 10 mm long, young twigs and bud scales ± densely hairy . . . . *Acer maximowiczianum*
1* Buds to 6 mm long, twigs and buds only slightly hairy, bark of older twigs and stems coming off in rolls . . *Acer griseum*

Acer maximowiczianum MIQ., Nikko maple
[*Acer nikoense* hort., non (MIQ.) MAXIM.]

Terminal buds to c. 10 mm long, narrowly ovoid to blunt-conical, with 6–8 pairs of outer bud scales. **Lateral buds** smaller, basally spreading from the twig and near the tip bent towards the twig. Bud scales reddish-brown to dark violet-brown, dorsally more hairy and on the lateral margins almost glabrous. **Twigs** ochre-brown to dark red-brown, with sparse long hairs. Rarely planted tree to 10 m high. Origin: Japan, central China.

Acer griseum (FRANCH.) PAX, paperbark maple

Buds c. 5 mm long, conical with 5–6 pairs of outer bud scales. Bud scales dark brown to grey, margin pale-ciliate and especially towards the tip with shaggy indument. Bud scales on the back with few hairs. **Twigs** towards the tip slightly hairy, when young ochre-brown to red-brown and on the weather side pale grey due to dead epidermis, with many small lenticels. Older twigs with bark which is grey to grey-brown and soon tears and flakes, underneath ochre to brown. **Bark** smooth cinnamon-brown, coming off in narrow rolls. **Fruits** mostly persistent over winter, 3–3.5 cm long, thick, thinly hairy, wings at a narrow angle or at right angles. Rarely planted, shrub or small tree. Origin: China.

Section Lithocarpa

Acer macrophyllum PURSH, bigleaf maple

Contain milky juice (white dots, visible with a magnifying glass in cross section). **Buds**: terminal buds large, often replaced by terminal infructescences; lateral buds appressed, 6–8 mm long, surrounded by (1–)2 pairs of bud scales, basally with short greenish 'stalk', above olive- or yellowish green to wine-red and dark violet-brown, glabrous, only the margins very thinly ciliate. **Twigs** glabrous, green towards the tip, on the sun side also reddish; initially with few, later with many longitudinally tearing, brown lenticels. Older growth: glossy, greenish to violet-brown, with grey expanses of dead epidermis. **Leaf scars** surrounding the buds and abutting, with (5–)7 traces of vascular bundles, with a wavy margin. **Fruits** hairy, spreading. Rarely planted, tree to 30 m high. Origin: North America.

Acer griseum
Twig

Acer macrophyllum
Terminal bud

Section Platanoidea

The species have 4–6 visible bud scales of wine-red, brown, green to reddish violet hue. Their buds are of medium size, compact and ± blunt. Flowers in corymbose panicles, appearing with or before the leaves. Contain milky sap. Distributed in Europe and Asia.

Key to section *Platanoidea*

1 Terminal buds large (> 7 mm), wine-red . . . .*Acer platanoides*

1* Terminal buds smaller 2

2 Twigs and buds red-green, twigs with waxy bloom . . . . . . **Acer cappadocicum**

2* Twigs brown to grey-brown 3

3 Buds greenish. 4

3* Buds brown, twigs sometimes with corky ridges . . . .**Acer campestre**

4 Twigs dark brown to olive-green. . . . .*Acer ×zoeschense*

4* Twigs ochre-brown . . . . *Acer pictum*

Acer platanoides L., Norway maple

Buds with few dark, initially wine-red bud scales or, in strong shadow, also with yellowish to greenish hue. Terminal buds short-ovoid, 8–12 mm long, relatively broad and compact (width × height. 1:< 1.5); lateral buds smaller, flat-appressed to twig. **Twigs** olive to dark brown, partly with grey, dead epidermis. **Lenticels** pale, tearing open longitudinally. **Bark** dark, fissured longitudinally and slightly scaly. **Flowers** in April before the leaves, in upright corymbs. Very frequent tree to 30 m high. Origin: Europe to the Caucasus.

Acer cappadocicum GLED., Cappadocian maple
[*Acer laetum* C. A. MEY.]

Buds globose-ovoid, c. 4–6 mm long, greenish violet or ± reddish. **Twigs** greenish, on the light side pale reddish-violet, glabrous, glossy for several years, often ± white-blueish. **Bark** dark grey, smooth, with long narrow pale brown fissures and oblique lenticels. Tree to 12(–20) m high with strong suckers. Occasionally planted, with many forms. Origin: Caucasus to eastern Asia.

Acer pictum THUNB. the painted maple,
is similar. Can be distinguished by the pale ochre to orange-red-brown first

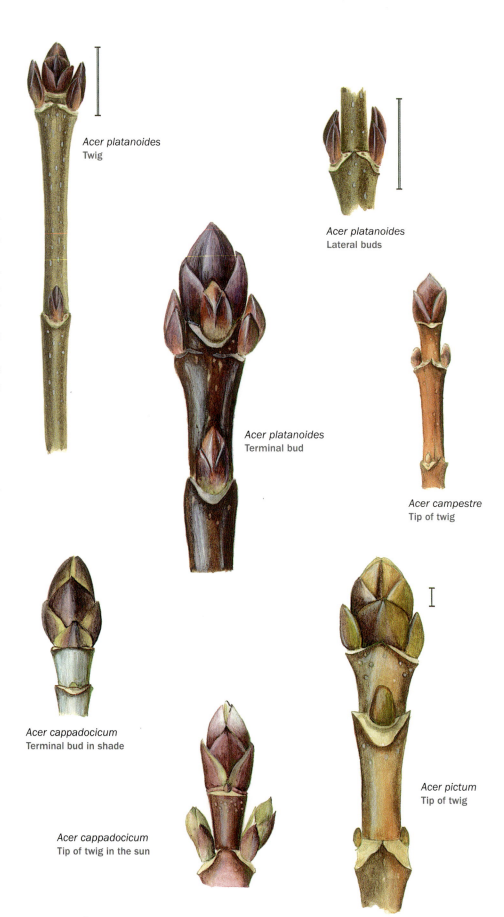

Acer platanoides
Twig

Acer platanoides
Lateral buds

Acer platanoides
Terminal bud

Acer campestre
Tip of twig

Acer cappadocicum
Terminal bud in shade

Acer cappadocicum
Tip of twig in the sun

Acer pictum
Tip of twig

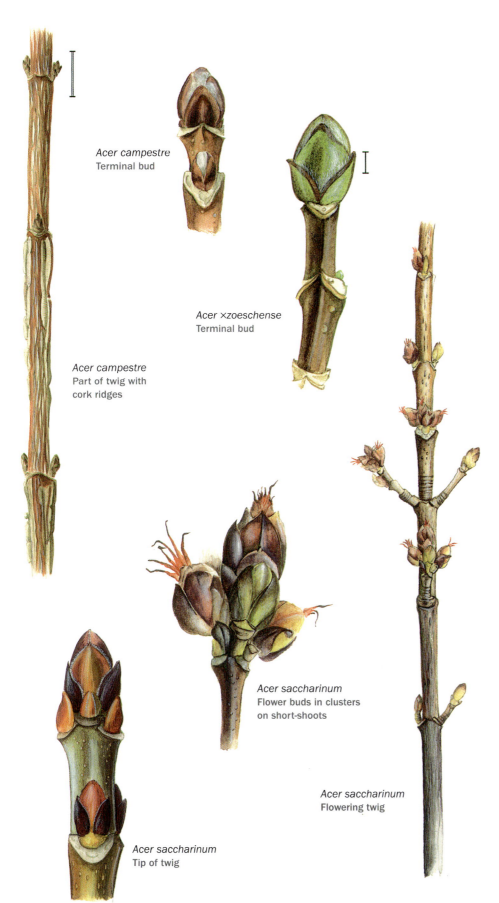

Acer campestre
Terminal bud

Acer ×zoeschense
Terminal bud

Acer campestre
Part of twig with
cork ridges

Acer saccharinum
Flower buds in clusters
on short-shoots

Acer saccharinum
Flowering twig

Acer saccharinum
Tip of twig

year's growth twigs, which are not glossy. Two-year-old twigs are rough, slightly longitudinally furrowed. **Bark** smooth and grey. Rarely planted, tree reaching to 15 (–20) m high. Origin: Asia, Manchuria, Japan, China and Korea.

Acer campestre L., field maple

Buds globose-ovoid. Terminal buds 5–6 mm, lateral buds c. 4 mm long, appressed to twig or slightly spreading, with 4–6 visible bud scales. Scales cinnamon red-brown to olive grey-brown with oblique dark stripes, white-ciliate at the margin. **Twigs** dirty yellow-ochre to red-brown, on the weather side and on older twigs grey-brown, often with cork ridges: the form *suberosum* (Dumort.) Rogow. **Lenticels** few to many. The bark begins to tear longitudinally early; old **bark** brown-grey; divided by longitudinal and cross-fissures into ± rectangular scales. Very frequent, tree to 20 m high, sometimes only shrubby. Origin: Europe to north Africa.

Acer ×*zoeschense* Pax [*Acer* ×*neglectum* Lange], a hybrid with *Acer cappadocicum* subsp. *lobelii* (Ten.) A. E. Murray is also occasionally planted. **Buds** globose to short-ovoid, c. 4 mm long, surrounded by 2–3 pairs of olive-green pairs of bud scales.

Section Rubra

This North American section is mainly characterised by the flower buds which occur in groups on short-shoots. The flowers appear long before the leaves. By the time the tree is in leaf, the typical maple keys have already developed. **Buds** with 4–6(–8) visible bud scales, sometimes those of the first pair of lateral buds deciduous. Due to elongation of the lowest internode, lateral buds often slightly stalked.

Acer saccharinum L., silver maple

Buds: the outer pair of bud scales often becomes deciduous. Terminal buds 4–6 mm long with 4–6 scales, lateral bud appressed to twig, 3–4 mm long, short-stalked with 4 bud scales (always when outer scales are lost, minus those). Outer bud scale ochre green, later often dry to dark violet-brown. The subsequent bud scales red on the sun side and on the shaded side reddish to yellowish-green. Bud scales densely ciliate at margin. Flower buds globose, many grouped in the axils of the first bud scales of the lateral buds. **Twigs** towards the tip olive-green to ochre-brown, older parts red-brown to grey-brown, glabrous. **Lenticels** twig-coloured, few to many. **Flowers** reduced: without petals, in March, long before the leaves. **Bark** initially ash-grey and smooth, later darker and flaking in oblong strips. Often planted, tree to 40 m high.

Acer rubrum L., red maple

Buds and twigs wine-red at their tips. Leaf- and flower buds different: flower buds more rounded, several grouped on short-shoots, with 2–6 outer bud scales, the first pair often deciduous. Margins of bud scales pale hairy. **Bark** of older twigs smooth, pale ash-grey, on stronger stems darker and flaking. Tree to 20 m high, but in its native area (North America and Canada) reaching 40 m.

Section Negundo

Buds with few bud scales, the outer pair often enclosing the bud entirely. Terminal buds flanked by large lateral buds.

Acer negundo L., box elder

Buds 5–10 mm long, surrounded by two outer dark wine-red to violet or also greenish scales, mostly with blueish waxy bloom, spreading. Inner bud leaves densely hairy, pale, brownish to grey-green. **Twigs** smooth, green to wine-red-violet, blueish-white, with waxy bloom layer that can be wiped off. **Flowers** in March to April, on male or female plants, pendulous: female flowers in long panicles, males in a dense cluster. Very frequently planted, large shrub or to 15 m high tree, often several-stemmed; with many forms. Origin: North America.

Acer cissifolium (SIEBOLD & ZUCC.) K. KOCH

Buds ovoid, enclosed by two wine-red, whitish-hairy bud scales which become partly glabrous. Lateral buds often acute, flat-appressed to twig. **Twigs** wine-red-brown, densely white-felty, later grey-brown. Rarely planted, small tree to 12 m high, often only shrubby. Origin: Japan and China (as *Acer henryi* PAX).

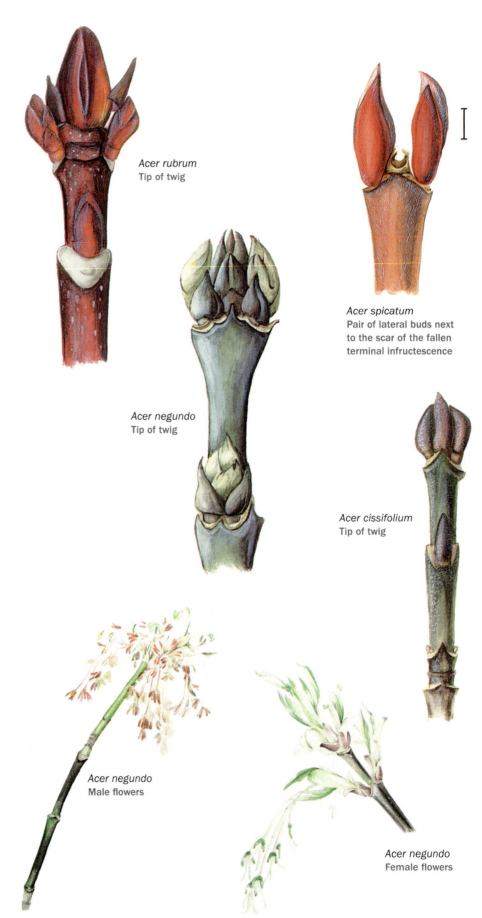

Acer rubrum
Tip of twig

Acer spicatum
Pair of lateral buds next to the scar of the fallen terminal infructescence

Acer negundo
Tip of twig

Acer cissifolium
Tip of twig

Acer negundo
Male flowers

Acer negundo
Female flowers

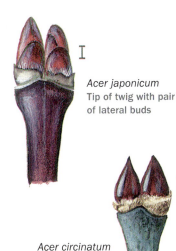

Acer palmatum
Twig

Acer shirasawanum
Twig

Acer shirasawanum
Lateral buds surrounded
by remnants of the leaf

Acer japonicum
Tip of twig with pair
of lateral buds

Acer circinatum
Tip of twig

Section Parviflora

Acer spicatum LAM., mountain maple

Buds 3–5(–8) mm long, outer bud scales red: the outer pair mostly enclosing the bud entirely, slightly hairy or more densely so towards the tip, inner pair of bud scales densely white-hairy. **Terminal bud** often missing, because end of shoot mostly used up by flowering. **Twigs** blunt, reddish, finely white-hairy with some raised lenticels which later tear open broadly and relatively deeply. **Leaf scars** narrow, 3-traced. Rarely planted, shrub or tree to 10 m high. Origin: North America.

Section Palmata

Lateral buds at the base enclosed, collar-like, by remnants of the leaf bases. Twigs and buds wine-red to dark purple, more rarely (in the shadow or in green-twigged species) ± green. The number of externally visible pairs of bud scales is 1 to 3. Frequently planted species. Origin: Asia and North America.

Key to section *Palmata*

1 Bud scales ± ciliate 2
1* Bud scales not ciliate. . . ***Acer palmatum***
2 Twigs with waxy bloom . . . .
. . ***Acer circinatum*** and *A. shirasawanum*
2* Twigs without waxy bloom. 3

3 Buds (almost) completely enclosed by pair of outer bud scales . . . .
. . . . *Acer shirasawanum*
3* Mostly with 2–3 outwardly visible pairs of bud scales . . . . ***Acer japonicum***

Acer palmatum THUNB., palmate maple

Buds with 2, more rarely 4, bud scales which often spread slightly at the tip. All parts glabrous, apart from the densely hairy insides of the leaf base remnants. **Twigs** slender and smooth, dark wine-red to pure green, without visible lenticels. Very frequently planted shrub or small tree to 10(–15) m high. Origin: Japan, where it has been in cultivation for centuries; it is extraordinarily rich in forms. In winter the cultivars with coloured twigs make an impact, e.g. '**Corallinum**' with bright coral red twigs. **Acer sieboldianum** MIQ., Siebold's maple, is similar. It too comes from Japan, but is only rarely planted. It has initially whitish-hairy twigs.

Acer shirasawanum KOIDZ.,
Shirasawa maple

Buds: mostly lateral buds only, these ovoid, wide at base, 5–7 mm long, almost completely enclosed by the two outer bud scales, strongly wine-red to dark violet, ± glossy, glabrous on the margins or very sparsely ciliate. The persistent leaf base encloses the lateral buds to $^2/_3$ or $^3/_4$ of their length. **Twigs** strongly wine-red on the sun side, green in the shade, slightly glossy, partly with waxy bloom and glabrous. Tree to 15 m high. Only the cultivar '**Aureum**', golden Shirasawa maple, is frequently planted. Origin: Japan.

Acer japonicum THUNB., Japanese maple

Buds mostly with more than 2 visible, 4–6 bud scales, ± hairy on the margins (2 bud scales only on slightly developed shoots or basal buds). Frequently planted, shrub or small tree to 7 m high. Origin: Japan.

Acer circinatum PURSH, the vine maple, is difficult to distinguish in winter. Its twigs often have a waxy bloom.

Section Macrantha, snake bark maples

Buds stalked; often without terminal bud and always surrounded by 2 bud scales, the margins of which fit tightly; often without terminal bud, and then with two lateral buds at the end of the shoot. The bark is distinctive: it tears open in strips, resulting in unmistakable patterns. In winter the species are very hard to tell apart.

Key to section *Macrantha*

1 Buds small, to 4 mm long (measured without stalk) 4

1* Buds mostly more than 5 mm long (measured without stalk) 2

2 Twigs ± with waxy bloom 3

2* Never with waxy bloom, buds large, predominantly (wine)-red . . . . *Acer pensylvanicum* and *Acer davidii*

3 Twigs and buds predominantly green, buds large. . . . . *Acer rufinerve*

3* Twigs and buds predominantly red, buds medium-sized . . . . *Acer capillipes*

4 (1) Buds and tips of twigs very dark violet-brown . . . . *Acer crataegifolium*

4* Buds and tips of twigs wine-red . . . . *Acer micranthum*

Acer crataegifolium Siebold & Zucc., hawthorn maple

Buds blackish-violet, even darker than tips of twigs. **Twigs** glabrous when young, slender, at the tip dark purple, when older dark green, also reddish on the sun side with pale green stripes and few white lenticels, or whitish interrupted stripes. Rarely planted, shrub to 10 m high. Origin: Japan.

Acer micranthum Siebold & Zucc.

Buds relatively small, strongly wine-red. **Twigs** slender, slightly lighter wine-red to orange-brown. Rare shrub or small tree. Origin: Japan.

Acer capillipes Maxim., red snake bark maple

Buds twig-coloured. **Twigs** smooth, green to brown-green, initially whitish with waxy bloom, older twigs with white oblong stripes on a green base. **Bark** dark grey, corky, tearing open longitudinally, forming green and white stripes. Tree to 15 m high. Origin: Japan.

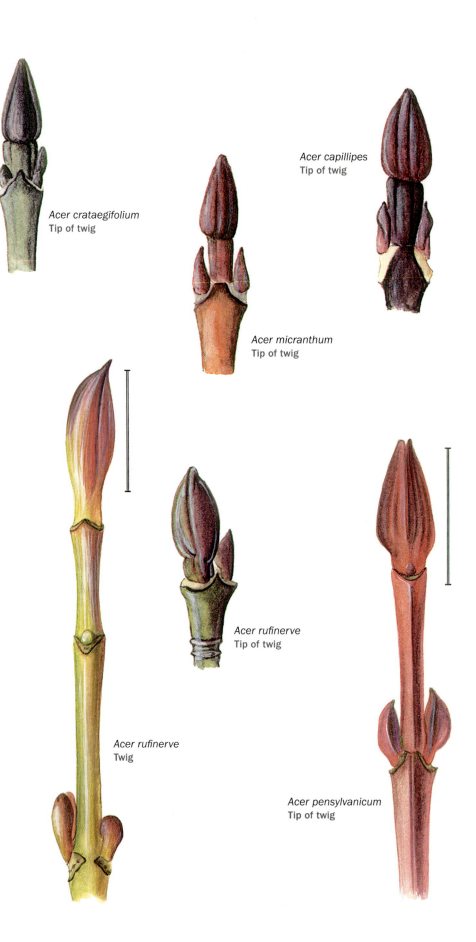

Acer crataegifolium
Tip of twig

Acer capillipes
Tip of twig

Acer micranthum
Tip of twig

Acer rufinerve
Tip of twig

Acer rufinerve
Twig

Acer pensylvanicum
Tip of twig

Acer glabrum
Tip of twig

Ailanthus altissima
Twig

Ailanthus altissima
Small lateral bud above
large leaf scar

Acer rufinerve SIEBOLD & ZUCC., grey-budded snake bark maple

Twigs mainly green, on the sun side also light reddish, initially matte, whitish with waxy bloom, older twigs glossy with scattered whitish stripes. **Bark** when older with few, hardly broader stripes, not as beautiful as in other species. Tree to 15 (–20) m. Origin: Japan.

Acer pensylvanicum L., moosewood

Buds narrowly ovoid, stalked, mostly reddish: orange-red to wine-red-violet, yellowish-green only on the shaded side; as in the twigs, never with waxy bloom. **Twigs** wine-red to brown-red, olive-green on the shaded side, from the second year with lighter stripes. **Bark** with very dense white stripes, on an initially reddish and greenish, later only greenish background. Small tree, in its native country often a shrub, to 12 m high. Origin: North America.

Section Glabra

Acer glabrum TORR., the rock maple, is rarely planted. Twigs and buds are similar to the snake bark maples: buds with 2 outer bud scales, these, like the twigs, dark wine-red. Terminal bud usually distinctly larger than the lateral buds. This species mostly grows as a shrub but can develop into a small tree to 6 m high. Origin: North America.

Picrasma quassioides
Terminal bud

Picrasma quassioides
Lateral bud

Simaroubaceae
tree of heaven family

Key to Simaroubaceae

1 Terminal buds yellow, twigs dense with yellow lenticels. . . . *Picrasma quassioides*

1* Lateral buds only . . **Ailanthus altissima**

Ailanthus altissima (MILL.) SWINGLE, tree of heaven

Buds only hemispherical lateral buds, c. 6 mm broad and 4 mm thick; surrounded by 2 outer, mostly dark wine-red, almost glabrous prophyll scales and 2–3 subsequent bud scales, these are often thinly hairy and dull wine-red to olive-green or yellowish-green. **Twigs** very thick: first year's growth 1 to more than 2 cm diameter, olive-brown to brown, near the leaf scars also reddish-brown, with many small, pale lenticels. Bark of twig finely patterned by longitudinal fissures. **Leaf scars** shield-shaped, pale grey-brown almost like the twig, only below the tip of the twig on large cushions with many traces, mostly with 5–7 traces or groups of traces. **Bark** of stem typically with rhomboid-oblong pale pattern on grey background. Shoots often end in **fruit-panicles** which become overtopped by lateral twigs, often already in the same year. **Fruits** oblong winged nutlets 3–4 cm long, pale brown to orange-red. Very frequently planted tree to 25 m high (growing wild in places). Origin: northern China.

Picrasma quassioides (D.DON) BENN., quassia

Buds without bud scales, distinctly dense yellowish-brown hairy. Thick through the in-rolled leaflets of the pinnate leaf. Terminal buds compact, broader (to 8 mm) than long (to 6 mm) and often also broader than the twig. Lateral buds slightly spreading. Leaf scars relatively large, pale ochre-grey, with 3 pale, indistinct traces. **Twig** slightly hairy only below the terminal bud, otherwise glabrous, covered very densely with many pale ochre-white to pale yellow lenticels. The latter contrast strongly with the dark brown bark of the twig. **Pith** solid, relatively thick, white (2.5 mm by 4 mm diameter). Tree to 10 m high. Origin: eastern Asia.

Meliaceae
neem family

Species of this family, mainly consisting of tropical woody plants, are only rarely grown in temperate regions.

Toona sinensis (JUSS.) M. ROEM., Chinese mahogany
[*Cedrela sinensis* A. JUSS.]

Buds without bud scales, finely ochre-brown hairy. **Terminal buds** conical-ovoid, to c. 10 mm long. These are surrounded by thick-stalked leaves which are slightly pinnate at the upper end. **Lateral buds** small, roundish. **Twigs** thick, initially brown-hairy, grey below with many rounded lenticels; lateral twigs grey and lenticels brown. **Leaf scars** shield-shaped with 5 traces. **Bark**: strongly fissured bark which comes off in strips. Tree to 15 m high. Origin: China.

Rutaceae
citrus family

A very varied family. Most species have oil glands and are more or less strongly aromatic. Mostly large shrubs or trees. Buds opposite or alternate, with or without bud scales. The fruits are multiple fruits with follicles (*Orixa*, *Tetradium*, *Zanthoxylum*), winged nuts (*Ptelea*), berries (*Citrus*) or stone fruits (*Phellodendron*). The remnants of the dry fruits are often retained into the winter. Many species are dioecious and then only female plants bear fruit. *Citrus* has hermaphrodite flowers, whilst *Ptelea* and *Zanthoxylum* have a tendency towards polygamous flowers.

Key to Rutaceae

1 Buds (sub-)opposite 2
1* Buds alternate 3
2 Buds without bud scales. 6
2* Buds with bud scales . . . . *Orixa japonica*
3 Twigs armed. 5
3* Twigs not armed. 4
4 Buds without bud scales. . . . .
. . . . **Ptelea trifoliata**
4* Buds with bud scales . . . . *Orixa japonica*
5 Twigs green, one prophyll spiny. . . . . . .
. . . . **Citrus trifoliata**

5* Twigs with prickles (often in pairs at the nodes) . . . . *Zanthoxylum*
6 Only lateral buds, these flat, hemispherical, enclosed by leaf scar. . . .
. . . . *Phellodendron*
6* With terminal buds, lateral buds, ovoid, taller than broad . . **Tetradium daniellii**

Orixa japonica THUNB.

Buds with many scales, differentiated into lateral and terminal buds 5–7 mm long. Lateral buds alternate in 4 lines on the twig: sometimes looking sub-opposite, slightly spreading from twig, 3–5 mm long. Bud scales decussate, basally greenish or dark violet-brown towards the margin. **Twigs** thickest at base (> 5 mm), becoming more slender towards the tip (2.5–4 mm), grey-green with many small, distinctive greenish to brownish lenticels; towards the tip dark and brownish. Strongly aromatic. **Leaf scars** with one trace, pale ochre-brown. Rarely planted shrub to 3 m high. Origin: eastern Asia, Japan, South Korea and China.

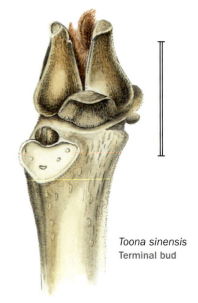

Toona sinensis
Terminal bud

Orixa japonica
Lateral bud

Orixa japonica
Tip of twig

Orixa japonica
Lateral bud

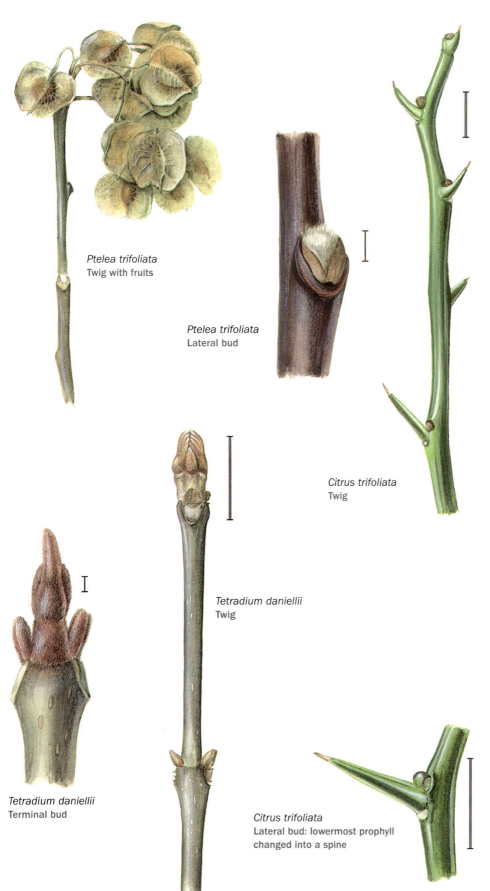

Ptelea trifoliata
Twig with fruits

Ptelea trifoliata
Lateral bud

Tetradium daniellii
Twig

Citrus trifoliata
Twig

Tetradium daniellii
Terminal bud

Citrus trifoliata
Lateral bud: lowermost prophyll
changed into a spine

Ptelea trifoliata L., hop tree

Buds: only alternate lateral buds, without bud scales, c. 1–1.5 mm broad, flat, slightly sunk into twig, tightly appressed, grey-white hairy. **Twigs** slender when young: 2–3 mm, dark brown, slightly glossy, with many small twig-coloured lenticels. **Leaf scars** 3-parted, encircling the buds almost completely except for the top edge. The dry **fruits** persist throughout the winter, grouped in panicles, flat and round, winged with a seed at the centre, similar to elm fruits. Often planted, small shrub or tree to 8 m. Origin: North America.

Citrus trifoliata (L.),
Japanese bitter orange
[*Poncirus trifoliata* (L.) RAF.]

Buds alternate, globose, 2–3 mm thick, with some white-ciliate, olive-green to wine-red-violet bud scales; one prophyll usually spiny and 1–3 cm long, green with short, dry brown tip. **Twigs** distinctly green when young, slightly glossy, smooth and flattened (with a lens the very close oil glands are visible). **Lenticels** noticeable only from the second year, grey longitudinally fissuring and merging. **Fruit** a globose, 3–5 cm thick, densely hairy, yellowish citrus fruit. Occasionally planted, 2–4 m high shrub. Origin: eastern Asia.

Tetradium daniellii (BENN.) T. G. HARTLEY, bee-bee tree
[*Euodia hupehensis* DODE, *Euodia daniellii* (BENN.) HEMSL.]

Buds alternate, without bud scales, densely velvety glossy reddish-brown hairy. Terminal buds 10–15 mm long, with 2 visible pairs of leaves: those of the outer pair curved and divided into petiole and pinnate leaf; lateral buds smaller, to 5 mm long. **Twigs** olive-grey to grey-brown, glabrous or hairy, with many ochre-brown lenticels. **Leaf scars** shield-shaped, hardly encircling lateral buds. **Fruits** in large, broad panicles: 4–5 small, 2-valvate, initially reddish, often finely warty follicles with small black glossy seeds. **Bark** dark grey, remaining thin. Occasionally planted, tree to 20 m high. Origin: central China.

Phellodendron amurense RUPR., Amur cork tree

Buds tight and delicate, appressed dark red brown to shining orange-brown hairy. Only lateral buds, sub-opposite, in relation to the twigs relatively small: 2–3 mm thick; flat, hemispherical, encircled by large horse-shoe-shaped leaf scars. **Twigs** glabrous, shiny orange-brown, with many flat, small, oblong pale ochre-brown lenticels. Older twigs matte, grey to grey-brown with warty lenticels. **Leaf scars** with 3 traces of vascular traces or groups of traces, especially the central trace often of 3 parts. **Bark** thickly corky and deeply furrowed. Tree to 15 m high. Origin: eastern Asia.

The Japanese cork tree, *Phellodendon japonicum* MAXIM. and the Sakhalin cork tree *Phellodendron sachalinense* (F. SCHMIDT) SARG. are more rarely planted. They have red-brown twigs and a thinner bark, which is only slightly furrowed.

Zanthoxylum L., prickly ash

Shrubs or small trees armed with prickles. Most of the species are evergreen and distributed throughout the tropical and subtropical regions of both hemispheres. Only a few, rarely planted deciduous species from the temperate zones of Eastern Asia and North America.

Zanthoxylum simulans HANCE., Chinese pepper
[*Zanthoxylum bungeanum* MAXIM.]

Buds globose, 1–2 mm long, surrounded by few bud scales. **Twigs** mostly glabrous, olive-green, finely oblong-fissured with many, initially round, later oblong lenticels. **Prickles** many, to 20 mm long, mostly in pairs on the nodes and scattered on the internodes, initially brownish with a broad base, narrowly acute; remaining on older twigs and stems on corky cushions. **Bark** olive-grey, with oblong lenticels and many cork cushions; partly unarmed, but usually with prickles. A tall and wide shrub to 3 m high. Origin: China.

Zanthoxylon americanum MILL., common prickly ash (*Zanthoxylum fraxineum* WILLD.), a tree to 8 m high from North America is rarely planted. **Buds** without bud scales, small, red-brown hairy, terminal bud

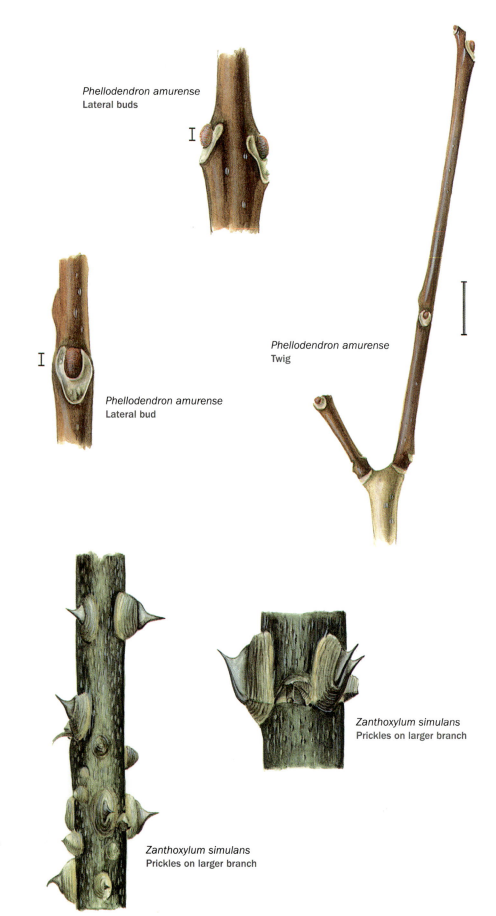

Phellodendron amurense
Lateral buds

Phellodendron amurense
Twig

Phellodendron amurense
Lateral bud

Zanthoxylum simulans
Prickles on larger branch

Zanthoxylum simulans
Prickles on larger branch

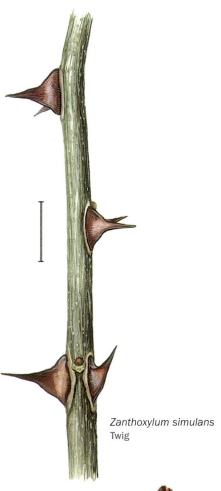

Zanthoxylum simulans
Twig

present. Young **twigs** quickly grey-brown (without green hues), finely longitudinally fissured with many small lenticels. Prickles in pairs on the nodes, relatively small, absent from older twigs and stems.

Tamaricaceae
tamarisk family

Shrubs with upright slender branches. Buds in the axils of alternate scale-shaped leaves, developing in spring into short-shoots and shed as one unit at the end of the growing season. Thus, at the tip of the twig, we find buds in the axils of scale leaves, and at the base of the twig the scars of last year's shoot with a circular bundle trace. Above the trace or bud there is a little pale swelling.

Key to Tamaricaceae

1 Pith narrow, excentric; older twigs dark red-brown to blackish. . . . .***Tamarix***

1* Pith wide, central; twigs ochre-brown to red-brown. . . . . *Myricaria germanica*

Tamarix L., tamarisk

The cultivated species cannot be distinguished in winter. **Buds** alternate, surrounded by 3 outer bud scales, which are pale grey-brown and also reddish on the light side. At the tips of twigs buds mostly solitary in the axils of the persistent scale-shaped leaves. Basal buds usually with 2 enrichment buds axillary to the prophylls. **Twigs** straight, slender, ending whip-like: basally 3–4 mm wide, at the tip less than 1 mm wide; greenish to dark red-brown, older ones black-brown. Shrubs to 5 m high, or trees to 10 m. Origin: From western south-east Europe to eastern Asia.

Myricaria germanica (L.) Desv., German or false tamarisk
[*Tamarix germanica* L.]

Buds small, globose-ovoid, mostly surrounded by the scale-shaped leaves. **Twigs** red-brown to pale ochre-brown, from the small distinct leaf cushion narrow ridges run down the twig. On the twigs one can often see the dried remnants of the assimilation shoots. Rarely planted, shrub to 2 m high. Origin: Europe to central Asia.

Zanthoxylum americanum
Twig

Tamarix
Towards twig tip lateral buds almost entirely surrounded by scale-shaped leaves

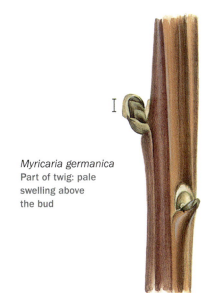

Tamarix
At the twig base the buds are larger with enrichment buds

Myricaria germanica
Part of twig: pale swelling above the bud

Plumbaginaceae
thrift family

Ceratostigma willmottianum STAPF, Chinese plumbago

Buds: only **lateral buds**, slightly spreading from twig; almost completely enclosed on one side by the first attenuate, ochre-brown, 6–8 mm long bud scale. The subsequent scales with shorter hairs, denser and longer. **Twigs** narrowing towards the tip, slightly zig-zag, with narrow longitudinal ridges, these violet-brown on the light side and green on the shaded side, with long, ochre-brown hair and delicate white dot-shaped resin glands. At the tip of the twig with globose (35 mm diameter) remnants of infructescences. **Leaf scars** narrow with many traces. **Pith** compact, white. Rarely planted, shrub to 1 m high. Origin: Tibet and Western China.

Polygonaceae
knotweed family

Several successive tubular stipule sheatings (ochreas) protect the buds. The laminas belonging to the outer envelopes are ± undeveloped.

The family occurs world wide but the centre of distribution is in the temperate zone; herb species are predominant and there are only few woody taxa, such as lianas and small shrubs. The dry remnants of some herbaceous species are distinctive in the winter, e.g. the c. 2 m high Japanese knotweed, *Reynutria japonica* HOUTT., an invasive species from Japan.

Key to Polygonaceae

1 Twigs armed, erect shrub . . . . *Atraphaxis spinosa*

1* Not armed, climbing or prostrate 2

2 Twigs pale, more than 2 mm thick, climbing several m high . . . . *Fallopia baldschuanica*

2* Twigs dark, less than 1 mm thick, prostrate or slightly rising. . . . . *Muehlenbeckia axillaris*

Atraphaxis spinosa L., buckwheat

Buds: only lateral buds, appressed to twig. **Twigs** grey-brown with narrow pale grey ridges, these are particularly distinct on long-shoots. Coming from the long-shoots, are gradually narrowing short-shoots which, however, do not end in a sharp point. Leaf cushion reaching far beyond buds with remnants of stipules enveloping the twig. **Leaf scar** indistinct. Rarely planted, shrub to almost 1 m high. Origin: eastern Europe and Asia Minor, but also to north-west China and Mongolia.

Muehlenbeckia axillaris (HOOK. f.) ENDL., creeping wire vine

Buds small, less than 1–2 mm long. Lateral buds appressed to twig. Terminal buds on short-shoots which are spreading from the long-shoot. **Twigs** very slender, c. 0.5 mm diameter, black-brown, finely hairy to glabrous, slightly winding, at the nodes with a stipule envelope which is slightly paler than ochre-brown. Rarely planted, densely branching cushion-forming shrub, prostrate or slightly climbing. Origin: New Zealand.

Fallopia baldschuanica (REGEL) HOLUB, Russian vine
[*Fallopia aubertii* (L. HENRY) HOLUB, sometimes placed in *Polygonum* or *Bilderdykia*]

Buds acute-conical, at the base with some dried-up grey-brown slightly spreading bud envelopes. Lateral buds almost appressed to obliquely spreading, on short-shoots near the base of the long-shoot, with terminal buds that are spreading, sometimes nearly at right-angles, to the twig. **Twigs** grey-green or grey-brown to brown, with grey epidermis which partly tears open. **Leaf scar** indistinct, small and round, dry brown, above the leaf scar the narrow stipule scar surrounds the twig. Frequent climbing shrub to 15 m high. Origin: central Asia and China.

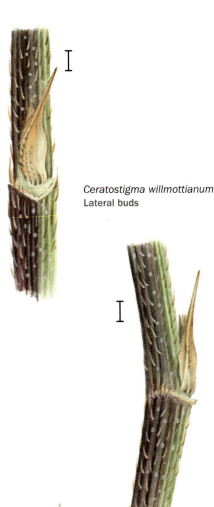

Ceratostigma willmottianum
Lateral buds

Atraphaxis spinosa
Twig ending in a thorn

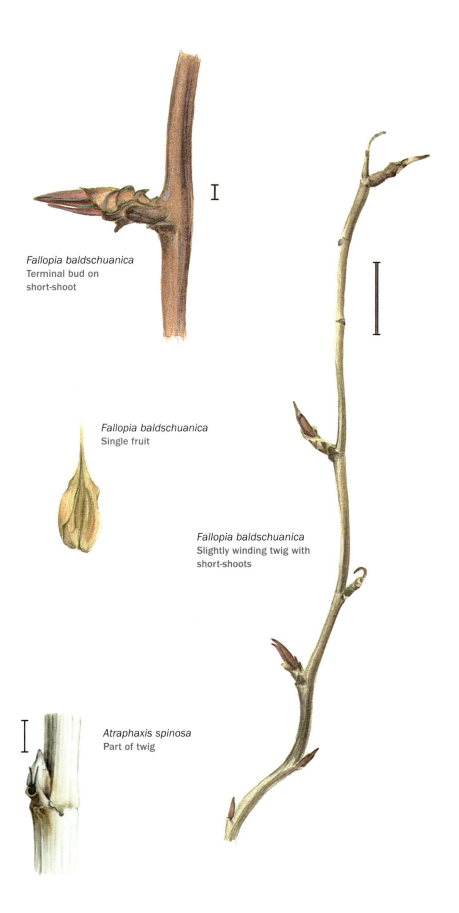

Fallopia baldschuanica
Terminal bud on
short-shoot

Fallopia baldschuanica
Single fruit

Fallopia baldschuanica
Slightly winding twig with
short-shoots

Atraphaxis spinosa
Part of twig

ORDER Cornales

Leaves or leaf scars opposite (Hydrangeaceae
and most of *Cornus*), more rarely alternate
(subfamily Nyssoideae, *Alangium* as well
as *Cornus alternifolia* and *C. controversa*).
Opposite leaf scars are connected by a line.
Whilst the fleshy fruits of the Cornaceae are
mostly no longer present in winter, the dry
capsules of the Hydrangeaceae often persist
the entire winter.

Cornaceae
dogwood family

Deciduous and evergreen trees and shrubs,
more rarely herbaceous perennials. Very
common in the temperate latitudes of the
Northern Hemisphere; in South East Asia
from the Indo-Malasian region to Australia
and from North to Central America, with a
few representatives in Central Africa.

Key to Cornaceae

1 Leaf scars alternate. 2
1* Leaf scars opposite; buds without or with
   bud scales. . . . . **Cornus**
2 Twigs with terminal buds, buds with bud
   scales 3
2* Twigs without terminal buds, buds
   without bud scales . . . . *Alangium*
3 Terminal buds to 5 mm long, bud scales
   with brown margin. . . . . *Nyssa*
3* Terminal buds longer 4
4 Shoot tips continuously overtopped by
   lateral axes, twigs medium-sized, buds of
   same colour as branches. . . . .
   . . **Cornus alternifolia** and **C. controversa**
4* Twigs grey-green, relatively thick, buds
   glossy dark wine-red . . . . **Davidia**

SUBFAMILY Nyssoideae

Five genera with some 30 species; two
of these in cultivation: the monotypic
handkerchief tree *Davidia* is seen in parks,
the American tupelo tree, genus *Nyssa*, with
12 species in North America and in south-
east Asia, is slightly rarer.

Nyssa sylvatica MARSHALL, tupelo

Terminal buds c. 4 mm long, with 3–4 visible bud scales. **Lateral buds** on the long-shoots, c. 2 mm long, surrounded externally by 2 opposite prophyll scales. **Bud scales** red-brown to greenish on the shaded side with pale brown dry margin, and, like the ends of twigs, loosely pale-hairy. **Twigs** brown, towards the tip paler ochre-brown; on short-shoots usually only terminal buds. **Leaf scars** with 3 vascular bundles on prominent leaf cushions. **Lenticels** inconspicuous. **Pith** wide and pale. Rarely planted, tree, narrowly conical, with thick branches. In Central Europe to 20 m high. Origin: North America from north-east USA to southern Mexico.

Davidia involucrata BAILL., handkerchief tree

Buds acute-ovoid, with wide base. Terminal buds c. 9–11 mm and lateral buds 6–8 mm long, few bud scales: glossy dark wine-red with a narrow, pale rim, the lowest occasionally slightly ciliate. **Twigs** glabrous, relatively thick and rigid, olive to grey-brown or grey-green with small, ochre-brown lenticels. **Leaf scars** broad, mostly with 3 traces, or through splitting of the outer ones, up to 5 traces. **Bark** grey-brown and flaking. Occasionally planted, broad tree to 20 m high. Origin: Western China.

SUBFAMILY Cornoideae

Two genera, *Cornus* and *Alangium*. The buds are mostly without bud scales. In *Cornus* the tendency towards bud scales can be observed in several sections.

Alangium LAM.

About 20 mostly evergreen trees, mainly in the (sub)tropics of the Old World. Only a few species reach the temperate latitudes. Two of these are sometimes cultivated:

Alangium platanifolium (SIEBOLD & ZUCC.) HARMS

Buds: lateral buds without bud scales, mostly forming a compact unit with a decending accessory bud; squat (3 mm long and just as thick), spreading-hairy, ochre-brown, completely surrounded by a circular 5–7-traced leaf scar. **Twigs** pale brown, fine- and densely hairy with few oblong lenticels. A small tree. Origin: Japan and China.

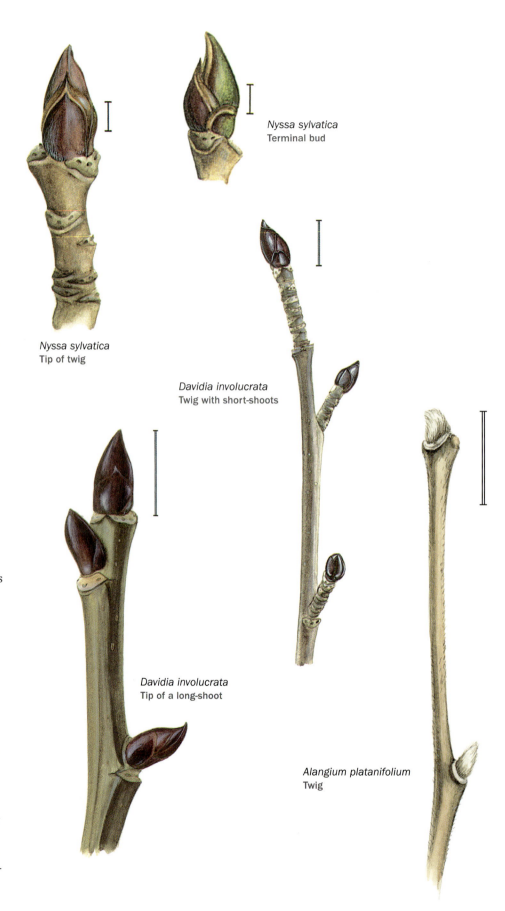

Nyssa sylvatica
Terminal bud

Nyssa sylvatica
Tip of twig

Davidia involucrata
Twig with short-shoots

Davidia involucrata
Tip of a long-shoot

Alangium platanifolium
Twig

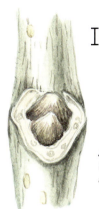

Alangium chinense
Lateral bud with
accessory bud

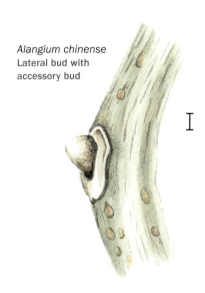

Alangium chinense
Lateral bud with
accessory bud

Alangium platanifolium
Lateral bud at the
shoot tip

Alangium chinense (LOUR.) HARMS

Buds similar to the above species, with closely appressed indument. **Twigs** (long-shoots) strongly zig-zag, relatively pale grey-green with fine narrowly fissured structure, brown lenticels and fine, inconspicuous indument. Leaf scar with 5 to 7 traces surrounding the bud, but not connected at the rear of the bud. Medium-sized tree. Origin: Vietnam via China to India and Africa.

Cornus, dogwood

Mostly deciduous, predominantly opposite, but can be alternate. Mostly shrubs, more rarely trees or herbaceous perennials. In Europe two native and several planted species, quite common in gardens. Twigs and buds mostly with typical 2-branched hairs (visible with hand lens).

Key to *Cornus*

1 Leaf scars alternate: buds with bud scales 7

1* Leaf scars opposite; at least leaf buds mostly without bud scales 2

2 Flower buds covered in bud scales, at least twice as wide as the twig 3

2* Flower buds without bud scales 8

3 Flower bud yellowish green to honey-brown, globose above the middle, hardly broader than high; twigs pale hairy, never with waxy bloom; flowers before the leaves in March 4

3* Buds and twigs different; flowers after the leaves 5

Cornelian cherries

4 Bark of twig hardly flaking, flower buds globose . . . . ***Cornus mas***

4* Twigs and stems with strongly flaking bark, flower bud more slender . . . . *Cornus officinalis*

flowering dogwoods

5 First year's growth twigs green to dark wine-red. Flower buds stalked, broader than high 6

5* First year's growth twigs brown to the tip (periderm), with many lenticels. Flower buds narrowly to the tip . . ***Cornus kousa***[1]

6 Inflorescence buds completely enclosed by four involucral bracts. . . . . ***Cornus florida***[1]

6* Inflorescence buds open, individual floral buds visible . . . . ***Cornus nuttallii***[1]

[1] In recent years more hybrids between these three species are being planted.

pagoda dogwood

7 (1) Buds dark red-brown-violet; twigs dark brown-violet with waxy bloom; below the tip more than 3 mm diameter . . . . ***Cornus controversa***

7* Buds and twigs wine-red-violet to ochre-brown, twigs more slender. . . . .*Cornus alternifolia*

other dogwoods

8 (2)Twigs with distinctly bright, intensive colour (yellow, orange to red or wine-red) 9

8* Twigs reddish to red-green or grey-brown. 11

9 Older twigs with uniform grey bark, the young twigs orange-red on one side more deeply reddish . . . . ***Cornus sanguinea*** — cultivars

9* Older twigs without such grey bark (though not showing the same brilliance of colour as first year's growth), first year's growth twigs the same colour on all sides. 10

10 Twigs bright yellow to greenish-yellow (without red hues). . . . .***Cornus alba*** subsp. ***stolonifera*** 'Flaviramea'

10* Twigs all-round strongly coral- or dark wine-red. . . . . ***Cornus alba*** — cultivars

11 First year's growth twigs with continuous grey to ochre-brown bark, remnants of inflorescence paniculate. . . . . *Cornus racemosa*

11* First year's growth twigs without such bark, with reddish and greenish hues 12

12 First year's growth twigs with oblong streaks of colour around the lenticels, greenish on the shaded side, to salmon-red in the sun. . . . . *Cornus rugosa*

12* Without coloured streaks around the lenticels 13

13 Older twigs with uniform dark bark, younger twigs at least in the shade (lower side) greenish. . . . .***Cornus sanguinea***

13* Older twigs (c. 10–15 mm diameter) without such bark, ± greenish or reddish; young twigs mostly reddish all-round, also in the shade. . ***Cornus alba***

Subgenus Cornus

Section Cornus

Cornus mas L., Cornelian cherry

Buds densely fine-hairy, the colour varying from yellow-ochre-brown to green olive-brown. **Inflorescence buds** terminal and axillary, globose, shortly acute to blunt, after 2 pairs of scale-shaped basal leaves follow 2 pairs of hemispherical bud scales, closed, only the outer pair visible. **Leaf buds** slender and without bud scales. **Twigs** green, on the light side partly with a red-violet hue, densely short-hairy. **Flowers** end of February to March, in heads. Single flowers small, tetramerous. **Bark** small-scaled, flaking. Very frequent shrub or small, to 5 m high, spreading tree. Origin: central and southern Europe.

The much rarer **Cornus officinalis** SIEBOLD. & ZUCC., Japanese cornelian cherry, is relatively similar and difficult to distinguish in winter. Origin: eastern Asia. **Inflorescence bud** ovoid, gradually acute, **bark** on several-year-old twigs coming off in rolls, on stems flaking in larger plants.

Sections Benthamia and Cynoxylon

Both sections are distinct in their infructescences, but also by the colour of their twigs and the shape of the buds. Hybrids between these subgenera, with large flowers, have come into cultivation in recent years. The hybrids have vigorous first year's shoots, which are almost completely grey-brown, as in *Cornus kousa*, and which show the greenish-red colour of *Cornus florida* or *Cornus nuttallii* only at the tips of the twigs.

Section Benthamia

Cornus kousa F. BUERGER ex MIQ., Korean dogwood

Buds red-brown to brown-violet, finely hairy, differentiated into leaf- and flower buds; mostly terminal, with only few, weakly developed lateral buds. **Flower buds** on short, dichasially branched shoots. Outer pair of the bud scales often deciduous basally, and remaining like a hood on the bud. About 7–8 mm long and 3 mm thick. **Leaf buds** mostly smaller and darker, acute-conical. **Twigs** on short-shoots red-brown to grey-brown, on long-shoots grey-brown to brown, and violet reddish towards the tip.

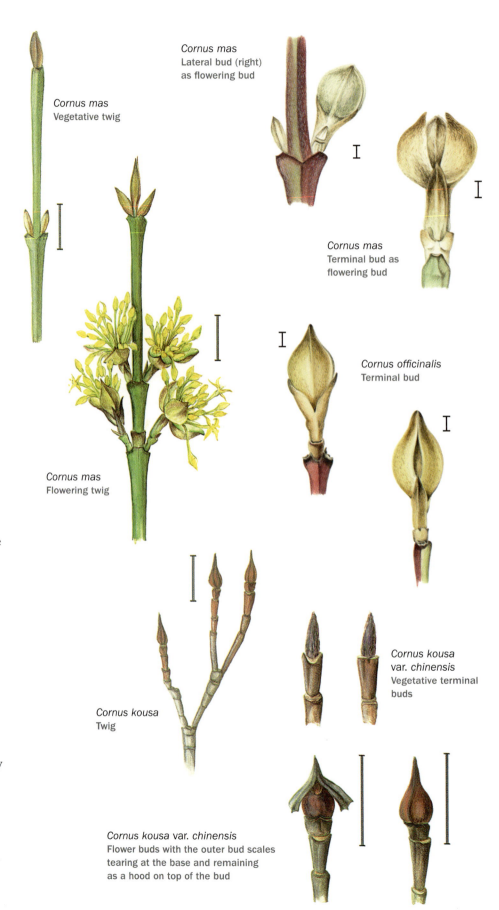

Cornus mas
Vegetative twig

Cornus mas
Lateral bud (right)
as flowering bud

Cornus mas
Terminal bud as
flowering bud

Cornus mas
Flowering twig

Cornus officinalis
Terminal bud

Cornus kousa
Twig

Cornus kousa
var. *chinensis*
Vegetative terminal
buds

Cornus kousa var. chinensis
Flower buds with the outer bud scales
tearing at the base and remaining
as a hood on top of the bud

Cornus florida
Vegetative twig

Lenticels very dense and many on long-shoots. Common, shrub with branches at right angles, or to 7 m high tree. Origin: Eastern Asia, Japan and Korea.

Cornus kousa var. ***chinensis*** OSBORN, Chinese dogwood, is more vigorous than the typical species: flower buds more compact and shorter-acute, c. 6–7 mm long and 4 mm thick. Origin: China.

Section Cynoxylon

Cornus florida L.,
eastern flowering dogwood

Buds mostly terminal, axillary ones weak or not developed. Leaf buds often small, 2–4(–7) mm long and 1–2 mm broad. Flower buds broadly globose, stalked: 12–13 mm long (without stalk 5–6 mm) and 7–8 mm wide. On the stalk are one or two pairs of basally connate scales. Two pairs of wider bud scales, the involucre bracts, surround the inflorescence bud, of which the outer pair almost completely encloses the bud. Twigs and buds finely appressed white-hairy or with waxy bloom. Twigs matte wine-red, in the shade sometimes entirely green. Hardly any lenticels. Frequent, 5(–8) m high shrub, occasionally also small tree. Origin: eastern North America: Canada to Mexico.

Cornus nuttallii AUDUBON,
Pacific or mountain dogwood

Buds and twigs similar to *Cornus florida*. Inflorescence buds terminal, 10–12 mm diameter, the basal narrow scales on the bud-stalk mostly persistent; the bowl-shaped 4–6 involucral bracts form an open basket (in *Cornus florida* closed) and surround the visible flower buds. Twigs dark violet-red, in the shade occasionally a little greenish, especially towards the tip with fine down. Rarely planted large shrub, in western North America a tree to 25 m high. ***Cornus* 'Eddie's white wonder'** [*Cornus florida* × *Cornus nuttallii*] is a hybrid of both North American dogwoods. It is occasionally planted, and in winter it looks like *Cornus nuttallii*.

Cornus florida
Flowering bud

Cornus florida
Twig with flower buds

Cornus florida
Leaf buds

Cornus nuttallii
Twig with typically open inflorescence bud

Cornus nuttallii
Scale-like leaves below the inflorescence bud

Subgenus Kraniopsis
Section Thelycrania

Cornus sanguinea L., common dogwood

Buds red-brown, finely hairy and without bud scales. **Leaf buds** 5–8 mm long and 1–2 mm thick. **Flower buds** to 10 mm long and 2–4 mm thick. **Twigs** matte wine-red, but last year's twigs underneath and where shaded, green. Older twigs, relatively quickly with grey, smooth bark (periderm), when over 6–8 mm diameter. Occasionally planted cultivars such as 'Midwinter Fire' with young orange-red twigs must be regularly pruned to make new young growth. Commonly planted shrub to 4 m high. Origin: Europe.

Cornus alba L., Siberian dogwood

[***Cornus sibirica*** LODD., ***Cornus tatarica*** MILL.]

Buds grey-brown and partly, particularly in the red-branched species, slightly reddish; densely and relatively long-hairy. **Twigs** ± glossy and usually all-round reddish, at the tip with deciduous fine hairs or with waxy bloom. Few grey to grey-brown lenticels. Older twigs, over 2 cm diameter, remain green to red. Very frequently planted, erect shrub without runners. In winter some cultivars stand out, such as **'Sibirica'** with coral-red twigs and **'Kesselringii'** with glossy dark wine-red twigs. Origin: eastern Asia.

subsp. *stolonifera* (MICHX.) WANGERIN, red osier dogwood

[***Cornus sericea*** hort. non L.]

This subspecies cannot be distinguished without fruit. Branches curving and looping and often taking root; or initially prostrate and later ascending. Frequent, spreading shrub to 2 m high. Origin: North America. Particularly the cultivar **'Flaviramea'** is often planted and easily recognised: **twigs** distinctly bright yellow to yellow-green, with few grey-brown lenticels.

Cornus rugosa LAM., round-leaved dogwood

Buds c. 10 mm long. Bud leaves in the lower part scale-like, only at the tip with rudimentary blade, greenish to reddish,

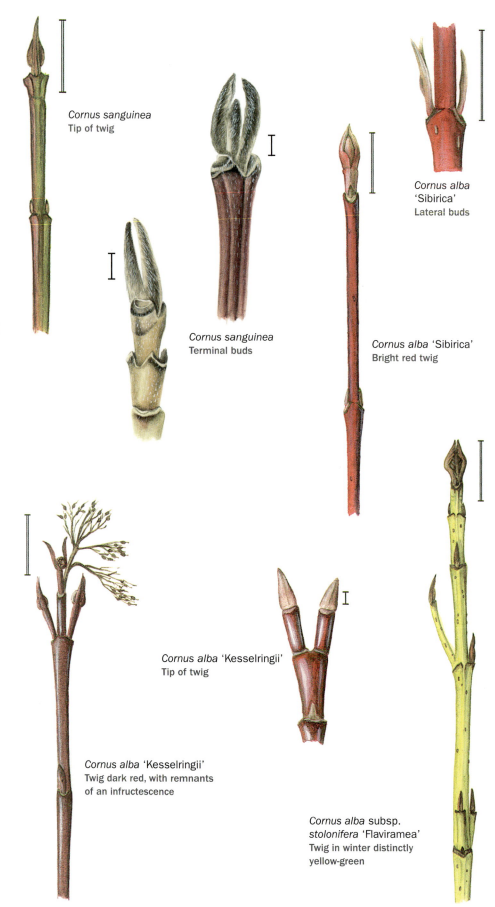

Cornus sanguinea
Tip of twig

Cornus sanguinea
Terminal buds

Cornus alba
'Sibirica'
Lateral buds

Cornus alba 'Sibirica'
Bright red twig

Cornus alba 'Kesselringii'
Tip of twig

Cornus alba 'Kesselringii'
Twig dark red, with remnants of an infructescence

Cornus alba subsp.
stolonifera 'Flaviramea'
Twig in winter distinctly yellow-green

Cornus rugosa
Part of twig: the lenticels
form in the colour streaks

Cornus rugosa
Terminal bud

Cornus racemosa
Terminal bud (periderm
formed up to tip of twig)

Cornus alternifolia
Terminal bud on
short-shoot

Cornus alternifolia
Terminal bud on the long-shoot
overtopped by lateral branch (right)

Cornus alternifolia
Short-shoots

finely hairy. **Twigs** green, salmon-pink in the sun over ochre hues at the tips of twigs, with oblong darker streaks in which the warty lenticels form. Rarely planted, shrub to 3 m high. Origin: north eastern North America.

Cornus racemosa LAM., panicled dogwood

Buds relatively small, c. 5 mm long, grey-brown, finely and inconspicuously hairy. **Twigs** to the tip pale grey on the light side, ochre-brown on the shaded side. In this feature it differs from all other species of shrubby dogwoods which, at least below the terminal bud, are coloured from green to red. (Only the flowering dogwood *Cornus kousa*, easily distinguished by its distinctive flower buds, also has twigs with a cork layer up to their tips; but this is mostly darker brown and has many lenticels). Rarely planted up to 5 m high shrub. Origin: central and eastern North America.

Section Bothrocaryum

This differs from the other species of *Cornus* by its alternate phyllotaxy and by the buds surrounded by bud scales. The buds are not differentiated into leaf- and flower buds. Almost every long-shoot terminates after a few internodes in a terminal bud and is overtopped by a lateral twig, which ends just as quickly when the next lateral twig takes over. This happens repeatedly during a given season. Only series of short-shoots grow permanent monopodially from the terminal bud.

Cornus alternifolia L.f., pagoda dogwood

Buds to 8 mm long and 2.5–4 mm thick. In the sun wine-red to violet-brown, in the shade ochre-coloured. The 5–6 visible bud scales darker towards the margin and the tip. Basal bud scales with their tips often slightly spreading. **Twigs** dark wine-red, long remaining glossy; in the shade below the branch also changing to olive-green. Below the terminal bud 2.5–3 mm thick. Growth form just as *Cornus controversa*, to 8 m high. Occasionally planted. Origin: eastern North America.

Cornus controversa HEMSL., wedding cake tree

Similar to *Cornus alternifolia*. **Terminal buds** narrowly ovoid, slightly acute, 6–9 mm long and 2.5–4 mm thick. Colour in the sun very dark wine-red-violet to violet-brown ± glossy. The 7–8 visible bud scales follow a spiral phyllotaxy. **Twigs** very dark violet-brown, glossy; immediately below the bud, matte wine-red. **Lenticels** few, tearing longitudinally. Several-year-old twigs mostly ± matte. Diameter of the twigs below the terminal bud 3–4 mm. Occasionally planted, large shrub or to 15 m high tree with the branches in distinctive tiers. Origin: eastern Asia.

Hydrangeaceae
Hortensia family

The family includes many often planted ornamental shrubs. They are characterised by their whorled, mostly opposite phyllotaxy in which the leaf scars of a node abut or are connected by a line.

Key to Hydrangeaceae

1 Buds concealed below the leaf scars. . . . . *Philadelphus*
1* Buds distinctly visible above the leaf scars 2
2 Buds and twigs with stellate hairs . . . . *Deutzia*
2* Buds and twigs with simple hairs or glabrous 3
3 Buds with a dense cover of long white hairs . . . . *Jamesia americana*
3* Buds glabrous or slightly hairy, when indument more dense, it is reddish-brown. . . . . **Hydrangea**

Hydrangea L., hortensia, hydrangea

Shrubs, more rarely small trees or climbing shrubs. In winter rather distinctive by the growth form, the indument and the type of infructescence. Of the 80 species which are native to North and South America or Asia, some are deciduous and are planted frequently.

Key to *Hydrangea*

1 Erect shrubs 2
1* Climbing shrubs 8
2 Long-shoots with dense sturdy spreading hairs . . . . *Hydrangea aspera* and **H. sargentiana**
2* Glabrous or different (finely) hairy . . . 3
3 Infructescences flat or globose corymbs 4
3* Infructescences often conical panicles, often partly 3-whorled . . . . *Hydrangea paniculata*
4 Terminal bud more than 10 mm long. . 5
4* Terminal buds shorter 6
5 Twigs slender . . . . *Hydrangea macrophylla* subsp. *serrata*
5* Twigs thick . . . *Hydrangea macrophylla* subsp. *macrophylla*
6 Bark of twig not flaking 7
6* Bark of twig red-brown, fissured and flaking . . . . *Hydrangea heteromalla* 'Bretschneideri'
7 Twigs grey-brown. . . . . *Hydrangea heteromalla*
7* Twigs red-brown. . . . . *Hydrangea arborescens*
8 Buds almost surrounded by two greenish bud scales which reach to the tip, pith chambered, twigs with adhesive rootlets . . . . *Hydrangea petiolaris*
8* Pith solid, buds without bud scales or with more red-brown bud scales 9
9 Buds without bud scales, red to red-brown, twigs with adhesive rootlets. *Hydrangea barbara*
9* Buds with bud scales . . . .*Hydrangea hydrangeoides*

Hydrangea petiolaris
Twig with adventitious rootlets and with flaking bark

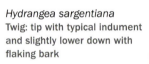

Hydrangea sargentiana
Twig: tip with typical indument and slightly lower down with flaking bark

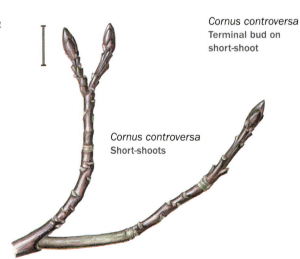

Cornus controversa
Short-shoots

Cornus controversa
Terminal bud on short-shoot

Hydrangea sargentiana
Terminal bud flanked by
large lateral buds

Hydrangea arborescens
Lateral buds

Hydrangea heteromalla
Twig

Hydrangea heteromalla
'Bretschneideri'
Twig with reddish-brown
bark, fissuring and lifting
in the first year

Hydrangea heteromalla
Twig remaining grey

Hydrangea sargentiana REHDER, Sargent's hydrangea
[*Hydrangea aspera* subsp. *sargentiana* (REHDER) E. M. McCLINT]

Buds velvety red-brown. Terminal buds surrounded by several lateral buds, relatively slender. Lateral buds acute-ovoid to c. 10 mm long. with 2 ± fused bud scales. **Twigs** thick, first year's growth 8–10 mm, towards tip densely rough, with spreading hairs. Red-brown **bark** of twigs fissuring into narrow rhomboids, underlying layer pale grey-ochre-coloured, later lifting off. **Leaf scars** large, broad, opposite scars abutting, with (5–) 7–9(–11) distinct traces. **Infructescences** flat corymbs to 25 cm wide. Laxly branching shrub to 3 m high. Origin: China.

Very similar, and not much different in winter, is the rough-leaved hydrangea, *Hydrangea aspera* D. DON.

Hydrangea arborescens L., sevenbark

Lateral buds spreading from twig, ovoid, 5–6 mm long, with some bud scales: the outermost dry grey-brown, the inner patchy-striped olive violet-red-brown, hairy mostly at the margins. **Twigs** red-brown to ochre-brown, glabrous or with fine remnants of hair. **Leaf scars** semi-circular to triangular with 3(–5) vascular traces. **Infructescences** globose, 5–10 cm thick. Shrub 2–3 m high. Origin: North America. Several subspecies and their cultivars frequently grown.

Hydrangea heteromalla D. DON, Himalayan hydrangea

Buds blunt-conical, thickest at the base. Terminal buds to 8 mm long, lateral buds at a wide angle, below the terminal bud even at right angles, to c. 5–6 mm long. Bud scales ochre-brown, glabrous or hairy towards the tip, occasionally early deciduous, which makes the firmly appressed greenish leaves visible; these are more hairy towards the tip. **Twigs** grey-brown, with many paler fissuring lenticels. Infructescences flat-domed, 10–30 cm broad corymbose panicles with some dry sterile flowers at the margin. Bark even in 2-year-old twigs not lifting. Frequently planted, shrub to 3 m high. Origin: China and Himalaya.

In the cultivar *Hydrangea heteromalla* 'Bretschneideri' [*Hydrangea bretschneideri* DIPPEL] the twigs are more vigorous, slightly glossy, violet-brown to ochre-brown; with fine, roundish lenticels and soon narrowly fissuring and lifting bark.

Hydrangea paniculata SIEBOLD, panicled hydrangea

Lateral buds often whorled in threes, small, blunt-conical, spreading from the twig, with some bud scales: the outer ones dark brown and the inner paler ochre-brown. **Twigs** mostly glabrous, ochre-brown to red-brown, with oblong, soon fissuring lenticels and then lifting bark. **Inflorescences** narrowly conical panicles to 20 cm long. In the common cultivar **'Grandiflora'** the panicles are to 30 cm long, consisting predominantly of sterile flowers. A shrub to 2 m high. In its native area a tree to 10 m high. Origin: Japan, south-east China and Sakhalin.

Hydrangea macrophylla (THUNB.) SER., bigleaf hydrangea, hortensia

Terminal buds very large: c. 15–20(–40) mm long and without bud scales: outer leaves distinctly visible, pinnate-nerved, partly surrounded by dried out larger foliage leaves. Bud leaves green (particularly in white-flowering species) to dark wine-red (in pink-flowering species). **Lateral buds** to c. 12 mm long, acute-ovoid or ovoid; basally with two outer brown prophyll scales. Of the next scales the abaxial one is mostly more developed and therefore envelops the entire tip of the bud. **Twigs** thick and glabrous, matte brown, partly olive-green or reddish. **Leaf scars** shield-shaped with 3 traces; abutting or connected by a line. Frequently planted, shrub 1–3 m high, many garden forms exist. Origin: Himalaya, Southern China, Japan.

Subsp. *serrata* (THUNB.) MAKINO [*Hydrangea serrata* (THUNB.) SER.] is also very common. The twigs of this subspecies are relatively slender. A small shrub usually less than 1 m high. Origin: mountain forests of Japan and South Korea.

Hydrangea petiolaris SIEBOLD & ZUCC., climbing hydrangea
[*Hydrangea anomala* subsp. *petiolaris* (SIEBOLD & ZUCC.) E. M. MCCLINT.]

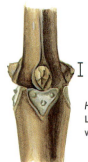

Hydrangea paniculata
Lateral buds, often whorled in threes

Hydrangea macrophylla
Twig of a dark-flowering species with reddish bud scales

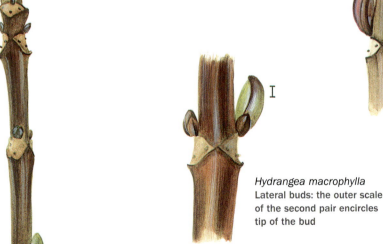

Hydrangea macrophylla
Lateral buds: the outer scale of the second pair encircles tip of the bud

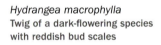

Hydrangea macrophylla
Twig of a pale-flowering species with green bud scales

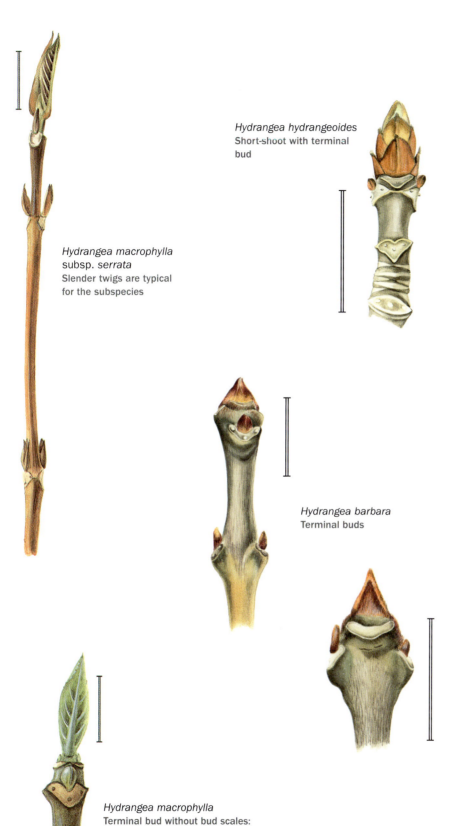

Hydrangea hydrangeoides
Short-shoot with terminal bud

Hydrangea macrophylla
subsp. *serrata*
Slender twigs are typical
for the subspecies

Hydrangea barbara
Terminal buds

Hydrangea macrophylla
Terminal bud without bud scales:
bud leaves distinctly show
pinnate venation

Buds ovoid, 6–10 mm long, almost completely surrounded by the first or second pair of bud scales. Bud scales green to reddish, acute. **Twigs** sturdy, red-brown with flaking bark; underneath pale ochre-grey. With adhesive roots. **Pith** chambered. **Infructescences** 15–20 cm broad corymbose panicles. Frequently planted, liana climbing to 10(–20) m high. Origin: eastern Asia.

Hydrangea hydrangeoides (SIEBOLD & ZUCC.) BERND SCHULZ, Japanese hydrangea vine
[*Schizophragma hydrangeoides* SIEBOLD & ZUCC.]

Terminal buds to c. 8 mm long, slightly narrowing towards the tip, relatively blunt, almost cylindrical, with 2–3 visible pairs of bud scales. **Lateral buds** 5–6 mm long. All buds with broad base. Bud scales rusty-red, flat, almost glabrous, but more densely hairy at tip with distinct hair tufts. Bud scales of former years remain at growth boundaries, dark violet-brown. **Twigs** sturdy, c. 4 mm thick, grey-brown to cinnamon-brown, glabrous, with many fissuring lenticels, particularly at the nodes. **Pith** solid, relatively wide, white with greenish xylem. **Leaf scars** shield-shaped to narrow and wide, with 3 traces, opposite, connected. Rarely planted, climbing shrub to 10 m high. Origin: Japan.

Hydrangea barbara (L.) BERND SCHULZ, woodvamp
[*Decumaria barbara* L.]

Buds without bud scales, densely hairy with long appressed red-brown hairs. Terminal bud short, compact, c. 3–4 mm long, but distinctly wider, enclosed by few leaves (2 pairs). Lateral buds smaller, to 3 mm long, but only 1 mm wide, appressed to the twig. **Twigs** grey-brown to brown, towards the tip with reddish-brown hairs, but soon entirely glabrous. No visible lenticels, bark fissuring longitudinally, on two-year-old twigs the fissures join, the under-surface initially pale grey to dark grey, later also grey-green. **Pith** loose, continuous in the nodes, greenish. Opposite **leaf scars** with 3 traces, connected by V-shaped to horizontal line. Rarely planted, climbing shrub with adventive adhesive rootlets, to 10 m high. Origin: eastern USA.

Philadelphus L., mock orange

Some species which are difficult to tell apart, and many hybrids and cultivars are planted. Without terminal buds, the lateral buds are (in the species treated here) concealed below the scar of the petiole. The lowest part of the leaf base is persistent and protects the buds. Opposite scars are connected by a line.

Key to *Philadelphus*

1 Erect shrubs more than 2 m high 2
1* To 1.50 m high, dainty shrubs . . . .
. . . . **Philadelphus microphyllus**
                    and **hybrids**
2 Bark on older twigs flaking . . . .
. . . . **Philadelphus coronarius**
2* Bark not flaking . .*Philadelphus pubescens*

Philadelphus coronarius L., sweet mock orange

Lateral buds concealed below the pale, 3-parted leaf scars; they are visible only when they begin to swell, which in mild weather can be very early. **Twigs**: first year's growth deep red-brown with long stripes, soon tearing; on two-year-old branches all flaked off. Apart from the normal rather slender twigs on the shrub, there are many long upright, thick and unbranched twigs. **Fruits** 4-valvate top-shaped capsules, in racemes. Frequently planted, shrub to 3 m high. Origin Europe, exact origin unknown.

Philadelphus microphyllus A. GRAY is similar but much smaller in all parts and has particularly slender twigs.

Beside the normal species there exist many hybrids which are hard to distinguish in winter, e.g. **Philadelphus ×lemoinei** hort., a frequent hybrid between *Philadelphus coronarius* and *P. microphyllus*. It is intermediate between the parents.

Philadelphus pubescens LOISEL., broad-leaved mock orange

Lateral buds invisible, concealed below leaf scars. **Twigs** glabrous, the slender ones red-brown; the long, thick, unbranched shoots ochre to pale brown. Several-year-old twigs grey-brown, with persistent bark, which is slightly furrowed and partly (similar to lenticels) bulging. Rarely planted, shrub to 3 m high. Origin: North America.

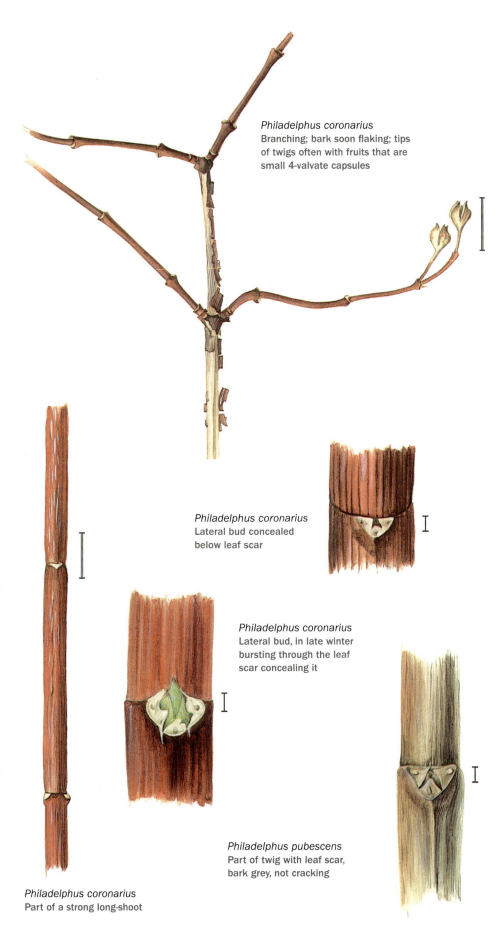

Philadelphus coronarius
Branching; bark soon flaking; tips of twigs often with fruits that are small 4-valvate capsules

Philadelphus coronarius
Lateral bud concealed below leaf scar

Philadelphus coronarius
Lateral bud, in late winter bursting through the leaf scar concealing it

Philadelphus pubescens
Part of twig with leaf scar, bark grey, not cracking

Philadelphus coronarius
Part of a strong long-shoot

Deutzia crenata
Lateral bud: tip of bud
scales spreading

Deutzia crenata
Lateral buds

Deutzia crenata
Lateral buds

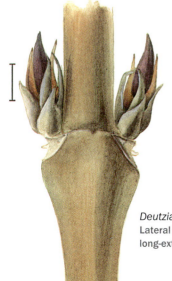

Deutzia gracilis
Lateral buds: bud scales with
long-extended spreading tips

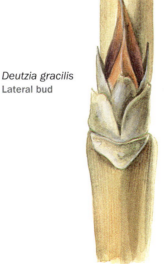

Deutzia gracilis
Lateral bud

Deutzia THUNB.

Visible buds with bud scales. **Twigs** mostly hollow or with loose pith, but solid at the nodes. Many hybrids, created by intensive breeding, cannot be distinguished in winter. Characters for identification are growth form and the size and shape of buds. **Fruits**: in panicles or racemes, globose 3–5-chambered capsules which disintegrate late.

Apart from *Deutzia crenata* (= *D. scabra* hort.) and *D. gracilis*, the wild species are not often planted. The more commonly planted hybrids combine the characters of their parents in different ways; in winter they cannot be identified with certainty.

Key to *Deutzia*

1 Buds without angles, bud scales often ± spreading at their tips. 2
1* Buds grey and 4-angular, bud scales appressed to bud . . . . *Deutzia parviflora*
2 Pith only in the nodes, brown; vigorous shrubs to 2 m high and more, fruits in narrow panicles with deciduous, short sepals (when sepals remain: hybrids). . . . . **Deutzia crenata**
2* Pith white also in the internodes (loose) 3
3 Shrubs less than 1 m high, fruits in narrow panicles, sepals short (to 1 mm) and deciduous (when persistent: hybrids) . . . . **Deutzia gracilis**
3* Fruits in broad, lax panicles, tips of sepals oblong (2–3 mm long), persistent . . . .*Deutzia discolor* (and others)

Deutzia crenata SIEBOLD & ZUCC.
[*Deutzia scabra* hort. non THUNB.]

Buds ovoid, 4–6 mm long, ochre-brown. Bud scales loose, long-acute and spreading at the tips. **Twigs** with stellate hairs, ochre-brown to orange red-brown, hollow, xylem brown. Bark flaking only slightly. **Fruits** in ± long racemose panicles, sepals short and deciduous. Stiffly erect shrub 2–3 m high. Origin: Japan, China.

This frequently planted species is often called *Deutzia scabra* THUNB., but this is another, very rarely cultivated species from Japan (cf. SCHULZ, in MDDG 2010). If sepals partly persist, the plant is *Deutzia ×magnifica* (LEMOINE) REHDER, a hybrid with *Deutzia discolor*.

Deutzia gracilis SIEBOLD & ZUCC.

Buds ovoid, 3–4 mm long, basal bud scales pale grey-brown, long-acute but often breaking off bluntly, slightly spreading from the bud; topmost bud scales ochre-brown to wine-red or red-brown with only a few (4–) 6–7 spreading hairs. **Twigs** weakly angular, slender, grey-brown to ochre-brown, with narrow longitudinal stripes. Very frequently planted, small and dense shrubs to 80 cm high. Origin: Japan.

Deutzia discolor HEMSL.

Buds ovoid, 3–4 mm long. Bud scales ochre, short-acute to dark brown and only slightly spreading or appressed. Bud scales with sparse stellate hairs, these hairs 8–12-rayed with broad base (appears as a dark dot at the centre of the hair). **Twigs** red-brown, initially with many stellate hairs, older twigs with strongly flaking dark brown bark, under-surface pale grey-brown. Pith loose, whitish to pale brown. **Leaf scars** narrow, triangular, with 3 indistinct traces. **Fruits** in loose panicles with persistent 2–3 mm long sepals. Shrub 1.5–2 m high with upright twigs hanging down at the tips. Origin: China.

Deutzia purpurascens (FRANCH. ex L. HENRY) REHDER is similar, a 1(–2) m high shrub with pendulous twigs. The stellate hairs of this species have mostly fewer than 8 rays. Origin: western China.

Deutzia parviflora BUNGE

Buds acute-ovoid, 3–5 mm long, distinctly 4-angular. Bud scales appressed, grey(!), only basal ones slightly brown, covered in many 6–9-stellate hairs. First year's growth **twigs** with dark, red-brown to violet-brown bark which fissures longitudinally and flakes; with few stellate hairs. **Pith** remnants at nodes and in the internodes: white. **Leaf scars** broadly triangular with 3, about equal-sized traces. Rarely planted shrub to c. 2 m high. Origin: China, Manchuria.

Jamesia americana TORR. & A. GRAY, cliffbush

Buds surrounded by two outer leaves, covered in long whitish hairs. Terminal buds 5–6 mm long, conical-cylindrical, with a broad base. Lateral buds smaller. **Twigs** ochre-brown to violet-brown to the tip, with ± dense pale indument, towards the base of the twig the darker, less hairy bark flakes off, showing a paler ochre-brown under-layer. **Leaf scars** narrow and wide, with 3 traces. Rarely planted, shrub to 1.5 m high. Origin: USA.

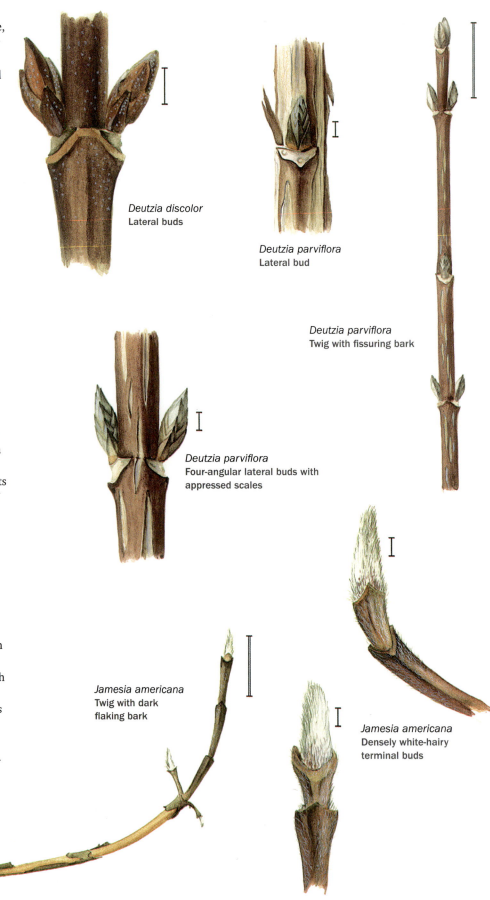

Deutzia discolor
Lateral buds

Deutzia parviflora
Lateral bud

Deutzia parviflora
Twig with fissuring bark

Deutzia parviflora
Four-angular lateral buds with appressed scales

Jamesia americana
Twig with dark flaking bark

Jamesia americana
Densely white-hairy terminal buds

ORDER Ericales

The deciduous woody plants in this order have leaf scars with one trace in common (leaf scars with a single trace are rarer than those with three traces): Ebenaceae, Theaceae, Symplocaceae, Styracaceae, Actinidiaceae, Clethraceae and Ericaceae.

Ebenaceae
persimmon family

Of this family, mainly from subtropical and tropical areas, only a few species of the genus **Diospyros** are planted in our climes. These are known for their wood (ebony) and fruit (kaki or persimmon).

Diospyros kaki
Twig with remnants of fruit stalks

Diospyros kaki
Tip of twig

Diospyros lotus
Lateral bud

Diospyros L., ebony, persimmon

About 500 mostly evergreen species, native to the tropics and subtropics. There are few deciduous plants, and these are rarely planted. Without terminal buds, lateral buds entirely enclosed by both prophyll scales. Fruits epigynous berries, the 4-merous calyx occasionally persistent for longer.

Key to *Diospyros*

1   Buds brownish, bluntly triangular; twigs ± hairy 2
1* Buds olive-brown, narrowly triangular, slightly acute; twigs becoming glabrous. . . . .*Diospyros lotus*
2   First year's growth twigs mostly less than 3 mm thick . . . .*Diospyros virginiana*
2* First year's growth twigs mostly thicker . . . . *Diospyros kaki*

Diospyros kaki THUNB., persimmon

Buds 4–5 mm high, bluntly triangular, surrounded by two brown bud scales with a narrow margin; loosely appressed-hairy like the twigs. The bud scales on the year's growth boundaries are long persistent. **Twigs** when young slightly angular and thickened at fruit insertion, ochre brown to grey-brown with many small raised lenticels. **Bark** with rough plates. **Fruits** shaped like tomatoes, orange-red, 5–8 cm. Tree 10–15 m high; frequently planted fruit tree in the Mediterranean region. Origin: China, Japan, South Korea.

Diospyros lotus L., European date plum

Buds acute-ovoid, 5–6 mm long, slightly glossy, olive-green at the base and dark violet-brown towards the tip; surrounded by two partly and slightly keeled bud scales with a white ciliate margin and sparsely hairy. **Twigs** olive-green to greenish-brown, becoming glabrous, glossy, 2–3 mm thick, with few round to oblong pale brown lenticels. **Bark** with small scales. **Fruits** smaller than in *Diospyros kaki* and blueish with waxy bloom. Tree 12–15 m high. Origin: western to eastern Asia, naturalised in the Mediterranean region.

Diospyros lotus
Tip of twig

Diospyros virginiana L.
American persimmon

Buds distichous, from above rounded-triangular, c. 3–4 mm high and wide, black-brown on the upper side, on the lower side olive-green to red-brown; surrounded by two bud scales, the first of these enveloping the second almost entirely. **Twigs** slightly zig-zag, slender to medium thick, from c. 2 mm diameter, due to dead epidermis pale grey on top, below this ochre to olive- or red-brown. Twigs finely spreading-hairy, becoming glabrous. **Lenticels** few, round to oblong, pale ochre to red-brown. **Fruits** in early winter partly persistent, 2–3.5 cm thick, globose, yellowish to orange berries, later only the fruit stalks remain. A tree, in its native area to 20 m high. Origin: eastern North America.

Theaceae
tea family

Only a few deciduous plants, rarely planted. Leaf scars with one trace. Terminal buds large.

Key to *Theaceae*

1 The first two bud scales of terminal buds dry, brown, contrasting with the subsequent green bud leaves . . *Stewartia*
1* The first bud leaves a little reddish, with dense indument; does not differ noticeably from the following . . . . *Franklinia*

Stewartia pseudocamellia MAXIM., deciduous camellia

Terminal bud oblong, 1–1.5 cm long, 4–5 mm broad and 3 mm thick; with distichously arranged **bud leaves** which are keeled and distinctly pinnate-nerved; the outer two bud leaves red-brown and sparsely hairy, covering the larger part of the bud, the inner ones olive-green with dense long white indument. Lateral buds alternate, small. **Twigs** initially brownish, later grey. **Bark** deciduous in plates. **Fruits** 5-angular, c. 2 cm long woody capsules. Rarely planted, erect shrub or small tree to 6(–15) m high. Origin: Japan.

Diospyros virginiana
Lateral bud

Diospyros virginiana
Twig with fruit and
fruit stalk scars

Stewartia pseudocamellia
Fruit

Stewartia pseudocamellia
Twig with large terminal bud

Franklinia alatamaha MARSHALL, Franklin tree

Buds between naked and covered (with bud scales). Bud leaves partly green, basally and on the light side also dark violet-red, towards the tip fine velvety-hairy. **Terminal buds** completely enclosed by 2 leaves, c. 15(–30) mm long, rarely also longer. **Lateral buds** mostly very small and inconspicuous. **Twigs** finely brown-hairy towards the tip, lower down with very fine lenticels (0.1× 0.3 mm), on older twigs larger and almost round. Bark of older twigs grey, fissured. **Leaf scars** ochre-brown, one-traced, looking like twig. Very rarely planted, tree to 10 m high. Origin: Georgia, but presumably extinct in the wild.

Franklinia alatamaha
Very long terminal bud

Franklinia alatamaha
Terminal bud

Symplocos paniculata
Part of twig

Symplocos paniculata
Tip of twig

Symplocaceae
sweetleaf family

Only one genus with some 100 species in the tropics and subtropics of America and Asia. Only a few deciduous species reach the temperate latitudes.

Symplocos paniculata (THUNB.) MIQ., sapphire berry

Buds only small, 1–1.5(–2) mm long lateral buds with (2–)6 outer, ochre-brown to red-brown bud scales, only the first two prophyll scales are darker grey-brown. **Twigs** grey-brownish to grey-greenish, ± angular, sparsely simply hairy with relatively large leaf cushions with ochre-brown margin and distinctly warty ochre-yellow lenticels. **Leaf scar** with one trace, mostly larger than the bud. Rarely planted shrub to 3 m high. Origin: northern China.

Styracaceae
storax family

The introduced species usually have descending accessory buds, at least in vigorous shoots. Their growing points are protected by some hardly differentiated leaves and this is the transition between buds with and without bud scales. The fruits which often persist to the winter are typical for the individual genera.

Key to *Styracaceae*

1 Twigs with terminal bud. 2
1* Only lateral buds present 4
2 Pith white, tearing open in places, fruits 4-winged . . . .***Halesia carolina***
2* Pith green, compact, fruits round and 5-winged or 10-angular 3
3 Fruit round in cross section, globose-ovoid, short-acute. . . . . *Sinojackia*
3* Fruits many in long, oblong panicles, with 10 fine ribs, densely long-hairy, more rarely slightly 5-winged. . . . .***Pterostyrax***
4 Buds densely felty-hairy, partly stalked, mostly with accessory buds, fruit round in cross section, superior . . . . ***Styrax***
4* Buds slightly hairy, lateral buds not noticeably stalked, fruit 2-winged, inferior . . . .*Halesia diptera*

Halesia J. ELLIS ex L., silverbell

Halesia in winter is marked by oblong, winged dry fruits. The fruits of the rarely planted *Halesia diptera* J. ELLIS have 2 wings, the fruits of the occasionally planted *Halesia carolina* have 4 wings. Origin: North America (several species) and China (1 sp.).

Halesia carolina L., Carolina silverbell, snowdrop tree
[*Halesia tetraptera* J. ELLIS]

Buds 5–6 mm long, acute, on the outside with 4–5 bud scales. The colour varies strongly: from predominantly green in the shade and on the lower side, to strongly wine-red in the sun and on the upper side. Mostly with small, descending accessory bud. **Twigs**, when young, deep red-brown and finely hairy. From the second year, the pale grey top layer of bark comes off in fibres and hangs down in shreds. Older **bark** typically scaly, breaking into small cubes. **Leaf scars** with one trace, the trace 'U' to 'V'-shaped. Older branches often distinctly fan-like in one plane. Frequently planted, shrub or small tree to 10 m high.

Similar, but upright and more vigorous, and always a tree, is the closely related mountain silverbell, **Halesia carolina** var. **monticola** REHDER [*Halesia monticola* (REHDER) SARG.].

Halesia diptera J. ELLIS, two-winged silverbell

Buds: only lateral buds, with two slightly hairy, slightly leafy, drying and deciduous prophylls. **Twigs** ochre to reddish brown, matte, relatively thickly hairy. **Pith** white, compact, 0.9–3.2 mm diameter. **Fruit** two-winged in one plane, thus flat, more than 5 cm long. Shrub to 5 m high. Origin: south-eastern USA.

Styrax japonicus SIEBOLD & ZUCC., Japanese storax

Buds slightly stalked, pale ochre-brown, with dense stellate hairs; with one to two descending accessory buds; apparently with a single scale, but surrounded by two outer bud leaves. **Twigs** zig-zag from bud to bud, slender, sparsely hairy, pale ochre-brown on top, slightly darker below; when several

Halesia carolina
Twig with fruits

Halesia carolina
Lateral buds

Styrax japonicus
Young twig

Styrax japonicus
Older twig, bark coming off in threads

Styrax japonicus
Lateral bud with descending accessory bud

Pterostyrax hispidus
Lateral bud (outer bud scales deciduous)

Pterostyrax hispidus
Tip of twig

Pterostyrax hispidus
Remnants of infructescence

Sinojackia xylocarpa
Tip of twig

Pterostyrax hispidus
Fruit: densely hairy stone fruit with long style

Sinojackia xylocarpa
Fruit

years old, rhombically fissuring and top bark coming off in threads. **Leaf scar** pale grey-brown with a dark black-brown trace of a vascular bundle. Frequent, 6–10 m high shrub to small tree. Origin: eastern Asia.

Pterostyrax hispidus SIEBOLD & ZUCC., fragrant epaulette tree

Buds oblong, slightly acute and slightly stalked; ± without bud scales, leaves distinctly pinnately veined; outer leaves (in the lateral buds, the two prophylls) dark grey to violet-brown, ± deciduous, leaving distinct scars, subsequent leaves greenish. Accessory buds almost always present. **Twigs** smooth when young, ochre to red-brown; two-year-old bark darker red-brown, fissuring longitudinally and coming off in strips. **Leaf scars** on distinct leaf cushions, with a horizontal to 'U'-shaped vascular bundle trace. In winter the species can be easily recognised by the oblong hairy fruits in long, hanging panicles. Shrub or medium-sized tree to 15 m high. Occasionally planted. Origin: eastern Asia.

Pterostyrax corymbosus SIEBOLD & ZUCC. with small, lax infructescences with 5-winged, only shortly hairy fruits, is rarely planted.

Sinojackia xylocarpa HU
[including *Sinojackia rehderiana* HU]

Buds without bud scales, basal hue greenish, covered in lax to dense brown tufted stellate hairs. Terminal bud c. 7 mm long, with rather dense indument. Lateral buds c. 6 mm long with smaller, descending accessory bud, often only sparsely hairy and green. **Twigs** slender, slightly zig-zag, initially ochre-brown to grey-brown, bark soon longitudinally fissured, turning grey and coming off in oblong shreds, under-surface smooth violet-brown. Pith greenish. **Fruits** are ovoid stone fruits, a few together in axillary panicles; they are to 2(–3) cm long, stalked and end in an attenuate point. Fruit stone woody, usually with a single seed. Shrub to 6 m high or small tree. Origin: eastern China.

Actinidiaceae
kiwifruit family

Actinidia LINDL., kiwifruit

Mostly dioecious climbing shrubs. Without terminal buds, **lateral buds** enclosed by thickened leaf base. On the lower half of the leaf base is the petiole scar, and above this the point where the bud comes through. A similar bud protection is found in the evergreen genus *Kalmia* in the family Ericaceae, belonging to the same order. **Pith** mostly chambered, more rarely solid. Origin: east and south-east Asia.

Key to *Actinidia*

1 Pith chambered 2
1* Pith solid . . . . *Actinidia polygama*
2 Pith mostly brown- or white-chambered, twigs glabrous 3
2* Twigs hairy, pith white-chambered. . . . . **Actinidia deliciosa**
3 Twigs thick, with many, pale, flat lenticels . . . . *Actinidia arguta*
3* Twigs slender, with warty lenticels . . . . *Actinidia kolomikta*

Actinidia deliciosa (A. CHEV.) C. F. LIANG & A. R. FERGUSON, kiwifruit

[*Actinidia chinensis* hort., non PLANCH.]

Buds concealed below globose-domed persistent leaf base, visible in longitudinal section; densely felty honey-brown hairy. **Twigs** grey-olive to grey-brown, when young densely red-brown bristly hairy, the female plants especially partly lose their indument, but usually remnants of this indument can be found. **Lenticels** few, pale and small. **Pith** chambered, only initially solid at the upper nodes, whitish to yellowish. Frequently planted climbing shrub to 8 m high. Origin China. As a fruit tree only suitable for warm-temperate habitats.

Actinidia arguta (SIEBOLD & ZUCC.) PLANCH. ex MIQ., hardy kiwi

Base of leaf covering the buds oblong and relatively flat. **Twigs** red-brown, slightly glossy and glabrous, with many pale, oblong small lenticels. **Pith** chambered, white to honey-brown. Frequently planted, liana to 7 m high and over 20 m wide. Selected strains of this frost-hardy species are used as fruit plants for small gardens. Origin: countries around the Sea of Japan.

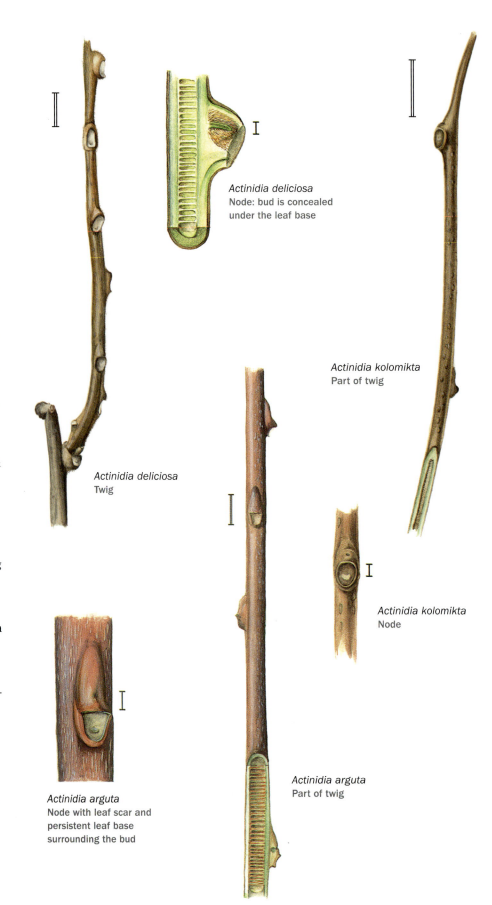

Actinidia deliciosa
Node: bud is concealed under the leaf base

Actinidia deliciosa
Twig

Actinidia kolomikta
Part of twig

Actinidia kolomikta
Node

Actinidia arguta
Node with leaf scar and persistent leaf base surrounding the bud

Actinidia arguta
Part of twig

Actinidia polygama
Part of twig

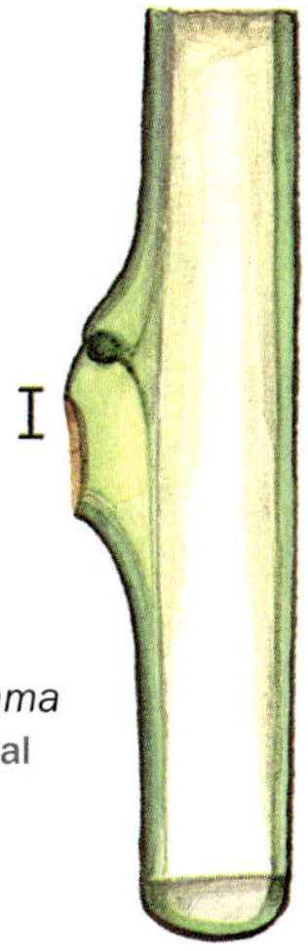

Actinidia polygama
Node, longitudinal
section

Actinidia kolomikta (RUPR. & MAXIM.) MAXIM., kolomikta

Twigs relatively slender: first year's growth less than 3(–4) mm thick, dark brown, glabrous, often only slightly climbing. **Lenticels** many, pale or dark, warty. **Pith** narrow, brown and chambered. Frequently planted. Erect shrub or, rarely, a liana climbing to 2(–7) m high. Origin: northern east Asia.

Actinidia polygama (SIEBOLD & ZUCC.) PLANCH. ex MAXIM, silver vine

Twigs brown and glabrous. **Pith** white, wide and solid(!). Liana winding to 5 m high. Only rarely planted. Origin: Western China, Manchuria, Korea, Sakhalin and Japan.

Clethraceae
lily-of-the-valley-tree family

Only one genus with about 70 species of which 4–5 are occasionally planted.

Clethra alnifolia L., sweet pepper bush

Buds without bud scales, oblong, 4–7 mm long, outer leaves narrow and irregularly spreading. Bud leaves hairy, initially reddish, later grey-brown. Lateral buds stalked, mostly at the end of an extended first internode. **Twigs** initially matte, pale grey-brown with fine stellate hairs and rounded-angular, later smooth, red brown to violet-brown and round. **Bark** tearing open in places and coming off in small flakes. **Leaf scars** small, relatively flat on the twig, with one trace. **Fruits** persist throughout the winter in terminal long narrow and ± upright racemes; they are small, dry, 3-valvate capsules which are surrounded by 5 free (not fused) sepals. Occasionally planted, erect shrub to 3 m high. Origin: North America.

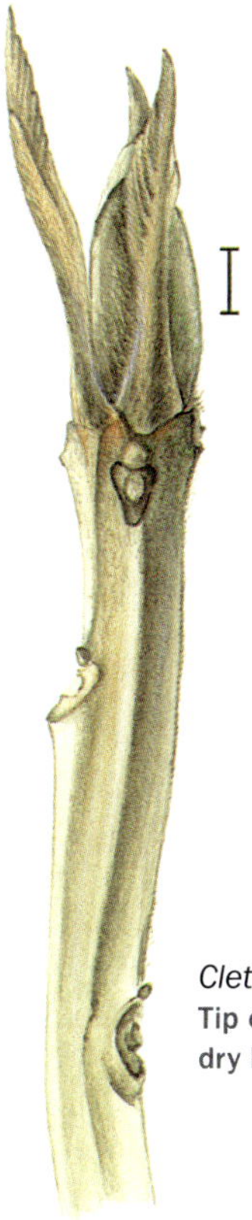

Clethra alnifolia
Tip of twig with
dry bud

Clethra alnifolia
Inflorescence:
an erect raceme

Clethra alnifolia
Terminal bud in
autumn

Clethra alnifolia
Single fruit

Ericaceae
heather family

A world-wide family with about 4000, mostly shrubby, species. Leaves mostly evergreen, more rarely deciduous. **Buds** usually in a spiral. **Leaf scars** of all species with a single trace.

Key to Ericaceae

1 Twigs without terminal buds 5
1* Twigs with terminal buds 2

Subfamily Arbutoideae

2 Prostrate dwarf shrub. . . . .
. . . .*Arctostaphylos alpina*
2* Erect shrubs 3
3 Buds glabrous or scales very long, rough-ciliate, scarcely more than 6 mm long . . 4

Subfamily Ericoideae

3* Buds often hairy, mostly over 7 mm long
. . . . **Rhododendron**
4 Buds dark, brownish. . . . .
. . . . *Rhododendron menziesii*

Subfamily Enkianthoideae

4* Buds greenish to reddish . . . .
. . . .***Enkianthus campanulatus***

Subfamily Vaccinioideae

5 Twigs orange-ochre, ± with waxy bloom, dead leaves long-persistent . . . .
. . . . *Zenobia pulverulenta*
5* Twigs not with waxy bloom 6
6 Leaf buds semi-orbicular, relatively flat. Twigs green to wine-red 7
6* Leaf buds longer than their diameter, twigs often edged, when round, brown-coloured 8
7 Inflorescences not initiated, fruits elongate . . . .*Oxydendrum arboreum*
7* Inflorescences already initiated, fruits orbicular . . . . *Eubotrys racemosa*
8 Fruits dehiscent or dried-up berries, buds, at least on the shaded side, green or completely brown . . . . **Vaccinium**
8* Fruits persistent capsules, buds all-round wine-red. 9
9 Buds with 2 scales, flat adjacent to twig, fruits orbicular 3.5 mm Ø, calyx lobes 1 mm long . . . . *Lyonia ligustrina*
9* Buds with several scales, patent, fruits 5-6 mm long, with 4-6 mm long calyx lobes . . . . *Lyonia mariana*

SUBFAMILY Enkianthoideae

Enkianthus campanulatus (MIQ.) G. NICHOLSON, redvein enkianthus

Buds 4–6(–8) mm long, compact, ovoid, acute. Bud scales pale ochre-brown, inner ones green, in the sun also reddish and ± outermost dry brown-glossy bud scales often coming off at the base and deciduous at the end of winter. **Twigs** ochre to orange-brown when young, later grey due to loosening epidermis, afterwards grey-brown. **Branching** with several lateral shoots developing below the tip of the shoot, which is usually used up by an inflorescence. **Infructescences**: long, dry racemes remaining on the plant. Frequently planted large shrub, in its native country tree to 10 m high. Origin: Japan.

SUBFAMILY Arbutoideae

Arctostaphylos alpina (L.) SPRENG., Alpine bearberry

Terminal buds present, often opening early. Dwarf shrub. Its decumbent twigs, to 50 cm long, ascend to 10 cm above ground. Origin: circumpolar.

SUBFAMILY Ericoideae

TRIBE Rhodoreae

Rhododendron L., azalea, rhododendron

The genus includes, depending on the author, several hundred to more than a thousand, predominantly evergreen species. The phyllotaxy on the shoots and of the bud scales in the terminal buds is spiral. The lowest bud scales of the terminal buds frequently show distinct similarities to foliage leaves. Two transverse simple prophyll scales often cover the mostly very much smaller lateral buds. Flower buds are mostly distinctly larger and thicker than leaf buds.

Two-year-old shoots often end in an infructescence. Below this, apparent whorls appear in dense proximity; the many first year's growth twigs then overtop the old tip of the shoot. These twigs form either from existing lateral buds below the terminal bud (subgenus *Hymenanthes*) or from the lowest axils of scale leaves on the terminal bud (subgenus *Azaleastrum*).

Origin: east and south Asia, Caucasus, up into the high mountains and arctic areas of Europe and North America.

Enkianthus campanulatus
Branching

Enkianthus campanulatus
Terminal bud in spring;
outer scales deciduous;
reddish on the sun side

The deciduous species flower long before the leaves (*Rhododendron mucronulatum*), shortly before or simultaneously with the leaves (*R. schlippenbachii, R. albrechtii, R. canadense, R. vaseyi, R. molle* and *R. luteum*) or after the leaves (*R. arborescens, R. calendulaceum* and *R. viscosum*). The wild species can mostly be identified fairly easily. Due to the long period in cultivation there are many hybrid groups which are difficult to tell apart. The main characters for identification are well formed terminal buds (particularly flowering buds) and twigs (colour and indument).

Key to *Rhododendron*

1   Twigs and buds without scurfy scales. . 2
1* (1) Twigs and buds with scurfy scales . . . . **Rhododendron mucronulatum**
2   Buds distinctly hairy, occasionally as if with waxy bloom 3
2* Top-most bud scales glabrous, only the margins finely ciliate 7
3   Buds densely long-hairy 5
3* Bud scales very finely appressed white-hairy (resembling waxy bloom), leaf scars slightly raised 4
4   Buds dirty pink to brown, bud scales with attenuate tip. . . *Rhododendron canadense*
4* Bud scales greenish to dark wine-red, the tip emarginate; hairy, resembling waxy bloom . . . . *Rhododendron vaseyi* and *Rhododendron albrechtii*

5 (3) Bud scales mostly with remnants of the lamina, low shrub. . . . . *Rhododendron yedoense*
5* Bud scales rarely with lamina, mostly taller shrubs 6
6   Buds ochre brown . . . . **Rhododendron schlippenbachii**
6* Buds red-brown to greenish grey-brown . . . . *Rhododendron reticulatum*
7 (2) Flower buds to 15 mm long, at least partly reddish. 8
7* Flower buds green with a brownish hue, longer than 15 mm. . . . . **Rhododendron luteum**
8   Twigs glabrous 9
8* Twigs with indument . . . . *Rhododendron calendulaceum*
9   Stalks of infructescences with glandular hairs . . . . **Rhododendron molle** subsp. **japonicum**
9* Infructescence stalks glabrous . . . . *Rhododendron arborescens*

Subgenus Rhododendron

The species of this predominantly evergreen subgenus have, in contrast to azaleas, scurfy scales. A few deciduous, very early-flowering species.

Rhododendron mucronulatum TURCZ.

Buds ovoid acute, 6–8 mm long; bud scales green at the base, but predominantly covered by reddish to orange-brown small, round scurfy scales. First year's growth **twigs** at the base orange-brown, slightly scaly below the tip, pale ochre to matte brown, also slightly violet. **Flowers** very early, often before March, solitary from buds aggregated at the end of shoots: 3 cm broad, funnel-shaped, purple-pink with 10 anthers. Densely branched shrub with short-shoots. Origin: north-east Asia.

Rhododendron dauricum L. is more frequent and very similar, mostly retaining some leaves; also the mostly evergreen garden hybrid with *Rhododendron ciliatum* HOOK. f.: **Rhododendron ×praecox** DAVIS.

Rhododendron mucronulatum
Tip of twig

Rhododendron mucronulatum
Flower in March

Enkianthus campanulatus
Terminal bud in autumn.
Shaded side mainly green

Subgenus Hymenanthes

Section Pentanthera

The leafy shoots derive from buds initiated last year, closely below the terminal bud.

Rhododendron canadense (L.) TORR., rhodora

Buds relatively small, 6–8 mm long, ovoid, narrowing to and ending in a slightly rounded point. Bud scales 8–10, basic colour ochre, covered with a pink-wine-red hue. Toward the tip more violet, very finely hairy (almost appearing with waxy bloom) and densely short-ciliate. **Twigs** ascending, slender, below the bud slightly thicker; here and below the dark orange-brown leaf scars ochre-brown, further below pale grey-brown; finely furrowed, glabrous, only initially hairy and partly with waxy bloom. When two-years-old, with some long-fissured grey remnants of epidermis, underneath these smooth red-brown to violet-brown. **Fruits** oblong, 10–15 mm long, initially pink to wine-red, with waxy bloom, with short, spreading, partly glandular hairs and on a 3–6 mm long stalk with glandular hairs. Rarely planted, shrub 30–70 cm high. Origin: north-eastern North America.

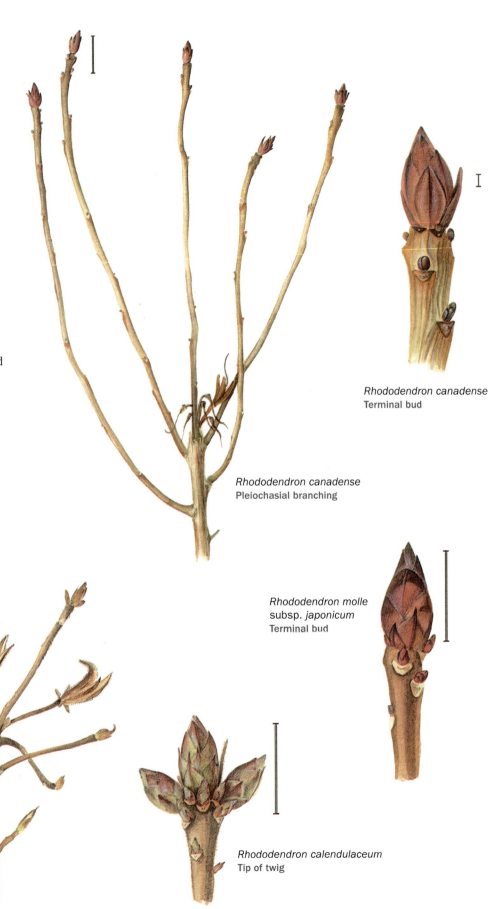

Rhododendron canadense
Pleiochasial branching

Rhododendron canadense
Terminal bud

Rhododendron molle
subsp. *japonicum*
Terminal bud

Rhododendron calendulaceum
Branching derived from buds below last year's terminal inflorescence bud

Rhododendron calendulaceum
Tip of twig

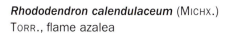

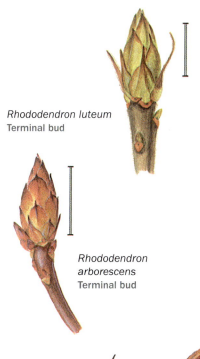

Rhododendron luteum
**Twig with remnants
of infructescence and
pleiochasial shoots**

Rhododendron luteum
Terminal bud

*Rhododendron
arborescens*
Terminal bud

Rhododendron viscosum
**Lateral buds below the
infructescence**

Rhododendron calendulaceum (MICHX.) TORR., flame azalea

Buds: 12–14 bud scales, similar to *Rhododendron arborescens*, but **twigs** long-hairy (not glandular), red-brown, darker grey-brown below the tip, spreading relatively wide (almost horizontal). Rarely planted, strongly branching shrub 1–1.5 m high. Origin: eastern USA.

Rhododendron molle (BLUME) G. DON, Chinese azalea

Twigs relatively upright, shaggy bristly-hairy when young. Many golden yellow, funnel-shaped flowers, appearing in May before the leaves. Very rarely cultivated in the pure form. Origin: mountains of central and eastern China.

Contributor to several hybrid groups: especially with American species the 'Knap Hill' cultivars; and 'Occidentale' cultivars with the long-budded American azalea (*Rhododendron occidentale* [TORR. & A. GRAY] A. GRAY), and 'Mollis' cultivars with the Japanese azalea (subsp. *japonicum*).

subsp. *japonicum* (A. GRAY) KRON, Japanese azalea

[*Rhododendron japonicum* (A. GRAY) SURINGAR]

Buds ovoid, acute, 10–12(–14) mm long, bud scales white-ciliate, reddish on the light side with darker violet-brown margin, lighter on the shaded side and greenish at the base. **Twigs** thick, ochre-brown and slightly glossy, glabrous to slightly bristly. **Leaf scars** with relatively broad vascular trace. **Flowers** before the leaves, in tufts of 6–10, with 6–8 cm broad funnel-shaped corollas: salmon-pink to orange-coloured, with large orange spot. Frequently in cultivation and involved in many crosses, shrub 1–2 m high. Origin: north and central Japan.

'Rustica' cultivars, hybrids with 'Gandavensis' cultivars, differ from this subspecies by more compact growth and slightly earlier flowering.

Rhododendron luteum SWEET, yellow azalea

Buds with 12–14 bud scales, the lowest ones long-acute with spreading indument, from the 4th bud scale onwards only ciliate. Flower buds 16–20 mm, leaf buds 8–10 mm long. **Twigs** initially with glandular-sticky indument. Frequently planted, wide and dense shrub 1–4 m high. Origin: eastern Europe to Caucasus and to the Black Sea. A species often employed in hybridisation.

Hybrids with the American species (*Rhododendron calendulaceum, R. viscosum* and others) are termed 'Gandavensis' cultivars (Ghent hybrids). They are very variable, the bud shape resembles, according to cultivars, various parent species, but seems to be typical for the individual cultivars.

Rhododendron arborescens (PURSH) TORR., smooth azalea

Buds 10–12 mm long, ovoid-acute, with 12–16 spiral slightly keeled bud scales, the lowest ones narrowly acute, the upper ones more rounded with a short apiculus; greenish at the base, red and dark violet-brown towards the margin; on the light side the darker hues predominate; glabrous, only the margins finely white-ciliate. **Twigs** slender, orange to red-brown, glabrous, sometimes with a slight waxy bloom; when two-years-old grey-brown, longitudinally fissuring. Flowering after the leaves. Rarely planted, erect irregularly branching shrub 3–4 m high. Origin: mountains of the eastern USA.

Rhododendron viscosum (L.) TORR., swamp azalea

Buds to 13(–15) mm long, fusiform and acute. Bud scales acute, densely white-ciliate, otherwise glabrous except the lowest ones; reddish on the sun side, yellowish-green on the shaded side. First year's growth **twigs** slender, orange-red-brown to dark brown and appressed rough-hairy, older twigs grey-brown with fissuring bark. **Fruits** in short racemose clusters at the end of the shoot: on stalks of c. 1 cm long, densely hairy and 1–3 cm long capsules with 4 cm long style remnants. A shrub 1–2 m high. Origin: swamps of eastern North America.

Subgenus Azaleastrum

New shoots emerge from the axils of the bud scales of terminal buds.

Section Tsutsusi

Rhododendron yedoense var. *poukhanense*
(H. LÉV.) NAKAI, Korean azalea
[*Rhododendron poukhanense* H. LÉV.]

Buds surrounded by 5–6 leaves/bud scales, outwardly densely long-hairy and sticky on the inside. Outer leaves of terminal buds often with developed lamina, brownish to grey-green, partly reddish on the light side, in the shade paler green, mainly on the underside of the lamina. About 1 cm long blade with recurved tips to 3 cm long. **Twigs** slender; first year's growth red to orange-brown, towards the tip darker brown with loosely appressed fine indument; two-years-old glabrous, grey-brown. **Flowers** before or with the leaves, 1–3 together, 4 cm broad, purple-lilac, dark spotted inside. Rarely planted, densely branching shrub, c. 1 m high. Origin: Japan, Korea.

In the occasionally planted Japanese garden form var. *yedoense*, the flowers are double and without stamens, so that no fruits are formed.

Section Sciadorhodion

Rhododendron reticulatum D. DON ex. G. DON

Buds with 10–12 outer bud scales, narrowly ovoid, with fine appressed indument, mainly green to pale brown at base, otherwise mostly orange-brown, sometimes darker towards the tip: grey, brown to wine-red. **Twigs** orange-brown when young, sparsely to densely hairy, becoming glabrous; two-year-old twigs fine rhomboidal, grey-brown. **Flowers** 1–2(–4), before the leaves, April to May. Rare, shrub c. 1–5 m high. Origin: Japan.

Rhododendron schlippenbachii MAXIM.
Schlippenbach's azalea

Buds 10–12(–15) mm long, narrowly ovoid with 7–9 bud scales: basally sometimes becoming glabrous, greenish, otherwise very dense and fine-hairy, yellow-ochre to honey-brown, towards the slightly spreading tips a little pale grey. **Twigs**

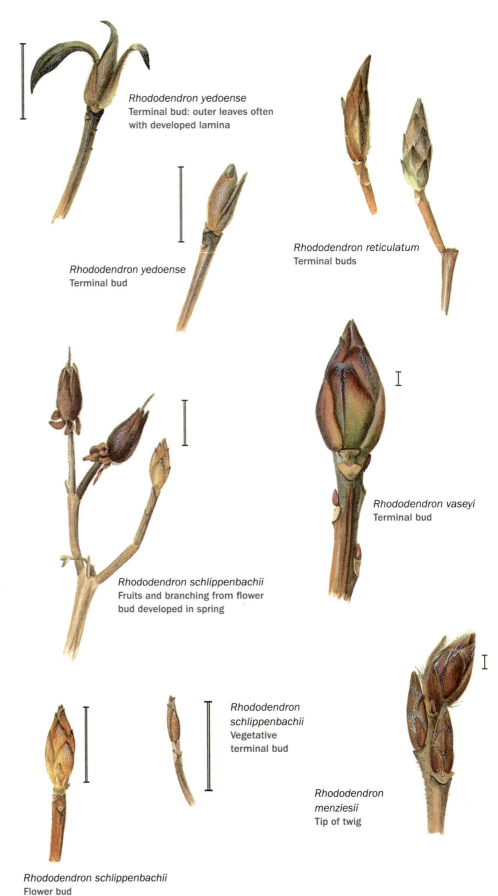

Rhododendron yedoense
Terminal bud: outer leaves often with developed lamina

Rhododendron yedoense
Terminal bud

Rhododendron reticulatum
Terminal buds

Rhododendron vaseyi
Terminal bud

Rhododendron schlippenbachii
Fruits and branching from flower bud developed in spring

Rhododendron schlippenbachii
Vegetative terminal bud

Rhododendron menziesii
Tip of twig

Rhododendron schlippenbachii
Flower bud

brown, initially with sparse glandular indument, at two-years-old glabrous, grey, longitudinally fissured. **Fruits** dark red-brown, with 5 slits, compact to 1.5 cm long capsules with dry sepals and a stalk to 1.5 cm long with densely glandular indument. **Flowers** with or shortly before the leaves. Shrub, branching irregularly and divaricate, to 2(–4) m high. Frequent. Origin: Central Japan, Korea, north-east Manchuria.

Rhododendron vaseyi A. Gray, pink-shell azalea

Buds acute-ovoid with c. 8 outer bud scales, greenish at the base, above dark wine-red to violet-brown, at the upper margin distinctly crenate, with very fine white hairs resembling a white waxy bloom cover. **Twigs** glabrous or finely hairy when young, orange red-brown to pale brown with grey expanses of dead epidermis, two-years-old grey-brown, fissuring longitudinally. **Flowers** 5–8, before the leaves in April to May, with deeply 5-lobed pale rose-red maculate, 3 cm broad corolla and mostly with 7 stamens. Frequent, shrub to 2 m high. Origin: North America.

Rhododendron albrechtii Maxim., a more rarely planted species is very similar and in winter not easily distinguished. Origin: forests of north and central Japan.

Rhododendron menziesii Craven, rusty leaf

[*Menziesia ferruginea* Sm.]

Buds differentiated into terminal 6–7 mm long and 4–5 mm thick flower buds and, below the tip of the twig, 5–6 mm long and only c. 2–3 mm thick foliage buds. **Bud scales** spiral, basally violet-red to ochre brown, toward the margin darker brown, distinctly long-ciliate and on the central vein with long, bristly hairs. **Twigs** slender, less than 2 mm thick, matte ochre to red-brown, densely covered by grey-black glandular hairs. The fruits, 3–7 mm long, 4–5-valvate capsules are grouped in terminal corymbs. Shrub to 1.5(–2) m high. Origin: North America.

SUBFAMILY Vaccinioideae

TRIBE Oxydendreae

Oxydendrum arboreum (L.) DC., sorrel tree

Buds shallowly semi-globose, c. 2 mm broad, covered for $^2/_3$ by the first two bud scales. Bud scales glabrous, margins sometimes ciliate, wine-red to olive-brown. **Twigs** glabrous, in the sun strongly wine-red, only in the shadow greenish,

with few small grey-brown lenticels. Older growth twigs red-brown, fissuring longitudinally, bark rusty-brown, when old deeply furrowed. **Leaf scars** half-round to triangular, with a central trace, mostly slightly wider than the buds. Rarely planted, shrub up to 5 m. In its native area a tree to 25 m high. Origin: North America.

TRIBE Gaultherieae

Eubotrys racemosa (L.) Nutt.

[*Leucothoë racemosa* (L.) A. Gray]

Flower buds overwinter glabrous, in 5-15 cm long racemes; each one in the axil of a lanceolate bract, which is often persistent at the tip of the inflorescence. Short, stalked c. 1 mm long, and to 3 mm long, wine-red on the lit, green on the shaded side. **Foliar buds** semi orbicular 1–1.5 mm long, with few brown bud scales. **Twigs** glabrous, initially wine-red to green, grey-brown when older. Leaf scars with one trace, elongate. At the tips of twigs often with racemose remnants of the infructescences. **Fruit** a globose capsule (c. 2.5 mm diameter) surrounded by calyx lobes and crowned by the remnant of the style. Rarely planted, shrub to 2 m high. Origin: eastern USA.

TRIBE Andromedeae

Zenobia pulverulenta (W. Bartram ex Willd.) Pollard, dusty zenobia

Buds: only lateral buds, in the lower half of the twig small and flat, almost completely covered by two brown scales. In Germany bud expansion begins from November up to the tip of the twig, the buds to 3 mm long, ovoid, spreading from twig, multi-scaled; bud scales here entirely (lower ones) or only at the margin (top ones) ochre-brown, deeper violet-wine-red. Glabrous and not ciliate. **Twigs** ochre yellow to reddish-orange-brown, partly blueish-white with waxy bloom. **Pith** solid, white to pale green. Remnants of the tufted infructescences persist during winter, with 5-valvate capsules. Leaves long persistent, half evergreen. Shrub c. 1 m high. Origin: USA — Virginia, north and south Carolina.

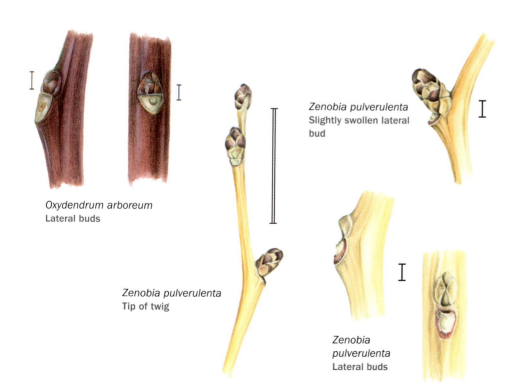

Oxydendrum arboreum
Lateral buds

Zenobia pulverulenta
Tip of twig

Zenobia pulverulenta
Slightly swollen lateral bud

Zenobia pulverulenta
Lateral buds

TRIBE Vaccinieae

Vaccinium L.

No terminal buds: sympodial branching. Lateral buds surrounded by two transverse bud scales (*Vaccinium myrtillus*) or also by the next simple scales, going from opposite to distichous phyllotaxy (*Vaccinium corymbosum*). Evergreen or deciduous shrubs and half-shrubs.

Key to *Vaccinium*

1   Small shrubs, rarely exceeding 50 cm in height 2
1* Larger, 1–2(–4) m high shrub . . . .
. . . . ***Vaccinium corymbosum***
2   Twigs ± round, grey-brown . . . .
. . . . ***Vaccinium uliginosum***
2* Twigs with ridges, green . . . .
. . . . ***Vaccinium myrtillus***

Vaccinium corymbosum L., blueberry

Buds distichous, acute-ovoid, 5–6 mm long; acute, wine-red bud scales on the sun side, in the shade they are of a paler pink, partly with narrow brown margin and especially the lowest bud scales slightly keeled. **Twigs** ± round, with fine pale dots, predominantly wine-red, greenish on the shaded side, with fine grey hairs descending in ridges. Often used fruit shrub. Origin: North America.

Vaccinium myrtillus L., bilberry

Buds spiral due to torsion of the twigs, sometimes appearing to be distichous; 2–4 mm long, flat-appressed to twig, narrowly triangular with rounded tip; mostly completely enveloped by the two pale greenish-yellow prophyll scales which sometimes have a reddish hue. **Twigs** green and strongly angular: from both sides of the leaf scar a ridge descends, the first merges with the opposite side of the leaf scar below, the second ends blind between the nodes. Mainly below the tip of the shoot between the ridges very deeply furrowed. A dwarf shrub which forms runners. Origin: Europe to the Caucasus and northern Asia, also in western North America.

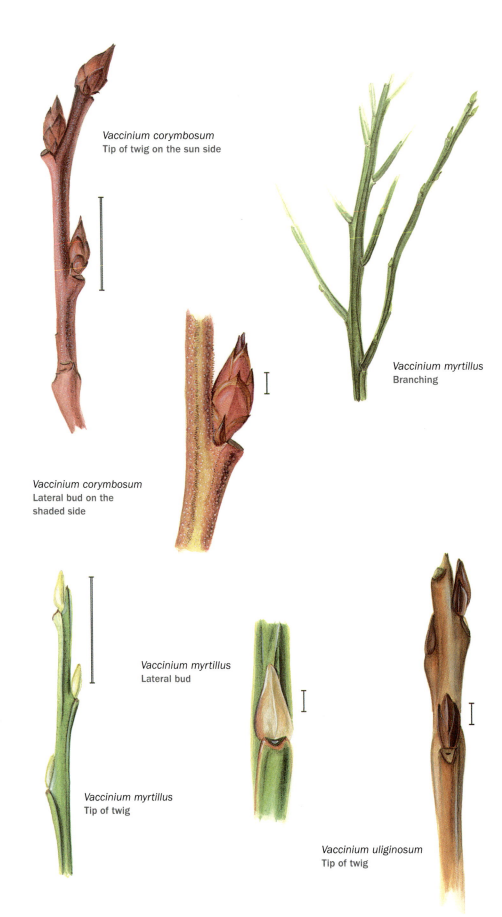

Vaccinium corymbosum
Tip of twig on the sun side

Vaccinium corymbosum
Lateral bud on the
shaded side

Vaccinium myrtillus
Branching

Vaccinium myrtillus
Tip of twig

Vaccinium myrtillus
Lateral bud

Vaccinium uliginosum
Tip of twig

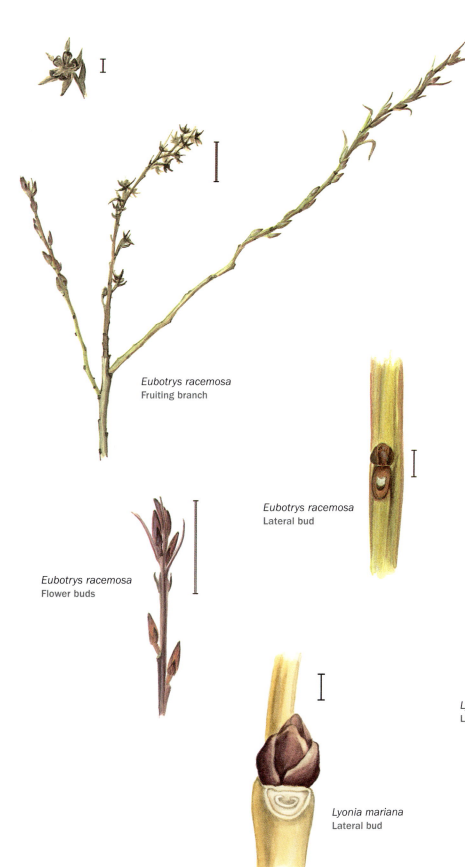

Eubotrys racemosa
Fruiting branch

Eubotrys racemosa
Flower buds

Eubotrys racemosa
Lateral bud

Lyonia mariana
Lateral bud

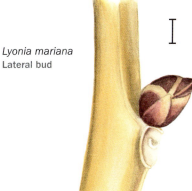

Lyonia mariana
Lateral bud

Vaccinium uliginosum L., bog bilberry

Buds 2–3 mm long, violet-brown to (wine-)red-brown, surrounded by 4 outer, distinctly keeled bud scales. **Twigs** round, grey-brown to orange-brown, slender, ± glossy, very finely hairy, becoming glabrous. Frequent shrub to 90 cm high. Origin: circumpolar.

TRIBE Lyonieae

Lyonia ligustrina (L.) DC., he-huckleberry, male berry

Lateral buds elongate, 4-5 mm long, adjacent to twig, completely enclosed by 2 wine-red bud scales. **Twigs** with brownish remainders of indumenta, leaf scars triangular with one trace. **Infructescences** short, to 5 cm long racemes, fruits orbicular, c. 3.5 mm Ø, stalked c. 3.5 mm long, bushily addorsed. Calyx lobes short, c. 1 mm long. Rarely planted, shrub to 4 m high. Origin: eastern North America.

Lyonia mariana (L.) D. Don., stagger bush

Buds distinctly wine-red, c. 5 mm long, squat-ovoid, acute; scales with a brown margin, glabrous but for the ciliate margin. **Twigs** ochre yellow, glabrous. **Leaf scars** dark brown, with one trace at the upper rim. **Pith** solid, dark green. Rarely planted, shrub to 2 m high. Origin: eastern North America.

Eucommiaceae
Chinese rubbertree family

Eucommia ulmoides OLIV.
Chinese rubbertree, gutta-percha tree

Buds acute-ovoid, 3–4 mm long and 2–2.5 mm thick, with 6–8 bud scales. The two opposite prophyll scales are followed by a few scales which transition to spiral placing. The lower bud scales dark brown and the upper ochre-brown, the lower ones mainly ciliate and sparsely white-hairy. **Twigs** grey-brown, moderately glossy, slightly long-fissured even when young. Lenticels few, pale ochre-brown. **Pith** chambered. **Leaf scars** on ± developed leaf cushions with a central semi-circular trace. When the bark is removed, fine gutta-percha threads (a sort of rigid rubber) and a dark inside become visible. Rarely planted tree 10 m high, in its native area to 20 m high. Origin: China.

Rubiaceae
coffee family

Cephalanthus occidentalis L.,
button bush

Buds opposite to 4-whorled, usually three lateral buds at a node. These are very small with few bud scales. As they are slightly embedded in the bark tissue, only a tiny bump is visible above the leaf scars. **Twigs** glabrous, thick, with wide, pale brown pith. Bark of twig grey-brown, with slightly glossy, grey epidermis fissured lengthwise. Lenticels few, large, narrowly rhomboid, very pale ochre. **Leaf scars** relatively large, trace circular, open above and rolled inwards. The leaf scars of a node are connected by narrow scars of fused stipules, which occasionally persist. Rarely planted, shrub to 2 m high. Origin: North America.

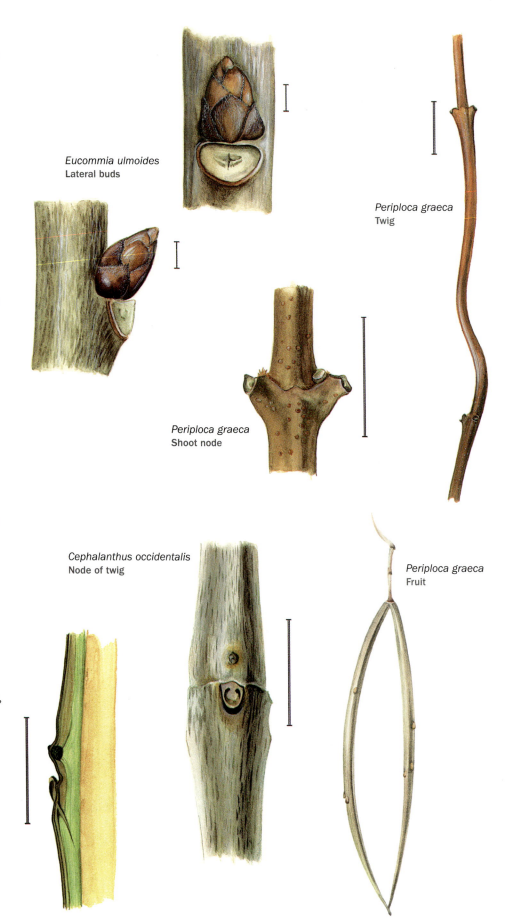

Eucommia ulmoides
Lateral buds

Periploca graeca
Twig

Periploca graeca
Shoot node

Cephalanthus occidentalis
Node of twig

Periploca graeca
Fruit

Cephalanthus occidentalis
Longitudinal section through lateral bud and leaf scar

Apocynaceae
periwinkle family

SUBFAMILY Asclepiadoideae

Periploca graeca L., silk vine

Buds opposite, very small and inconspicuous, sunk in the axils of the persistent leaf base of the leaves; with thick, ochre-brown indument. **Twigs** sturdy, 3–4 mm thick, glabrous, olive-brown to orange-brown, with many small, paler lenticels. **Pith** tearing apart, oblong, fibrous and twigs soon hollow. **Leaf scars** on protruding leaf cushion; dark grey with a pale, narrow semi-circular trace. Opposite leaf cushions connected by a line. The **fruit** consists of two oblong, separate follicles which often stay connected at the tip. Left-winding (counterclockwise) liana, to 15 m high. Origin: south-eastern Europe.

Oleaceae
olive family

Small and large shrubs to large trees. Native to the tropics and the temperate zone. Phyllotaxy rarely alternate, mostly opposite, but opposite leaf scars not connected to one another and often becoming sub-opposite. **Leaf scars** occasionally with a compound trace, composed of several small traces.

Key to Oleaceae

1 Pith solid 2
1* Pith hollow or chambered, flowers very early 3

Tribe Fontanesieae

2 Twigs very slender, first year's growth to 2 mm thick, without terminal bud . . . . *Fontanesia*[1]
2* First year's growth twigs thicker and/or with terminal bud. 4

Tribe Forsythieae

3 (1) Flowers yellow from buds covered by scales . . . . *Forsythia*
3* Flower buds without bud scales, surrounded only by the purple sepals. . . . . *Abeliophyllum*

Tribe Jasmineae

4 Twigs pure green and angular, shrubs, opposite or alternate . . . .*Jasminum*
4* Twigs rounded or grey to red-brown, shrubs and trees, opposite 5

Tribe Oleeae

5 Bud scales with very dense felty or scaly indument, buds at most with 2 outer pairs of scales, mostly trees . . . *Fraxinus*

5* Bud scales glabrous or only ciliate, when more hairy, with more than 2 outer bud scale-pairs; shrubs, more rarely small trees 6
6 Twigs slender 8
6* Twigs sturdy, when slender, mostly without terminal bud 7
7 Always with terminal bud, this is to 3 mm long and less than 1.5 times as high as broad . . . . *Chionanthus*
7* Without terminal buds or these 1.5 times as long or longer than broad and often longer than 3 mm. . . . . *Syringa*
8 Terminal buds 2–3 mm long, with 5 or more bud scale-pairs, coloured yellow-ochre . . . . *Forestiera*
8* Buds usually slightly larger, with fewer bud scales. 9
9 Fruits dry capsules . . . . *Syringa*
9* Fruits black, berry-like stone-fruits. . . . . *Ligustrum*

[1]Compare with *Ligustrum* and *Syringa*

TRIBE Fontanesieae

Fontanesia phillyreoides LABILL., Syrian privet
[**Fontanesia angustifolia** DIPPEL]

Buds small and compact: to 2 mm high and broad, with two pairs of outer bud scales. **Twigs** slender, initially ochre-orange-red-brown to grey, later entirely grey, mostly drying up towards the tip. **Leaf scars** on relatively large cushions: small, dark with a trace composed of several smaller traces. Rarely planted, densely branching shrub 1.5–2 m high. Origin: south-eastern Sicilia, Anatolia and Syria.

Fontanesia phillyreoides
Twig

Fontanesia phillyreoides
Lateral buds

TRIBE Forsythieae

Abeliophyllum distichum NAKAI, white forsythia

Buds 1–2 mm long, lateral buds spreading from the twig, short-ovoid and acute, surrounded by few bud scales. Occasionally with small descending accessory buds. **Flower buds** of the early appearing flowers (March to April) in terminal globose-ovoid racemes, only enclosed by the dark wine-red calyx lobes. **Twigs** 4-angular, pale brown to pale grey-brown, finely longitudinally striped. Rachis of inflorescences wine-red like the flower buds. **Pith** initially chambered, later hollow. **Leaf scars** rounded-off triangular, with a central trace. **Flowers** from March, white, funnel-shaped, with 3–4 mm basal tube and with four 8–10 mm long lobes. Frequent shrub, divaricate, to 1.5 m high. Origin: Korea.

Forsythia VAHL f., forsythia

Buds: the lateral buds often with descending serial accessory buds and/ or lateral enrichment buds in the axils of prophylls. These serve mainly in the formation of flowers. **Twigs** hollow or with chambered pith. **Leaf scars** on strong leaf cushions and, as in most genera of the family, with only one trace of a vascular bundle. Flowers yellow, appearing before the leaves. Very frequently planted shrubs, few species. Origin: eastern Asia and south-eastern Europe.

Key to *Forsythia*

1 Twigs with solid pith at the nodes, internodes hollow or pith partly chambered 2
1* Pith of twig chambered in the nodes and in the internodes 3
2 Twigs between nodes completely hollow, pendulous. . . . . **Forsythia suspensa**
2* Pith between nodes slightly chambered, twigs ± upright **Forsythia ×intermedia**
3 Individual buds mostly over 7 mm long, twigs mostly distinctly 4-angular . . . . **Forsythia viridissima**
3* Lateral buds mostly less than 7 mm long, twigs angular only at the tip. . . . . *Forsythia europaea* and *Forsythia ovata*

Abeliophyllum distichum
Flowering twig

Forsythia ×intermedia
Twig

Abeliophyllum distichum
Lateral bud

Abeliophyllum distichum
Lateral buds

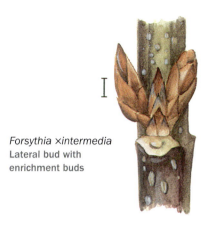

Forsythia ×intermedia
**Lateral bud with
enrichment buds**

Section Forsythia × Giraldianae

Twigs intermediate between the sections:
Pith solid at the nodes, hollow in the
middle of the internodes and chambered in
between.

Forsythia ×intermedia ZABEL,
border forsythia
[*Forsythia suspensa* × *Forsythia viridissima*?]

Buds with opposite, ± strongly keeled,
ochre-brown bud scales. Lateral buds
narrowly ovoid, acute, 6–10 mm long.
Mostly with descending accessory buds
and enrichment buds in the axils of the
lowermost bud scales. Because of this there
are often several buds in dense groups
above the leaf scar. **Twigs** 4-angular, due to
ridges running downward from the margins
of leaf scars, ochre yellow and olive green
to violet, with few small to many large and
warty grey-white to ochre-brown lenticels.
Flowers in April; deep yellow, corolla
basally fused into a short cup and ending
in four long lobes. Upright and spreading
shrub, 2–3 m high. Most of the cultivated
Forsythia cultivars are in this group.

Section Forsythia

Twigs hollow in the internodes, with solid
pith at the nodes.

Forsythia suspensa (THUNB.) VAHL,
golden bell

Buds fusiform, bud scales 6–8 mm long,
pale brown with a slightly darker margin.
Twigs pale brown to ochre, slightly angular,
hollow in the internodes, with solid pith
at the nodes. **Lenticels** many, grey-brown.
Flowers in the first half of April with,
depending on the form, campanulate or
spreading corolla. Shrub 2.5–3 m high with
arching or pendulous branches. Origin:
China and Japan. After the border forsythia,
this is the most commonly cultivated species.

Forsythia suspensa
**Pendulous twig with flowers
and last year's fruit capsules**

Forsythia suspensa
Twig

Section Giraldianae

Pith of twig chambered at the nodes and in the internodes.

Forsythia viridissima LINDL.

Buds 7–9 mm long, fusiform, at the base slender and stalk-like. Lowest bud scales ochre brown to red-brown, the next ones basally and the upper ones entirely dark wine-red to violet-brown. **Twigs** on the shaded side olive greenish, on the light side dark red-brown; filled entirely with chambered pith; only at the base of vigorous long-shoots occasionally hollow; growing strongly erect. **Flowers** arranged in 1–3s with narrow reflexed corolla lobes; flowering, as last forsythia, in the 2nd half of April. Occasionally planted erect shrub to 2 m high. Origin: China.

The other, only rarely planted species of the section are difficult to tell apart: e.g. the Albanian forsythia **Forsythia europaea** DEGEN & BALD., from Albania and the early-flowering Korean forsythia, **Forsythia ovata** NAKAI, two slightly more dainty species with buds to 7 mm long.

TRIBE Jasmineae

Fruit divided in two halves by constriction of the tip, seeds erect in the ovary chambers.

Jasminum L., jasmine

Buds opposite or alternate. No terminal buds. A genus mostly from the tropics and subtropics, with many evergreen species. Few deciduous species cultivated:

Jasminum nudiflorum LINDL., winter jasmine

Buds: opposite lateral buds differentiated into flower buds and foliage buds. Leaf buds small, c. 3–5 mm long with few bud scales, flower buds narrowly ovoid to c. 1 cm long, surrounded by several pairs of scales. Bud scales (especially lower ones) acute, green, partly reddish, at the tips and margins dry and brown. **Twigs** 3–4 mm diameter, glabrous, 4-angular, dark green, finely white-spotted. **Pith** white, rounded-rectangular, loose and porous. **Leaf cushions** with orange-brown border, grey, with a shrivelled dried-up scar with one trace. **Flowers** in mild weather, from December to April,

yellow, funnel-shaped with a basal tube and with (5–)6 lobes. Frequently planted, low shrub, pendulous or on a trellis to 3 m high. Origin: northern China.

Some alternate-leaved and (in mild climates) evergreen species are rarely cultivated: for example the yellow jasmine **Jasminum fruticans** L., a shrub growing to just over 1 m high. Origin: Mediterranean area and central Asia: **twigs** green, slender, long, rod-shaped, angular, furrowed. Leaf cushions dried up, grey-brown. **Buds** c. 2 mm long, ± surrounded by 2 grey-brown prophyll scales. The Italian jasmine **Jasminum humile** L., grows slightly taller, the twigs are green as well and a bit more vigorous, slightly 6-angular and glabrous. **Buds** with 5–6 visibly ciliate bud scales.

Forsythia viridissima
Twig

Forsythia europaea
Twig

Jasminum nudiflorum
Flowers appear during
mild winter weather

Forsythia ovata
Lateral buds

Jasminum nudiflorum
Lateral buds

Jasminum nudiflorum
Transverse section of twig

Jasminum nudiflorum
Flower bud

Jasminum humile
Latyeral bud

TRIBE Oleeae

Shrubs with slender twigs to trees with large branches, with dry fruits such as capsules (*Syringa*), winged nuts (*Fraxinus*) or stone fruits (*Chionanthus*, *Ligustrum* and *Forestiera*).

SUBTRIBE Ligustrinae

Syringa L., lilac

Buds globose to narrowly ovoid, mostly acute with several pairs of bud scales. **Leaf scars** with one trace. **Fruits** two-chambered, opening by 2 valves, woody dry capsules in terminal or lateral panicles. The genus contains about 30 species. Origin: South Eastern Europe to Eastern Asia.

Most common is the lilac, ***Syringa vulgaris***, with many cultivars. Looking at the bud one can already guess the flower colour. In reddish- or violet-flowering species rich in anthocyanin (a violet-red substance frequent in plant cells), the buds have a violet-red hue, while white-flowering species without anthocyanin have pure green buds.

The many cultivated wild species are difficult to tell apart. The slender-twigged species, moreover, are similar to the deciduous species of the genus *Ligustrum* which belongs, according to molecular phylogenetic data, to the genus *Syringa sensu lato*. The fruits which, in both genera, mostly persist in winter can be helpful in identification.

Key to *Syringa*

1   Twigs regularly with terminal buds . . . 4
1* In place of the dead terminal bud, two terminal lateral buds at the tip of the twig; very rarely single terminal buds (in young plants or on coppice shoot) 2
2   Bark very smooth or coming off in thin rolls; large shrubs or small trees. 9
2* Bark not coming off in rolls, but scaly or coming off in threads 3
3   Buds more than 4 mm long, twigs ± thick (> 3 mm) or glabrous, fruit capsules ± smooth 7
3* Buds up to 4 mm long, twigs thinner, often hairy, fruit capsules warty 11
4   Twigs thick 12

Subgenus *Syringa*, series *Syringa*
4* Twigs slender 5
5   Terminal buds with 2–3 pairs of bud scales . . . . *Syringa pinnatifolia*
5* Terminal buds with more bud scales . 6
6   All long-shoot tips with terminal bud. . . . .**Syringa protolaciniata**
6* Only some long-shoot tips with terminal bud. . . . . **Syringa ×persica**
7   Buds mostly more than 7 mm long . . 8
7* Buds smaller, brown. Bud scales keeled . . . . **Syringa ×persica**
8   Buds predominantly brown. Fruit capsule long-acute . . . . . *Syringa oblata*
8* Buds green to ± violet-red. Fruit capsule short-acute . . **Syringa vulgaris**

Subgenus *Ligustrina*
9 (2) Twigs hairy, buds very small (<2 mm) . . *Syringa reticulata* subsp. **pekinensis**
9* Twigs and buds ± becoming glabrous 10
10 Buds globose, only a little longer than broad . . . .*Syringa reticulata* subsp. **reticulata**
10* Buds ovoid, distinctly longer than broad . . *Syringa reticulata* subsp. **amurensis**

Subgenus *Syringa*, series *Pubescentes*
11 (3) Twigs densely hairy . . **S. microphylla**
11* Twigs sparsely hairy and becoming glabrous . . . . *Syringa meyeri*

Subgenus *Syringa*, series *Villosae*
12 (4) Terminal buds red-brown, compact, with 3 pairs of bud scales. . . . . *Syringa emodi*
12* Mostly more scales: ± broad with a grey margin; terminal buds at least twice as long as thick 13
13 Fruits in pendulous panicles. . . . . **Syringa reflexa** and **hybrids**
13* Fruits in erect to horizontal, rarely pendulous infructescences 14
14 Fruit infructescences dense, narrow, conical . . . . *Syringa josikaea* and *Syringa villosa*
14* Fruit infructescences lax, very wide . . . . . *Syringa tomentella* subsp. *sweginzowii*

Subgenus Syringa

Series Villosae

Shrubs, twigs mostly relatively robust, with terminal buds.

Syringa reflexa C. K. SCHNEID., nodding lilac

[*Syringa komarowii* subsp. *reflexa* (C. K. SCHNEID.) P. S. GREEN & M. C. CHANG]

Twigs grey to grey-brown, relatively robust with pale lenticels. Older twigs pale grey. **Fruits** in narrow, pendulous cylindrical infructescences, c. 12 mm long, smooth to slightly warty. Frequently planted shrub 3–4 m high. Origin: central China.

Many hybrids with: *Syringa villosa* (**Syringa ×prestoniae** MCKELVEY), *Syringa josikaea* (**Syringa josiflexa** PRESTON ex J. S. PRINGLE) and **Syringa ×swegiflexa** J. S. PRINGLE. Because of further breeding but also by unintentional hybridisation in botanical gardens, pure species are found only rarely in horticulture. *Syringa reflexa* has become involved in most cultivars which belong to this section; these, therefore, have more or less distinctly pendulous inflorescences.

Syringa josikaea JACQ. f. ex RCHB. f., Hungarian lilac

Buds: ovoid, acute, terminal buds to 10 mm long and 4–5 mm broad, with 4–5 pairs of scales. Bud scales brown-red, partly basally olive-greenish, often with a dark margin and hairy at least towards the tip. **Twigs** thick, slightly angular, glossy grey-brown to olive-brown, glabrous or with remnants of fine indument. **Lenticels** few, oblong, pale ochre-whitish. **Fruits** in narrow, conical dense clusters: 10 mm long, smooth and acute. Rarely planted, stiff, erect shrub 3–4 m high. Origin: Hungary to Ukraine.

The late lilac, **Syringa villosa** VAHL is very similar: **buds** ovoid, slightly acute, with 5–6 outer pairs of scales. Terminal buds 8–12, lateral buds to 10 mm long. Bud scales turning glabrous, ± keeled, orange-brown to dark red-brown, at the margin with distinct grey margin. **Twigs** robust, slightly angular, violet-brown to grey-brown, with many small and almost round lenticels. **Fruits** in pyramidal panicles, 10–15 mm long, almost smooth. Rarely planted shrub, 3–4 m high. Origin: northern China.

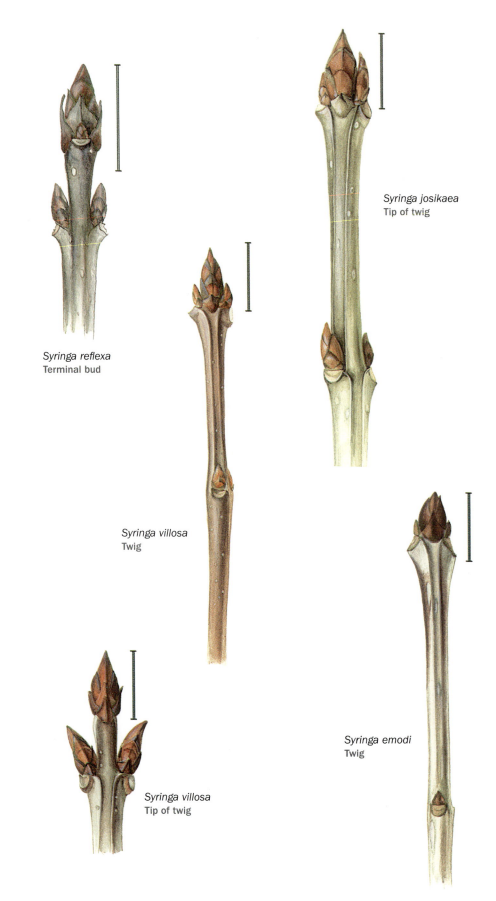

Syringa reflexa
Terminal bud

Syringa villosa
Twig

Syringa josikaea
Tip of twig

Syringa villosa
Tip of twig

Syringa emodi
Twig

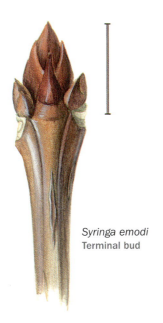

Syringa emodi
Terminal bud

Syringa tomentella subsp. **sweginzowii** (KOEHNE & LINGELSH.) JIN Y. CHENG & D. Y. HONG, Chengtu lilac
[*Syringa sweginzowii* KOEHNE & LINGELSH.]

Twigs glabrous, smooth, purple- to red-brown with pale lenticels. **Fruits** in broad panicles, often joined by lateral panicles, c. 10 mm long and smooth. Occasional, wide-branched shrub, higher than 3 m. Origin: north-western China.

Syringa emodi WALL. ex G. DON, Himalayan lilac

Buds almost evenly red-brown, bud scales almost glabrous but ciliate. Terminal buds with 3 outer pairs of scales, compact-ovoid, acute, 7–9 mm long and 4–6 mm wide. Lateral buds mostly relatively small with 1–2 pairs of scales. **Twigs** when young smooth and glabrous; grey-brown to olive-brown, below the tip of the twig occasionally orange-red-brown. **Lenticels** oblong, whitish to grey-brown; often strongly pad-like and oblong-fissured even on the youngest twigs. Older twigs also fine-fissuring between the lenticels. Rarely planted, upright growing shrub 2–5 m high. Origin: Central Asia.

Syringa microphylla DIELS, very little leaf lilac

Buds ovoid, acute 3–4 mm long. Mostly only lateral buds. Lowest bud scales small with ochre-brown margin, upper bud scales which cover the bud larger, violet-brown, glabrous but slightly ciliate. **Twigs** slender, grey-brown, long persistent with dense felty hair. **Leaf scar** on distinct cushion. **Fruits** in small 4–7 cm long panicles, c. 12 mm long, often curved and warty. Frequently planted, shrub 1–1.5 m high. Origin: northern China.

Korean lilac is similar, **Syringa meyeri** C. K. SCHNEID. A shrub to 1.5 m high. **Twigs** slightly 4-angular, soon becoming glabrous. **Fruits** in panicles to 10 cm long, 12–20 mm long, warty. Origin: northern China.

Series Syringa

Twigs usually without terminal buds. In young coppice shoots or in seedlings, some of the shoot tips may have true terminal buds.

Syringa oblata LINDL., early lilac

Lateral buds globose-ovoid. Bud scales ochre to red-brown, under a paler border dark purple-brown, at the base often slightly greenish. **Twigs** glabrous, round, yellow- or red-grey. **Leaf scars** vertical on a faint cushion. Rare, erect, wide-branching shrub, 2–4 m high. Origin: northern China.

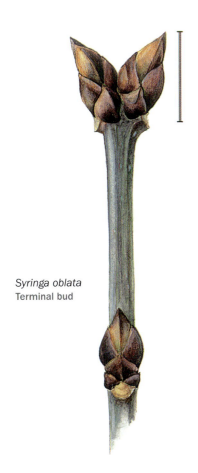

Syringa oblata
Terminal bud

Syringa microphylla
Part of twig

Syringa vulgaris L., lilac, common lilac

Lateral buds globose-ovoid, acute, green to dark red; bud scales rounded, ± keeled. White-flowering varieties have entirely green buds, whilst the buds of red- to violet flowering varieties have a ± reddish hue. A true **terminal bud** is found occasionally in young plants and following coppicing. In these cases resembling some species of maple (*Acer*) but distinguished by the single-trace leaf scar and sub-opposite phyllotaxy (maple has a three-trace leaf scar, the scars positioned exactly opposite). **Twigs** round, initially yellow-grey to olive-green, older twigs grey. **Bark** finally rough, fissured, and coming off longitudinally. **Pith** of young twigs relatively wide. Erect widely branching shrub, 2–5 m high, more rarely a small tree to 10 m high. Most frequently planted species of lilac with very many varieties. Origin: southern Europe to south-western Asia.

Syringa protolaciniata P. S. GREEN & M. C. CHANG

[*Syringa afghanica* hort. p.p., non SCHNEID., *Syringa ×laciniata* hort. p.p., non MILL.]

Bud scales glabrous, glossy ochre-brown, especially the lower ones strongly keeled and long-acute. Terminal bud ovoid, to c. 5 mm long with 5–6 visible pairs of bud scales. **Lateral buds** slightly spreading from the twig, to 4 mm long. **Twigs** relatively slender (c. 2 mm), glabrous, glossy (olive) brown, with many warty cork pores. From each of the two leaf scars of one node, two sharp ridges and one blunt, central edge, run down the twig. The twig is thus 6-angular in transverse section. Shrub to 3 m high, often planted. Origin: south-western China.

Syringa ×persica 'Laciniata' [*Syringa ×laciniata* MILL.], presumably a back-cross of *S. ×persica* with *S. protolaciniata*, is very similar and forms many terminal buds too. Even during the growth period this more rarely cultivated form is difficult to tell apart from its ¾ parent.

The pinnate lilac, *Syringa pinnatifolia* HEMSL. is very rarely planted: bud scales wine-red to dark brown, glabrous, mostly 2–3 visible pairs. Scales of a pair occasionally basally connate. A shrub to 3 m high. Origin: south-western China.

Syringa vulgaris
Buds of a dark-flowering cultivar

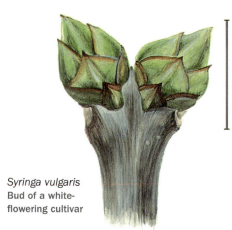

Syringa vulgaris
Bud of a white-flowering cultivar

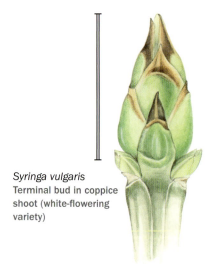

Syringa vulgaris
Terminal bud in coppice shoot (white-flowering variety)

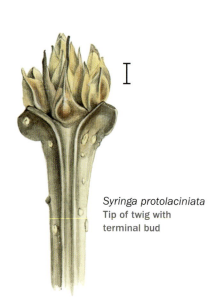

Syringa protolaciniata
Tip of twig with terminal bud

Syringa pinnatifolia
Tip of twig with terminal bud

Syringa ×persica 'Laciniata'
Atrophied tip of twig, without terminal bud

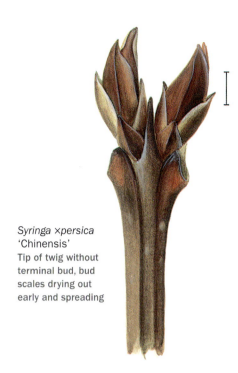

Syringa ×*persica*
'Chinensis'
Tip of twig without
terminal bud, bud
scales drying out
early and spreading

Syringa ×*persica* L., Persian lilac

[*Syringa afghanica* hort. p.p., non. SCHNEID., *Syringa* ×*laciniata* MILL., *Syringa* ×*chinensis* WILLD.]

[*Syringa protolaciniata* × *Syringa vulgaris*]

Buds acute, globose to ovoid, to 5 mm long, mostly without terminal bud. Bud scales plain red- to ochre-brown, distinctly keeled. **Twigs** slender: youngest member of shoot < 2 mm diameter, slightly angular, yellow-ochre to grey-brown, and often silvery grey due to dead epidermis. **Pith** relatively wide, white, ellipsoid. Bushy shrub 1.5–2 m high. A hybrid, cultivated for centuries in different cultivars. The most frequent is **'Chinensis'** the Chinese lilac, probably a back-cross with *S. vulgaris*: usually only lateral buds, to 9 mm long, bud scales spreading even in the growth period, as if dried up, keeled, dry-skinned, brown and glabrous. **Twigs** glossy, olive-green, partly brown, towards the tip ± angular, with many lenticels. **Pith** rounded-rectangular. Erect bushy shrub to 4 m high with down-curving slender twigs.

Syringa ×*persica*
Tip of twig occasionally
with terminal bud

Syringa reticulata
subsp. *pekinensis*
Lateral buds at
tip of twig

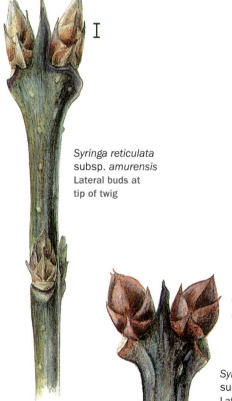

Syringa reticulata
subsp. *amurensis*
Lateral buds at
tip of twig

Syringa reticulata
subsp. *reticulata*
Lateral buds at
tip of twig

Subgenus Ligustrina

Presently amalgamated into one species, *Syringa reticulata*, a group of multiple forms of tree lilacs from eastern Asia. The smooth bark, coming off a bit in rolls, with horizontal bands of lenticels on stronger twigs and stems is distinctive. Terminal buds are mostly lacking (but are formed now and then on vigorous shoots).

Syringa reticulata (BLUME) H. HARA subsp. *pekinensis* (RUPR.) P. S. GREEN & M. C. CHANG, Peking lilac

[*Syringa pekinensis* RUPR.]

Buds hairy, with 2–3 pairs of bud scales, these distinctly ciliate. **Twigs** round, relatively slender, hairy, grey-olive-green, ± brown; older twigs dark grey, hardly glossy. **Lenticels** few. **Bark** of larger branches very smooth, finely fissured. Shrub similar to *Ligustrum*, 2.5–5 m high. Lateral branches spreading at right angles, wide and partly slightly pendulous. Occasional. Origin: northern China.

subsp. *reticulata*, Japanese lilac

[*Syringa amurensis* var. *japonica* (MAXIM.) FRENCH. & SAV.]

Buds glossy, brownish-yellow, ± globose; with keeled bud scales which gape slightly and are ciliate at the margin. **Twigs** round, initially glossy red-brown to grey; two-year-old twigs form distinct ringed bark; older twigs with the outer bark coming off in rolls, below this smooth and ± olive-green. **Leaf scars** oblique, on very large cushions. Occasional, erect shrub to 3 m high, or small tree with short stem, branches at right angles to trunk and round crown. Origin: northern Japan.

subsp. *amurensis* (RUPR.) P. S. GREEN & M. C. CHANG, Amur lilac

[*Syringa reticulata* var. *mandshurica* (MAXIM.) H. HARA, *Syringa amurensis* RUPR.]

Similar to subsp. *reticulata*, but the **buds** are narrowly ovoid and not glossy. **Twigs** grey. Shrub 3–4 m high or small tree, wide with branches at right angles to main stem. Origin: Manchuria and northern China.

Ligustrum L., privet

Buds opposite and small. **Leaf scars** with one trace. Dense deciduous-and evergreen shrubs or small trees with delicate twigs. **Fruits** black, berry-like stone fruits. Often used for hedges. Without fruits difficult to tell apart from slender-twigged species of *Syringa*.

Key to *Ligustrum*

1   Twigs glabrous . . . . . ***Ligustrum vulgare***
1* Twigs hairy 2
2   Twigs delicate short hairy. . . . .
  . . . . ***Ligustrum obtusifolium***
2* Twigs shaggy long-hairy. . . . .
  . . . . *Ligustrum quihoui*

Ligustrum vulgare L., wild privet

Buds 2–4 mm long. Terminal bud often undeveloped. Lateral bud broadly spherical, c. 2 mm broad and high. Bud scales greenish to wine-red-violet. **Twigs** slender, at the end of the shoot 1–1.5 mm diameter, grey-olive to grey-brown, with pale ochre or twig-coloured lenticels. **Fruits** persistent into winter: black, globose, glossy, 6–7 mm long. Very common, particularly for hedges: deciduous, some forms also evergreen; shrub to 5 m high. Origin: south-western Europe to eastern central Europe.

Ligustrum obtusifolium SIEBOLD & ZUCC.

Buds very small, ovoid. Terminal buds 1–1.5(–2) mm long, ochre-brown. **Twigs** slender, slightly 4-angular, grey, very delicately dark-hairy (lens needed). **Fruits** blackish, occasionally with slight waxy bloom, to 5–6 mm diameter. Frequently planted, spreading shrub 2–3 m high. Origin: Japan.

Ligustrum ibota SIEBOLD is rarely planted, twigs only initially hairy, soon becoming glabrous. Fruits roundish-ovoid, 7–8 mm long, a few united in short panicles. Origin: Japan.

Ligustrum quihoui CARRIÈRE, waxyleaf privet

Buds very small, 1–1.5 mm long. Bud-scales ochre-brown to dirty red-brown, sparsely white-hairy. **Twigs** slender, grey, round or slightly angular, with long, shaggy hair. Rarely planted, shrub to 2 m high, with branches at right angles to stem. Origin: China

Another rare species is *Ligustrum sinense* LOUR., with twigs with dense, felty indumentum and small, globose black-blue fruits to 4 mm thick. Origin: China.

Ligustrum quihoui
Twig

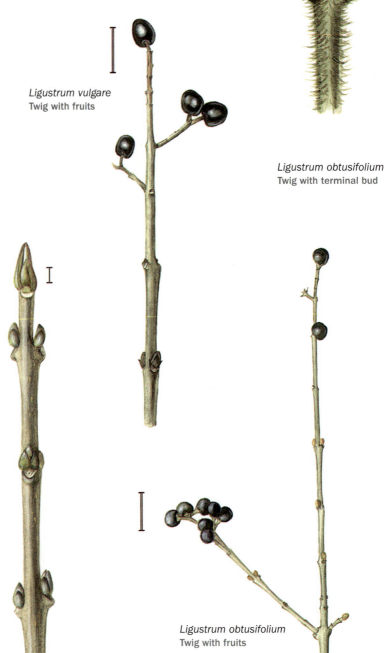

Ligustrum vulgare
Twig with fruits

Ligustrum obtusifolium
Twig with terminal bud

Ligustrum vulgare
Tip of twig

Ligustrum obtusifolium
Twig with fruits

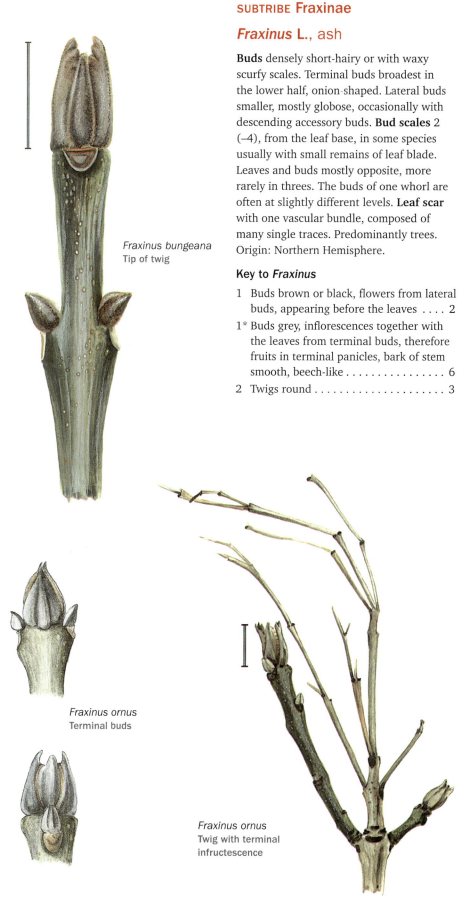

Fraxinus bungeana
Tip of twig

Fraxinus ornus
Terminal buds

Fraxinus ornus
Twig with terminal
infructescence

Fraxinus L., ash

Buds densely short-hairy or with waxy scurfy scales. Terminal buds broadest in the lower half, onion-shaped. Lateral buds smaller, mostly globose, occasionally with descending accessory buds. **Bud scales** 2 (–4), from the leaf base, in some species usually with small remains of leaf blade. Leaves and buds mostly opposite, more rarely in threes. The buds of one whorl are often at slightly different levels. **Leaf scar** with one vascular bundle, composed of many single traces. Predominantly trees. Origin: Northern Hemisphere.

Key to *Fraxinus*

1 Buds brown or black, flowers from lateral buds, appearing before the leaves 2
1* Buds grey, inflorescences together with the leaves from terminal buds, therefore fruits in terminal panicles, bark of stem smooth, beech-like 6
2 Twigs round 3

Section *Dipetalae*
2* Twigs sharply 4-angular . . . .
. . . . *Fraxinus quadrangulata*

True ashes (Section *Fraxinus*)
3 Buds deep black . . . . .*Fraxinus excelsior*
3* Buds red-brown to dark-brown 4
4 Outer bud scales of terminal buds usually with primordial lamina, flower buds often 3-whorled . .*Fraxinus angustifolia*
4* Outermost scales of terminal buds usually without primordial lamina, always opposite 5

American ashes (Section *Melioides*)
5 Leaf scar rather twig-like, hardly surrounding lateral buds; terminal bud higher than wide; shoots glabrous or hairy . . . . *Fraxinus pennsylvanica*
5* Leaf scar on large leaf cushion, clasping lateral buds; terminal bud mostly wider than high; twigs always glabrous . . . . . .
. . . .*Fraxinus americana*

Flower-ashes (Section *Ornus*)
6 (1) Buds with waxy surface layers . . . . . .
. . *Fraxinus chinensis* subsp. *rhynchophylla*
6* Buds fine velvety hairy . .*Fraxinus ornus*

Section Ornus, flower-ashes

Buds grey to brown. Terminal fruit panicles from inflorescences following the leaves.

Fraxinus ornus L., manna ash

Buds fine felty-hairy. Terminal buds 7–12 mm long, with 2 pairs of bud scales: outer pair appressed or slightly spreading with brownish hairy margin. Lateral buds mostly much smaller, globose to short-ovoid. **Twigs** greenish-grey, glabrous or initially slightly hairy. **Bark** grey, long remaining smooth. **Fruits** in terminal panicles: 2.5–3.5 cm long winged samaras, the wing extending down to about the middle of the seed-bearing part. Frequently planted, shrub or small tree to 15 m high. Origin: southern Europe to Asia Minor.

Species such as *Fraxinus bungeana* A. DC. are rarely encountered: a shrub 1.5–2 m high. Its **buds** are dark grey-brown, ± hairy and the terminal bud acute-ovoid. **Twigs** grey, ± finely hairy, with delicate lenticels. Origin: northern China.

Fraxinus chinensis subsp. *rhynchophylla*
(HANCE) A. E. MURRAY

[*Fraxinus rhynchophylla* HANCE]

Buds grey with whitish surface layers of
wax, bud scales brown-hairy along the
margins. Terminal buds c. 7 mm long. The
two outer bud scales, with the tips mostly
turned outwards, envelop the significantly
smaller subsequent pair of scales. Lateral
buds smaller with only two bud scales which
are ± closed at the margins. **Twigs** glabrous,
occasionally hairy just below the buds.
When young, brownish-yellow to pale grey-
green, with pale, ochre-coloured lenticels,
later twigs darker grey. **Bark** long remaining
smooth, later small-scaly. A rare tree,
15–25 m high, usually encountered only in
collections. Origin: northern and southern
east China.

Section Dipetalae

Fraxinus quadrangulata MICHX., blue ash

Buds with thick, felty indument, pale grey
with reddish or ochre-yellow hue. Terminal
buds 5–6 mm long, with two pairs of
bud leaves; the outer pair enveloping the
bud with well-developed pinnate lamina.
Lateral buds small with one opposite pair of
scales, the opposing leaf scars connected by
a line. **Twigs** sharply four-angular, slightly
winged, when young pale orange-grey to
red-brown, later grey. Tree 35–40 m high.
Origin: north America.

Fraxinus anomala TORR. ex S. WATSON,
mostly only shrubby, also has four-angular
twigs. Both species are not planted very
often, but are distinct by the shape of their
twigs. Origin: Western North America.

Section Melioides

Fraxinus americana L., white ash

Buds scaly-glandular, brown-hairy. **Terminal
buds** dark brown, globose, c. 5–6 mm
broad and 4–5 mm high. **Lateral buds**
smaller, globose to obtuse-conical, to 3 mm
high. **Twigs** glabrous, first olive-green to
brown-green, glossy with pale remnants of
epidermis and with whitish lenticels; older
twigs pale grey to grey-brown. **Leaf scars**
on large cushion, the opposing ones often

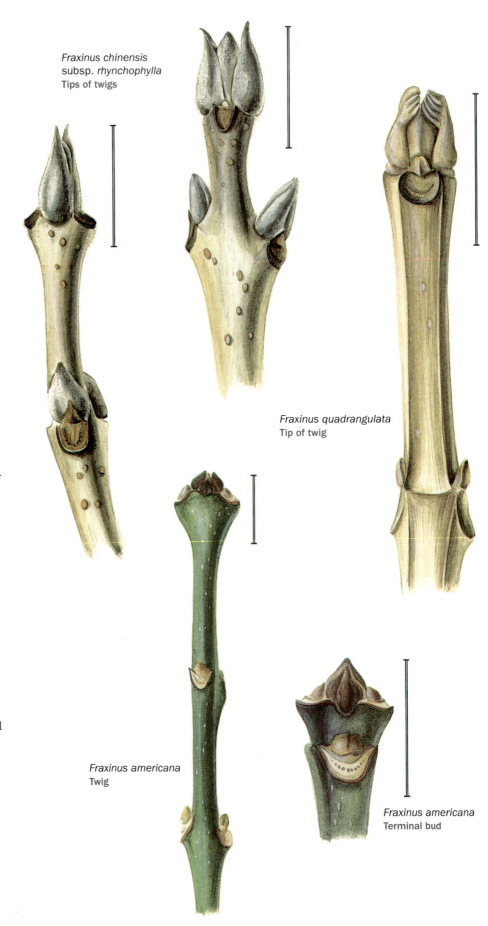

Fraxinus chinensis
subsp. *rhynchophylla*
Tips of twigs

Fraxinus quadrangulata
Tip of twig

Fraxinus americana
Twig

Fraxinus americana
Terminal bud

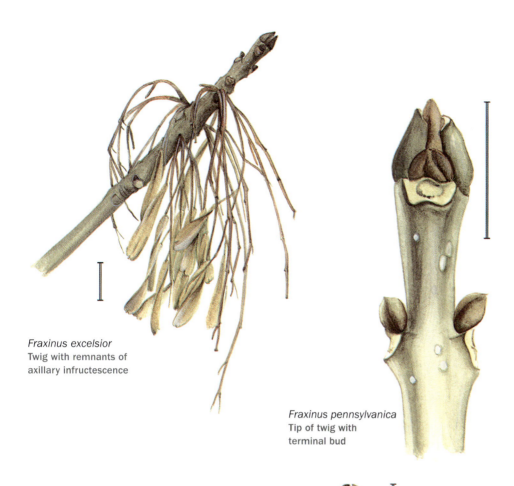

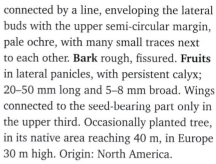

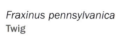

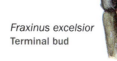

Fraxinus excelsior
Twig with remnants of
axillary infructescence

Fraxinus pennsylvanica
Tip of twig with
terminal bud

Fraxinus pennsylvanica
Twig

Fraxinus excelsior
Tip of twig

Fraxinus excelsior
Terminal bud

connected by a line, enveloping the lateral buds with the upper semi-circular margin, pale ochre, with many small traces next to each other. **Bark** rough, fissured. **Fruits** in lateral panicles, with persistent calyx; 20–50 mm long and 5–8 mm broad. Wings connected to the seed-bearing part only in the upper third. Occasionally planted tree, in its native area reaching 40 m, in Europe 30 m high. Origin: North America.

Fraxinus pennsylvanica MARSHALL, green ash

Buds with pale brown to dark grey brown indument. **Terminal bud** 4–7 mm long, globose to conical-ovoid, mostly higher than broad. **Lateral buds** smaller, compact, relatively broad, not surrounded by leaf scars. **Twigs** hairy or glabrous, grey to grey-brown, with small, whitish lenticels. **Leaf scars** relatively large, grey to pale brown with a central large leaf trace. **Bark** rough and fissured. Frequently planted, tree to 20 m high, with a round crown. Origin: North America. According to the indument of the twigs two varieties (with intermediates present) are recognised: var. *pennsylvanica*, with dense, felty indument, and var. *lanceolata* (BORKH.) SARG., with glabrous twigs.

Section Fraxinus, true ashes

Buds brown to black, rarely grey. **Flowers** before or with the leaves, lateral in axillary panicles, often dioecious. **Fruits** with or without calyx, fruit wings fused with the seed-bearing part for various lengths.

Fraxinus excelsior L., common ash

Buds deep black, rarely in the shade only brown-black. Bud leaves with felty-velvety surface. **Terminal buds** hemispherical to conical and acute or obtuse, occasionally also with slightly spreading bud leaf tips, 6–8 mm long and up to 10 mm wide. **Lateral buds** ovoid, globose or, most of all in flower buds, also broader than high, 3–7 mm long and 2.5–9 mm broad. **Twigs** pale grey-green to dark olive-green, occasionally with a violet hue, with few to many pale lenticels. **Leaf scars** semi-circular, ochre to brown with a linear to circular trace which is composed of many smaller traces. Old bark shallowly fissured. **Fruits** in panicles, without calyx, 25–50 mm long and 7–11 mm broad, the wings $\frac{1}{3}$ to entirely fused with the seed-bearing part. Tree to 40 m high, the most frequent ash in Europe. Origin: Europe to western Asia.

There are several cultivars which are distinct in winter: **'Aurea'** a small tree to 8 m high and **'Jaspidea'**, a robust tree to 20 m high, both with yellow twigs; **'Nana'**, a ball-shaped cultivar usually grafted on a high stem, and **'Pendula'**, large trees with pendulous twigs. The rarely planted *Fraxinus mandshurica* RUPR., Japanese ash, also has black buds, and is a tall tree. The **buds** are more black-brown, rarely deep black as in the common ash. Immediately below the terminal bud is a pair of lateral buds. **Twigs** dark yellowish-red-brown, blunt 4-angular, warty, with distinct lenticels. Origin: Manchuria and north and central Japan.

Fraxinus angustifolia VAHL, narrow-leaved ash

Buds dark brown, finely felty-hairy. Terminal buds conical-ovoid, 5–6 mm long, outer scale leaves with leaf remnants. Lateral buds smaller, spreading from the twig, particularly flower buds often 3-whorled on the twig. **Twigs** glabrous, ± green-brown, glossy, later grey, with whitish lenticels. **Bark** of old stems rough, deeply furrowed. **Fruits** 2–6 cm long, winged samaras, wing fused to the base or to the lower third of the seed-bearing part. Frequent tree, to 30 m high. Origin: southern Europe from Spain to Transcaucasia.

SUBTRIBE Oleinae

One-seeded stone fruits. Genera belonging here are, apart from the olive (*Olea*), the fringe tree and the desert olive.

Fraxinus excelsior 'Aurea'
Twig with yellowish bark

Fraxinus angustifolia
Twig with opposite leaf buds

Fraxinus angustifolia
Twig with 3-whorled flower buds

Fraxinus angustifolia
Tip of twig with terminal bud

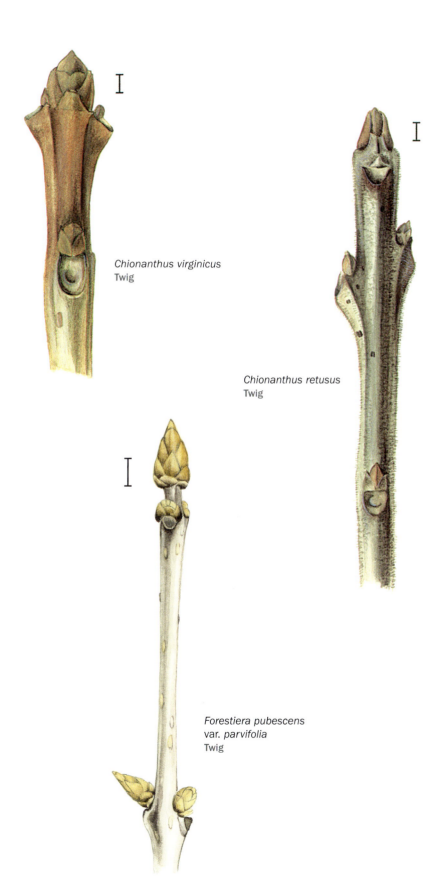

Chionanthus virginicus
Twig

Chionanthus retusus
Twig

Forestiera pubescens
var. parvifolia
Twig

Chionanthus L., fringe tree

Buds with a few pairs of leaf base scales. Large shrubs or small trees: two species.

Key to Chionanthus

1   Twigs glabrous . **Chionanthus virginicus**
1* Twigs fine-hairy . . . .*Chionanthus retusus*

Chionanthus virginicus L., white fringe tree

Buds with brown bud scales. The latter are often distinctly keeled and acute, with sparse indument. Terminal buds ovoid, with 3–4 pairs of scales. Lateral buds smaller, globose. **Twigs** ochre-brown to dark grey-brown, robust, slightly angular, initially hairy but mostly becoming glabrous. Frequently planted; large shrub or small tree to 10 m high. Origin: North America.

Chionanthus retusus LINDL. & PAXTON, Chinese fringe tree

Buds conical. Terminal buds c. 3 mm, lateral buds 1.5–2 mm long. Bud scales 3–4 pairs: the lowermost dark grey-brown and hairy, the subsequent red-brown and the uppermost ochre-brown with sparse indument. **Twigs** grey and only below the leaf scars red-brown; especially below the tip densely hairy, slightly 4-angular and relatively thick. **Leaf scars** on distinct cushion, oblong-semi-circular and slightly sunken, with a round trace at the centre. **Lenticels** few to many, warty, small, ovate. Rarely planted shrub 2–3 m high. Origin: Korea, China and Taiwan.

Forestiera pubescens var. parvifolia
(A. GRAY) G. L. NESOM, desert olive
[*Forestiera neomexicana* A. GRAY]

Terminal buds ovoid, acute, 2–3 mm long. **Lateral buds** smaller. **Bud scales** opposite, in 5–7 pairs, the lowermost grey-brown. Margin ciliate, otherwise glabrous. Uppermost bud scale honey-yellow to ochre-coloured. Surface partly with dead, grey-white lifting epidermis. **Twigs** very slender, below the terminal bud <1 mm diameter, grey to grey-brown, with few pale dot-shaped lenticels. Bark later shallow rhomboid-furrowed. Very rarely planted shrub to 3 m high. Origin: south-western North America.

Scrophulariaceae
figwort family

Buddleja L., butterfly bush

Perennial herbs, shrubs and trees, mostly native to the tropics to subtropics. **Buds** opposite or alternate, without bud scales, just like twigs usually with stellate hairs.

Key to *Buddleja*

1 Alternate, leaves deciduous . . . .
. . . . *Buddleja alternifolia*
1* Opposite, evergreen, often dying back
   under severe frost. . . . . *Buddleja davidii*

Buddleja alternifolia MAXIM., alternate-leaved butterfly bush

Buds ovoid, 2–3 mm long, without bud scales, enveloped by several pale leaves with fine, velvety indument: the outer ones occasionally slightly spreading and also a little longer than the bud. **Twigs** long, rod-like, slender, narrowing towards the tip, narrowly striped and almost glabrous; ochre-brown to grey-brown. **Bark** of older twigs narrowly rhomboid-fissuring. **Leaf scar** on distinct cushion with a central, larger, mostly faint trace. Frequently planted, high and wide overhanging shrub 2–4 m high. Origin: China.

Buddleja davidii FRANCH., common butterfly bush

Buds not formed, mostly lateral shoots in various stages of development. Silver-felted leaves drying out and entire plants often dying back with frost. **Twigs** thick, round, matte ochre-brown, initially hairy. **Infructescences** at the tip of overhanging branches, dense cylindric-conical panicles 10–25 cm long. Very frequently planted shrub, 3–5 m high. Origin: China. Spreading as an invasive plant in temperate regions. Globose infructescences are formed by the occasionally planted hybrid ***Buddleja ×weyeriana*** WEYER, Weyer butterfly bush (*Buddleja davidii* × *Buddleja globosa* HOPE).

Callicarpa bodinieri
Tip of twig

Buddleja davidii
Twig

Buddleja alternifolia
Twig

Buddleja alternifolia
Lateral bud

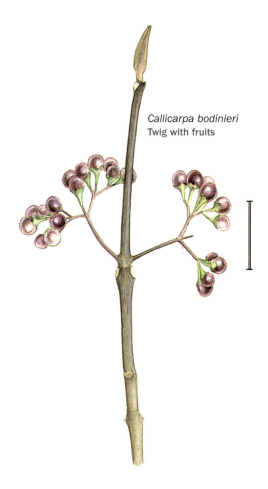

Callicarpa bodinieri
Twig with fruits

Callicarpa bodinieri
Lateral bud

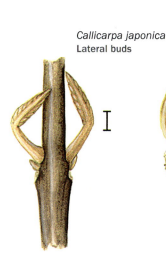

Callicarpa japonica
Terminal bud

Callicarpa japonica
Lateral buds

Lamiaceae
[Labiatae *nom. alt.*] mint family

Many genera formerly in the Verbenaceae now belong to this large globally distributed family. Herbs, perennial herbs, sub-shrubs, shrubs, lianas and trees which are represented in the temperate zones by relatively few taxa. The c. 7000 species are distributed among 230 genera in seven subfamilies. About 10 genera, among them *Callicarpa*, have not yet been placed in any subfamily. There are not many woody taxa in the family. Besides the species treated here, many sub-shrubs are in cultivation, as well as mostly evergreen species such as thyme (*Thymus*), hyssop (*Hyssopus*), lavender (*Lavandula*), sage (*Salvia*) and savory (*Satureja*). **Buds** opposite or whorled. **Twigs** aromatic, at least initially 4-angular. **Leaf scars** of a node connected by lines, with one trace.

Key to Lamiaceae

1   Buds without bud scales, usually densely hairy 4
1* At least the first pair of leaves of lateral bud distinctly scale-like. Plant rare, reaching a height of 1.5 m 2
2   Twig with simple hairs, not continuous 3
2* Twigs with very dense white stellate hairs . . . . *Perovskia*
3   Upper bud scales with velvety ochre-brown simple hairs . . . . . *Elsholtzia stauntonii*
3* Upper bud scales with white stellate hairs . . . . *Perovskia scrophularifolia*
4   Buds compact to ovoid, mostly only lateral buds 5

4* Buds oblong, stalked. Terminal buds present . . . . *Callicarpa*
5   Twigs towards tip with very dense white stellate hairs . . . . *Perovskia*
5* Indument different 6
6   Young twigs to 2 mm diameter, persistent fruit calyces like parchment . *Caryopteris*
6* Twigs distinctly thicker 7
7   Buds with wine-red indument . . . . *Clerodendrum*
7* Buds with ochre-brown indument . . *Vitex*

Callicarpa L., beautyberry

Buds pale ochre-brown, with dense stellate hairs. **Terminal bud** always formed. **Lateral buds** opposite, stalked, often with descending accessory buds. **Twigs** slender, with stellate hairs. **Fruits** persistent into early winter, globose, violet berry-like stone fruits (c. 4–6 mm diameter) in dichasial-branching heads. Many evergreen (sub) tropical species of the genus exist. Origin: eastern Asia to Australia. Several deciduous species from eastern Asia and North America are cultivated, and can be difficult to tell apart.

Callicarpa bodinieri H. LÉV. var. *giraldii* (HESSE ex REHDER) REHDER, Bodinier's beautyberry

Buds without bud scales, with oblong outer bud leaves which are clearly pinnate with dense stellate hairs. Terminal buds 5–10 mm long, slim, a little oblique, non-symmetrical and slightly bifid at the tip. Lateral buds stalked, mostly smaller, c. 6 mm long, appressed to slightly spreading, outer leaf blades often slightly spreading. **Twigs** slender, 1–1.5 mm diameter, grey-brown to green-brown, sparsely stellate-hairy. **Leaf scars** small, ± round with a slightly raised trace of a vascular bundle. **Inflorescences** short, stalked for 6–8 mm. Shrub 1.5–2 m high. Origin: central to western China.

The terminal buds of **Callicarpa japonica** THUNB., Japanese beautyberry, are slightly longer, 8–12 mm long and slim. The lateral bud, to 10 mm long is long-stalked. The up to 5 mm long stalk is often clearly spreading, whilst the actual bud turns again towards the stalk with a kink. **Inflorescences** are slightly longer with a c. 1.2 cm long stalk. A shrub with pendulous twigs to 2 m high. Origin: Japan.

SUBFAMILY Viticoideae

Vitex agnus-castus L., chaste tree

Buds opposite, 1 × 1.5 mm, without bud scales, with dense ochre- to gold-brown indument, often with a tiny descending accessory bud. **Twigs** 4–6 mm diameter; especially when young ± 4-angular, later mostly only slightly flattened, ovoid in transverse section; grey-brown, sparsely hairy, with occasional glands. **Leaf scar** with a large central trace and a small faint trace on either side. **Pith** wide and white. **Bark** aromatic with very small, faint lenticels. In winter the remnants of the terminal, to 20 cm long, panicle are visible. **Fruit** a black-brown 4-locular stone fruit, 3–4 mm diameter. Shrub to 3 m high. Origin: southern Europe (Mediterranean countries) and western Asia.

SUBFAMILY Ajugoideae

Clerodendrum trichotomum THUNB., harlequin glorybower

Buds small, c.2.5 mm broad and 1.5–2 mm high, some with tiny descending accessory bud; naked, with densely dark wine-red to aubergine-coloured indument. **Twigs** slightly 4-angular, 3–5 mm diameter; only initially densely felty red-brown hairy, later more sparsely so and ochre-brown to grey and finally becoming glabrous. **Leaf scar** relatively large, 2.5–3 mm, ovate and turned a little inward at the upper margin, with a semi-circular vascular bundle which consists of several small vascular bundles which merge. **Fruits** in 12–25 cm broad corymbs: soon deciduous, blue stone fruits, surrounded by lilac-red sepals. Shrub or tree to 8 m high. Origin: China.

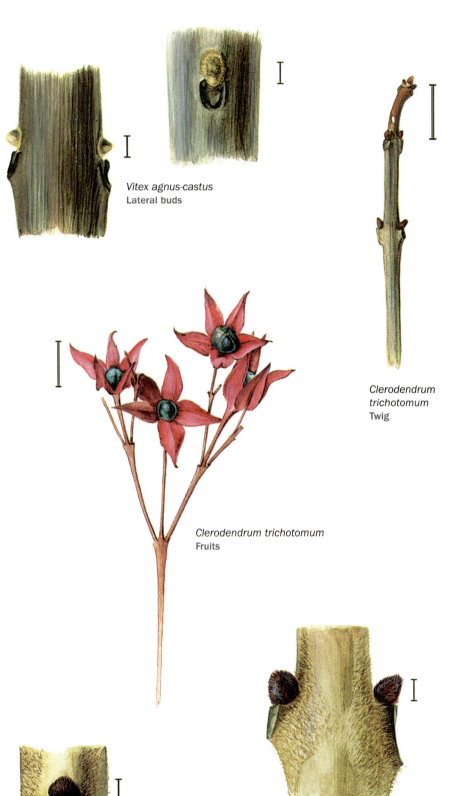

Vitex agnus-castus
Lateral buds

Clerodendrum trichotomum
Twig

Clerodendrum trichotomum
Fruits

Clerodendrum trichotomum
Lateral bud

Clerodendrum trichotomum
Lateral buds

Caryopteris incana
Tip of twig with remnants of fruit

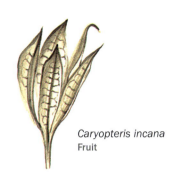

Caryopteris incana
Fruit

Caryopteris incana
Lateral bud

Elsholtzia stauntonii
Lateral buds

Caryopteris BUNGE, bluebeard

A few species of low shrubs. Origin: eastern Asia, Himalaya and Mongolia.

Caryopteris incana (THUNB. ex HOUTT.) MIQ.

Buds small and opposite, 1–2 mm long, without bud scales, grey-green and densely appressed-hairy; basally often with ± developed leaves, which can be several times larger than the buds. **Twigs** slender, brown, with sparse white hair. **Leaf scars** small, with a central trace. **Fruits** in axillary corymbs, the bell-shaped, parchment-like 5-lobed calyx persisting. Rarely planted shrub, a little over 1 m high. Origin: eastern Asia.

The hybrid with *Caryopteris mongholica* BUNGE, which can hardly be distinguished in winter: ***Caryopteris* ×*clandonensis*** hort. ex REHDER is more often planted.

SUBFAMILY **Nepetoideae**

Elsholtzia stauntonii BENTH., mint bush

Buds ovoid to narrowly ovoid, 3–10 mm long, appressed to twig. Outer bud scales brown, almost glabrous, inner scales pale ochre-brown, velvety hairy. **Twigs** 4-angular, with longitudinally striped, brown, hairy bark. Older twigs grey-brown, **bark** flaking. **Pith** white, wide and 4-angular. **Leaf scars** with one trace. Frequently planted shrub to 1.5 m high. Origin: northern China.

Perovskia KAR.

Aromatic subshrubs, their main axis straight from the ground, with a few weak lateral branches; these only more numerous in the terminal paniculate inflorescence. Origin: central Asia.

Key to *Perovskia*

1 Twigs very dense, covered with white, stellate hairs . . . .***Perovskia abrotanoides*** and ***Perovskia atriplicifolia***

1* Twigs 4-angular, with sparse simple hairs . . . . *Perovskia scrophulariifolia*

Perovskia atriplicifolia BENTH., Russian sage

Buds globose-ovoid, 3–4 mm long, occasionally slightly stalked; with 2–3(–4) visible, very densely woolly-felty hairy bud scales. The indument consists of tiny stellate hairs which can only be seen with a strong lens [×10, ×20]. **Twigs** aromatic, straight, to 1.5 m high, often falling over, almost round, finely longitudinal-furrowed, densely grey-white (looking like mould), stellate-haired. **Leaf scars** with a faint central large trace, the opposite ones connected by a line which is often concealed by the indument. Frequently planted bushy shrub to 1.5 m high.

Perovskia abrotanoides KAR., occasionally planted, is similar. It is a short shrub, to 1 m high. When old, it has prostrate-ascendent main axes.

Perovskia scrophulariifolia BUNGE

Buds are opposite lateral buds, ovoid, c. 3 mm long; lowermost bud scale brownish, very sparsely hairy, the subsequent ones ± thick densely white stellate-hairy. **Twigs** relatively thick, distinctly four-angular, with remnants of simple indument, becoming glabrous. Rarely planted, upright aromatic shrub, over 1 m high. Origin: central Asia.

Paulowniaceae
empress-tree family

Paulownia tomentosa (THUNB.) STEUD., foxglove tree

No **terminal buds** formed. **Lateral buds** opposite, but often slightly displaced from one another, c. 2 × 3 mm with densely ochre to olive-brown indument. Sometimes with descending accessory bud. **Flower buds** without bud scales, in upright, to 30 cm long opposite-branched panicles: globose, c. 10 mm long and 8 mm thick; surrounded only by the calyx lobes which are densely ochre-brown hairy. **Twigs** thick, 8–12 mm diameter, olive-brown to grey-brown, with many round to oblong, pale grey lenticels, which tear open vertically.

Perovskia atriplicifolia
Lateral bud

Perovskia abrotanoides
Lateral buds

Perovskia abrotanoides
Lateral bud

Perovskia scrophulariifolia
Lateral buds

Paulownia tomentosa
Lateral bud

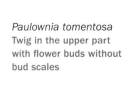

Paulownia tomentosa
Twig in the upper part
with flower buds without
bud scales

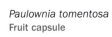

Paulownia tomentosa
Fruit capsule

Paulownia tomentosa
Seed

Paulownia tomentosa
Longitudinal twig section,
chambered pith

Pith chambered, 2–3 mm in diameter. **Fruits** persist on the tree over winter: globose-ovoid, sharply acute capsule, opening by 2 valves (c. 35 mm long and 20–25 mm thick) with many small (2–3 mm) winged seeds. Frequently planted, tree to 20 m high. Origin: central- to northern China.

Bignoniaceae
trumpet vine family

Widely distributed family from the tropics and subtropics. A few deciduous genera reach temperate latitudes. **Fruits**: in the treated species oblong 2-valvate capsules which open late, with many flat, 2-winged seeds. **Leaf scar** with one trace or with several individual traces which, together, make up a united shape, a circle or semi-circle.

Key to Bignoniaceae

1 Lateral buds 3-whorled. Erect tree or shrub 2

1* Lateral buds opposite. Often climbing shrubs with adhesive roots . . . . . ***Campsis***

2 Tree, leaf scars ovate, rounded at the top . . . . ***Catalpa***

2* Shrub or small tree, leaf scars shield shaped, flat at the top. . . . . . . ×*Chitalpa*

Campsis Lour., trumpet vine

Two species. These can be confused with climbing *Hydrangea* species such as *Hydrangea petiolaris,* which also has adhesive roots and opposite leaves, but which differs in significantly larger buds, flaking bark, 3-trace leaf scars and inflorescences in corymbs. Origin: North America and eastern Asia.

Key to *Campsis*

1 With many adhesive roots, climbing strongly . . . . ***Campsis radicans***

1* More shrubby or with few adhesive roots 2

2 Erect hardly climbing shrub with adhesive roots . . ***Campsis*** ×*tagliabuana*

2* Winding, without or with very few adhesive roots, only in warm spots. . . . . *Campsis grandiflora*

Campsis radicans (L.), BUREAU,
American trumpet vine

Buds opposite, flat-appressed to twig,
rounded-off triangular, with two branch-
coloured prophyll scales which cover the
buds for about two-thirds. **Twigs** ochre
to orange-brown, slightly longitudinally
fissured and with many adhesive roots. **Pith**
solid, also fissuring and chambered in the
nodes. **Leaf scar** with a 'C' to 'U' shaped
trace. Shrub climbing to 10 m high. Origin:
south-eastern USA.

The frost-sensitive Chinese trumpet vine
Campsis grandiflora (THUNB.) K. SCHUM.
is only encountered in more clement areas.
Adhesive roots almost entirely absent.
Winding to 6 m high. In some species
the hybrid between both species is often
planted: *Campsis ×tagliabuana* (VIS.)
REHDER, a more upright shrub, only slightly
climbing with adhesive roots.

Catalpa SCOP.

Erect, rough-branched, medium-sized to
large trees. The lateral buds are in three
verticils and are small in relation to the
robust branches and large leaf scars.
Terminal buds absent. Bud scales simple and
like the buds themselves mostly 3-whorled.
The first whorl, however, consists of two
prophyll scales. Noticeably long fruit
capsules with many bilaterally winged seeds.
Origin: North America and Eastern Asia.

Key to *Catalpa*

1 Buds glabrous, mainly at the tip of the
   twig with long acute bud scales, seeds
   approx. 10 mm wide[1]. . . . .*Catalpa ovata*

1* Buds with blunt or hardly acute, ciliate
   bud scales. Seeds wider than 15 mm . . 2

2 Fruit capsules 6–8 mm thick, seeds with
   acute wings . . . . . **Catalpa bignonioides**

2* Fruit capsules 8–15 mm thick, seeds with
   rounded-off wings . . . .**Catalpa speciosa**

[1]Measured without fringe

Campsis radicans
Twig

Campsis radicans
Fruits

Campsis radicans
Node of twig with
lateral bud and
adventitious adhesive
roots

Campsis radicans
Seed

Campsis grandiflora
Lateral bud and leaf scar
with distinct trace

Campsis grandiflora
Node with lateral buds

Catalpa bignonioides
Node of shoot

Catalpa bignonioides
Seed

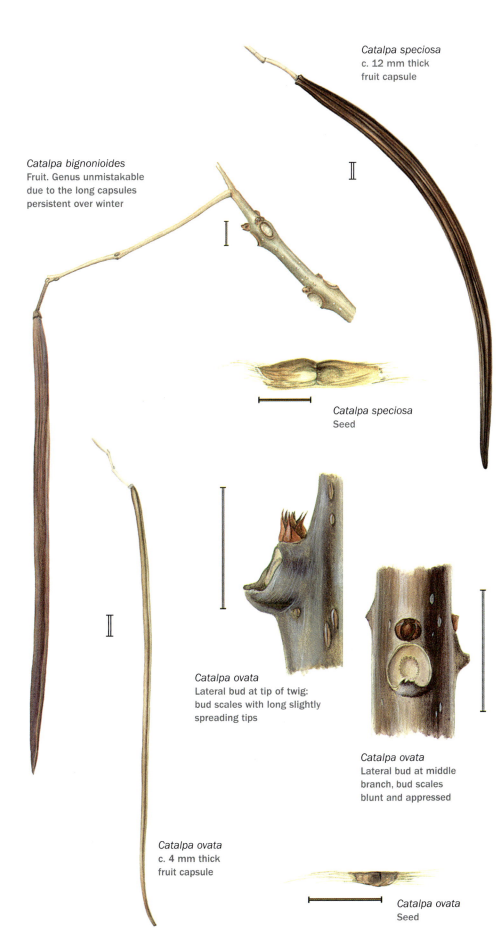

Catalpa speciosa
c. 12 mm thick
fruit capsule

Catalpa bignonioides
Fruit. Genus unmistakable
due to the long capsules
persistent over winter

Catalpa speciosa
Seed

Catalpa ovata
Lateral bud at tip of twig:
bud scales with long slightly
spreading tips

Catalpa ovata
Lateral bud at middle
branch, bud scales
blunt and appressed

Catalpa ovata
c. 4 mm thick
fruit capsule

Catalpa ovata
Seed

Catalpa bignonioides WALTER, southern catalpa, Indian bean tree

Buds globose, with c. 8 visible brown, loosely appressed, acute and fine glandular-ciliate bud scales; on large leaf cushion. **Twigs** initially yellowish grey-brown, with solid wide whitish pith. Bark of twig, when rubbed, has an intense scent. **Lenticels** many and distinct. **Bark** pale brown, thin and small-scaly. **Leaf scars** very large. **Fruit** capsules 6–8 mm thick, thin-walled. **Seeds** with two acute wings which terminate in tufts of hair. Without fringes c. 25 mm broad. Tree to 25 m high, the most often cultivated *Catalpa*. Origin: south-eastern USA.

Catalpa speciosa WARDER ex ENGELM., northern catalpa

Buds similar to *Catalpa bignonioides*, on raised leaf cushion. **Twigs** sturdy, reddish-brown, occasionally with waxy bloom. **Pith** whitish, initially solid, only at the nodes partly chambered, later disappearing. **Fruit** capsules 8–12(–15) mm thick and to 45 cm long. **Seeds** with rounded wings ending in broad fringes. Measured without fringes: c. 20 mm broad. **Bark** thick and deeply furrowed. Frequent, tree over 30 m high. Origin: south-eastern USA.

Catalpa ovata G. DON, Chinese catalpa

Buds 2–3 mm long, glabrous, red-brown; with c. 5–8 visible, acute bud scales, distinctly spreading in buds at the tip of the twig. Lower down the twig the buds are smaller with appressed bud scales, buds often absent from the twig base. **Twigs**, when young, pale-grey to violet-brown on the shaded side, glabrous or with occasional stiff hairs. **Lenticels** round to oblong, pale ochre-grey. **Leaf scars** large, round and sunken, on the lower edge often with balcony-like processes. Leaf cushion at the base of the twig weak and only near the tip more pronounced. **Fruit** capsules 3–4 mm thick and to 30 cm long. Seeds significantly smaller than in both American species. Without the relatively long fringes to 10 mm broad. Occasionally planted, tree 10–15 m high with a broad crown. Origin: China.

Also occasionally planted is ***Catalpa ×erubescens*** CARRIÈRE, the hybrid with *Catalpa bignonioides*.

×*Chitalpa tashkentensis* T. S. ELIAS & WISURA, summer bells

[*Catalpa bignonioides* × *Chilopsis linearis*]

Buds: only lateral buds, these very small and flat, 3-whorled as in *Catalpa*, more rarely tending towards alternate as in *Chilopsis*. **Twigs** medium robust, pale brown to ochre-brown, particularly in upper reaches, with paler lenticels. Appearing glabrous, but with sparse tiny hairs. **Leaf scar** on distinct leaf cushion with a vascular bundle trace. Shrub to 6 m high.

This generic hybrid is intermediate between the parents. *Chilopsis linearis* (CAV.) SWEET, the desert willow, from the southern USA and Mexico has relatively slender, slightly hairy grey twigs, the leaf scars are alternate, but also tend towards whorled. The leaf cushion is raised far above the twig and clasps the base of the lateral bud.

×Chitalpa tashkentensis
3-whorled node of shoot

Lycium barbarum
Short-shoot

Solanaceae
nightshade family

Buds alternate, relatively small with few bud scales. **Leaf scars** with one trace. Mostly native to the tropics and subtropics: herb, shrub and tree species.

Key to Solanaceae

1 Winding, half-shrubby liana . . . .
. . . .*Solanum dulcamara*
1* Often armed shrubs . . . . *Lycium*

Solanum dulcamara L., bittersweet

Buds alternate, with a few outer red-brown to grey-brown hairy bud scales. **Twigs** glabrous, angular, greenish, olive to brownish, with an unpleasant smell. **Leaf scar** dark, with one trace. **Fruits** occasionally persistent in early winter, many (poisonous!) red berries, with often longer persistent 5-toothed calyces. Infructescences apparently originate from the twig far above the scar of the leaf. The axis of the infructescence is often fused in its lower part with the twig (concaulescence). Climbing half-shrub to 2 m high. Origin: Eurasia.

Solanum dulcamara
Twig with remnants of fruit

Solanum dulcamara
Lateral bud

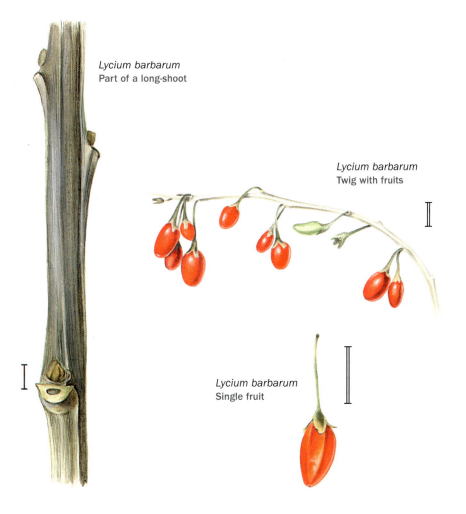

Lycium barbarum
Part of a long-shoot

Lycium barbarum
Twig with fruits

Lycium barbarum
Single fruit

Lycium L., box-thorn

Medium-sized, pendulous or slightly climbing shrubs 1–3 m high. Phyllotaxy alternate. Axillary short-shoots often thorny. The species are difficult to tell apart: characters mainly come from thorns and fruits. Origin: temperate to subtropical latitudes.

Lycium barbarum L., Chinese box thorn

Buds alternate, small, with few scales, sometimes with a tiny accessory bud. At the base of young shoots often laterally with two small buds (axillary buds of the prophylls). **Twigs** glabrous, pale grey or pale ochre-brown to olive-grey; slightly 5-angular due to fine ridges running down from the sides of the leaf scars. Lateral twigs branching off at right angles, often thorny. **Leaf scars** semi-circular, with a large, central, relatively dark trace. Shrub with initially upright, later pendulous twigs and ± strong development of runners. Fruit remnants are sometimes present: stalked for 17 mm, mostly solitary; only basally on the twig also in pairs. **Calyx** c. 4 mm long, divided halfway into 1–3 (mostly 2), blunt, 1.5–2 mm long lobes. Origin: China, in Europe naturalised here and there.

Helwingiaceae
flowering-rafts family

Helwingia japonica (THUNB.) F. DIETR.

Terminal buds 3–4 mm long, globose to short-ovoid, slightly flattened on one side, with 2 visible wine-reddish-greenish bud scales, outermost scale forming a hood over the top of the bud. **Lateral buds** alternate, small, strongly wine-red, with 2 acute bud scales, spreading at their tips. **Twigs** glabrous, slightly angular and curved, matte green, with a reddish-brown hue on the lighter side, beneath the leaf scars also stronger violet-brown. **Leaf scars** roundish, on distinct cushions, with a central trace and at the sides with small, spreading stipule lobes. Rare, dense shrub to 1.5 m high, producing runners. Origin: Japan.

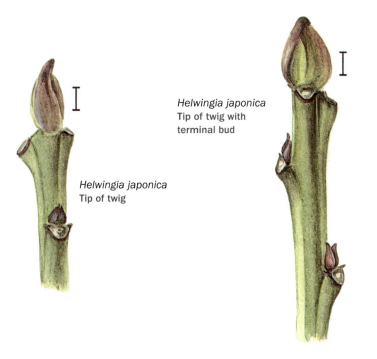

Helwingia japonica
Tip of twig

Helwingia japonica
Tip of twig with terminal bud

Aquifoliaceae
holly family

Woody plants, mainly in the tropics and subtropics. The largest genus is *Ilex* with approx. 400 species.

Ilex L., holly

Mostly dioecious trees and shrubs. Leaves alternate. Several evergreen and only a few deciduous species planted. The slender twigs of the introduced deciduous species with terminal buds. Leaves and bud scales have stipule lobes, the bud scales occasionally with a remnant of a leaf blade. Occasional red stone-fruits with 2–7 stone kernels, which persist long into winter.

Key to *Ilex*

1 Twigs grey-brown to ochre-brown, not noticeably dark. 2
1* Twigs dark violet-brown, fruit short-stalked with 4–7 smooth stones . . . .
. . . . ***Ilex verticillata***
2 Twigs sparsely hairy or almost glabrous 3
2* Twigs initially felty-hairy . . . . *Ilex serrata*
3 Bud scales of 2 colours, appressed to bud, fruits long-stalked (stalk at least 2–3 times as long as the fruit) with 3–5 woody stones . . . .*Ilex mucronata*
3* Bud scales narrowly acute, points slightly spreading. Twigs stiff, almost glabrous. Fruits short-stalked with 4 ribbed stones . . . . *Ilex decidua*

Ilex verticillata (L.) A. Gray, winterberry
Terminal buds short-ovoid, to 2 mm long.
Lateral buds appressed to twig, mostly less than 1 mm long. Bud scales dirty ochre to reddish-brown, darker at the margin, in the terminal buds with two stipule lobes.
Twigs slender, glabrous, very dark at first year's growth, olive to violet; at two years grey-white in places due to dead epidermis.
Lenticels few, small and pale. **Leaf scars** with one trace. **Fruits** long-persistent in winter: with short stalks, often in pairs, bright red and c. 6 mm thick, with 4–7 smooth, white stones. Frequently planted, divaricate shrub to 3 m high. Origin: North America. In winter, berried twigs are sold as Christmas decoration.

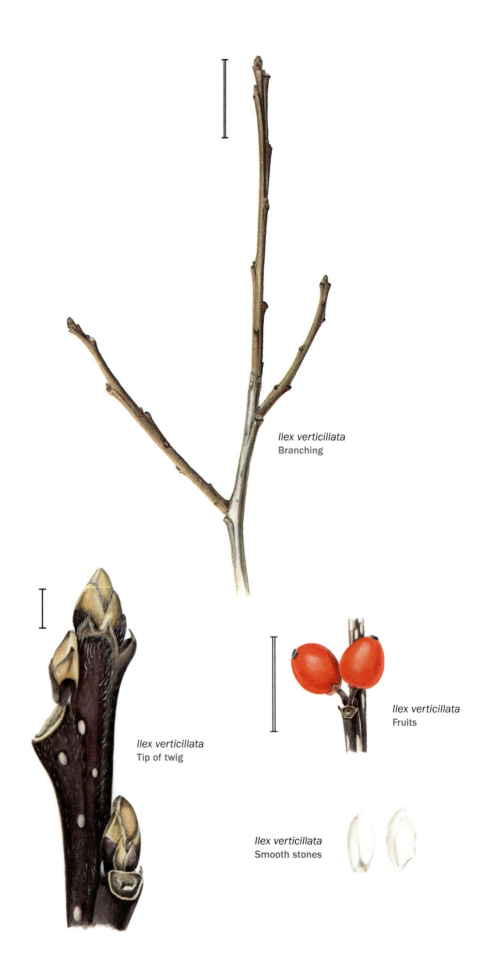

Ilex verticillata
Branching

Ilex verticillata
Tip of twig

Ilex verticillata
Fruits

Ilex verticillata
Smooth stones

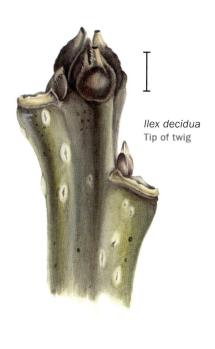

Ilex decidua
Tip of twig

Ilex decidua
Fruits on short-shoot

Ilex decidua
Twig with fruits

Ilex decidua
Ribbed stones

Ilex serrata
Tip of twig

Ilex mucronata
Tip of twig

The following species are rarely planted:

Ilex decidua Walter

Buds 1.5 to 2 mm long, short-ovoid. **Bud scales** dark wine-red to violet-black, on terminal buds with brownish blunt point and stipule lobes, on lateral buds with central ochre-brown point and inconspicuous stipule lobes. **Twigs** glabrous, olive-green in the shade to dark wine-red-brown in the sun. **Fruits** globose with 1 cm long stalks, solitary or in clusters on short-shoots, red, 7–8 mm diameter, with four ribbed stones. Rarely planted shrub, 1–2 m high. Origin: North America.

Ilex serrata Thunb.

Buds 1–2 mm long, globose-ovoid with some ochre- to dark brown bud scales with whitish indument; often with descending accessory buds. **Twigs** red-brown to pale grey-green, towards the tip relatively dark and hairy. **Fruits** in short clusters, 4–5 mm, red, with 2–4 stones, long-persistent in large numbers. At least stalks of fruit persistent over winter. Rarely planted shrub, 3–5 m high. Origin: eastern Asia.

Ilex mucronata (L.) M. Powell, Savol. & S. Andrews
[*Nemopanthus mucronatus* (L.) Trel.]

Terminal buds to 2.5 mm long, blunt-ovoid. The bud scales towards the tip paler ochre-brown, at the base discontinuous, glossy dark red-brown to violet-brown. **Lateral buds** smaller, spreading from the twig, sometimes with small descending accessory bud. **Twigs** sparsely hairy, slender (at most 2 mm diameter), tip dark brown, on the shaded side pale ochre, with green and grey hues, with small lenticels. **Leaf scar** with one trace. **Fruits** red, with 3–5 woody stones. Shrub to 3 m high, generating runners. Origin: north-east of North America.

Asteraceae
[Compositae *nom alt.*]
daisy family

The family occurs world-wide and contains approx. 1000 genera. It consists mainly of annual and perennial herbs. Some species of sub-shrubs are sometimes planted.

Baccharis halimifolia L., bush groundsel

Buds ovoid to orbicular, 4–6 mm long, covered by many small, fleshy, greenish-brown, sticky-resinous glossy bud scales. **Twigs** fine-angular, furrowed, whitish resinous with remnants of fruits at the end: of the heads the outer two small involucral bracts remain. Twigs usually with many dried brown and delicately white-spotted leaves. **Leaf scars** narrow, with 3 traces. Shrubs to 3 m high. Origin: southern and central America; naturalised in France and Spain.

Artemisia abrotanum L., southernwood

Buds with few, grey-green to ochre-brown scales, densely hairy, particularly towards tip. Terminal buds 2–3 mm long, globose-ovoid; lateral buds c. 2 mm long, ovoid, particularly towards tip of twig in the axil of a scaly leaf-remnant. **Twigs** aromatic and fragrant, ochre-brown to dark brown, hairy towards tip of twig. A sub-shrub to 1 m high. Distributed from southern Europe to Asia Minor and the Himalayas.

The true absinth wormwood *Artemisia absinthium* L., is similar but more densely hairy. Distributed from Europe to western Siberia and Asia Minor.

Escalloniaceae
currybush family

Escallonia virgata (RUIZ & PAV.) PERS.

Terminal buds 2–3 mm long, green with a slight reddish tint, occasionally with some persistent leaves. Bud-leaves at the margin with brown-red glands. **Lateral buds** smaller, spreading from the twig, lower down on the twig with some leaf scars at the base. **Twigs** at first year's growth c. 1.5 mm diameter, red-brown to ochre, hairy; at two years old, grey-brown. **Pith** solid, green. **Leaf scar** with one trace, on a distinct cushion,

Baccharis halimifolia
Lateral buds

Artemisia abrotanum
Lateral buds on the twig

Baccharis halimifolia
Twig with persistent
dry leaves

Artemisia abrotanum
Twig

Escallonia hybrid
Tip of twig

Sambucus racemosa
Twig

Sambucus nigra
Parts of twig

on both sides with ridges running down the twig a little. Very rarely planted shrub to 1 m high. Origin: Chile and Argentina.

Adoxaceae
elder family

Infructescences usually much branched umbel-like corymbs, with deciduous fleshy berries or stone fruits. Twigs usually robust. In the genus *Sambucus* pith very wide.

Key to Adoxaceae

1 Twigs thick, with warty lenticels and with wide pith, glabrous or with simple hairs, terminal bud often missing, or if present often half open. Lateral buds with more than 4 bud scales. Often with small stipule remnants between the leaf scars . . . . *Sambucus*

1* Lenticels and pith normal, glabrous, with simple hairs or often with scales or stellate hairs, mostly with terminal bud, with or without bud scales . . . *Viburnum*

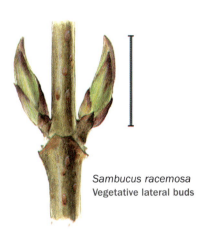

Sambucus racemosa
Vegetative lateral buds

Sambucus racemosa
Flower buds

Sambucus L., elder

Buds opposite, opposing leaf scars abut or connected. **Twigs** thick, with wide pith. One perennial herb and two common shrubs native to Europe. World-wide, about 25 species.

Key to *Sambucus*

1 Bud scale greenish to wine-red or brown, enclosing the bud, pith red-brown. . . . .*Sambucus racemosa*

1* Bud scales wine-red to violet-brown, mostly slightly spreading from early in season, pith white . . . . *Sambucus nigra*

Sambucus nigra L., common elder

Buds 5–9(–20) mm long, narrowly ovoid, at base with some small brown scales. The next bud leaves loose, dark wine red, often spreading and glabrous. Due to bud leaves emerging from the bud, always appearing well-advanced, and in mild weather also with bud growth even in midwinter. **Twigs** thick, glabrous, olive or brown to grey, slightly angular with many large, raised grey-brown lenticels. **Pith** very wide, loose, white. **Leaf scars** on distinct cushions, with 3 traces. Frequent shrub or small tree to 7 m high. Native to Europe.

In botanical collections closely related taxa may be found, these can be seen as species or subspecies: *Sambucus caerulea* RAF. and *Sambucus canadensis* L.

Sambucus racemosa L., red-berried elder

Buds glabrous, 10–12 mm long, usually only lateral buds: narrow fusiform leaf buds and ± globose flower buds. The buds at the base relatively slender (almost stalked) with a pair of small, brown scales, next are 2–3 green to wine-red pairs of bud scales which envelop the bud. **Twigs** glabrous, relatively thick, brown to olive. **Pith** wide, mostly distinctly brown-red. **Lenticels** many, relatively large, ± brown. **Leaf scars** on cushions with 3(–5) traces. Shrub to 4 m high. Origin: Europe.

Viburnum L.

Shrubs with opposite 3-trace leaf scars, connected by a line. There are taxa without bud scales, with free, and with fused bud scales. In some species stellate hairs occur, others have simple hairs or are glabrous. May be confused with species of maple and dogwood.

Key to *Viburnum*

1 Buds without bud scales. Inflorescence bud terminal, very broad 2
1* Buds with bud scales 3
2 Twigs dark grey-brown with stellate hairs, glossy, centre of inflorescence bud composed of individual buds (partial inflorescences) . . . .
. . . . ***Viburnum carlesii*** and **hybrids**
2* Twigs matte, pale grey to ochre grey-green, centre of inflorescence-bud forming a compact unit . . . .
. . . . ***Viburnum lantana***
3 Buds 2–3 times as long as thick, outermost pair of bud scales completely fused and enveloping the bud, (green-) red, glabrous . . . . ***Viburnum opulus***
3* Outermost pair of scales at most fused at base or buds longer, often hairy 4
4 Terminal bud over 3 times as long as thick 5
4* Terminal buds more compact 7
5 Buds brown to blue-grey due to peltate hairs or wax layers 6
5* Buds greenish, often with a strong red hue, if brown with stellate hairs 7
6 Buds without peltate hairs . . . .
. . . .*Viburnum lentago*
6* Buds densely covered with peltate hairs (flat, roundish, brown-silvery, slightly darker at the centre) . . . .
. . . . *Viburnum cassinoides*
7 Flowers with or after the leaves 8
7* Flower buds well-developed and in mild temperatures flowering in winter . . . . . .
. . . .*Viburnum farreri* and
Viburnum ×bodnantense
8 Bud scales and twigs brown with stellate hairs. Terminal buds to 12(–15) mm, first pair of bud scales mostly as long as the bud, often spreading rather wide . . . . . .
. . . . ***Viburnum plicatum***
8* Buds to 10 mm long, first pair of scales not reaching length of buds 9

9 Stone fruits black, mostly deciduous . . .
. . . .*Viburnum dentatum*
9* Fruits in large infructescences and persistent during winter, red . . . .
. . . .*Viburnum betulifolium*

Section Viburnum (including Lentago)

Viburnum carlesii HEMSL., Korean spice viburnum

Buds dark grey-brown with felty stellate indument. Floral buds broad, basket-shaped, composed of several individual buds (partial inflorescence), between these small scale-like bracts. Inserted below the entire inflorescence bud is a (rarely deciduous) pair of leaves, which consists of two pinnate-veined leaves which reach over the bud and which, folded up, show their lower (abaxial) side. **Twigs** red brown to grey-brown, initially densely felty-stellate, gradually becoming glabrous and the glossy surface of the twig becoming visible. Frequently planted, lax shrub which contributes to many hybrids. Origin: Korea.

Beside the wild species, some often planted hybrids: **Viburnum ×carlcephalum** BURK. ex R. B. PIKE (×*Viburnum macrocephalum* FORTUNE) and the semi-evergreen **Viburnum ×burkwoodii** BURKW. & SKIPW. (×*Viburnum utile* HEMSL.).

Viburnum lantana L., wayfaring tree

Buds densely pale grey to ochre or pale green felty-stellate. Foliar buds oblong. **Inflorescence bud** compact, surrounded by relatively broad, flat involucral bracts. A division into individual buds is, therefore, not visible. The inflorescence bud is flanked by a pair of leaves that overtop it. **Twigs** matte, pale grey-brown with dense stellate felty indument, slightly angular, striped. **Lenticels** few, becoming distinct only on older twigs. **Bark** grey-brown and fissured lengthwise. Frequently planted shrub, erect, strong growing, to 5 m high. Origin: Europe and Asia Minor.

Viburnum carlesii
Twig

Viburnum lantana
Twig

Viburnum carlesii
Tip of twig with
inflorescence bud

Viburnum lentago
Twig with terminal
inflorescence bud

Viburnum lantana
Terminal inflorescence
bud

Viburnum lentago
Twig with terminal
inflorescence bud

Viburnum lentago
Lateral buds

Viburnum lentago
Terminal bud

"Section Lentago" (with Section Viburnum)

Viburnum lentago L., sheepberry

Buds ochre to red-brown on the shaded side, in the light darker, in large part covered by pale grey wax. Terminal bud long and narrow, acute, outer pair of scales enveloping the bud, usually fused at base. Inflorescence bud c. 2 cm long, widest below the middle, long-attenuate above. **Twigs** smooth, slightly glossy, glabrous, olive-green to grey-brown, with many distinct lenticels towards the tip. Rarely planted shrub or small tree to 10 m high. Origin: North America.

Viburnum cassinoides L., the Appalachian tea tree is similar. **Buds** densely covered in peltate hairs (flat, roundish, brownish-silvery, slightly darker at the centre), scales of the inflorescence buds in the top-half long, narrowed, at base dark brown, above also grey. Inflorescence buds similar, but the first pair of scales does not cover the wider part, often two-tipped. **Twigs** below the tip brownish, lower down predominantly grey, matte, with fine dots (peltate scales), lenticels initially faint. Rarely planted shrub to 2 m high. Origin: North America.

Viburnum cassinoides
Terminal inflorescence bud

Section Opulus

Viburnum opulus L., Guelder rose

Buds glabrous, glossy, red to red-brown, at the base ± greenish, in the shade too predominantly greenish. Terminal buds c. 7 mm long, but mostly lacking; like the lateral buds, surrounded by a bud scale, originating from two fused scales. Floral buds globose, foliar buds oblong. **Twigs** glabrous, ochre to red-brown, in the shade also grey-green. **Lenticels** few to many. **Bark** pale grey, glossy. **Fruits** red, globose-ovoid, in flat corymbs and often long persistent. Shrub 2–4 m high, common in moist woods and cultivated since ancient times. Origin: Europe to north Africa.

Viburnum sargentii KOEHNE is very similar to *Viburnum opulus*. **Buds** red-brown, glabrous and slightly glossy, 7–9 mm long. **Twigs** initially red-brown to grey-brown, with remnants of indument. **Bark** slightly corky and thick. **Fruits** almost globose, 1 cm in diameter, pale red. Shrub 2–3 m high. Origin: north-east Asia.

Section Pseudopulus

Viburnum plicatum THUNB., Japanese snowball

Buds matte, wine-red to grey-brown with sparse stellate hairs; with an outer pair of bud scales which initially envelops the buds completely, but which fall off later, particularly from the terminal bud. Lateral buds 4–6 mm, oblong, narrow. Terminal buds 6–12(–15) mm long, slim, ovoid in floral buds. **Twigs** initially felty-stellate, grey-brown to ochre-brown, with round, warty ochre-brown lenticels. **Leaf scars** ochre-brown with 3 traces of vascular bundles. Only in the wild form var. *tomentosum* MIQ. with red, later with blackish fruits. Frequently planted, shrub to 3 m high. Origin: China and Japan; cultivated there since ancient times.

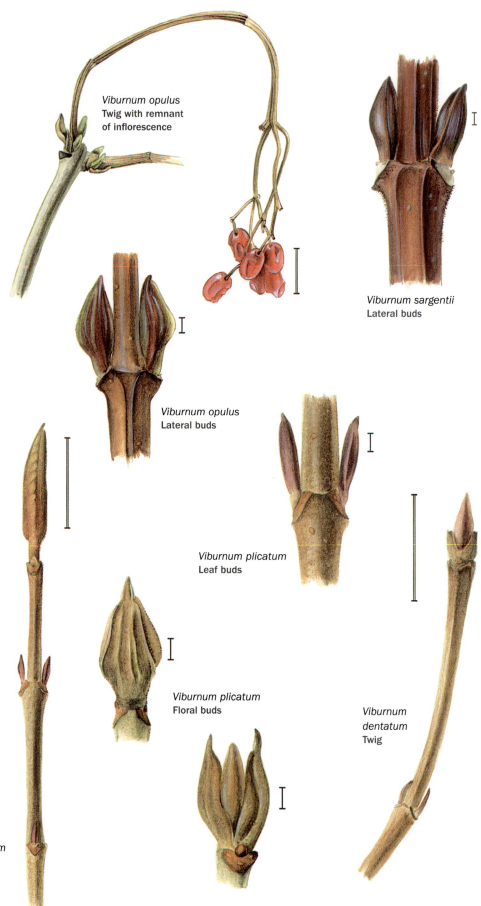

Viburnum opulus
Twig with remnant of inflorescence

Viburnum sargentii
Lateral buds

Viburnum opulus
Lateral buds

Viburnum plicatum
Leaf buds

Viburnum plicatum
Floral buds

Viburnum dentatum
Twig

Viburnum plicatum
Twig

Viburnum dentatum
Terminal bud

Viburnum dentatum
Lateral buds

Viburnum
betulifolium
Lateral buds

Viburnum
betulifolium
Terminal bud

Viburnum farreri
Inflorescence

Viburnum farreri
Individual flower

Viburnum farreri
Terminal bud

Viburnum farreri
Tip of twig

Viburnum ×bodnantense
Inflorescence

"Section Odontotinus"

[The Old- and New World species grouped here have a certain relationship, but do not necessarily belong together.]

Viburnum dentatum L., arrowwood viburnum

Buds 4–6 mm long, surrounded by two pairs of bud scales: basally two rough, grey-brown slightly spreading and hairy scales which reach only little over half the height of the bud; subsequently two wine-red to brown smooth ± glabrous scales which envelop the conical tip. Terminal bud acute-ovoid, lateral buds slimmer, when the terminal bud is absent two near-terminal lateral buds are more developed. **Twigs** with matte grey-brown to light red-brown bark, partly with stellate hairs. **Lenticels** few, orange-brown to grey-brown, warty. **Fruits** blue-black. Rarely encountered, very variable shrub, to 1.5 m high. Origin: North America, with various varieties in dry to moist habitats.

Viburnum betulifolium BATALIN

Buds ovoid, with 2 pairs of scales, the lower one about half as long as the bud. **Twigs** glabrous, red-brown to grey. **Fruits** only in older shrubs: red, in heavy, pendulous corymbs. Rarely planted shrub to 4 m high. Origin: central and north China.

Section Solenotinus

Viburnum farreri STEARN, Farrer's viburnum

Buds with 2–3 visible greenish-red pairs of bud scales. **Twigs** upright, red-brown. Densely branched shrub to 3 m high. The flowers are distinctive, because they sometimes appear from November, but mostly from February to April. In the floral bud they are pink, when flowering white and strongly scented. Origin: north China.

This species can only be confused with a hybrid with *Viburnum grandiflorum* WALL. ex DC. as the other parent, which is also planted frequently: ***Viburnum ×bodnantense*** ABERC. ex STEARN. It is characterised by slightly larger flowers, which in the floral buds are pink to rose-red.

Caprifoliaceae
honeysuckle family

Leaf scars opposite (rarely 3- or 4-whorled), leaf scars of a node connected by a line or abutting. Usually 3, more rarely 5 or more traces of vascular bundles per leaf scar. Inferior ovaries, where the calyx lobes are always at the top of the fruit, are typical for the entire order of Dipsacales. Predominantly medium-sized shrubs, almost no trees. Can be confused with maple (*Acer*), species of *Viburnum* or species of *Deutzia*. The fruits, which in some species are regularly available in winter, are useful for identification.

Key to Caprifoliaceae

1 Twigs hollow[1] . . . .
. . . . **Subfamily Caprifolioideae**

1* Twigs with solid pith 2

2 Lateral buds closely appressed to twig, fruits 2-chambered, oblong, woody capsule . . . . . . **Subfamily Diervilloideae**

2* Lateral buds not appressed to twig (or not visible) 3

3 Fruits indehiscent, persistent in winter . 4

3* Fruits paired and mostly deciduous berries . . . . *Lonicera*

4 Fruits with enlarged sepals. Uppermost node of the strong twigs (at the base 3 mm diameter) died off . . . . *Heptacodium*

4* Fruits different or youngest branches at the base to 2 mm diameter . . . .
. . . . **Subfamily Linnaeoideae**

[1]Some very seldom cultivated *Abelia* species, not dealt with here, also have hollow twigs.

SUBFAMILY **Diervilloideae**

The genera have in common: acute lateral buds which are closely appressed to the twig, and dry, long-persistent fruit capsules. There are dIfferences in the formation of the terminal bud, the shape of the capsule, and in the seeds.

Key to Diervilloideae

1 Terminal bud absent, fruit capsules 6–15 mm long, low shrubs to 1 m high . . . . *Diervilla*

1* Terminal buds present, fruit capsules 15–30 mm long, shrubs usually well over 1 m high 2

2 Terminal buds naked . . . .
. . . . *Weigela* Section *Weigela*

2* Terminal buds covered in bud scales . . 3

3 Terminal bud pale straw-coloured, lowest scales almost reaching to top of bud. Fruit capsule without central column, seeds long-winged on both sides . . . . . .
. . . . *Macrodiervilla middendorffiana*

3* Terminal bud deep brown, lowest scales shorter. Capsule with central column, seeds without wings . . . .
. . . . *Weigela*, Section *Calysphyrum*

Weigela THUNB., weigela

Mostly 2–3 m high erect shrubs with 1.5–3 cm high, 2-chambered fruit capsules which have a central column and open by 2 valves. Origin: eastern Asia. The traditionally narrow species delimitation leads to species which are very close to each other and which are difficult to tell apart. The cultivated species can mostly only be identified to section. This is because the plants hybridise easily and also because breeding has created many cultivars.

Key to *Weigela*

1 Terminal buds with bud scales. Fruit: calyx lobes broad-triangular, fused for half their length into a tube . . . .
. . . . Section *Calysphyrum*

1* Terminal buds without bud scales, or tip of twig dying. Fruit: calyx lobes narrow, linear, separate down to the base, deciduous . . . . Section *Weigela*

Weigela
Fruiting twig

Weigela florida
Tip of twig with terminal bud

Weigela florida
Twig

Weigela florida
Lateral buds

Weigela florida
Lateral bud

Weigela section Weigela
Shoot tip

Macrodiervilla middendorffiana
Tip of twig with terminal bud

Section Calysphyrum

Weigela florida (BUNGE) A. DC.
(incl. var. **praecox** (LEMOINE) Y. C. CHU).

Buds acute-ovoid. Terminal buds 8–10 mm, lateral buds appressed to twig, 6–8 mm long. Bud scales acute and partly slightly keeled, ochre-brown, darker towards the margin and long white-ciliate. **Twigs** light ochre-brown to orange-brown, with slight oblong stripes and little fissuring; with two rows of hairs which descend between the leaf scars. **Leaf scars** pale, with 3 traces almost lying in one plane with the twig. **Fruits** 3–4 glabrous axillary capsules, without a common stalk, calyx lobes broadly triangular. Seeds not winged. The most frequently planted species of the genus. Shrub to over 3 m high. Origin: northern China and Korea.

Section Weigela

Weigela floribunda (SIEBOLD. & ZUCC.) K. KOCH

Twigs slender, at least initially hairy in two rows. Terminal buds without bud scales or tip of shoot dying. **Fruits** mostly hairy; cylindrical, often a little curved, to 2.5 cm long, solitary in leaf axils and in small groups on short lateral shoots. Seeds with wing-like margin. Shrub to 3 m high. Origin: Japan.

Some other species of the section are similar, like **Weigela coraeenensis** THUNB. with glabrous twigs and fruits. The latter is less common in cultivation. Origin: Japan.

Macrodiervilla middendorffiana (CARRIÈRE) NAKAI

Buds: with terminal buds, scales loose, long-acute, partly with visible pinnately-veined structure, straw-coloured, pale brown, only a few green areas. **Twigs** below the tip relatively thick (3 mm diameter), matte, pale ochre-grey with two hairy vertical ridges. **Fruit** without central column, opening by two valves, calyx deciduous. **Seed** long-winged on both sides. Very rarely planted, shrub c. 1.5 m high. Origin: eastern Asia, from Japan and Korea towards the north into the Amur region and also on Sakhalin.

Diervilla MILL., bush honeysuckle

Twigs and buds very similar to the weigelas, but plants differ by lower growth (to 1 m high) and shorter, 6–15 mm long, 2-valvate fruit capsules. Dense shrubs with short stolons. Origin: North America.

Key to *Diervilla*

1 Twigs hairy, often on longitudinal ridges
. . . . *Diervilla sessilifolia*

1* Twigs entirely glabrous . . . .
. . . . *Diervilla lonicera*

Diervilla sessilifolia BUCKLEY, southern bush honeysuckle

Buds long-ovoid, acute, lateral buds appressed to twig. **Twigs** slightly 4-angular, ridges and nodes hairy. **Fruits** many in corymbs, lateral ones with 5–7, terminal with 15 or more fruits. The most frequently planted species of the genus.

Diervilla rivularis GATT. [*Diervilla sessifolia* var. *rivularis* (GATT.) H. E. AHLES] is very similar, but differs in rounder twigs which are hairy throughout.

Diervilla lonicera MILL., northern bush honeysuckle

Differs from the other species of the genus by round, glabrous twigs and few-fruited cymes: axillary cymes with 3 and terminal ones with 5 fruits. Shrub producing stolons.

Heptacodium miconioides
Node with lateral buds

Symphoricarpos albus
var. *laevigatus*
Lateral bud with
enrichment bud

Symphoricarpos albus
var. *laevigatus*
Lateral bud

Symphoricarpos albus
var. *laevigatus*
Twig with white
stone fruits

Diervilla sessilifolia
Lateral buds

Diervilla sessilifolia
Twig with remnants
of fruit

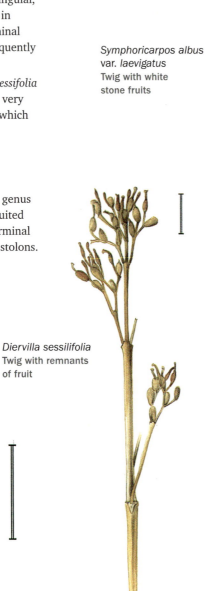

Symphoricarpos
×*chenaultii*
Fruiting twig

Symphoricarpos occidentalis
Twig with fruits that quickly
turn brown

SUBFAMILY Caprifolioideae

Fruits usually fleshy berries or stone fruits; only in the genus *Heptacodium*, rather isolated in the subfamily, a dry nut.

Key to Caprifolioideae

1 Twigs hollow 2
1* Twigs with solid pith 5
2 Lianas . . . *Lonicera*, Subgenus *Lonicera*
2* Erect shrubs 3
3 Berry-like stone fruits in terminal spikes, twigs brown or green 4
3* Berries in axillary pairs, buds often with ascending accessory buds . . . .
. . . . . *Lonicera*, Section *Coeloxylosteum*
4 Twigs brown, buds to 2(–3) mm long, often with enrichment buds . . . .
. . . . *Symphoricarpos*
4* Twigs green, thick buds almost concealed under petiole remnants . . . . *Leycesteria*
5 Fruits paired, mostly deciduous berries, buds often with ascending accessory buds
. . . . *Lonicera*
5* Fruits dry oblong nut, crowned by large calyx lobes, buds solitary . . *Heptacodium*

Heptacodium miconioides REHDER, seven son flower tree

[*Heptacodium jasminoides* AIRY SHAW]

Buds only narrowly ovoid, acute, glabrous lateral buds with 5–6 visible ochre-brown pairs of scales. **Twigs** stiff, 3–4 mm diameter, slightly angular, larger below the leaf scars, from where weak ridges descend, almost to the next node. Bark on one side very pale-ochre-coloured with green hues, on the other side a little darker with an epidermis which is red-brown, but part of it dies and turns grey. Lenticels many, small, ochre-coloured like the twig. The uppermost part of the shoot is dead, without buds. **Pith** solid, white, at the node interrupted by a compact green wall. **Leaf scars** opposite, connected by a fine line, blackish, with three traces. **Fruits** hypogynous, a nut crowned by 5 persistent, c. 1 cm long, narrow calyx lobes (similar to *Abelia*). Rarely planted shrub to 7 m high. Origin: China.

Symphoricarpos DUHAMEL., snowberry

Buds relatively small, often with enrichment buds from the axils of the bud scales. **Twigs** slender and hollow. **Fruits**: white to red berry-like stone fruits in terminal spikes. Frequently planted ornamental shrubs.

Key to Symphoricarpos

1 Twigs glabrous, fruits white . . . .
. . *Symphoricarpos albus* var. *laevigatus*
1* Twigs hairy, fruits reddish, if white, soon turning brown 2
2 Fruits reddish 3
2* Fruits initially greenish-white, soon a dirty brown . . . .
. . . . *Symphoricarpos occidentalis*
3 Fruits completely red . . . .
. . . . *Symphoricarpos orbiculatus*
3* Fruits only red on one side . . . .
. . . . . *Symphoricarpos* ×*chenaultii* and
Symphoricarpos ×*doorenbosii*

Symphoricarpos albus (L.) S. F. BLAKE var. laevigatus (FERNALD) S. F. BLAKE, snowberry

Buds 1.5 mm long, glabrous, with 2–4 outer grey to red-brown pairs of bud scales, towards the tip occasionally green. On larger twigs often one to two enrichment buds in the axils of prophylls, the primary axillary bud partly wasting away. **Twigs** grey-brown, ± glossy, glabrous, often slightly zig-zag. In winter the species can be easily recognised by the 1.5 cm thick stone fruits. This frequently planted variety attains 2 m in height. Origin: California to Alaska.

The typical variety, var. *albus*, has more slender and erect finely hairy twigs. This shrub, growing to 1 m high, is rare in cultivation. Origin: North America.

Symphoricarpos ×chenaultii REHDER, pink snowberry

[*Symphoricarpos microphyllus* KUNTH × *Symphoricarpos orbiculatus*]

Buds obliquely spreading (45°), on relatively large leaf cushion, 1–1.5 mm long. Bud scales, especially the two prophyll scales, spreading at the tips. Buds the same colour as the twig: red-brown to grey-brown, densely but not continuously spreading-hairy. **Twigs**, when young, short and finely hairy, matte ochre- to red-brown, 1.5–2 mm diameter. **Stone fruits** reddish on the light side with white dots, on the shaded side white with reddish dots. Dainty, laxly branching shrub to over 2 m high. In the coralberry, *Symphoricarpos orbiculatus* MOENCH., the parent species from North America, which is also planted frequently, the stone fruits are purple-red and 4–6 mm thick. An upright 1–2 m high shrub. A further hybrid with white fruits which are reddish on one side is *Symphoricarpos* ×*doorenbosii* KRÜSSM. [*Symphoricarpos albus* var. *laevigatus* × *Symphoricarpos* ×*chenaultii*], a vigorous shrub to 2 m high.

Symphoricarpos occidentalis HOOK., wolfberry

Twigs hairy, slightly pendulous. **Stone fruits** greenish-white, soon turning brown, c. 1 cm thick. Occasionally encountered, 1–1.5 m high, erect shrub with overhanging branches. Origin: USA.

Symphoricarpos ×*chenaultii*
Lateral buds

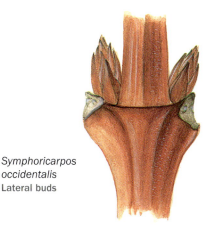

Symphoricarpos occidentalis
Lateral buds

Lonicera L., honeysuckle

Many planted and several native species. Common characters are the long-persistent bud scales of the previous year, and the buds that often spread from twig (except for the *Lonicera fragrantissima* group).

While the plants are fairly easily assigned to groups, determination to species within the groups is often difficult or impossible. Helpful hints can be the type of growth (shrub, tree or liana), round or angular twigs, pith (solid or twigs hollow), the presence of ascending(!) accessory buds, indument and waxy bloom, shape and size of buds as well as the bark.

Key to *Lonicera*

1 Winding lianas 2
1* Erect shrubs 3

Subgenus Lonicera

2 On the twigs at least blackish dots of original indument, bracts separate . . . . .
. . . . *Lonicera periclymenum*
2* Twigs completely glabrous, bracts fused at tip of twig. .*Lonicera caprifolium* and many often planted species and hybrids

Subgenus Chamaecerasus

3 Twigs with solid pith 4
3* Twigs hollow 10
4 Buds in transverse section 4-angular with many bud scales (> 6 pairs) 13
4* Buds in transverse section ± round or with few bud scales 5

Basal Groups

5 Buds with 1(–2) visible pairs of bud scales 6
5* Buds with more pairs of visible bud scales
. . . . 8
6 Buds and twigs bluish with waxy bloom, twigs with terminal buds . . . .
. . . .*Lonicera caerulea*
6* Without waxy bloom, twigs mostly without terminal buds 7
7 Growing almost during the entire winter and flowering white . . . .
. . . .*Lonicera fragrantissima*
7* Dormant. . . . . *Lonicera ferdinandi*
8 Twigs angular. . . . *Lonicera involucrata*
8* Twigs round, bark splitting lengthwise into fibres or tearing open lengthwise . 9
9 Terminal buds c. 10 mm long, twigs relatively robust . . . . .*Lonicera alpigena*
9* Terminal buds c. 3 mm long, twigs slender . . . . *Lonicera pyrenaica*

Section Coeloxylosteum

10(3) Buds globose to ovoid, fruit stalks over 5 mm long 11
10* Buds narrowly ovoid to fusiform or fruit stalks under 5 mm long 12
11 Twigs glabrous . . . . . *Lonicera tatarica*
11* Twigs densely hairy *Lonicera morrowii*
12 Buds fusiform, twigs soon becoming glabrous . . . . *Lonicera xylosteum*
12* Buds narrowly ovoid, twigs hairy for longer, fruit stalk less than 5 mm long .
. . . . *Lonicera maackii*

Section Rhodanthae

13(4) Buds sharply 4-angular, bud scales not oblong . . . .*Lonicera maximowiczii*
13* Buds rounded 4-angular, bud scales with long extended tip 14
14 Twigs with fine indument, fruit stalk usually over 2 cm long . *Lonicera nigra*
14* Twigs glabrous, fruit stalk less than 1.5 cm long . . . .*Lonicera caucasica*

Subgenus Lonicera

Mostly climbing by winding (clockwise), more rarely erect shrubs, with hollow twigs. Fruits mostly persistent into early winter, mostly in groups of 6 in pseudo-whorls at the twig ends, in contrast to the pairs of fruits of the *Chamaecerasus* group. Very frequently planted species and hybrids, which are difficult to tell apart.

Lonicera caprifolium L., perfoliate honeysuckle

Buds 8–10 cm long, spreading from the twig. Outer bud scales ochre-brown to grey-brown. Inner leaves, particularly when developing, greenish. **Twigs** ochre-brown to red-brown and occasionally with waxy bloom, finely striped lengthwise, later tearing open. Initially sparsely hairy, becoming glabrous. Towards the tip of twig usually with remnants of the disk-like fused bracts. **Fruits** deciduous, in the axils of the upper bracts. Frequent shrub, winding to 5 m high. Origin: eastern central Europe to southern Europe.

Lonicera caprifolium
Lateral buds

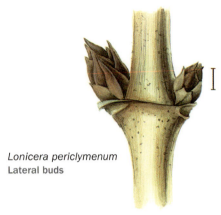

Lonicera periclymenum
Lateral buds

Lonicera ×*brownii*
Lateral buds

Lonicera ×*heckrottii*
Lateral buds

Lonicera periclymenum L., common honeysuckle, woodbine

Buds ovoid, to c. 3 mm long with a few acute brown and ± glabrous bud scales. **Twigs** pale grey-brown to orange-brown, with fine blackish dots and remnants of indument. **Leaf cushions** prominent, with indistinct leaf scar. Bracts not fused. Frequently planted shrub, winding to 6 m high. Origin: Central to Western Europe, in the South spreading to North Africa.

The hybrids of *Lonicera sempervirens* L., trumpet honeysuckle, with glabrous twigs and fused bracts are frequently planted. They can be identified only when fruit remnants are present. Two fruit whorls immediately above one another are typical for the yellow-flowered hybrid ***Lonicera ×tellmanniana*** MAGYAR ex H. L. SPÄTH, Tellmann's honeysuckle [× *Lonicera tragophylla* HEMSL.], with olive-green shoots, whilst the fruiting whorls of the other two hybrids are distinctly separate. In ***Lonicera ×brownii*** (REGEL) CARRIÈRE [× *Lonicera hirsuta* EATON] there are only small bracts of the first order next to the ovary. In ***Lonicera ×heckrottii*** hort. ex REHDER [? × *Lonicera ×americana* (MILL.) K. KOCH], a relatively slightly winding shrub in which, next to the ovary, bracts of the second order are also developed.

Lonicera fragrantissima
Twig with flowers

Lonicera fragrantissima
Lateral buds

Subgenus Chamaecerasus

Mainly erect shrubs, more rarely climbing lianas. The classical division into sections by REHDER (1903) does not portray the actual relationships. Therefore the species of the "basal groups" are combined here and contrasted with the sections *Coeloxylosteum* and *Rodanthae*. The group sequence reflects an increase in the number of bud scales. Fruits: berries in axillary pairs, very rarely (*Lonicera iberica* M. BIEB.) also at the ends of shoots.

Basal groups

The species have a solid pith. Buds often with accessory buds. Terminal buds present or absent. The sequence mirrors the position in the phylogeny (THEIS 2008). The buds of the first species (*L. fragrantissima*, *L. caerulea* and *L. ferdinandi*) are mostly widely enveloped by the first pair of bud scales.

Lonicera fragrantissima LINDL. & PAXTON, sweetest honeysuckle
[*Lonicera standishii* CARRIÈRE, *Lonicera ×purpusii* REHDER]

Terminal buds mostly absent, lateral buds with only two outer bud scales: the upper ones almost always ± enlarged, basal lateral buds less often so: c. 4 mm long, relatively flat, with two acute, ochre-brown glabrous but ciliate prophyll scales. **Twigs** glabrous or slightly hairy, ochre-brown to pale grey. **Leaf scars** indistinct 3-traced, on distinct cushions. Shrubs flowering in March, or in mild weather from December onwards. **Flowers** strongly scented, white, when fading cream-coloured or with a pink hue, c. 1.5 cm long with short, fused crown tube, pleated-buckled at the base, and a 2-lipped corolla. Occasionally planted, lax shrub to 2 m high. Origin: China.

Lonicera fragrantissima
Pair of flowers in February

Lonicera caerulea L., blue honeysuckle

Buds dark wine-red to violet-brown, on one side white-blue with waxy bloom. Terminal buds onion-shaped: basally globose, acute, 7–9 mm long, outer bud scales spreading at the tips, lateral buds ± spreading, narrowly ovoid with 2 keeled bud scales. Often with several accessory buds. **Twigs** on first year's growth wine-red to dark brown, white-blue with waxy bloom above; older growth with flaking grey-brown bark, pale ochre-brown below. Erect shrub 1–1.5 m high. Origin: very widely distributed across the entire Northern Hemisphere; variable shrub with many varieties; among these some with larger edible fruits, which are often regarded as a separate species: *Lonicera kamtschatica* Pojark.

Lonicera ferdinandi Franch.

Buds are lateral buds spreading from the twig with a visible pair of scales, glabrous on the outside, like the twigs ochre- to grey-brown, with dense indument inside the bud-scales. **Twigs** with stiff, long hairs on little pedestals. On vigorous shoots winged-to disk-shaped dilations between leaf stalks (formation of auricles), strongest on shoots from the base of the stem. **Fruits** very short-stalked, axillary and less often also terminal. Large shrub. Origin: Mongolia and northern China.

Lonicera pyrenaica L.

Buds blunt-conical, basally grey to grey-brown, uppermost bud scales pale ochre-brown, lateral buds 2–3 mm, terminal buds 3–4 mm long. **Twigs** with solid pith; glabrous, pale grey, brownish below the leaf scars, longitudinally fissured, later lifting off in fine fibres. Small shrub to 1 m in height. Origin: Pyrenees.

Lonicera involucrata (Richardson) Banks ex Spreng., Californian honeysuckle

Buds grey to grey-brown, terminal buds ovoid, to 4(–5) mm long, blunt or slightly acute, with c. 3–4 pairs of scales; lateral buds small, ± triangular seen from above, with 1–2 pairs of outer scales. Bud scales keeled and slightly acute. **Twigs** glabrous, matte grey-brown to glossy ochre-brown and at the tip of twigs strongly 4-angular,

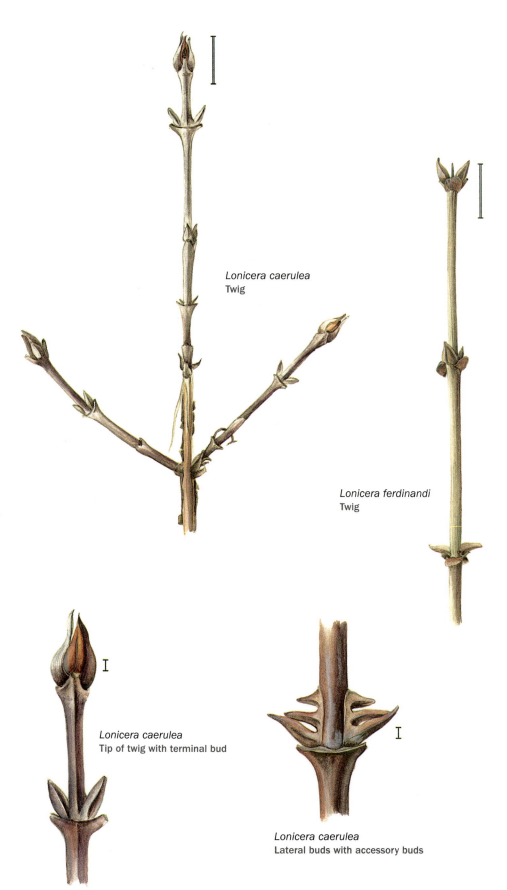

Lonicera caerulea
Twig

Lonicera ferdinandi
Twig

Lonicera caerulea
Tip of twig with terminal bud

Lonicera caerulea
Lateral buds with accessory buds

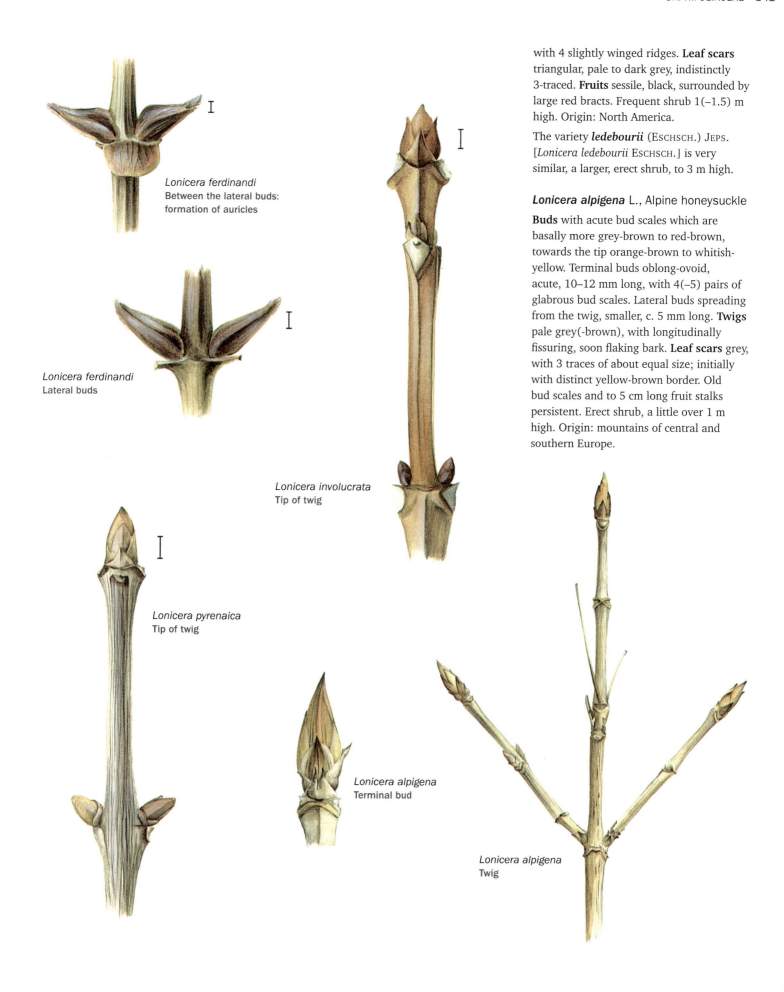

Lonicera ferdinandi
**Between the lateral buds:
formation of auricles**

Lonicera ferdinandi
Lateral buds

Lonicera pyrenaica
Tip of twig

Lonicera involucrata
Tip of twig

Lonicera alpigena
Terminal bud

Lonicera alpigena
Twig

with 4 slightly winged ridges. **Leaf scars** triangular, pale to dark grey, indistinctly 3-traced. **Fruits** sessile, black, surrounded by large red bracts. Frequent shrub 1(–1.5) m high. Origin: North America.

The variety ***ledebourii*** (ESCHSCH.) JEPS. [*Lonicera ledebourii* ESCHSCH.] is very similar, a larger, erect shrub, to 3 m high.

Lonicera alpigena L., Alpine honeysuckle

Buds with acute bud scales which are basally more grey-brown to red-brown, towards the tip orange-brown to whitish-yellow. Terminal buds oblong-ovoid, acute, 10–12 mm long, with 4(–5) pairs of glabrous bud scales. Lateral buds spreading from the twig, smaller, c. 5 mm long. **Twigs** pale grey(-brown), with longitudinally fissuring, soon flaking bark. **Leaf scars** grey, with 3 traces of about equal size; initially with distinct yellow-brown border. Old bud scales and to 5 cm long fruit stalks persistent. Erect shrub, a little over 1 m high. Origin: mountains of central and southern Europe.

Section Coeloxylosteum

Erect shrubs with hollow twigs. In the genus *Lonicera* hollow twigs are only found outside this section in winding lianas (subgenus *Lonicera*) and in the evergreen section *Nintooa* (e.g. *Lonicera japonica*).

Lonicera tatarica L.

Buds spreading from the twig, globose to ovoid, blunt or short-acute, to 5 mm long; with ascending accessory buds. Bud scales grey-brown, appressed or slightly spreading at the margins, glabrous, only at the margins slightly ciliate. **Twigs** glabrous, initially pale ochre-brown to silvery grey. **Bark** later grey, its flaking hardly noticeable. **Leaf scars** with 3 traces. Many infructescence stalks persist, if not broken off 1.5–2 cm long. Many cultivars and hybrids, very frequently planted, shrub, 2–4 m high. Origin: western Europe to western Asia.

Lonicera korolkowii STAPF is similar; a shrub to 3 m high with hairy twigs. Origin: central Asia.

Lonicera xylosteum L., fly honeysuckle

Buds slender, fusiform, to c. 8 mm long, lateral buds spreading with ascending accessory buds, these often persistent above lateral branching. Covered by 5–7(–9), red or ochre-brown to dark brown coloured pairs of bud scales which have, towards the tip, a long and whitish indument. **Twigs** initially sparsely spreading-hairy; slender, usually zig-zag; pale grey, grey-brown to ochre and red-brown. Pith thinly hollow, xylem brown. **Leaf scars** with 3 traces, on a slightly developed leaf cushion. Very frequently planted shrub, to 2.5 m high. Origin: Europe to central Asia.

The occasionally planted **Lonicera chrysantha** TURCZ. ex LEDEB. is similar, but more vigorous in all parts: a 2–4 m high shrub. **Buds** larger: lateral buds to 10 mm, terminal buds to 12 mm long. Origin: northeast Asia to central China.

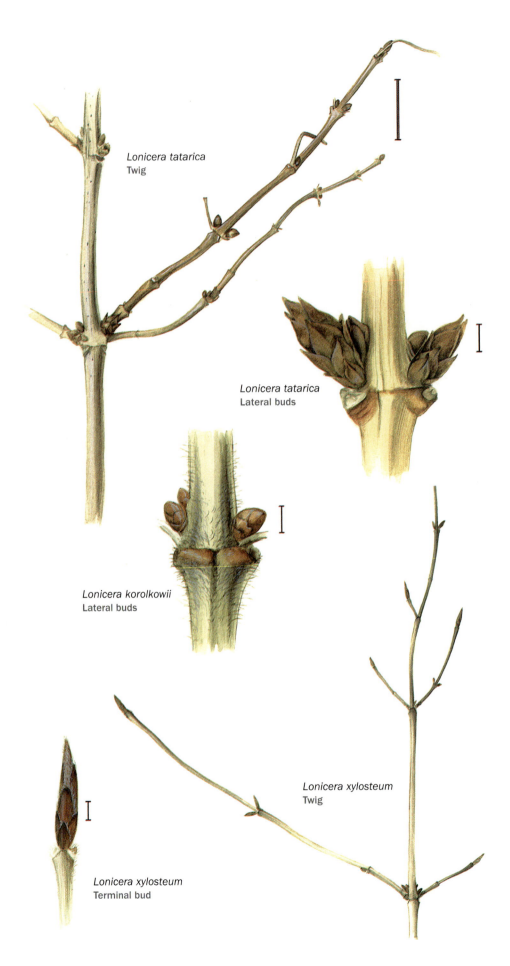

Lonicera tatarica
Twig

Lonicera tatarica
Lateral buds

Lonicera korolkowii
Lateral buds

Lonicera xylosteum
Terminal bud

Lonicera xylosteum
Twig

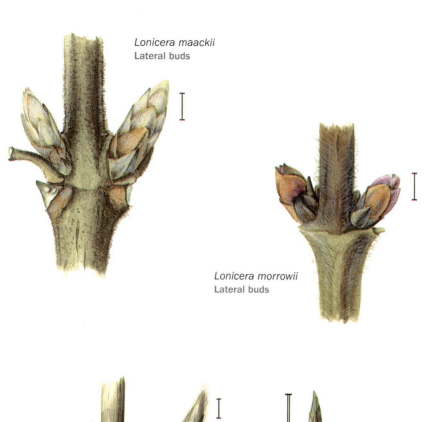

Lonicera maackii
Lateral buds

Lonicera morrowii
Lateral buds

Lonicera nigra
Twig

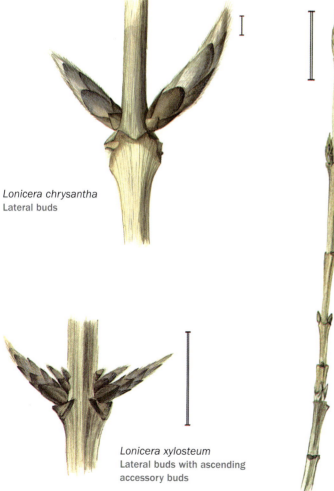

Lonicera chrysantha
Lateral buds

Lonicera xylosteum
Lateral buds with ascending
accessory buds

Lonicera nigra
Tip of twig

Lonicera maackii (Rupr.) Maxim.

Buds ovoid, to 5(–8) mm long, spreading from the twig at a wide angle (but not at right angles), occasionally with accessory buds. Bud scales to 7 pairs, brown, apart from lowest 1–2 pairs densely long white-hairy; old bud scales long persistent, black-grey. **Twigs** grey, very densely fine-hairy, older bark grey to grey-brown, narrowly fissured and flaking. **Leaf scars** with 3 traces on leaf cushions with brown margins. **Stalks of infructescence** short, 3–7 mm long. Erect wide shrub to 5 m high. Origin: eastern Asia.

Lonicera morrowii A. Gray

Buds 2–3 mm, terminal buds can be to 4 mm long, short-ovoid with 4(–5) pairs of bud scales. Lowest scales dark grey-brown, densely hairy, central scales ochre to red-brown and uppermost ones greenish, often also with violet-red hues and only sparsely hairy. Sometimes with small, ascending accessory buds. **Twigs** matte grey-brown, finely shaggy-hairy. **Leaf scars** on spreading cushions, with 3 traces which are small and faint. Rarely planted shrub to 2 m high. Origin: Japan.

Section Rhodanthae

Buds acute, oblong, distinctly 4-angular with very many pairs of scales.

Lonicera nigra L.,
black-berried honeysuckle

Buds 4–7 mm long, acute-ovoid, glabrous. Lateral buds often with ascending accessory buds. **Terminal bud** with 6–7 visible pairs of bud scales, these distinctly keeled, long-acute, lowest one ochre-brown, upper ones paler, more slender. **Twigs** ochre-grey, slightly glossy and very finely hairy. **Leaf scars** faint with 3 traces. Stalks of infructescence 2–3 cm long. Frequent shrub to 1.5 m high. Origin: from the Pyrenees to the Karpathian Mountains.

Lonicera maximowiczii (RUPR.) REGEL

Buds 8–10 mm long, sharply 4-angular through c. 8 strongly keeled pairs of bud scales. In contrast to *Lonicera nigra* and *L. caucasica*, the buds are only short-acute, glabrous, ochre-brown to grey-brown, on the weather side with dead grey-white epidermis. **Twigs** glabrous, first year's growth ochre-brown to red-brown, older twigs grey to grey-green. At the border between years' growth, the old bud scales can be persistent. **Leaf scars** small, faintly 3-traced. **Stalks of infructescence** c. 15 mm long. Rarely planted, shrub to 3 m. Origin: Korea, Manchuria.

Lonicera caucasica PALL.
[*Lonicera orientalis* LAM.]

Buds distinctly 4-angular, but not as sharp as in *Lonicera maximowiczii*. Well-developed terminal bud with c. 11 pairs of bud scales, these ochre-brown and darker brown towards the margin. **Twigs** glabrous, ochre-grey, slightly glossy. Occasionally planted shrub to 2.5 m high. Origin: Turkey to the Caucasus.

Leycesteria formosa WALL.,
Himalayan honeysuckle

Buds small, 2–3 mm long lateral buds, hidden below the remnants of leaf stalks. **Twigs** grass-green, glabrous, almost round (weak ridges between the nodes); thick and hollow, occasionally with waxy bloom. No lenticels, but under high magnification dense white dots are visible. **Leaf scar** absent, instead dry brownish remnants of leaf stalks abutting. Stiff erect shrub to 2 m high. Origin: Himalaya.

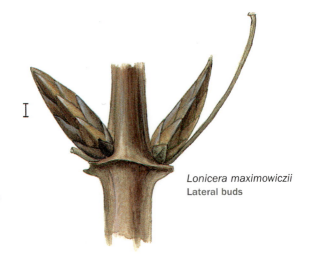

Lonicera maximowiczii
Lateral buds

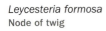

Leycesteria formosa
Node of twig

Abelia chinensis
Lateral buds

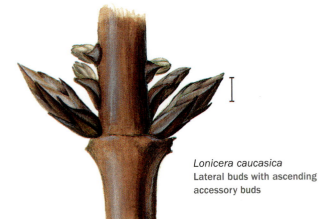

Lonicera caucasica
Lateral buds with ascending accessory buds

Zabelia mosanensis
Node of twig

Zabelia mosanensis
Lateral buds initially hidden under leaf scars

Dipelta floribunda
Twig

Kolkwitzia amabilis
Lateral buds

SUBFAMILY Linnaeoideae

Frequently with distinctly flaking bark. Twigs and buds differ in size. Terminal buds mostly absent, but occasionally formed on a twig. The fruits, which mostly persist into winter, are useful in identification.

Key to Linnnaeoideae

1 Flowers/fruits in pairs, firmly connected at the base. Ovary long and densely spreading brown-hairy with long calyx tube . . . . ***Kolkwitzia***

1* Without oblong floral tube between ovary and calyx lobe 2

2 Calyx lobes persistent, enlarging at fruit ripening 3

2* Fruit between two distinctly enlarged bracts . . . . *Dipelta*

3 Buds visible normally, with bud scales . . . . *Abelia*

3* Buds hidden below leaf bases, often developing early. . . . . *Zabelia*

Abelia chinensis R. Br., Chinese abelia

Twigs with remnants of foliage and fruits at the dying shoot tips. **Buds**: c. 1 mm long lateral buds: surrounded by 1–2 bud scales with ciliate margin. The first pair keeled, ochre-brown, the second pair green. **Twigs** very slender, the youngest branches at base to 1.5 mm thick. From pale grey (dead) to pale ochre-brown, hairy. **Leaf scars** (apart from the small size) distinctly visible, with three traces of vascular bundles. **Pith** white, solid at the nodes, green as the wood. Rarely planted shrub to 2 m high. Origin: China.

Zabelia mosanensis (Chung ex Nakai) Hisauti & Hara
[*Abelia mosanensis* T. H. Chung ex Nakai]

Lateral buds only; these initially hidden below the leaf cushions of the leaves, but larger buds always stand out. With dark wine-red, glabrous bud scales, at most 2 pairs visible, of these the outer one enveloping the buds for two-thirds. **Twigs** slender, youngest branches at base to 2 mm thick, very pale brown to ochre-grey, with scattered relatively long (c. 1 mm) appressed bristly hairs, slightly denser at the nodes. Due to the persistent leaf bases, the twigs are knotty and no leaf scar is visible. **Pith** white, at the node compact, green. Still relatively seldom planted shrub, to 2 m high. Origin: Korea.

Dipelta floribunda Maxim., rosy dipelta

Buds narrowly ovoid, 4–6 mm long, spreading from the twig. The many bud scales are wine-red on the light side and green on the shaded side, and slightly paler towards the margin. **Twigs** hollow, red-brown to grey-brown, densely hairy. Several-year-old twigs with strongly flaking bark. **Leaf scars** indistinct, with 3- (or 5-) traces. **Fruit** flanked by two distinctly enlarged (25 × 20 mm) dry shield-shaped bracts; fruits slightly octagonal, hairy, at the tip with calyx lobes. Occasionally planted, 2–3(–5) m high shrubs. Origin: central China.

Kolkwitzia amabilis Graebn., beauty bush

Buds opposite, spreading from the twig, ovoid, 2–4 mm long; with 4–5 acute, brown, densely white-hairy pairs of bud scales. **Twigs** ochre-brown to red-brown, initially finely hairy. Bark of twig soon tearing open lengthwise, pale ochre-grey below. **Fruits** in compact panicles, always in pairs connected at base, densely bristly hairy, c. 6 mm long, the thin-tubular calyx ending in 5 equal lobes, persisting on the shrub throughout the winter. **Bark** in large sheets, coming off in thin flakes. Very frequently planted 2–3 m high shrub; unmistakable due to the fruits. Origin: western China.

Kolkwitzia amabilis
Fruits typically growing in pairs

Araliaceae
ivy family

Shrubs and trees, little-branched with coarse branches, often armed with prickles. Leaf scars alternate, narrow and broad and with many traces, usually almost enveloping the twig. Remnants of fruit occasionally present in winter: infructescences compound-umbellate.

Key to Araliaceae

1 Lateral buds hemispherical, glossy violet-brown, trees . . . ***Kalopanax septemlobus***

1* Lateral buds different . . . .

2 Twigs covered very densely with slender prickles, these almost covering the terminal bud . . . . *Oplopanax*

2* Prickles sparser, and not covering the terminal bud 3

3 First year's growth twigs to 5 mm thick, moderately branching shrubs . . . . ***Eleutherococcus***

3* Twigs mostly more than 1 cm thick, very slightly branched . . . . ***Aralia***

Aralia L., angelica tree

Little-branched, many-stemmed shrubs due to runners, mostly armed with prickles. In winter the species are hard to tell apart. The growth form and the spines are helpful in identification.

Aralia elata (MIQ.) SEEM., Japanese angelica tree
[*Aralia mandshurica* RUPR. & MAXIM.]

Buds conical, narrower than the 10–15 mm thick twigs, surrounded in a spiral by partly spreading brown-red to dark brown bud scales of the leaf-base. **Twigs** differentiated into long and short-shoots, first year's growth long-shoots pale brown to grey-brown, with sparse 2–5 mm long prickles. **Leaf scars** with many traces of vascular bundles (more than 20–35). Stiff, slightly branched 4–5(–6) m high shrubs forming stolons. Origin: eastern Asia.

Aralia chinensis, the Chinese angelica tree, is encountered less often. Usually with only a few prickles. Shrub to 3 m high. Origin: China.

Aralia spinosa L., angelica tree, is larger and beset with many prickles. A large shrub or small tree to 8 m high, this is also rare. Origin: southern North America.

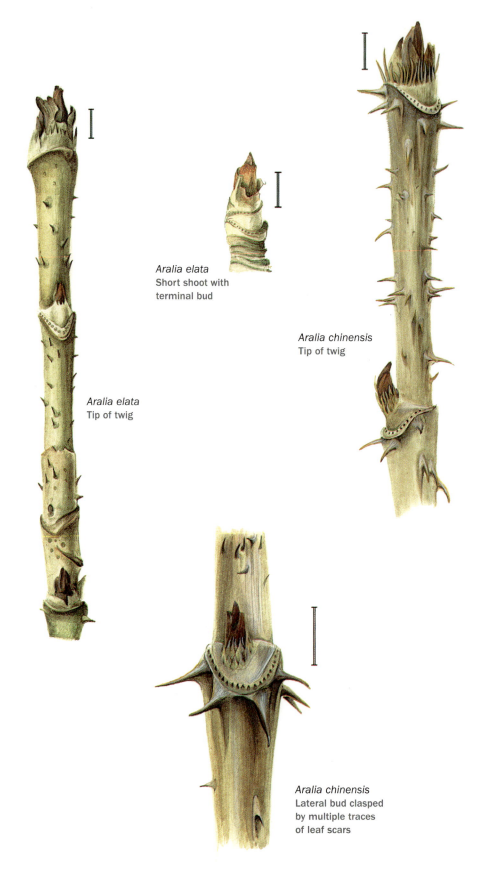

Aralia elata
Short shoot with terminal bud

Aralia chinensis
Tip of twig

Aralia elata
Tip of twig

Aralia chinensis
Lateral bud clasped by multiple traces of leaf scars

*Kalopanax
septemlobus*
Lateral bud

Kalopanax septemlobus
Short-shoot

*Kalopanax
septemlobus*
Twig

Kalopanax septemlobus (THUNB. ex MURRAY) KOIDZ., haragiri

[*Kalopanax pictus* (THUNB.) NAKAI]

Buds broadly sessile, with smooth, dark-violet-brown bud scales: the two outer ones wide-clasping the bud. **Terminal buds** hemispherical, bud scales often attenuate to a small point or lobe, c. 5–6 mm high and as wide. **Lateral buds** from very flat to almost hemispherical, 3–4(–5) mm broad and thick as well as 2–3 mm long (high), with rounded to only slightly acute prophyll scales. **Twigs** thick and stiff, matte, dark brown-green to pale ochre-brown, with many light ochre-coloured lenticels; later twigs with deep longitudinal tears, grey-brown to grey-green. **Prickles** on young individuals dense and strong, later few, red-brown, short, broad-based and mostly ± blunt. **Remnants of infructescence**: short terminal axes, towards winter's end usually with only scars of the fruit-stalk. **Leaf scars** very narrow with many (7–9) traces of vascular bundles aligned next to each other. Occasionally planted, laxly branched, small, single-stemmed tree. Origin: eastern Asia.

Eleutherococcus MAXIM., finger aralia

Trees and shrubs native to Eastern Asia from the Himalayas to the Philippines. Some shrubby species planted frequently.

Key to *Eleutherococcus*

1. First year's growth twigs c. 3 mm thick, with a prickle below the leaf scar . . . .*Eleutherococcus sieboldianus*
1* Twigs thicker 2
2 Twigs bristly with dense prickles . . . .*Eleutherococcus senticosus*
2* Twigs (almost) unarmed. . . . .*Eleutherococcus sessiliflorus*

Eleutherococcus sessiliflorus (RUPR. & MAXIM.) S. Y. HU

[*Acanthopanax sessiliflorus* (RUPR. & MAXIM.) SEEM.]

Buds with a few grey-brown appressed bud scales. Lateral buds 3–4 mm long, ovoid, spreading from the twig. Terminal buds mainly at the end of short-shoots, short-conical, distinctly thinner than the shoot. **Twigs** mostly unarmed, pale grey, slightly brownish, with some round lenticels tearing open. Spreading shrub to 4 m high. Origin: Manchuria, northern China and Korea.

Eleutherococcus sessiliflorus
Short-shoot

Eleutherococcus sessiliflorus
Twig

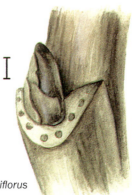

Eleutherococcus sessiliflorus
Lateral bud

Eleutherococcus sieboldianus (MAKINO) KOIDZ., Siebold eleutherococcus

[*Acanthopanax sieboldianus* MAKINO]

Buds grey-brown, to 3 mm long. Lateral buds ± appressed to twig. Terminal buds hardly longer, ovoid. **Twigs** at first year's growth pale olive-grey, below the leaf scars with 1(–3), 5–10 mm long, red-brown to dark brown prickles, several-year-old long-shoots pale grey-brown. **Leaf scars** narrow, encircling the bud, with 5(–7) traces. **Lenticels** warty, ochre-brown. **Bark** fissuring longitudinally, with many short-shoots. Frequently planted shrub, 1–3 m high. Origin: Japan and China.

Eleutherococcus senticosus (RUPR. & MAXIM. ex MAXIM.) MAXIM., Siberian ginseng

[*Acanthopanax senticosus* (RUPR. & MAXIM. ex MAXIM.) HARMS]

Buds with red to dark brown bud scales, at the margin slightly lighter. Terminal buds 8–10 mm long. Lateral buds smaller. **Twigs** ochre-brown to grey-brown, densely covered in fine, slender, c. 5 mm long prickles. Older twigs grey-brown, fine rhomboid-fissured lengthwise, with round ochre-brown lenticels. **Leaf scars** narrow, with c. 9 traces (7–11). Only rarely planted shrub, 2.5 m high. Origin: China and Manchuria.

Oplopanax horridus (SM.) MIQ., Devil's club

Lateral and terminal buds hidden behind dense prickles. **Twigs** pale, ochre-brown, thick; at the tip c. 8–10 mm diameter, with very dense, fine prickles; also around the leaf scars with dense prickles which cover the lateral buds. At the boundary of the year's growth with persistent dark brown bud scales. **Pith** wide, white. **Leaf scar** with 10–16 traces of vascular bundles, but due to the prickles not easily seen. Thick, spiny shrub, up to 1.5 m, rarely 3 m in height. Origin: north-west America.

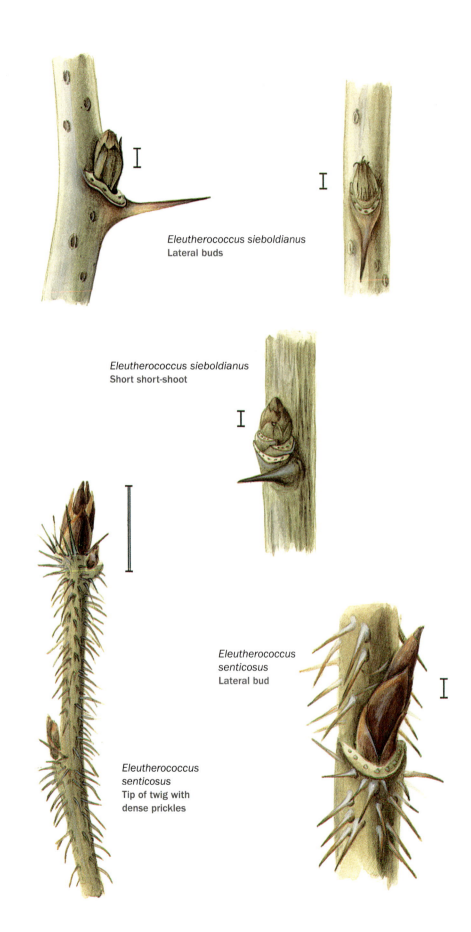

Eleutherococcus sieboldianus
Lateral buds

Eleutherococcus sieboldianus
Short short-shoot

Eleutherococcus senticosus
Lateral bud

Eleutherococcus senticosus
Tip of twig with dense prickles

References and further reading

Phylogeny and Systematics

In the past two decades many new discoveries concerning the relationships of organisms have been made with the help of molecular biological comparisons of their genomes. The numerous individual studies cannot be listed here. Their titles can, however, be found easily online using a search engine by connecting the taxon i.e. "*Betula*" or "*Betulaceae*" with key words such as "Phylogeny". A fundamental survey of the systematic botany of angiosperms can be found in APG IV (2016). References to additional literature can be accessed in STEVENS (2001 onwards).

APG IV (2016). An update of the Angiosperm Phylogeny Group classification for the orders and families of flowering plants: APG IV. *Bot. J. Linn. Soc.* 181: 1–20.

CHRISTENHUSZ, M. J. M., FAY, M. E. & CHASE, M.W. (2017). *Plants of the World: An illustrated encyclopedia of vascular plants.* Royal Botanic Gardens, Kew.

STEVENS, P. F. (2001 onwards). Angiosperm Phylogeny Website. Version 12 July 2012. http://www.mobot.org/MOBOT/research/APweb/

THEIS, N., DONOGHUE, M. J. & LI, J.-H. (2008). Phylogenetics of the Caprifolieae and *Lonicera* (Dipsacales) based on nuclear and chloroplast DNA sequences. *Syst. Bot.* 33 (4): 776–783.

Anatomy and morphology of buds

BRICK, E. (1914). Die Anatomie der Knospenschuppen in ihrer Beziehung zur Anatomie der Laubblätter. *Beih. Bot. Centralbl.* 31, Abt. 1.

BUCHHEIM, G. (1953). *Knospenbau, Sproßge-staltung und Verzweigung der Magnoliaceae.* PhD thesis. Freien Universität zu Berlin.

CADURA, R. (1886). *Physiologische Anatomie der Knospendecken dicotyler Laubbäume.* Breslau.

HENRY, A. (1847). Knospenbilder. Ein Beitrag zur Kenntnis der Laubknospen und Verzweigungsart der Pflanzen. In: *Nov. Actorum Acad. Caes. Leop.-Carol. Nat. Cur.* 22 (1): 169–342.

LAEGAARD, S. (1973–1978). Morfologiske undersogelser af vegetative vinterknopper hos traer og buske. (3. Teile) *Dansk Dendrol. Arsskr.*, I 1973, II 1975, III 1978.

LOEFLING, P. (1749). Gemmae arborum. In: *Amoen. Acad.* 1: 1–32. Uppsala.

LUBBOCK, J. (1899). *On buds and stipules.* K. Paul, Trench, Trübner & Co., London.

MALPIGHI, M. (1675–79). *Anatome plantarum.* J. Martyn, Londini.

MONACHINO, J. (1960). The Internal Characteristics of Winter Buds. *Bull. Torrey Bot. Club* 87 (6): 419–421.

ROLOFF, A. (1986). Morphologische Untersuchungen zum Wachstum und Verzweigungssystem der Rotbuche (*Fagus sylvatica* L.). *Mitt. Deutsch. Dendrol. Ges.* 76: 5–47.

SANDT, W. (1925). Zur Kenntnis der Beiknospen. *Bot. Abh.* 7.

SCHULZE, M. G. (1965). Vergleichend-morphologische Untersuchungen an Laubknospen und Blättern australischer und neuseeländischer Pflanzen. *Feddes Repert. Beih.* 76.

SCHÜTZSACK, U. (1965). Die morphologischen Grundlagen der Blührhythmik der Hamamelidaceae. *Willdenowia Beih.* 3: 1–147.

WARD, H. M. (1904). *Trees. 1 Buds and Twigs.* Cambridge University Press, Cambridge.

WILDE, L. (1988). Untersuchungen über den Gerbstoff- und Cyanidgehalt der Winterknospen einiger mitteleuropäischer Gehölze. *Mitt. Inst. Allg. Bot. Hamburg* 22: 234–255.

Identification

BEENTJE, H. (2010). *The Kew Plant Glossary – an illustrated dictionary of plant terms.* Royal Botanic Gardens, Kew.

BLAKESLEE, A. F. (1931). *Trees in Winter. Their Study, Planting, Care and Identification.* MacMillan, New York.

BÖHNERT, E. (1952). *Die wichtigsten Erkennungsmerkmale der Laubgehölze im winterlichen Zustande.* Ulmer, Stuttgart.

BOOM, B. K. (1982). *Flora der cultuurgewassen van Nederland.* Teil 1 *Nederlandse dendrologie.* H. Veenman, Wageningen.

BRAUN, E. L. (1923). *A key to the deciduous trees of Ohio, native and planted, in winter condition.* Cincinatti.

ČERVENKA, M. & CIGÁNKOVÁ K. (1989). *Klíč k určzování dřevin-podle pupenů a větviček.* Prague. [Key to leafless woody plants – buds and twigs]

CORE, E. L. & AMMONS, N. P. (1999). *Woody Plants in Winter. Identification of Southeastern Trees in Winter.* West Virginia University Press.

LANCE, R. (2004). *Woody Plants of the Southeastern United States: A Winter Guide.* University of Georgia Press, Athens, Georgia.

NOWIKOW, A. L. (1959). *Opredelitiel derewoew i kustarnikow w bezlistnom sostojanii.* Kiev. [Identification of trees and shrubs in the leafless state]

SANTA, S. (1966–67). Identification hivernale des plantes ligneuse de la Flore de France. *Naturalia Monspel., Sér. Bot.* 19.

SCHMIDT, P. A. & SCHULZ, B. (eds) (2017). *Fitschen Gehölzflora.* 13th edn. Quelle & Meyer, Wiebelsheim.

SCHNEIDER, C. (1903). *Dendrologische Winterstudien.* Fischer, Jena.

SCHULZ, B. (2014). *Taschenatlas Knospen und Zweige.* 2nd edn. Ulmer, Stuttgart.

SCHULZ, B. (2018). Sommergrüne Gehölze im Winter. In: A. Roloff, & A. Bärtels, *Flora der Gehölze.* 5th edn. Ulmer, Stuttgart.

SHIRASAWA, H. (1895). Die Japanischen Laubhölzer im Winterzustande. *Bull. Coll. Agric. Tokyo Imp. Univ.* 2 (5): 229–300.

SZYMANOWSKI, T. (1974). Rozpoznawanie drzew i krzewów ozdobnych w stanie bezlistnym (Identification of ornamental trees and shrubs in the leafless state). Warschau.

TRELEASE, W. (1931). *Winter Botany: An Identification Guide to Native and Cultivated Trees and Shrubs.* 3rd edn. Urbana.

WILLKOMM, M. (1859). *Deutschlands Laubhölzer im Winter.* Dresden.

ZUCCARINI, J. G. (1829). *Charakteristik der deutschen Holzgewächse im blattlosen Zustand.* München.

Index to scientific and common plant names

Page numbers in italic relate to synonyms.

Index to botanical terms

Acknowledgements

Author's acknowledgements

The main thanks are due to my wife Renata for her love and patience, her active help and faith in me. I dedicate this book to her.

Two other ladies unknowingly set the course: my biology lecturer Dr Margret Bemmann (Schwerin) suggested the subject and the book illustrator, the late Dagmar Elsner-Schwintowsy strengthened my preferences in illustration. Additionally, two professors working in Tharandt: Andreas Roloff gave the final push by referring me to the publisher Eugen Ulmer, after Peter A. Schmidt kept my dream of publishing in book form alive for a long time.

Before the first edition, my then colleagues at the Institute of Botany of the TU Dresden encouraged me. I would particularly like to mention Dr Bernd Egger, and remember the then Director of the Institute, Prof. Werner Hempel, who died in 2012.

My special thanks go to Mr Robert Ulmer who, without seeing any references, wanted to publish the book from sight of the first sketches. I am also very grateful for the recognition after the publication of the first edition: kind reviews and the distinction of receiving the Book Prize of the German Horticultural Society (Deutsche Gartenbaugesellschaft) 1999 and the Prix Redouté, mention botanique 2000.

Obtaining suitable branches was occasionally difficult. My first point of call was the Botanical Gardens of the TU Dresden. Other material was obtained from the following botanic gardens to whom I extend thanks: Berlin-Dahlem, Bayreuth, Bonn, Pallanza (Villa Taranto, Italy), Strasbourg (France), Ulm, Hillier Gardens (England) and the collections of the Julius-Kühn Institute, Dresden-Pillnitz.

My heartfelt thanks go to all my colleagues and friends who helped with material, above all the many garden colleagues not mentioned by name. I also thank Andreas Bärtels (Waake), Matthias Bartusch (Dresden), Wolfgang Bopp (England), Dr Barbara Ditsch (Dresden), Sigfried Gand (Mainz), Eike Jablonski (Kruchten), Jan De Langhe (Ghent, Belgium), Volker Meng (Göttingen), Dr Ullrich Pietzarka (Tharandt), Jörg Schröder, Rudolf Schröder (Dresden), Dr Mirko Schuster (Dresden-Pillnitz), Antje Verstl (Leipzig) and Ullrich Würth (Westerstede).

The second edition was thoroughly revised. Numerous findings of molecular phylogenetic systematics, which have an important rôle in the teaching and science of our institute, were added to the book. Here I thank Prof. Christoph Neinhuis and all colleagues.

Discussions and talks with gardeners, botanists and environmental architects were of great help. These included Eike Jablonski (Kruchten), Dr Christiane Ritz (Görlitz), Dr Friedrich Ditsch (Dresden) and Prof. Siegfried Sommer (Dresden). Seminars with students on the knowledge of woods in winter, as well as the Winter Seminars founded by the German Dendrological Society in 1998, provided as ever a rich exchange of information and new thinking for which I thank all participants deeply.

Last but not least, I thank the Publishing House Eugen Ulmer, and particularly the publisher Matthias Ulmer and my editor Ina Vetter for the realisation of the book and its beautiful layout.

Publisher's acknowledgements

Kew Publishing would like to thank the following for the time, assistance and advice given in the preparation of this English edition: Sigrun Wagner at Eugen Ulmer KG; Henk Beentje, Tony Kirkham, Kevin McGinn and Monika Shaffer-Fehre at the Royal Botanic Gardens, Kew.

Most of all we thank Bernd Schulz and Friedrich Ditsch for their diligent checking procedures and improvements made to the final English text.

Appendix — Quick reference key

Note: This quick reference key is taken from *Taschenatlas Knospen und Zweige* (Schulz, 2014,) which treats only the 270 most common species in Central Europe, significantly less than half of the species featured in this book. All species, including the rarer cultivated ones, can be identified with the more detailed keys at the beginning of this book (page 26).

1 Number of leaves or leaf scars per node?

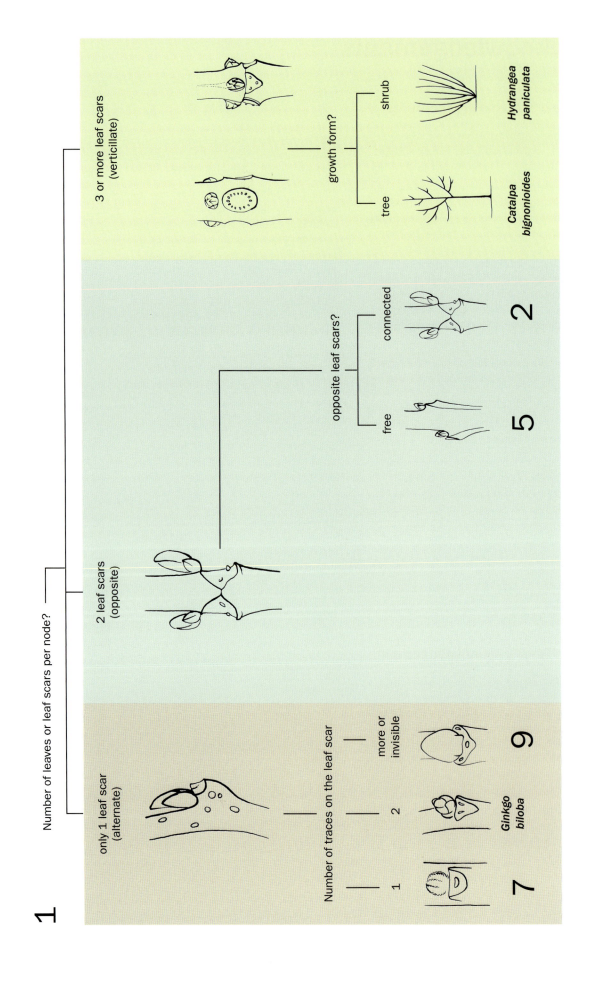

only 1 leaf scar (alternate)

Number of traces on the leaf scar
- 1
- 2
- more or invisible

7

Ginkgo biloba

9

2 leaf scars (opposite)

opposite leaf scars?
- free
- connected

5

2

3 or more leaf scars (verticillate)

growth form?
- tree
- shrub

Catalpa bignonioides

Hydrangea paniculata

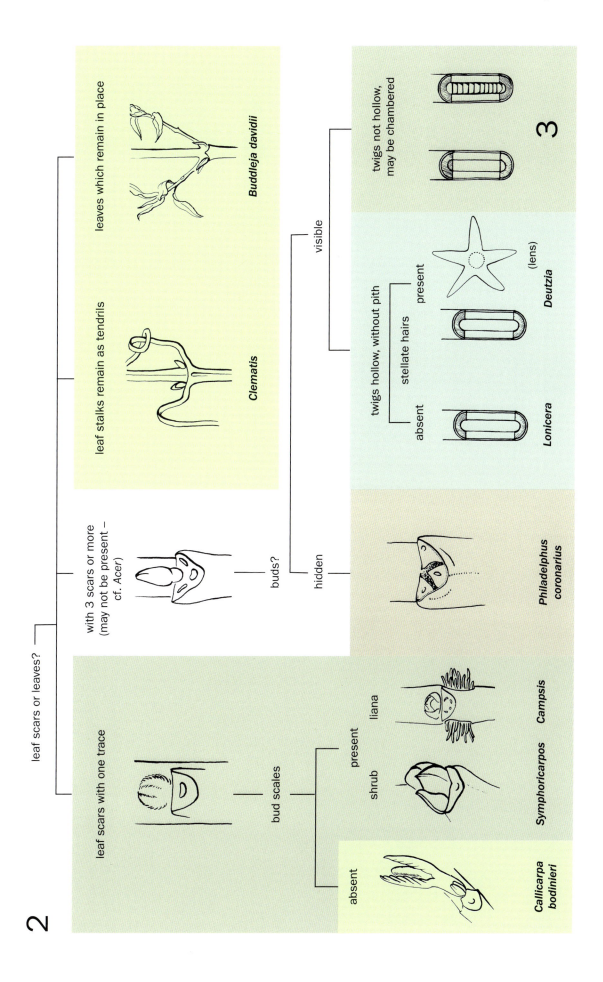

2

leaf scars or leaves?

with 3 scars or more
(may not be present –
cf. *Acer*)

buds?

visible

hidden

leaf stalks remain as tendrils

Clematis

leaves which remain in place

Buddleja davidii

twigs hollow, without pith

twigs not hollow,
may be chambered

3

stellate hairs

present

absent

(lens)

Deutzia

Lonicera

**Philadelphus
coronarius**

leaf scars with one trace

bud scales

present

absent

liana

shrub

Symphoricarpos

Campsis

**Callicarpa
bodinieri**

3

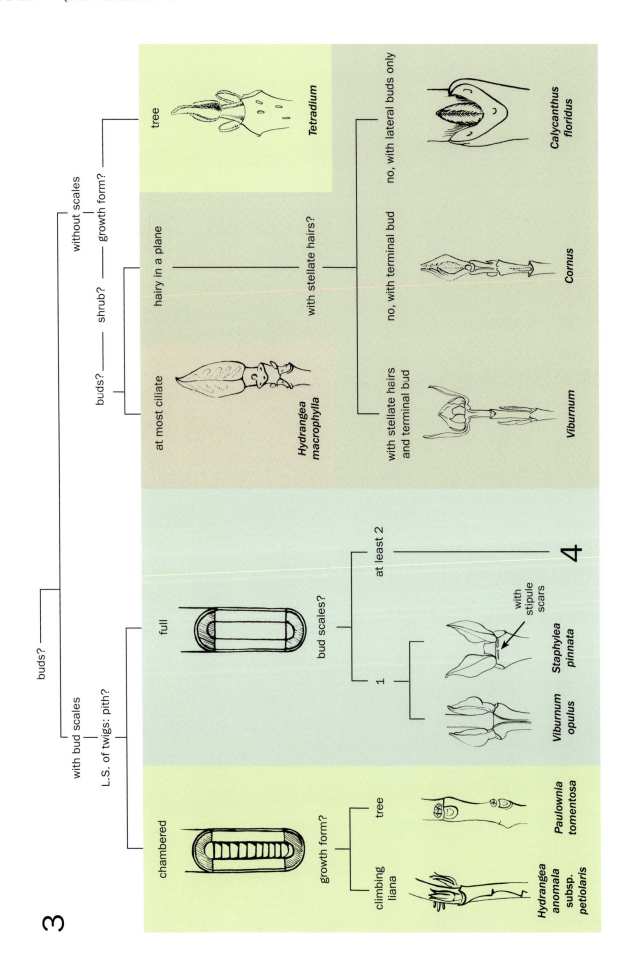

4

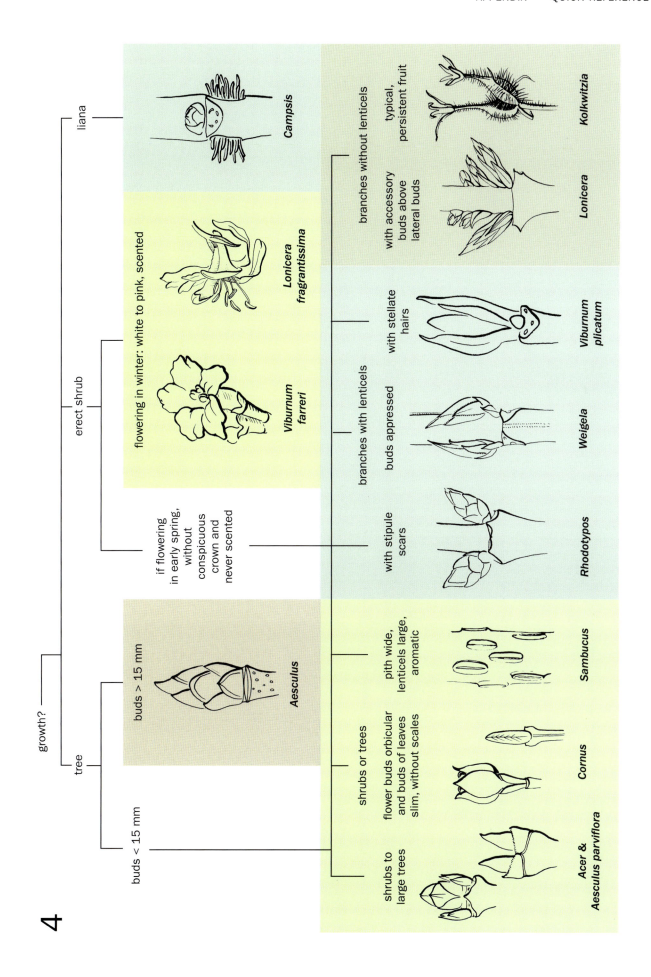

growth?

tree

liana

erect shrub

buds > 15 mm

Aesculus

buds < 15 mm

flowering in winter: white to pink, scented

Campsis

Lonicera fragrantissima

Viburnum farreri

branches without lenticels

typical, persistent fruit

Kolkwitzia

with accessory buds above lateral buds

Lonicera

if flowering in early spring, without conspicuous crown and never scented

branches with lenticels

with stellate hairs

Viburnum plicatum

buds appressed

Weigela

with stipule scars

Rhodotypos

shrubs or trees

pith wide, lenticels large, aromatic

Sambucus

flower buds orbicular and buds of leaves slim, without scales

Cornus

shrubs to large trees

Acer & Aesculus parviflora

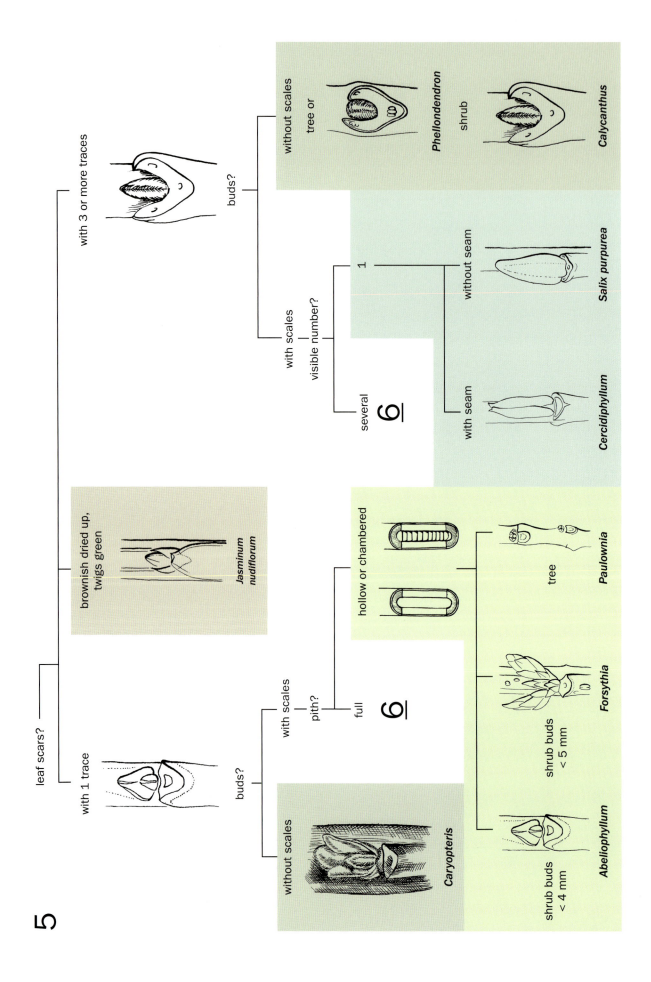

5

leaf scars?

with 1 trace

brownish dried up, twigs green

Jasminum nudiflorum

with 3 or more traces

buds?

without scales

Caryopteris

with scales

pith?

full

6

hollow or chambered

tree

Paulownia

shrub buds < 5 mm

Forsythia

shrub buds < 4 mm

Abeliophyllum

buds?

without scales
tree or shrub

Phellondendron

Calycanthus

with scales
visible number?

several

6

1

with seam

Cercidiphyllum

without seam

Salix purpurea

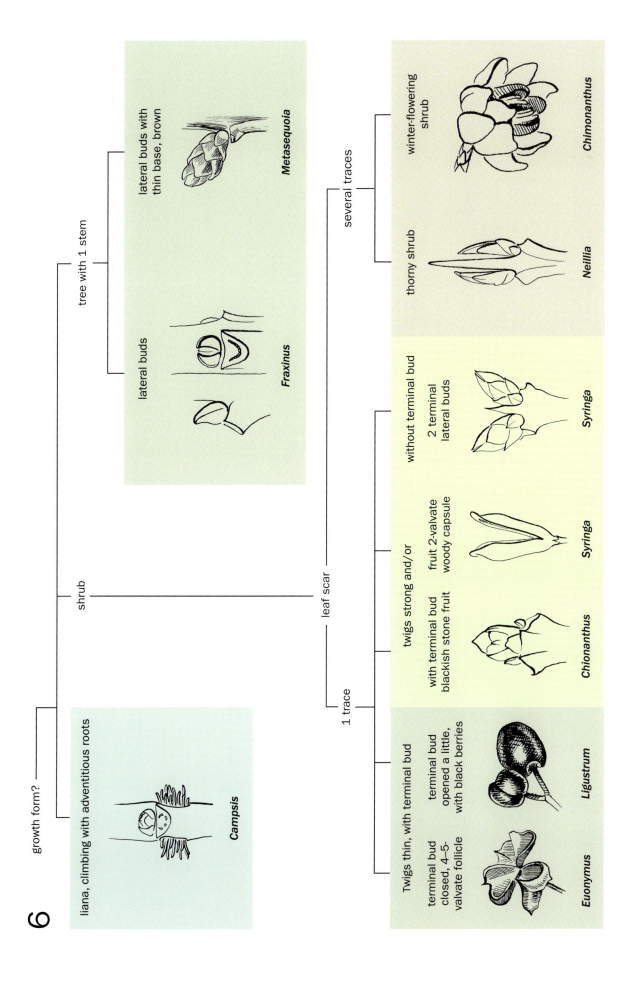

6

growth form?

liana, climbing with adventitious roots

Campsis

shrub

tree with 1 stem

lateral buds

Fraxinus

lateral buds with thin base, brown

Metasequoia

leaf scar

several traces

thorny shrub

Neillia

winter-flowering shrub

Chimonanthus

1 trace

twigs strong and/or

without terminal bud

with terminal bud blackish stone fruit

Chionanthus

fruit 2-valvate woody capsule

Syringa

2 terminal lateral buds

Syringa

Twigs thin, with terminal bud

terminal bud closed, 4–5-valvate follicle

Euonymus

terminal bud opened a little, with black berries

Ligustrum

7

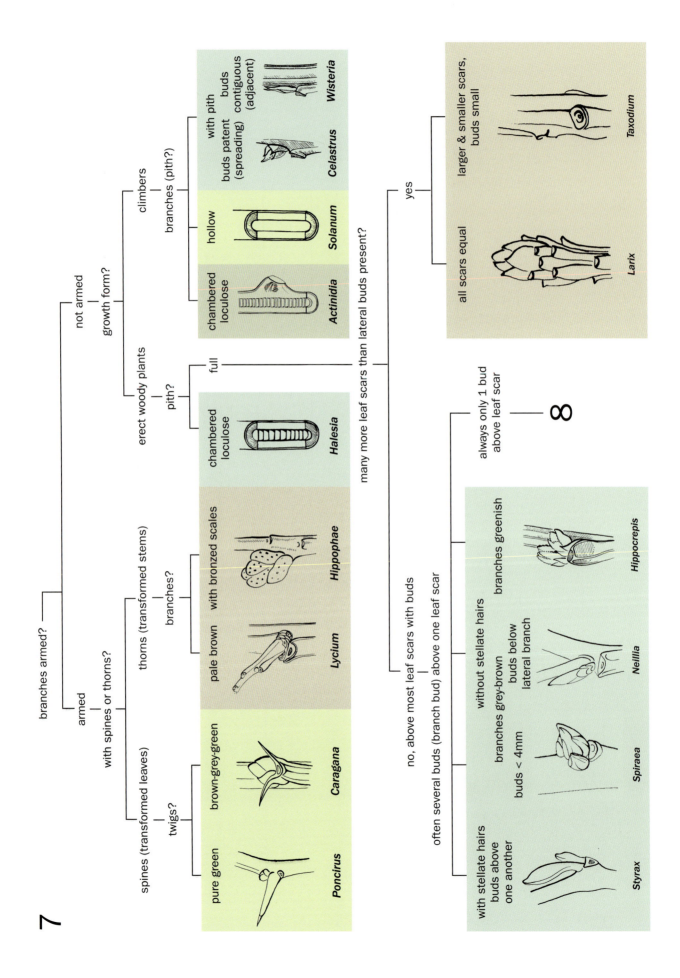

branches armed?

armed

with spines or thorns?

spines (transformed leaves)

twigs?

pure green — *Poncirus*

brown-grey-green — *Caragana*

thorns (transformed stems)

branches?

pale brown — *Lycium*

with bronzed scales — *Hippophae*

not armed

growth form?

erect woody plants

pith?

chambered loculose — *Halesia*

full

climbers

branches (pith?)

chambered loculose — *Actinidia*

hollow — *Solanum*

with pith

buds patent (spreading) — *Celastrus*

buds contiguous (adjacent) — *Wisteria*

many more leaf scars than lateral buds present?

yes

all scars equal — *Larix*

larger & smaller scars, buds small — *Taxodium*

no, above most leaf scars with buds

often several buds (branch bud) above one leaf scar

with stellate hairs buds above one another — *Styrax*

without stellate hairs

branches grey-brown

buds < 4mm — *Spiraea*

buds below lateral branch — *Nelllia*

branches greenish — *Hippocrepis*

always only 1 bud above leaf scar — ∞

8

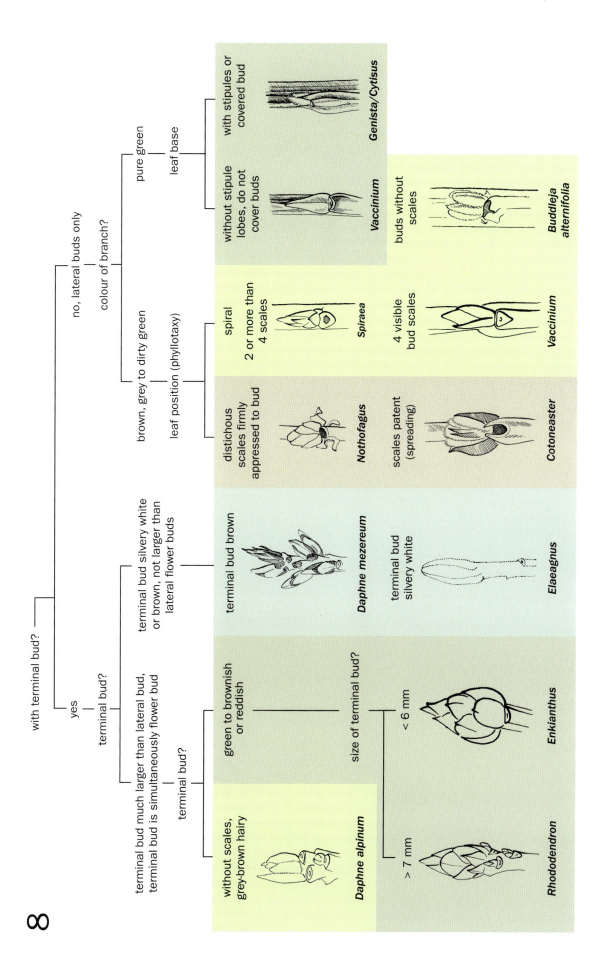

with terminal bud?

yes

no, lateral buds only
colour of branch?

terminal bud?

pure green
leaf base

brown, grey to dirty green
leaf position (phyllotaxy)

with stipules or covered bud

Genista/Cytisus

without stipule lobes, do not cover buds

Vaccinium

buds without scales

Buddleja alternifolia

spiral
2 or more than 4 scales

Spiraea

4 visible bud scales

Vaccinium

distichous
scales firmly appressed to bud

Nothofagus

scales patent (spreading)

Cotoneaster

terminal bud much larger than lateral bud, terminal bud is simultaneously flower bud

terminal bud silvery white or brown, not larger than lateral flower buds

terminal bud?

terminal bud brown

Daphne mezereum

terminal bud silvery white

Elaeagnus

green to brownish or reddish

size of terminal bud?

< 6 mm

Enkianthus

> 7 mm

Rhododendron

without scales, grey-brown hairy

Daphne alpinum

9

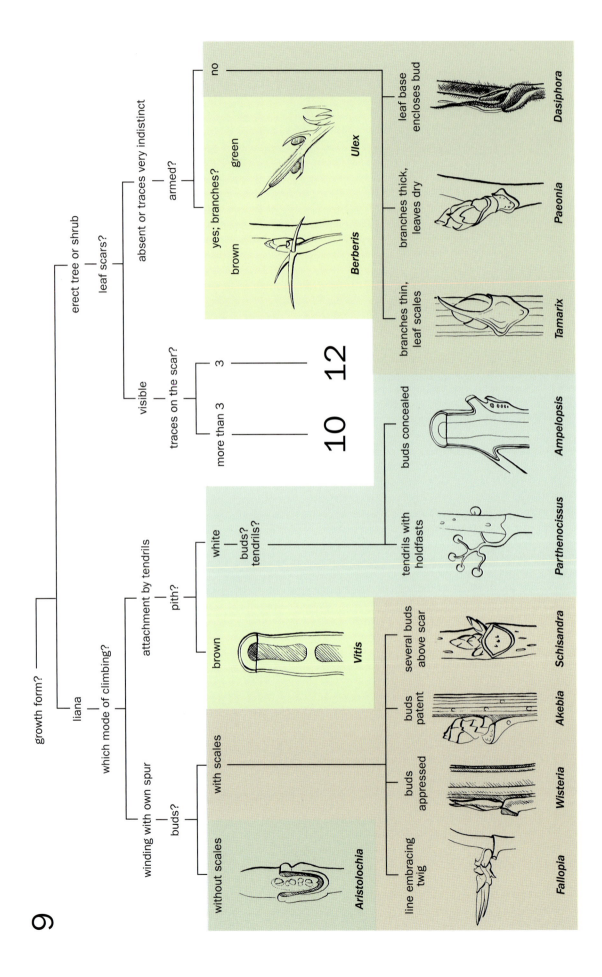

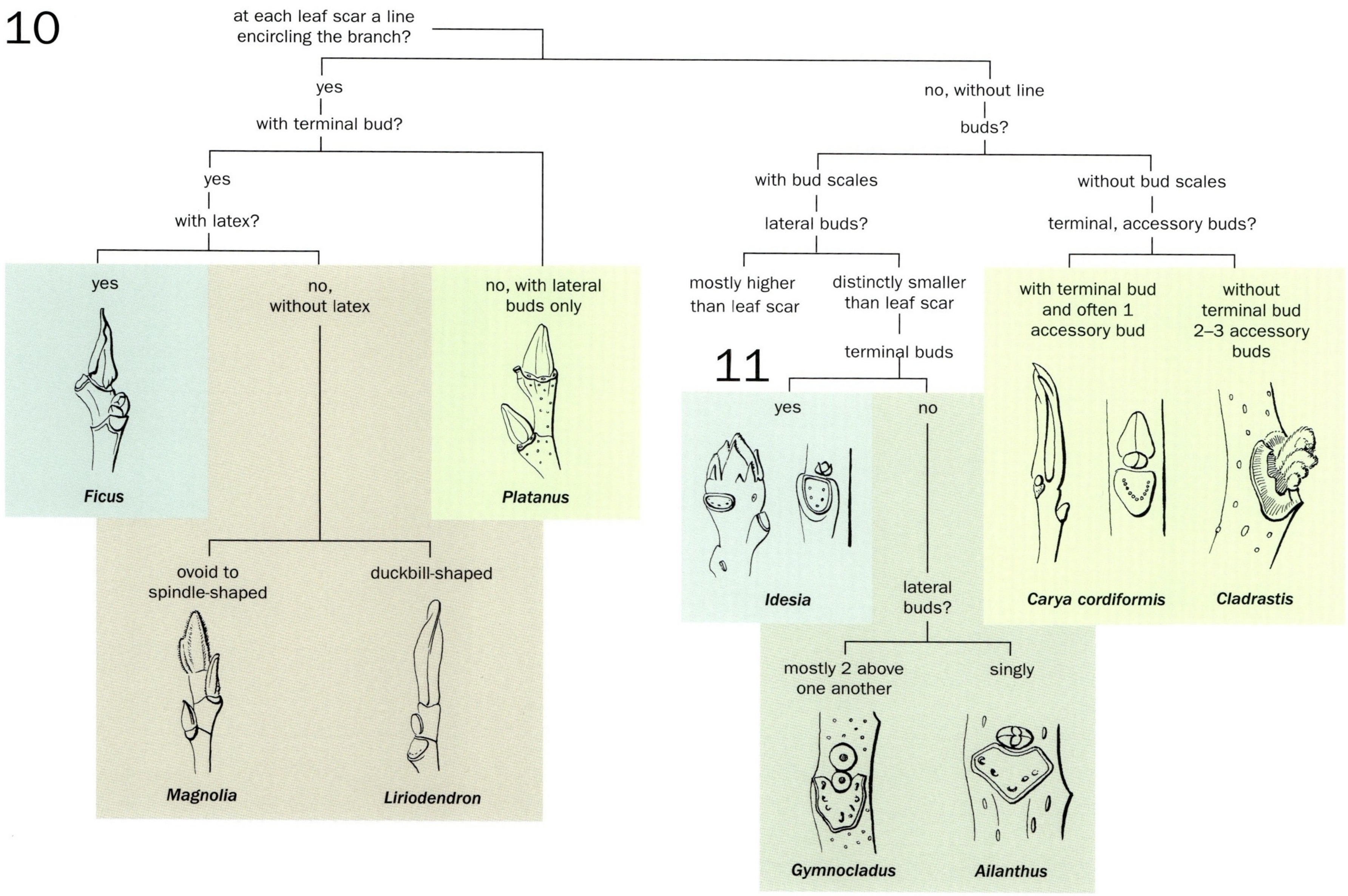

10

at each leaf scar a line encircling the branch?

yes

with terminal bud?

yes

with latex?

yes

Ficus

no, without latex

duckbill-shaped

Liriodendron

ovoid to spindle-shaped

Magnolia

no, with lateral buds only

Platanus

no, without line buds?

with bud scales

lateral buds?

mostly higher than leaf scar

distinctly smaller than leaf scar

terminal buds

yes

11

Idesia

no

lateral buds?

mostly 2 above one another

Gymnocladus

singly

Ailanthus

without bud scales

terminal, accessory buds?

with terminal bud and often 1 accessory bud

Carya cordiformis

without terminal bud 2–3 accessory buds

Cladrastis

11

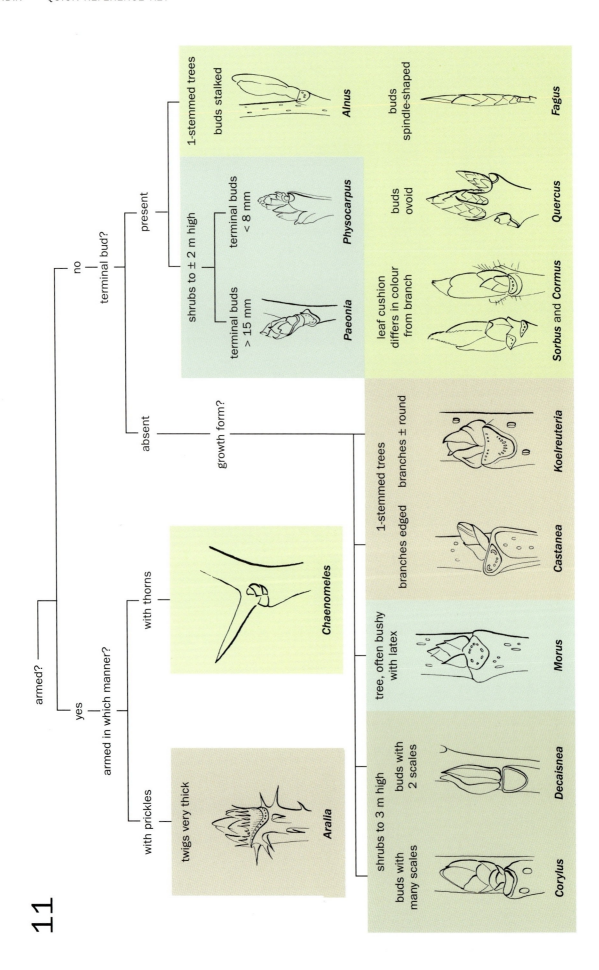

armed?

yes

no

armed in which manner?

terminal bud?

with prickles

with thorns

present

absent

twigs very thick

Aralia

Chaenomeles

1-stemmed trees
buds stalked

Alnus

shrubs to ± 2 m high

terminal buds
> 15 mm

Paeonia

terminal buds
< 8 mm

Physocarpus

buds
spindle-shaped

Fagus

buds
ovoid

Quercus

leaf cushion
differs in colour
from branch

Sorbus and *Cornus*

growth form?

1-stemmed trees
branches edged   branches ± round

Koelreuteria

Castanea

tree, often bushy
with latex

Morus

shrubs to 3 m high
buds with
2 scales

Decaisnea

buds with
many scales

Corylus

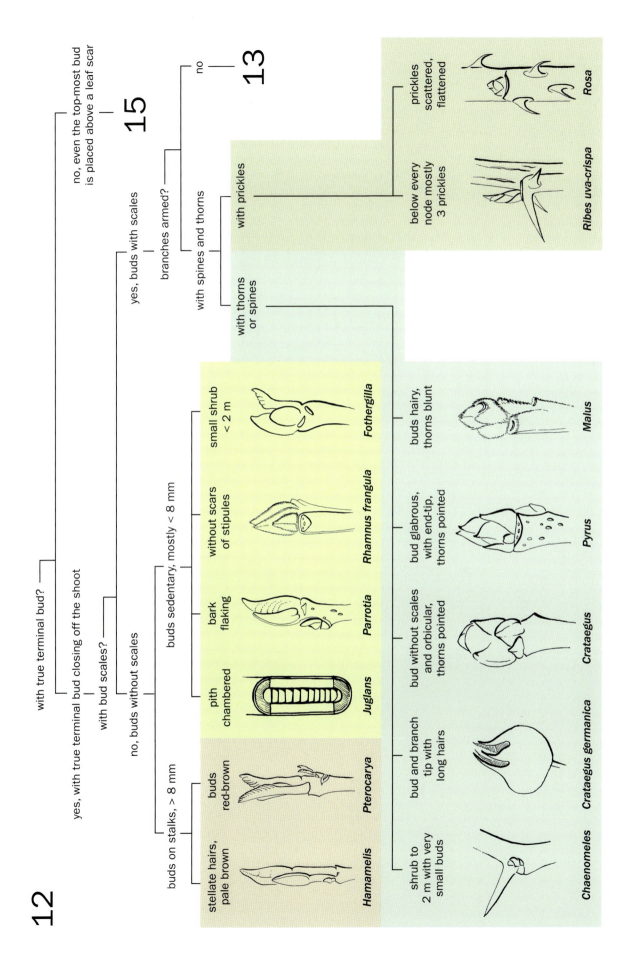

12

with true terminal bud?

yes, with true terminal bud closing off the shoot

with bud scales?

no, buds without scales

buds sedentary, mostly < 8 mm

pith chambered	bark flaking	without scars of stipules	small shrub < 2 m
Juglans	*Parrotia*	*Rhamnus frangula*	*Fothergilla*

buds on stalks, > 8 mm

stellate hairs, pale brown	buds red-brown
Hamamelis	*Pterocarya*

no, buds without scales

buds sedentary, mostly < 8 mm

yes, buds with scales

branches armed?

no — **15**

no, even the top-most bud is placed above a leaf scar

with spines and thorns

with prickles

with thorns or spines

13

below every node mostly 3 prickles	prickles scattered, flattened
Ribes uva-crispa	*Rosa*

shrub to 2 m with very small buds	bud and branch tip with long hairs	bud without scales and orbicular, thorns pointed	bud glabrous, with end-tip, thorns pointed	buds hairy, thorns blunt
Chaenomeles	*Crataegus germanica*	*Crataegus*	*Pyrus*	*Malus*

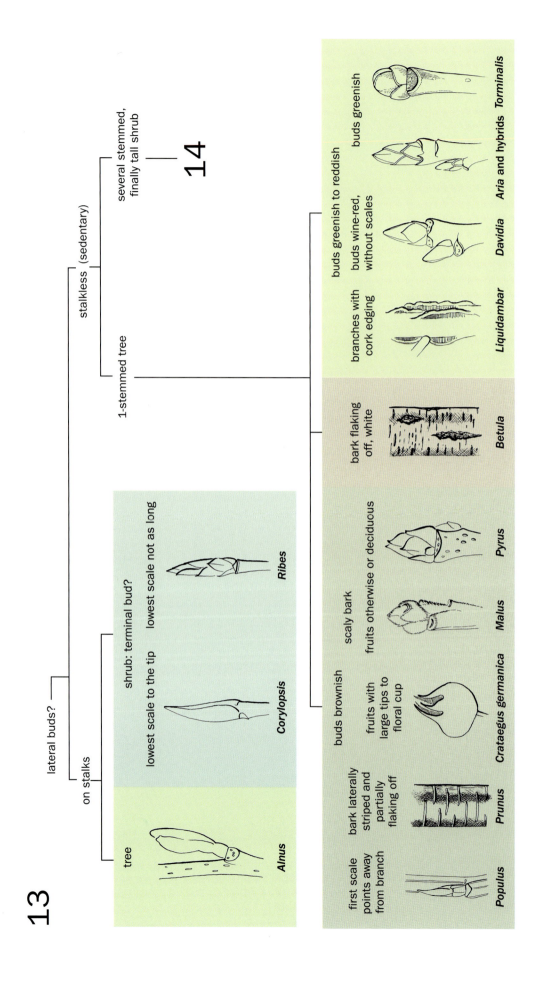

13

lateral buds?

on stalks

tree — *Alnus*

first scale points away from branch

shrub: terminal bud?

lowest scale to the tip — *Corylopsis*

lowest scale not as long — *Ribes*

stalkless (sedentary)

1-stemmed tree

several stemmed, finally tall shrub — 14

buds brownish

fruits with large tips to floral cup

bark laterally striped and partially flaking off — *Prunus*

Crataegus germanica

scaly bark

fruits otherwise or deciduous

Malus

Pyrus

bark flaking off, white — *Betula*

branches with cork edging — *Liquidambar*

buds greenish to reddish

buds wine-red, without scales — *Davidia*

Aria and hybrids

buds greenish — *Torminalis*

Populus

14

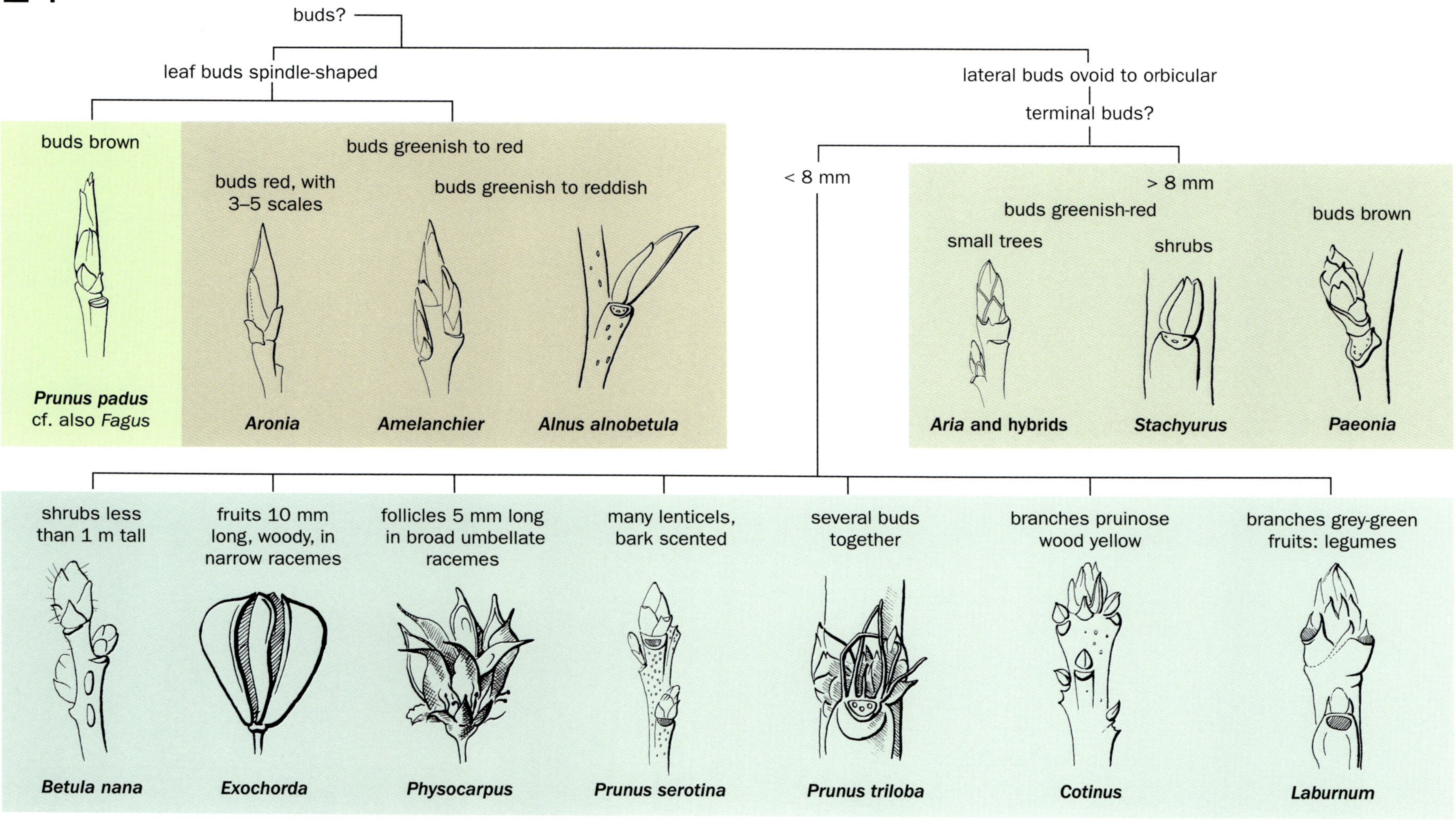

buds?

leaf buds spindle-shaped

lateral buds ovoid to orbicular terminal buds?

buds greenish to red

buds greenish to reddish

buds brown
Prunus padus
cf. also *Fagus*

buds red, with 3–5 scales
Aronia

Amelanchier

Alnus alnobetula

> 8 mm

< 8 mm

buds greenish-red   shrubs
small trees
Stachyurus

Aria and hybrids

buds brown
Paeonia

shrubs less than 1 m tall
Betula nana

fruits 10 mm long, woody, in narrow racemes
Exochorda

follicles 5 mm long in broad umbellate racemes
Physocarpus

many lenticels, bark scented
Prunus serotina

several buds together
Prunus triloba

branches pruinose wood yellow
Cotinus

branches grey-green fruits: legumes
Laburnum

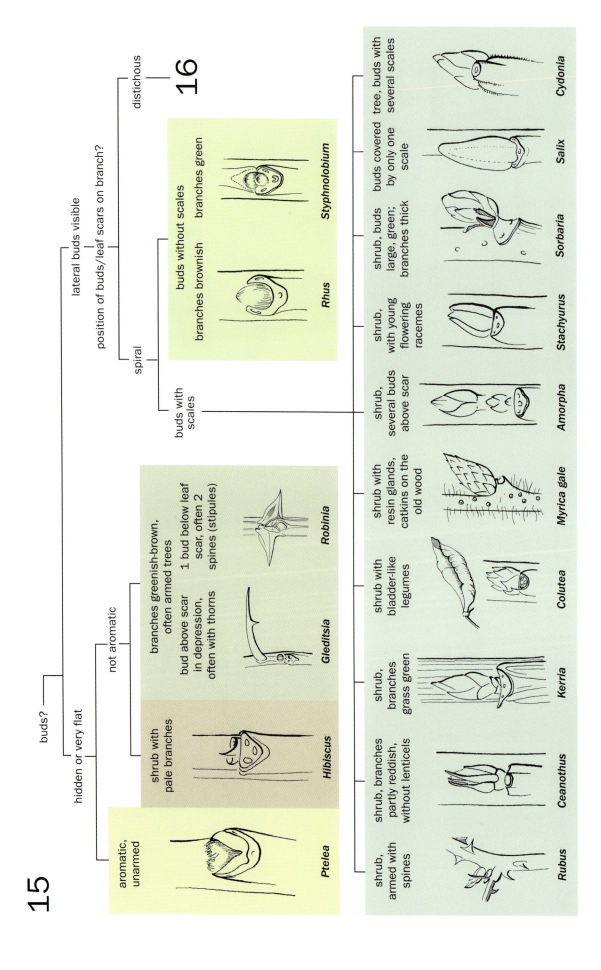

15

buds?

hidden or very flat

not aromatic

branches greenish-brown, often armed trees

1 bud below leaf scar, often 2 spines (stipules)

Robinia

bud above scar in depression, often with thorns

Gleditsia

shrub with pale branches

Hibiscus

aromatic, unarmed

Ptelea

lateral buds visible

position of buds/leaf scars on branch?

distichous

16

spiral

buds without scales   branches green

Styphnolobium

branches brownish

Rhus

buds with scales

shrub, several buds above scar

Amorpha

shrub, with young flowering racemes

Stachyurus

shrub, buds large, green; branches thick

Sorbaria

tree, buds with several scales

Cydonia

buds covered by only one scale

Salix

shrub with resin glands, catkins on the old wood

Myrica gale

shrub with bladder-like legumes

Colutea

shrub, branches grass green

Kerria

shrub, branches partly reddish, without lenticels

Ceanothus

shrub, armed with spines

Rubus

16

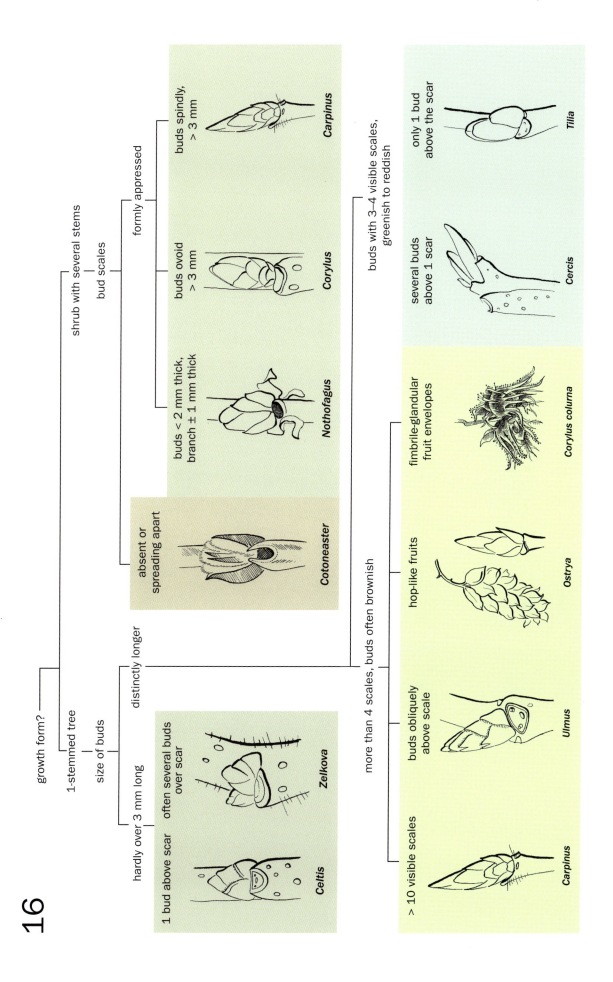

growth form?

1-stemmed tree

size of buds

distinctly longer

hardly over 3 mm long

1 bud above scar

often several buds over scar

Zelkova

Celtis

more than 4 scales, buds often brownish

> 10 visible scales

Carpinus

buds obliquely above scale

Ulmus

hop-like fruits

Ostrya

fimbrile-glandular fruit envelopes

Corylus colurna

shrub with several stems

bud scales

formly appressed

buds spindly, > 3 mm

Carpinus

buds ovoid > 3 mm

Corylus

buds < 2 mm thick, branch ± 1 mm thick

Nothofagus

absent or spreading apart

Cotoneaster

buds with 3–4 visible scales, greenish to reddish

only 1 bud above the scar

Tilia

several buds above 1 scar

Cercis